ELECTRONIC PROPERTIES OF SYNTHETIC NANOSTRUCTURES

Previous Proceedings in the Series of
International Kirchberg Winterschools

	Year	Held in	Publisher	ISBN
XVII	2003	Kirchberg, Austria	AIP Conf. Proceedings Vol. 685	0-7354-0154-3
XVI	2002	Kirchberg, Austria	AIP Conf. Proceedings Vol. 633	0-7354-0088-1
XV	2001	Kirchberg, Austria	AIP Conf. Proceedings Vol. 591	0-7354-0033-4
XIV	2000	Kirchberg, Austria	AIP Conf. Proceedings Vol. 544	1-56396-973-4
XIII	1999	Kirchberg, Austria	AIP Conf. Proceedings Vol. 486	1-56396-900-9
XII	1998	Kirchberg, Austria	AIP Conf. Proceedings Vol. 442	1-56396-808-8
XI	1997	Kirchberg, Austria	World Scientific Publishers	981-02-3261-6

Other Related Titles from AIP Conference Proceedings

725   DNA-Based Molecular Electronics: International Symposium on DNA-Based Molecular Electronics
Edited by Wolfgang Fritzsche, September 2004, 0-7354-0206-X

696   Scanning Tunneling Microscopy/Spectroscopy and Related Techniques: 12th International Conference, STM'03
Edited by Paul M. Koenraad and Martijn Kemerink, December 2003, CD-ROM included, 0-7354-0168-3

640   DNA-Based Molecular Construction: International Workshop on DNA-Based Molecular Construction
Edited by Wolfgang Fritzsche, November 2002, 0-7354-0095-4

590   Nanonetwork Materials: Fullerenes, Nanotubes, and Related Systems, ISNM 2001
Edited by Susumu Saito, Tsuneya Ando, Yoshihiro Iwasa, Koichi Kikuchi, Mototada Kobayashi, and Yahachi Saito, October 2001, 0-7354-0032-6

To learn more about these titles, or the AIP Conference Proceedings Series, please visit the webpage **http://proceedings.aip.org**

ELECTRONIC PROPERTIES OF SYNTHETIC NANOSTRUCTURES

XVIII International Winterschool/Euroconference on Electronic Properties of Novel Materials

Kirchberg, Tirol, Austria   6 –13 March 2004

EDITORS
Hans Kuzmany
Universität Wien, Austria

Jörg Fink
Institut für Festkörperphysik
Dresden, Germany

Michael Mehring
Universität Stuttgart, Germany

Siegmar Roth
Max-Planck-Institut für Festkörperforschung
Stuttgart, Germany

SPONSORING ORGANIZATIONS
Bundesministerium für Bildung, Wissenschaft und Kultur, Austria
Phantoms, Spain
Verein zur Förderung der Internationalen Winterschulen in Kirchberg, Austria

Melville, New York, 2004
AIP CONFERENCE PROCEEDINGS ■ VOLUME 723

Editors:

SEP/AE
PHYS

Hans Kuzmany
Institut für Materialphysik
Universität Wien
Strudlhofgasse 4
A-1090 Wien
AUSTRIA
E-mail: kuzman@ap.univie.ac.at

Michael Mehring
2. Physikalisches Institut
Universität Stuttgart
Pfaffenwaldring 57
D-70550 Stuttgart
GERMANY
E-mail: sk2@physik.uni-stuttgart.de

Jörg Fink
Institut für Festkörperphysik
Postfach 270016
D-01171 Dresden
GERMANY
E-mail: j.fink@ifw-dresden.de

Siegmar Roth
Max-Planck-Institut für Festkörperforschung
Heisenbergstr. 1
D-70569 Stuttgart
GERMANY
E-mail: S.Roth@fkf.mpg.de

L.C. Catalog Card No. 2004111528
ISBN 0-7354-0204-3
ISSN 0094-243X
Printed in the United States of America

CLE
12/13/04

CONTENTS

QC176
.8
N35
±588
2004
PHYS

Preface xv
Table of Previous Kirchberg Winterschools xvi
Organizer, Patronage, Supporters, Sponsors xvii

FULLERENES, ENDOHEDRALS, AND FULLERIDES

What is Special about Endofullerenes? 3
    K.-P. Dinse and T. Kato
Distortion and Orientation of Fulleride Ions in A_4C_{60} 8
    G. Klupp, K. Kamarás, N. M. Nemes, C. M. Brown, and J. Leao
VIS-NIR, Raman and EPR Spectroscopy on Medium Cage Sized
Endohedral Fullerenes 12
    A. Grupp, O. Haufe, M. Hecht, M. Mehring, M. Panthöfer, and M. Jansen

CARBON NANOSTRUCTURE SYNTHESIS, PURIFICATION AND SEPARATION

Approaching the Rational Synthesis of Carbon Nanotubes 19
    R. Herges, D. Ajami, and S. Kammermeier
Controlled Growth of SWCNT on Solid Catalysts with
Narrow (n,m) Distribution 27
    D. E. Resasco, L. Balzano, J. E. Herrera, O. Matarredona, and L. Zheng
Blowing Carbon Nanotubes to Carbon Nanobulbs 32
    D. S. Su, Z. P. Zhu, Y. Lu, R. Schlögl, G. Weinberg, and Z. Y. Liu
On the Temporary Growth of Multi-Walled Carbon Nanaotubes 36
    K. Bartsch and A. Leonhardt
Synthesis, Properties and Possible Applications of Helical
Carbon Nanotubes 40
    A. Szabó, A. Fonseca, A. Volodin, C. Van Haesendonck, L. P. Biró,
    and J. B. Nagy
A Low Cost Method for the Synthesis of Carbon Nanotubes and
Highly Y-Branched Nanotubes 45
    O. T. Heyning, L. Kouwenhoven, P. Bernier, and M. Glerup
Raman-Spectroscopy on Metallic SWNTs Separated from
Semiconducting SWNTs with Dielectrophoresis 49
    F. Hennrich, R. Krupke, H. v. Löhneysen, and M. M. Kappes
Production of Aligned Carbon Nanotube Films and Nitrogen Doped
Carbon Nanotube Films from the Pyrolysis of Styrene 53
    Y. Z. Jin, W. K. Hsu, Y. Q. Zhu, P. C. P. Watts, Y. L. Chueh, L. J. Chou,
    H. W. Kroto, and D. R. M. Walton
Limited Number of Possible Mean Diameters in the Evaporation
Synthesis of Single-Walled Carbon Nanotubes 57
    O. Jost, W. Pompe, X. Liu, and J. Fink

Mass Production of Multiwalled Carbon Nanotubes by Chemical Vapor Deposition.....................61
    A. Magrez, C. Mikó, J. W. Seo, R. Gaál, and L. Forró

Electrochemical Synthesis of an Array of Aligned Polypyrrole Nanotubes.....................65
    P. R. Marcoux and L. Forró

Metallic/Semiconducting Nanotube Separation and Ultra-Thin, Transparent Nanotube Films.....................69
    Z. Chen, Z. Wu, J. Sippel, and A. G. Rinzler

Manipulating Carbon Nanotubes with Nucleic Acids.....................75
    M. Zheng

Direct Growth of Carbon Nanofibers on Plastic Substrates.....................81
    M. Cantoro, S. Hofmann, B. Kleinsorge, G. Csanyi, M. C. Payne,
    A. C. Ferrari, and J. Robertson

PROPERTIES OF SINGLE-WALL CARBON NANOTUBES

Changes in Carbon Nanotube Electronic Properties by Collisions with Inert Gases.....................87
    K. Bolton, A. Rosén, H. Romero, and P. Eklund

Thermal Measurements on Multi-Wall Nanotubes.....................91
    E. Brown, L. Hao, J. C. Gallop, and J. C. Macfarlane

Pathogenic Activity of 1D Nanocarbons: *In Vivo* Studies on Guinea Pigs.....................95
    A. Huczko, H. Lange, M. Sioda, M. Bystrzejewski, P. Nejman,
    H. Grubek-Jaworska, K. Czumińska, and J. Glapiński

Comparison of Current-Voltage Characteristics of Nanofibers and Nanotubes.....................99
    A. B. Kaiser, B. Chapman, U. Schlecht, and M. Burghard

Superconductivity in Long and Short Molecules.....................103
    A. Y. Kasumov, K. Tsukagoshi, M. Kawamura, T. Kobayashi, Y. Aoyagi,
    T. Kodama, V. T. Volkov, Y. A. Kasumov, I. I. Khodos, D. V. Klinov,
    M. Kociak, R. Deblock, S. Guéron, and H. Bouchiat

Effect of Irradiation on Aligned Carbon Nanotube Fibers.....................107
    C. Mikó, M. Milas, J. W. Seo, E. Couteau, N. Barišić, R. Gaál,
    and L. Forró

Luminescence from Pillar Suspended Single-Walled Carbon Nanotubes.....................111
    J. Lefebvre, P. Finnie, and Y. Homma

Strain-Induced Shifts of the Photoluminescence of Single-Walled Carbon Nanotubes in Frozen Aqueous Dispersions.....................116
    K. Arnold, S. Lebedkin, F. Hennrich, and M. M. Kappes

Mechanical Dynamics of Chiral Carbon Nanotubes: Magnetochyrodynamic Effects.....................121
    V. Krstić, G. Wagnière, and G. L. J. A. Rikken

CHARACTERIZATION OF CARBON NANOTUBES

Scanning Conductance Microscopy of Carbon Nanotubes and Polyethylene Oxide Nanofibers 129
C. Staii, N. J. Pinto, and A. T. Johnson, Jr.

Characterization and Gas Adsorption on Multi-Walled Carbon Nanotubes Before and After Controlled Chemical Opening 133
M. Rachid Babaa, E. McRae, C. Gommes, S. Delpeux, G. Medjahdi, S. Blacher, and F. Beguin

Wide Range Optical Studies on Transparent SWNT Films 137
F. Borondics, K. Kamarás, Z. Chen, A. G. Rinzler, M. Nikolou, and D. B. Tanner

Bulk Synthesis and Characteristic Properties of Boron Nitride Nanostructures: Nanocapsules and Nanotubes 141
E. Borowiak-Palen, M. H. Rümmeli, M. Knupfer, G. Behr, T. Gemming, R. J. Kalenczuk, and T. Pichler

Structure and Electronic Properties of Suspended Single Wall Carbon Nanotubes 145
A. Hassanien and M. Tokumoto

STM Investigation of Irradiated Carbon Nanotubes 149
Z. Osváth, G. Vértesy, G. Pető, I. Szabó, J. Gyulai, W. Maser, and L. P. Biró

A Resonant Raman Study of SWNTs under Electrochemical Doping 153
P. M. Rafailov, M. Stoll, J. Maultzsch, and C. Thomsen

Raman on Carbon Nanotubes Using a Tunable Laser and Comparison with Photoluminescence 157
A. Jorio, C. Fantini, M. de Souza, R. Saito, G. G. Samsonidze, G. Dresselhaus, M. S. Dresselhaus, and M. A. Pimenta

Local Phonon Modes of Single-Walled Carbon Nanotubes Observed by Near-Field Raman Spectroscopy 163
A. Hartschuh, A. J. Meixner, and L. Novotny

Vibrational Spectromicroscopy of Graphite and Carbon Nanotubes 168
L. Vitali, M. Burghard, M. A. Schneider, and K. Kern

Atomically Clean Integration of Carbon Nanotubes with Silicon 173
P. M. Albrecht and J. W. Lyding

FUNCTIONALIZATION OF CARBON NANOTUBES

NMR on Cesium Intercalated Carbon Nanotubes 181
M. Schmid, C. Goze-Bac, M. Mehring, S. Roth, and P. Bernier

Purification and Dispersion of Carbon Nanotubes by Sidewall Functionalization with Single-Stranded DNA 185
S. Taeger, O. Jost, W. Pompe, and M. Mertig

Transport Properties of Functionalized Single Wall Nanotubes Buckypaper 189
V. Skákalová, U. Dettlaff-Weglikowska, and S. Roth

Controlled Functionalization of Carbon Nanotubes by *in Situ*
Polymerization Strategy...193
    C. Gao, H. Kong, and D. Yan
Effect of Physical and Chemical Doping on Optical Spectra
of SWNT's...197
    K. Kamarás, H. Hu, B. Zhao, S. Niyogi, M. E. Itkis, and R. C. Haddon
Electrochemical Functionalization of Single-Walled Carbon
Nanotubes with Polyaniline Evidenced by Raman
and FTIR Spectroscopy...201
    S. Lefrant, M. Baibarac, I. Baltog, C. Godon, J. Y. Mevellec,
    and O. Chauvet
Covalent Interaction in Ba-Doped Single-Wall Carbon Nanotubes...........205
    X. Liu, T. Pichler, M. Knupfer, and J. Fink
Reaction of Single-Wall Carbon Nanotubes with Radicals...................209
    A. S. Lobach, V. V. Solomentsev, E. D. Obraztsova, A. N. Shchegolikhin,
    and V. I. Sokolov
A Raman Study of Potassium-Doped Double-Wall Carbon Nanotubes 213
    H. Rauf, T. Pichler, F. Simon, and H. Kuzmany
A Photoemission Study of Potassium-Doped Single Wall
Carbon Nanotubes 217
    T. Pichler, H. Rauf, M. Knupfer, J. Fink, and H. Kataura
Electronic Structure of Single-Wall Carbon Nanotubes and Peapods;
Photoemission Study..222
    H. Kataura, H. Shiozawa, S. Suzuki, Y. Achiba, M. Nakatake,
    H. Namatame, M. Taniguchi, H. Yoshioka, and H. Ishii

NANOTUBE FILLING AND DOUBLE-WALL CARBON NANOTUBES

Metal-Filled Nanotubes: Synthesis, Analysis, Properties
and Applications 229
    D. Golberg, Y. Bando, Y. B. Li, J. Q. Hu, Y. C. Zhu, Y. H. Gao,
    and C. C. Tang
The Growth Process of Nanotubes in Nanotubes.........................234
    M. Holzweber, C. Kramberger, F. Simon, R. Pfeiffer, M. Mannsberger,
    F. Hasi, H. Kuzmany, and H. Kataura
^{13}C NMR on Intercalated 2D-polymerised C_{60} and Modified Peapods........238
    T. Wågberg, C. Goze-Bac, R. Röding, B. Sundqvist, D. Johnels,
    H. Kataura, and P. Bernier
The Redox Behavior of Potassium Doped C_{60} Peapods 242
    M. Kalbáč, L. Kavan, H. Kataura, M. Zukalová, and L. Dunsch
Distinct Redox Doping of Core/Shell Nanostructures: Double Wall
Carbon Nanotubes 247
    L. Kavan, M. Kalbáč, M. Zukalová, M. Krause, H. Kataura, and L. Dunsch
Highly Diameter Selective ^{13}C Enrichment in Carbon Nanotubes 251
    C. Kramberger, F. Simon, R. Pfeiffer, M. Mannsberger, M. Holzweber,
    F. Hasi, and H. Kuzmany

Inserting Fullerene Dimers into Carbon Nanotubes: Pushing the Boundaries of Molecular Self-assembly 255
K. Porfyrakis, A. N. Khlobystov, D. A. Britz, J. J. L. Morton, A. Ardavan, M. Kanai, T. J. S. Dennis, and G. A. D. Briggs

Properties of N@C_{60}-Derived Peapods 259
A. Gembus, F. Simon, A. Jánossy, H. Kuzmany, and K.-P. Dinse

Interaction between Inner and Outer Tubes in DWCNTs 263
R. Pfeiffer, C. Kramberger, F. Simon, H. Kuzmany, and V. N. Popov

Single Wall Carbon Nanotube Specific ^{13}C Isotope Enrichment 268
F. Simon, C. Kramberger, R. Pfeiffer, and H. Kuzmany

Thin Films of C_{60} Peapods and Double Wall Carbon Nanotubes 273
F. Hasi, F. Simon, and H. Kuzmany

Raman Spectroscopy of PbO-Filled Single Wall Carbon Nanotubes 278
M. Hulman, P. Costa, M. L. H. Green, S. Friedrichs, and H. Kuzmany

NON-CARBONACEOUS NANOTUBES

Thermally Induced Templated Synthesis for the Formation of SiC Nanotubes and More 285
M. H. Rümmeli, E. Borowiak-Palen, T. Gemming, M. Knupfer, K. Biedermann, R. J. Kalenczuk, and T. Pichler

Raman Spectra of $B_xN_yC_z$–Nanotubes: Correlation between B, N–Content and Frequency Shifts of the G-band 289
T. Skipa, P. Schweiss, K.-P. Bohnen, S. Lebedkin, and B. Renker

EELS Measurements in Single Wall Boron Nitride Nanotubes 293
R. Arenal, O. Stephan, M. Kociak, D. Taverna, C. Colliex, A. Rubio, and A. Loiseau

Magnetic Properties of TiO_2 Based Nanotubes 298
R. Blinc, P. Umek, P. Cevc, D. Arčon, B. Zalar, Z. Jagličić, T. Apih, and J. Dolinšek

EPR Study of TiO_2 Based Nanotubes and NO_2 Adsorption 302
P. Cevc, P. Umek, R. Blinc, A. Jesih, B. Jančar, and D. Arčon

Mechanical Properties of Individual WS_2 Nanotubes 306
I. Kaplan-Ashiri, S. R. Cohen, K. Gartsman, R. Rosentsveig, V. Ivanovskaya, T. Heine, G. Seifert, H. D. Wagner, and R. Tenne

THEORY OF NANOSTRUCTURES

Entanglement of Spin States in $^{15}N@C_{60}$ 315
W. Scherer, A. Weidinger, and M. Mehring

Omniconjugation 321
M. H. van der Veen, H. T. Jonkman, and J. C. Hummelen

Quantum Chemical Study on $La_2@C_{80}$: Configuration of Endohedral Metals 326
H. Shimotani, T. Ito, A. Taninaka, H. Shinohara, Y. Kubozono, M. Takata, and Y. Iwasa

Raman Excitation Profiles for the (n_1, n_2) Assignment in Carbon Nanotubes 330
    H. Telg, J. Maultzsch, S. Reich, F. Hennrich, and C. Thomsen

Resonant Raman Spectroscopy of Nanostructured Carbon-Based Materials: The Molecular Approach 334
    M. Tommasini, E. Di Donato, C. Castiglioni, G. Zerbi, N. Severin,
    T. Böhme, and J. P. Rabe

Orientational Charge Density Waves and the Metal-Insulator Transition in Polymerized KC_{60} 339
    B. Verberck, A. V. Nikolaev, and K. H. Michel

First Principles Calculations for the Electronic Band Structures of Zone Folding Metallic Single Wall Carbon Nanotubes 343
    V. Zólyomi and J. Kürti

Stabilizing Y-Junctions and Ring Structures through Nitrogen Substitution 347
    A. C. M. Carvalho and M. C. dos Santos

Sticking Effect of Carbon Nanotube Y-Junction Branches 351
    L. A. Chernozatonskii and I. V. Ponomareva

How (and why) Twisting Cycles Make Individual MWCNTs Stiffer ... 355
    A. DiCarlo, M. Monteferrante, P. Podio-Guidugli, V. Sansalone,
    and L. Teresi

The Electronic Structure of Achiral Nanotubes: A Symmetry Based Treatment 359
    E. Di Donato, M. Tommasini, C. Castiglioni, and G. Zerbi

MD Simulations of Catalytic Carbon Nanotube Growth: Important Features of the Metal-Carbon Interactions 364
    F. Ding, A. Rosén, and K. Bolton

DFT Investigation of Nanostructured Binary Compounds 368
    S. Gemming, G. Seifert, and M. Schreiber

Electron-Phonon Interaction and Raman Intensities in Graphite 372
    A. Grüneis, R. Saito, J. Jiang, L. G. Cançado, M. A. Pimenta, A. Jorio,
    C. Fantini, G. G. Samsonidze, G. Dresselhaus, M. S. Dresselhaus, and
    A. G. Souza Filho

First Principles Calculations for the Electronic Band Structures of Zone Folding Non-Metallic Single Wall Carbon Nanotubes 377
    J. Kürti and V. Zólyomi

The Strength of the Radial-Breathing Mode in Single-Walled Carbon Nanotubes 381
    M. Machón, S. Reich, J. Maultzsch, P. Ordejón, and C. Thomsen

Quantum Mechanical Calculations of the Structure, Energetics, and Electronic Properties of the $(C_{60})_2$ and $(C_{60})_2^{2-}$ Fullerene Dimer ... 385
    O.E. Kvyatkovskii, I.B. Zakharova, A.L. Shelankov, and T. L. Makarova

Atomic Pseudopotential Model for Wave Packet Tunneling through a Carbon Nanotube 389
    G. I. Márk, L. P. Biró, L. Tapasztó, A. Mayer, and P. Lambin

Ab initio Approach to Superexchange Interactions in Alkali Doped Fullerides AC_{60} 393
    A. V. Nikolaev and K. H. Michel

Phonon Dispersion of Graphite **397**
J. Maultzsch, S. Reich, C. Thomsen, H. Requardt, and P. Ordejón
Electron Interactions and Excitons in Carbon Nanotube Fluorescence Spectroscopy **402**
C. L. Kane and E. J. Mele
Double Resonance Raman Spectroscopy and Optical Properties of Single Wall Carbon Nanotubes **407**
R. Saito, A. Grüneis, J. Jiang, A. Jorio, L. G. Cançado, C. Fantini,
M. A. Pimenta, G. G. Samsonidze, G. Dresselhaus, M. S. Dresselhaus,
and A. G. Souza Filho

NEW MATERIALS AND BIOLOGICAL NANOSTRUCTURES

Boomerang-Shaped VOx Nanocrystallites **415**
U. Schlecht, L. Kienle, V. Duppel, M. Burghard, and K. Kern
Conducting Properties of Single Bundles of $Mo_6S_3I_6$ Nanowires **419**
M. Uplaznik, A. Mrzel, D. Vrbanic, P. Panjan, B. Podobnik,
and D. Mihailovic
$Mo_6S_3I_6$ Nanowires **423**
D. Vrbanic, A. Meden, B. Jancar, M. Ponikvar, B. Novosel, P. Venturini,
S. Pejovnik, and D. Mihailovic
Synthesis and Magnetic Characterization of $Cu(OH)_2$ Nanoribbons **427**
P. Umek, J. W. Seo, L. Fórró, P. Cevc, Z. Jagličič, M. Škarabot, A. Zorko,
and D. Arčon
XPS Study of Carbyne-Like Carbon Films **431**
T. Danno, Y. Okada, and J. Kawaguchi
Isolation, Positioning and Manipulation of $Mo_6S_3I_6$ by (di)electrophoresis **435**
M. Ploscaru, A. Mrzel, D. Vrbanic, P. Umek, M. Uplaznik, B. Podobnik,
D. Mihailovic, D. Vengust, V. Nemanic, M. Zumer, and B. Zajec
Motions in Catenanes and Rotaxanes **439**
F. Zerbetto
Synthesis of Silicon Nanowires **445**
A. Colli, A. C. Ferrari, S. Hofmann, J. A. Zapien, Y. Lifshitz, S. T. Lee,
S. Piscanec, M. Cantoro, and J. Robertson
Electrical Properties of InAs-Based Nanowires **449**
C. Thelander, M. T. Björk, T. Mårtensson, M. W. Larsson, A. E. Hansen,
K. Deppert, N. Sköld, L. R. Wallenberg, W. Seifert, and L. Samuelson

NANOCOMPOSITES

CNF Re-Inforced Polymer Composites **455**
M. L. Lake, G. G. Tibbetts, and D. G. Glasgow
Carbon Nanotubes as Backbones for Composite Electrodes of Supercapacitors **460**
F. Béguin, K. Szostak, M. Lillo-Rodenas, and E. Frackowiak

Synthesis and Characterization of Carbon Nanotubes/Amylose Composites 465
  P. Bonnet, D. Albertini, C. Godon, M. Paris, H. Bizot, J. Davy, A. Buleon, and O. Chauvet
Route for Single-Walled Nanotube-Polymer Composites 469
  M. Holzinger, J. Steinmetz, D. Samaille, P. Bernier, V. Aboutanos, and M. Glerup
Investigations on Polycarbonate-Nanotube Composites 473
  B. Hornbostel, M. Dubosc, P. Pötschke, and S. Roth
Dispersion of Carbon Nanotubes into Thermoplastic Polymers Using Melt Mixing 478
  P. Pötschke, A. R. Bhattacharyya, I. Alig, S. M. Dudkin, A. Leonhardt, C. Täschner, M. Ritschel, S. Roth, B. Hornbostel, and J. Cech

APPLICATIONS

Field Emission from Individual Thin Carbon Nanotubes 485
  N. de Jonge, M. Allioux, M. Doytcheva, M. Kaiser, K. B. K. Teo, R. G. Lacerda, and W. I. Milne
Fundamental Aspects and Applications of Low-Field Electron Emission from Nano-Carbons 490
  A. N. Obraztsov, A. P. Volkov, A. A. Zakhidov, D. A. Lyashenko, Y. V. Petrushenko, and O. P. Satanovskaya
Light-Driven Molecular Motors 498
  R. A. van Delden and B. L. Feringa
Progress towards a Rotary Molecular Motor 503
  G. Rapenne, A. Carella, R. Poteau, J. Jaud, and J.-P. Launay
Catalytic CVD of SWCNTs at Low Temperatures and SWCNT Devices 508
  R. Seidel, M. Liebau, E. Unger, A. P. Graham, G. S. Duesberg, F. Kreupl, W. Hoenlein, and W. Pompe
Engineering Nanomotor Components from Multi-Walled Carbon Nanotubes via Reactive Ion Etching 512
  T. D. Yuzvinsky, A. M. Fennimore, and A. Zettl
Contact Resistance between Individual Single Walled Carbon Nanotubes and Metal Electrodes 516
  Y. Woo, M. Liebau, G. S. Duesberg, and S. Roth
Gate-Field-Induced Schottky Barrier Lowering in a Nanotube Field-Effect Transistor 520
  T. Brintlinger, B. M. Kim, E. Cobas, and M. S. Fuhrer
Fabrication of Field Effect Transistors Based on Carbon Nanotubes Made by LASER Ablation 524
  M. Dipasquale, P. Repetto, F. Gatti, D. Ricci, and E. Di Zitti
Emission Characteristics of CNT-Based Cathodes 528
  G. S. Bocharov, A. V. Eletskii, A. F. Pal, A. G. Pernbaum, and V. V. Pichugin

Application of Metal Coated Carbon Nanotubes to Direct Methanol
Fuel Cells and for the Formation of Nanowires. 532
    E. Frackowiak, G. Lota, K. Lota, and F. Béguin
Electrical Interconnects Made of Carbon Nanotubes 536
    M. Liebau, A. P. Graham, Z. Gabric, R. Seidel, E. Unger, G. S. Duesberg,
    and F. Kreupl
Freestanding Nanostructures for TEM-Combined Investigations
of Nanotubes 540
    J. C. Meyer, D. Obergfell, M. Paillet, G. S. Duesberg, and S. Roth
Controlling the Position and Morphology of Nanotubes for
Device Fabrication 544
    E. Lahiff, R. Leahy, A. I. Minett, and W. J. Blau
Conjugated Polymeric Donor — Fullerene Type Acceptor Systems for
Photoelectrochemical Energy Conversion 548
    A. Gusenbauer, A. Cravino, G. Possamai, M. Maggini, H. Neugebauer,
    and N. S. Sariciftci
Integration of Carbon Nanotubes with Semiconductor Technology by
Epitaxial Encapsulation 552
    J. Nygård, A. Jensen, J. R. Hauptmann, J. Sadowski and P. E. Lindelof
Electrical Transport in Dy Metallofullerene Peapods 556
    D. Obergfell, J. C. Meyer, P.-W. Chiu, Shi. Yang, Sha. Yang, and S. Roth
Simultaneous Deposition of Individual Single-Walled Carbon
Nanotubes onto Microelectrodes via AC-Dielectrophoresis. 561
    M. Oron, R. Krupke, F. Hennrich, H. B. Weber, D. Beckmann,
    H. v. Löhneysen, and M. M. Kappes
Carbon Nanotubes: Can They Become a Microelectronics
Technology? 565
    W. Hoenlein, F. Kreupl, G. S. Duesberg, A. P. Graham, M. Liebau,
    R. Seidel, and E. Unger
Suitability of Carbon Nanotubes Grown by Chemical
Vapor Deposition for Electrical Devices 574
    B. Babić, J. Furer, M. Iqbal, and C. Schönenberger
A Few Electron-Hole Semiconducting Carbon Nanotube
Quantum Dot. 583
    P. Jarillo-Herrero, S. Sapmaz, C. Dekker, L. P. Kouwenhoven,
    and H. S. J. van der Zant
Electrically Driven Vaporization of Multiwall Carbon Nanotubes for
Rotary Bearing Creation 587
    A. M. Fennimore, T. D. Yuzvinsky, B. C. Regan, and A. Zettl
Conducting Transparent Thin Films Based on Carbon
Nanotubes — Conducting Polymers 591
    N. Ferrer-Anglada, V. Gomis, Z. El-Hachemi, M. Kaempgen, and S. Roth
Flourine Effect on the Binding Energy of Nitrogen Atoms
Incorporated into Multiwall CNx Nanotubes 595
    L. G. Bulusheva, A. V. Okotrub, E. M. Pazhetnov, and A. I. Boronin

Author Index 599

PREFACE

The present book contains the proceedings of the 18th International Winterschool on Electronic Properties of Novel Materials in Kirchberg, Tirol, Austria. The winterschool was held from the 6th to the 13th of March, 2004 at the Hotel Sonnalp. The series of these schools started in 1985. Originally, the school was held every other year and was devoted to conducting polymers. After the discovery of high temperature superconductors, the periodicity changed to an annual format, and the topic alternated between conjugated polymers and superconductors. Since fullerenes are both conjugated compounds, and in some cases superconductors, it was tempting to choose fullerenes as the topic of the Kirchberg schools. The evident extension of this topic is carbon nanotubes and so the title changed from fullerenes via Fullerene Derivatives and Fullerene Nanostructures to Molecular Nanostructures. This gradual change enables us to keep a fairly large interdisciplinary scientific community together, and to stimulate numerous international cooperations. A compilation of the previous Kirchberg Winterschool is presented in the table at the end of this preface.

The term "synthetic nanostructures" implies the "bottom up" (synthetic) approach, as opposed to the "top down" (lithography and etching) techniques in nanostructure technology. As for the physics, we are in a field where solid state physics and molecular physics overlap. This is nicely illustrated with the example of carbon nanotubes. Perpendicular to their axis, nanotubes are molecular as their diameter is in the order of a few nanometers, and different diameters lead to different electronic structures, while along their axis they are extended solids.

Contributions to the 18th Winterschool focused on new nanostructured materials, with data presented on functionalized fullerenes and carbon nanotubes, filled and double wall nanotubes, non-carbon nanotubes such as BN and MoS_2 tubes, and new biological nanostructures. The direction of nanoelectronics research was explored in depth, and advancements in composite technology, and novel applications for nanotubes were discussed. Importantly, participants were updated on the theoretical and experimental determinations of structural and electronic properties, as well as on characterization methods for molecular nanostructures.

The meeting could not have taken place without the support of the Bundesministerium rur Wissenschaft und Forschung in Wien and the Verein fur Forderung der Winterschulen in Kirchberg, as well as from numerous industrial sponsors. Without their contribution, all the enthusiasm and dedication could be wasted and so we express our gratitude to the sponsors and supporters.

Finally, we are indebted, to the manager of the Hotel Sonnalp, Frau Edith Mayer, and to her staff for their continuous support and for their patience with the many special arrangements required during the meeting. Special thanks to Viera Skakalova for her efforts in editing and compiling the Winterschool volume 2004.

H. Kuzmany, J. Fink, M. Mehring, S. Roth
Wien, Dresden, Stuttgart 2004

Table of Previous Kirchberg Winterschools

Year	Title	Published By
2003	Structural and Electronic Properties of Novel Materials	AIP Conference Proceedings 685 (2003)
2002	Structural and Electronic Properties of Molecular Nanostructures	AIP Conference Proceedings 633 (2002)
2001	Electronic Properties of Molecular Nanostructures	AIP Conference Proceedings 591 (2001)
2000	Electronic Properties of Novel Materials - Molecular Nanostructures	AIP Conference Proceedings 544 (2000)
1999	Electronic Properties of Novel Materials - Science and Technology of Molecular Nanostructures	AIP Conference Proceedings 486 (1999)
1998	Electronic Properties of Novel Materials - Progress in Molecular Nanostructures	AIP Conference Proceedings 442 (1998)
1997	Molecular Nanostructures	World Scientific Publ. 1998
1996	Fullerenes and Fullerene Nanostructures	World Scientific Publ. 1996
1995	Physics and Chemistry of Fullerenes and Derivatives	World Scientific Publ. 1995
1994	Progress in Fullerene Research	World Scientific Publ. 1994
1993	Electronic Properties of Fullerenes	Springer Series in Solid State Sciences 117
1992	Electronic Properties of High- T_c Superconductors	Springer Series in Solid State Sciences 113
1991	Electronic Properties of Polymers - Orientation and Dimensionality of Conjugated Systems	Springer Series in Solid State Sciences 107
1990	Electronic Properties of High- T_c Superconductors and Related Compounds	Springer Series in Solid State Sciences 99
1989	Electronic Properties of Conjugated Polymers III - Basic Models and Applications	Springer Series in Solid State Sciences 91
1987	Electronic Properties of Conjugated Polymers	Springer Series in Solid State Sciences 76
1985	Electronic Properties of Polymers and Related Compounds	Springer Series in Solid State Sciences 63

ORGANIZER

Institut für Materialphysik Universität Wien, Austria

PATRONAGE

ELISABETH GEHRER
Bundesministerin für Bildung, Wissenschaft und Kultur, Austria

Magnifizenz
Univ. Prof. Dr. GEORG WINCKLER
Rektor der Universität Wien, Austria

HERBERT NOICHEL
Bürgermeister von Kirchberg, Austria

SUPPORTERS

BUNDESMINISTERIUM FÜR BILDUNG, WISSENSCHAFT UND KULTUR, AUSTRIA
PHANTOMS, SPAIN
VEREIN ZUR FÖRDERUNG DER INTERNATIONALEN WINTERSCHULEN IN
KIRCHBERG, AUSTRIA

SPONSORS

BRUKER OPTIC GmbH, Rudolf-Planckstrasse 23, D-76275 Ettlingen, Germany
ELECTRO V AC GmbH, Aufeldgasse 37-39, A-3400 Klosterneuburg, Austria
JOBIN YVON GmbH, Neuhofstrasse 9, D-64625 Bensheim, Germany
NANOCARBLAB, 1812-year str. 7, apt.6, 121170 Moscow, Russia
NANOCYL S.A., Rue de Seminaire 22, 5000 Namur, Belgium
OMICRON NanoTechnology GmbH, Limburger Strasse 75, D-65232, Taunusstein, Germany
PHYSICA STATUS SOLIDI, WILEY VCH. P.O.Box 101161, D-69451 Weinheim, Germany
VAKUUM-u. SYSTEMTECHNIK GmbH, Hohenauergasse 10, A-1190 Wien, Austria

The financial assistance from the sponsors and the supporters is gratefully acknowledged.

FULLERENES, ENDOHEDRALS, AND FULLERIDES

What Is Special About Endofullerenes?

K.-P. Dinse* and T. Kato**

*) Chem. Dept., Darmstadt University of Technology, Petersenstr. 20, D-64287 Darmstadt, Germany
**) Institute for Molecular Science, Okazaki 444-8585, Japan

Abstract. It is demonstrated that metallo-endofullerenes can exhibit close lying electronic levels with different spin multiplicity, thus serving as sensitive spin probes for charge transfer in carbon nanotubes. In contrast, $N@C_{60}$ is an inert spin label with exceptional sensitivity against local electric fields.

INTRODUCTION

Since trace amounts of metallo-endofullerenes (MEF) were detected more than a decade ago, scientific interest was focused to solve the problem of ion localization at a specific binding site and to determine the amount of charge transfer from the encased metal ion to the cage. Investigation of "pure" samples via powder diffraction techniques yielded information about binding site, cage topology, and the oxidation state of the ion or cluster. Compared to X-ray diffraction, application of less direct magnetic resonance methods like NMR and EPR also has advantages because they can be performed using highly diluted material. Even more important, the spin multiplicity can be determined, thus enabling to explore details of the molecular wave function. Envisioning the MEF as examples of internal charge transfer complexes, sign and size of the resulting exchange coupling between cage and ion will determine the properties of the ground state of the coupled system. Knowledge of the resulting effective electronic spin is mandatory when describing for instance MEF-based peapods. In contrast to MEF, group 15-derived endofullerenes can be classified as atoms freely suspended in a "chemical trap", thus indicating that the encased atom is localized in space with negligible interaction with its confinement. For this reason, these compounds are ideal sensors because spin or charge transfer is practically absent and the localized paramagnetic atom in its quartet spin ground state probes the multipole moments of the local charge distribution. In particular, deviations from spherical symmetry as expected for $N@C_{60}$ embedded in CNT, could in principle be detected.

EXPERIMENTAL

Soot containing $Gd@C_{82}$ was prepared by conventional arc vaporization of a rod composed of graphite and Gd_2O_3 in helium atmosphere. After soxhlet extraction, a highly purified sample of the major isomer of $Gd@C_{82}$ ($Gd@C_{82}(I)$ (C_{2v})) was obtained by employing high performance liquid chromatography (HPLC) as described

CP723, *Electronic Properties of Synthetic Nanostructures*, edited by H. Kuzmany et al.
© 2004 American Institute of Physics 0-7354-0204-3/04/$22.00

in a previous report [1]. The purity of the sample of more than 99% was confirmed by laser desorption time of flight (LD-TOF) mass spectrometry. $La_2@C_{80}$ was prepared and separated by the method reported in [2]. Mono anions of $La_2@C_{80}$ were produced by chemical reduction using 1, 5-diazabi-cycloundecene (DBU) as well as by electrochemical reduction. in well-dried o-dichlorobenzene (ODCB) with electrochemical grade tetra-n-butylammonium perchlorate (TBAP). $N@C_{60}$ was prepared by ion bombardment and purified as described elsewhere [3]. Using HPLC, the relative concentration of $N@C_{60}$ in C_{60} was increased to the 50 ppm level. $N@C_{60}$-doped SWCNT were provided by F. Simon (Universtity Wien). Standard continuous wave EPR powder spectra of HPLC pure $Gd@C_{82}$ and of La_2C_{80} mono anions were obtained using BRUKER 300 E, E500 (X-band) and E680 (W-band) spectrometers. EPR spectra of strongly microwave absorbing $N@C_{60}$ peapods were measured in a dielectric cavity with sample tubes of reduced diameter. Temperature was controlled by a helium flow cryostat.

RESULTS AND DISCUSSION

Spin Ground State of Gd@C$_{82}$

According to an analysis of EPR spectra of $La@C_{82}$, this MEF in first approximation can be described as a charge transfer state consisting of a $^1S_0$ state (La^{3+}) and a $^2S_{1/2}$ state ($C_{82}{}^{3-}$), resulting from a transfer of three electrons from metal to cage. Weak coupling results in just one state of spin multiplicity two as ground state, in which the absence of spin density at the metal leads to a very small Fermi contact hyperfine interaction (hfi) of only 3.2 MHz. In $Gd@C_{82}$, the half-filled $4f$ shell changes the metal ion sub state to $^8S_{7/2}$, allowing the formation of two different spin multiplets of $^9S_4$ and $^7S_3$. By identifying the effective spin of the ground state, the sign of the exchange interaction can be determined. Because no previous knowledge about sign and size of the exchange part of the Coulomb interaction of the charge transfer complex is available, EPR at low temperatures was invoked to observe the anticipated thermal population of a close-lying second spin multiplet via a change of the EPR spectrum. Although in general it is not an easy task to distinguish between $S_{eff} = 3$ and 4 if only powder EPR data are available, in case of a weakly coupled spin system, in which fine structure (FS) originates predominantly from subunit contributions, a simple relation can be used to predict the observable FS parameter D_{Seff} of the coupled spin system in terms of its constituents D_{Gd} and D_{cage} as follows

$$D_{S_{eff}} = c_{7/2}^{(S_{eff})} D_{Gd} + c_{1/2}^{(S_{eff})} D_{cage} \tag{1}$$

The coefficients in (1) can be derived from Clebsch-Gordan tables and are given as

$$
\begin{aligned}
S_{eff} &= 3: \quad c_{7/2}^{(3)} = 5/4; \quad c_{1/2}^{(3)} = -1/8 \\
S_{eff} &= 4: \quad c_{7/2}^{(4)} = 3/4; \quad c_{7/2}^{(4)} = +1/8
\end{aligned}
\tag{2}
$$

If the FS contribution from the high spin metal ion dominates as can be expected here, the ratio of the observable FS interactions of both multiplets is 5/3 in favor of the low spin configuration. In Fig. 1, EPR spectra of a low-concentrated solid solution of $Gd@C_{82}$ measured at 4 K and 20 K are depicted. At the higher temperature, additional lines emerge which can be attributed to the $S_{eff} = 4$ excited spin multiplet. Because of the rather large FS constant of approximately 10 GHz, spectral interpretation was only possible by performing the EPR experiments at the much higher Larmor frequency of 94 GHz. A detailed study of the temperature dependence revealed an energy separation of 11 cm^{-1} between both electronic states, thus allowing for the first time to determine the exchange coupling in a charge transfer complex involving a lanthanide ion encapsulated in a fullerene.

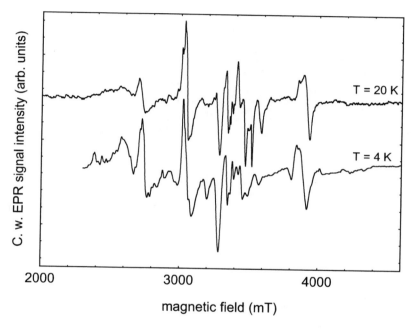

FIGURE 1. 94 GHz EPR spectra of $Gd@C_{82}$ in tetrachlorobenzene.

Charge Redistribution in La$_2$@C$_{80}$ Under Reduction

Apart from C_{60}, the only fullerene with I_h symmetry realized up to now is one of the C_{80} topoisomers. Caused by its topology, the electronic structure allows easy reduction to the hexa-negative state, because the highest occupied MO (in Hückel approximation) is 4-fold degenerate and only occupied by 2 electrons in its neutral ground state. In this respect it was not surprising that this particular cage is used to encapsulate the La_2^{6+} cluster. Any additional electron of the paramagnetic mono-anion of $La_2@C_{80}$ should then be accommodated either by an excited cage MO or by AO on the cluster, the latter implying a change of its oxidation state. Like in the case of

La@C_{82}, La hfi can be used to probe for the amount of spin density at the metal cluster. Changing from formal La$_2^{6+}$ to La$_2^{5+}$ would be equivalent of switching from a vanishing hfi to a value in the order of several 100 MHz. For this purpose, high-frequency (94 GHz) EPR was performed on a frozen solution of La$_2$@C_{80} mono-anions. W-band EPR was chosen to be able to disentangle the complicated multi-line EPR spectrum originating from hfi with two equivalent I = 7/2 nuclear spins. With the information obtained, it was possible also to analyze the overcrowded powder spectrum measured at the standard X-band spectrometer frequency of 9.4.GHz as shown in Fig. 2.

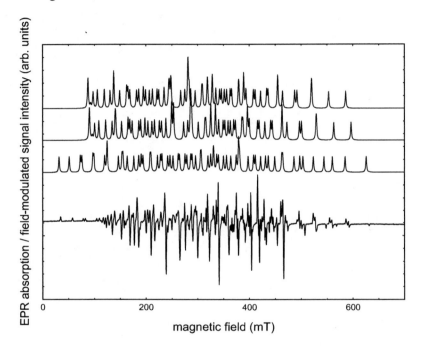

FIGURE 2. Comparison of fictitious single crystal EPR absorption spectra in three canonical orientations with the experimental field-modulated powder spectrum of La$_2$@C_{80}^{-1}. From line positions in the extreme field positions, g-matrix and hfi tensor elements could be obtained with high precision.

The g matrix (1.849, 1.887, 1.975) as well as the La hfi (918, 924, 1170 MHz) exhibited near axial symmetry. The very large La hfi clearly indicates that the spin density is localized at the La$_2$ cluster, thus confirming the assumption of "in cage" reduction.

Making Use Of N@C_{60} In The Nanotube World

If EPR spectra of N@C_{60} are measured either in liquid solution or in a solid C_{60} matrix, extremely narrow lines are observed. In contrast, after insertion into SWCNT the lines are broadened by more than a factor of 10 as shown in Fig. 3. The reason for this line broadening, which was observed in all samples studied is not clear yet but might be related to the presence of additional paramagnetic centers resulting from defects in the CNT. For further information about possible line broadening mechanisms see the contribution of Gembus et al. in this volume.

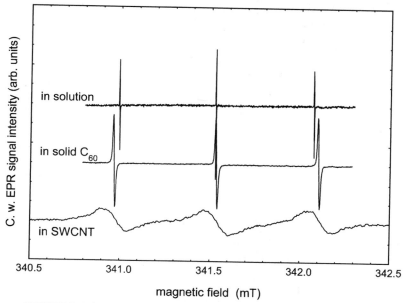

FIGURE 3. X-band EPR spectra of N@C$_{60}$ in various matrices at room temperature.

ACKNOWLEDGMENTS

Financial support by the Deutsche Forschungsgemeinschaft under various grants is gratefully acknowledged. German/Japanese collaboration was supported by a visiting professor fellowship and travel grants of the Institute for Molecular Science, Okazaki. We thank Dr. S. Stoll (ETH Zürich) for providing the EasySpin software package. The investigation of N@C$_{60}$ peapods is performed in collaboration with F. Simon (Universität Wien).

REFERENCES

1. Okubo, S., Kato, T., *Appl. Magn. Reson.* **23**, 481 (2003).
2. Suzuki, T., Maruyama, Y., Kato, T., Kikuchi, K., Achiba, Y., Kobayashi, K., Nagase, S., *Angew. Chem. Int. Ed. Engl.* **34**, 1094 (1995).
3. Jakes, P., Dinse, K.-P., Meyer, C., Harneit, W., and Weidinger, A., *Phys. Chem. Chem. Phys.* **5**, 4080-4083 (2003).

Distortion and orientation of fulleride ions in A_4C_{60}

G. Klupp*, K. Kamarás*, N. M. Nemes[†**], C. M. Brown[†**] and J. Leao[†]

*Research Institute for Solid State Physics and Optics, Hungarian Academy of Sciences, P. O. Box 49, H-1525 Budapest, Hungary
[†]NIST Center for Neutron Research, Gaithersburg, MD 20899-8562, USA
**Department of Materials Science & Engineering, University of Maryland, College Park, MD 20742, USA

Abstract.
A_4C_{60} compounds (A = K, Rb, Cs) are good candidates to exhibit the Mott-Jahn-Teller insulating state. We present near-IR and neutron scattering data to reflect molecular and crystal stucture changes with temperature. We show how the size of the cation affects the structural and electronic properties of these compounds.

The joint appearance of the Mott insulating state and the Jahn-Teller (JT) effect was first suggested for the A_4C_{60} (A = alkali metal) compounds [1]. In the C_{60}^{4-} molecule the Jahn-Teller distortion changes the molecular symmetry from I_h, either by a uniaxial distortion to D_{5d} or D_{3d} or by a biaxial distortion to D_{2h} [2]. From the splitting of the vibrational states found in mid-IR (MIR) experiments [3, 4] it is known that the distortion is temperature dependent: a transition from biaxial to uniaxial occurs on heating. The transition temperature depends on the cation: it is 400 K in Cs_4C_{60} and 270 K in K_4C_{60}. In this work we show that this molecular change can also be detected by transitions between the split electronic states in the near-IR (NIR) and from the splitting of high symmetry intramolecular vibrations observed in inelastic neutron scattering.

A_4C_{60} compounds were prepared by reacting stoichiometric amounts of the alkali metal and C_{60} at 350 °C in a dry box. The purity, checked by x-ray diffraction and Raman scattering was over 95% in all samples. Temperature dependent IR measurements were performed on KBr pellets in dynamic vacuum in a liquid nitrogen cooled flow-through cryostat with a Bruker IFS 66v/S spectrometer.

Temperature dependent neutron diffraction data were collected on the NCNR BT1 diffractometer using Cu(311) monochromator and a wavelength of $\lambda = 1.5403$Å. Inelastic neutron scattering spectra were collected on the NCNR BT4 beamline using the filter analyser neutron spectrometer for high energies and the triple-axis spectrometer for the librational studies. Details are given in Refs. [5, 6].

Rietveld analysis of powder diffraction data proved that in Cs_4C_{60} there is a structural transition from orthorhombic to tetragonal [7]. To look for a similar change in K_4C_{60} and Rb_4C_{60}, we have performed temperature dependent neutron diffraction measurements. We found that the structure of both K_4C_{60} and Rb_4C_{60} remains tetragonal ($I4/mmm$) between 6 K and 300 K (Figure 1).

CP723, Electronic Properties of Synthetic Nanostructures, edited by H. Kuzmany et al.
© 2004 American Institute of Physics 0-7354-0204-3/04/$22.00

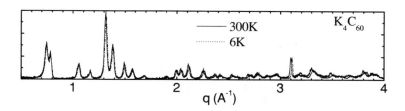

FIGURE 1. Neutron diffraction profiles of K_4C_{60} at 6 K and 300 K, showing only a small thermal contraction of the lattice but no change in symmetry or unit cell.

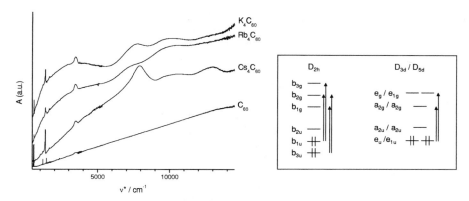

FIGURE 2. MIR and NIR spectrum of K_4C_{60}, Rb_4C_{60} and Cs_4C_{60} without background correction. Second panel shows the allowed transitions between the split molecular orbitals of the distorted fulleride ions leading to the observed peaks in the NIR. The ordering of the levels on this figure is arbitrary.

The ($I4/mmm$) structure is analogous to the high temperature Cs_4C_{60} structure: a disordered arrangement comprising two fulleride ion orientations [7, 8]. In contrast, the fulleride ions are orientationally ordered in the orthorhombic phase [9]. It was shown [7, 9] that this orientational order appears to avoid close Cs–C contacts, which would arise in the disordered structure. This frustration is released at high temperature. In K_4C_{60} and Rb_4C_{60} the orientational order does not appear on cooling, presumably because there is no frustration due to the smaller alkali–C contacts.

Intramolecular electronic transitions can provide information about changes in the molecular geometry. While C_{60} does not have an electronic transition in the 1000-14000 cm^{-1} region, fulleride ions show intramolecular $t_{1u} \rightarrow t_{1g}$ excitations due to the added electrons. The degeneracy of these states is lifted by the distortion, resulting in multiple spectral lines. In the solid, transitions between t_{1u} derived states on neighboring sites are also possible [10].

The room temperature spectra of K_4C_{60}, Rb_4C_{60} and Cs_4C_{60} are shown in Fig. 2. Above 6000 cm^{-1} intramolecular transitions between the split t_{1u} and t_{1g} states can be found [10]. This transition is fourfold split in Cs_4C_{60}, and twofold split in K_4C_{60} and Rb_4C_{60} (Fig.2). Based on group theory there are four dipole allowed transitions in the

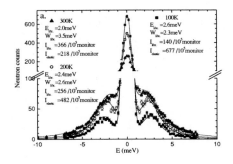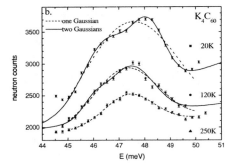

FIGURE 3.   a. Low energy inelastic spectra of K_4C_{60}. Squares: 300K, circles: 200K, triangles: 100K. Lines are fits to the data. Note the change in scale in the y-axis. b. Neutron vibrational spectrum of K_4C_{60} showing the temperature dependent splitting of the $H_u(1)$ mode.

case of a biaxially distorted ion, and two in the case of an uniaxially distorted one (see inset of Fig.2). Therefore the fulleride ion is biaxially distorted in Cs_4C_{60} and uniaxially in K_4C_{60} and Rb_4C_{60}. This finding is consistent with the MIR results [3, 4].

Upon heating, the spectrum of Cs_4C_{60} also develops two peaks, indicating a transition to uniaxial distortion, in accordance with MIR results [4]. MIR experiments found this biaxial–uniaxial transition to be also present in K_4C_{60} and Rb_4C_{60}, although at different temperature.

In the latter two compounds the molecular change is present even though the neutron diffraction measurements did not reveal a structural phase transition. This confirms that the driving force behind the molecular change is an interplay between the molecular Jahn-Teller effect and the crystal potential. In both the tetragonal and the orthorhombic phase, the symmetry of the environment (the alcali cation sublatice) distorts the fulleride ion biaxially. This distortion is dominant at low temperature [3, 4]. At higher temperature the molecular JT forces are dominant, resulting in a uniaxially distorted fulleride ion [3, 4].

the cations around a fulleride ion have such a symmetry, that they distort the fulleride ion biaxially. This distortion is dominant at low temperature [3, 4]. At higher temperature the molecular JT forces are dominant, resulting in a uniaxially distorted fulleride ion [3, 4].

Below 6000 cm^{-1} intermolecular transitions between the split t_{1u} states can be seen [10]. The transitions are weaker in Cs_4C_{60}, which indicates a more difficult electron hopping, due to the larger interfullerene separation in Cs_4C_{60}.

Low energy neutron inelastic scattering is a well established technique to study the rotational potential of C_{60} in pristine and doped fullerides [5]. The excitations observed are due to C_{60} molecules librating about their equilibrium position. Figure 3a shows the librational peaks in K_4C_{60} at selected temperatures. Similar data were collected on Rb_4C_{60}.

The main features are the elastic line and the librational peak at ≈ 2 meV. The intensity of the elastic line decreases with increasing temperature due to the Debye-Waller effect, while the width is resolution limited. The peak at ≈ 2 meV is a librational mode since the Q-dependence follows that expected for C_{60} with intensity maxima at $Q = 5.65$Å and $Q = 3.5$Å while the Q-dependence of the weaker features above 5 meV are flat indicating more translational character [5, 11]. The librations do not show any drastic change as a function of temperature that would indicate a change in the C_{60}^{4-} orientational potential. Following Ref. [11], we can estimate the magnitude of the orientational potential barrier. Using $E = 2.0$ meV for the measured librational energy at 300 K for K_4C_{60} we get 160 meV for hops between orientations related by 44.5° rotations and 630 meV for hops by 90° around (001). The librational energy in Rb_4C_{60} at 300 K was also found to be 2.0 meV.

In A_4C_{60} the MIR spectra show a splitting of the degenerate T_{1u} modes at 146 meV and 166 meV at low temperature [3, 4]. Neutron vibrational spectroscopy can yield similar information on all intramolecular modes without limitations from selection rules. We measured the NVS spectra for K_4C_{60} and Rb_4C_{60} between 25–150 meV in the temperature range 20 K to 300 K. Here, we focus on the peak centered at 47.5 meV at room temperature as this is the only mode that is observed to split within instrumental resolution (1.3 meV). By comparing it to pristine C_{60}, this peak can be assigned to the 5-fold degenerate $H_u(1)$ mode. Figure 3b shows the inelastic neutron spectra of K_4C_{60} at selected temperatures. We fit the spectra with either one or two Gaussians and a sloping background and found that the line is best described by a single Gaussian above 150 K while a split is revealed by the two-Gaussian fit at lower temperature. The discrepancy of the observed transition temperature between the neutron and optical work may be a result of different instrumental resolution.

In summary, the structure and rotational dynamics of K_4C_{60} and Rb_4C_{60} remain unchanged between 6 K and 300 K. Therefore the observed changes in the vibrational spectra are proof of Jahn-Teller distortion of the C_{60}^{4-} anions.

This work was supported by OTKA grant T 034198 and NSF-INT grant 9902050.

REFERENCES

1. Fabrizio, M., and Tosatti, E., *Phys. Rev. B* **55**, 13465 (1997).
2. Chancey, C. C., and O'Brien, M. C. M., *The Jahn-Teller effect in C_{60} and Other Icosahedral Complexes* (Princeton University Press, Princeton, 1997).
3. Kamarás, K., Klupp, G., Tanner, D. B., Hebard, A. F., Nemes, N. M., and Fischer, J. E., *Phys. Rev. B* **65**, 052103 (2002).
4. Klupp, G., Borondics, F., Oszlányi, G., and Kamarás, K., *AIP Conference Proceedings* **685**, 62 (2003).
5. Neumann, D. A., Copley, R. D., Reznik, D., Kamitakahara, W. A., Rush, J. J., Paul, R. L. and Lindstrom, R. M., *J. Phys. Chem. Solids* **54**, 1699 (1993).
6. Copley, R. D., Neumann, D. A., and Kamitakahara, W. A., *Canadian J. Phys.* **73**, 763 (1995).
7. Dahlke, P., and Rosseinsky, M. J., *Chem. Mater.* **14**, 1285 (2002).
8. Kuntscher, C. A., Bendele, G. M. and Stephens, P. W., *Phys. Rev. B* **55**, R3366 (1997).
9. Dahlke, P., Henry, P. F., and Rosseinsky, M. J., *J. Mater. Chem.* **8**, 1571 (1998).
10. Knupfer, M. and Fink, J., *Phys. Rev. Lett.* **79**, 2714 (1997).
11. Reznik, D., Kamitakahara, W. A., Neumann, D. A., Copley, R. D., Fischer, J. E., Strongin, R. M., Cichy, M. A. and Smith, A. B. III, *Phys. Rev. B* **49**, 1005 (1994).

VIS-NIR, Raman and EPR Spectroscopy on Medium Cage Sized Endohedral Fullerenes

Arthur Grupp[1], Oliver Haufe[2], Martin Hecht[1], Michael Mehring[1], Martin Panthöfer[2], and Martin Jansen[2]

1) 2. Physikalisches Institut, Universität Stuttgart, Pfaffenwaldring 57, 70550 Stuttgart, GERMANY
2) Max-Planck-Institut für Festkörperforschung, Heisenbergstr. 1, 70569 Stuttgart, GERMANY

Abstract. $Eu@C_{74}$ and both isomers of $Eu@C_{76}$ have been isolated from endohedral metal fullerenes containing raw soots produced by means of the radio-frequency method. $M@C_{74}$ (M = Ca, Sr, Ba, Eu) and $M@C_{76}$ (M = Sr, Eu) have been characterized by means of VIS and Raman spectroscopy. The VIS-NIR and Raman spectra are dominated by the spectral features of the fullerene cage. Small shifts of the band positions in the VIS spectra are mostly due to the variations of the ionization potentials of the respective metal. The analysis of the metal vs. cage vibration observed in the low frequency range of the Raman spectra depends on the atomic mass and the charge state of the endohedrally enclosed metal, irrespective of the type and molecular structure of the fullerene cage. Thus, the frequency of the metal vs. cage vibration is a first measure of the valence state of the endohedral metal atom. Preliminary W- and X-band ESR spectroscopic investigations on $Eu@C_{74}$ clearly show the presence of a $S = ^7/_2$ spin state of the endohedral Eu atom pointing to an inner salt $Eu^{2+}@C_{74}^{2-}$. The total fading of the ESR signal at temperatures above 15 K is a consequence of the thermal excitation of low lying spin states of the Eu^{2+} cation inside the C_{74} fullerene cage.

INTRODUCTION

The radio frequency method has proven its high applicability to the production of small to mid cage size empty fullerenes and divalent metal endohedral fullerenes in the gap of C_{70} to C_{80}. C_{74} is a prominent species in all raw soots produced by this method. Anyhow, due to its low band gap, this fullerene has exclusively been extracted from raw soot in the state of the exo- and endohedrally reduced dianion, i.e. C_{74}^{2-} and $M^{II}@C_{74}$ (M = Ca, Sr, Ba, Sm, Eu). In this work we present the isolation of $Eu@C_{74}$ and both isomers of $Eu@C_{76}$ from raw soot, VIS and Raman spectroscopic investigations on $M@C_{74}$ (M = Ca, Sr, Ba, Eu) and $M@C_{76}$ (M = Sr, Eu), the influence of the endohedral metal on the spectral features, and preliminary EPR spectroscopic investigations on $Eu@C_{74}$.

Experimental

Details on the synthesis and isolation of $M@C_n$ (M = Ca, Sr, Ba; n = 74, 76) have been reported elsewhere [1, 2], these apply to the isolation of $Eu@C_{74}$ and $Eu@C_{76}$, as well (see fig. 1). In addition, the late fraction from the first separation step showed the presence of the hitherto not observed species $Eu_2@C_{88}$ and $Eu@C_{90}$. It was not

CP723, *Electronic Properties of Synthetic Nanostructures*, edited by H. Kuzmany et al.

possible to isolate $Eu_2@C_{88}$, due to the small amounts of sample available at that stage of separation. Consecutive multistep separation enabled the enrichment of $Eu@C_{90}$ besides C_{94} (see fig. 1c).

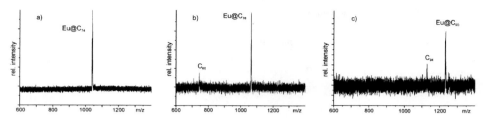

FIGURE 1. Mass spectra of pure fractions of $Eu@C_{74}$ (a), $Eu@C_{76}$ (b) and the $Eu@C_{90}$ enriched fraction (c).

Results and Discussion

VIS-NIR Spectroscopy

The VIS-NIR spectra of the endohedral metal fullerenes characterized so far are dominated by the spectral features of the fullerene cage. In case of $M@C_{74}$ (M = Ca, Sr, Ba, Eu) the endohedrally enclosed metal causes only slight shifts of the VIS-NIR band positions (see fig. 2), which are mostly due to the variations of ionization potentials of the respective metals. A characteristic absorption band is located at about 760 nm. Upon change of solvent, i.e. toluene instead of carbon disulfide, a hypsochromic shift of all bands is observed. This is typically found for π-π^* transitions. In contrast to $M@C_{76}$-I (M = Sr, Eu), the VIS-NIR spectra of $M@C_{76}$-II are quite unstructured. The spectra of both isomers of $Sr@C_{76}$ are almost identical to those reported for $Sm@C_{76}$-I, and $Sm@C_{76}$-II [3], respectively. The most prominent band of $Sr@C_{76}$-I is located at 672 nm.

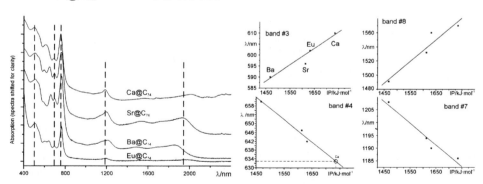

FIGURE 2. VIS-NIR spectra of $M@C_{74}$ (M = Ca, Sr, Ba, Eu; left) and dependency of the band positions on the ionization potential of the corresponding metal (right).

13

Analogous to the VIS-NIR spectra, the Raman spectra of endohedral fullerenes are dominated by the spectral features of the fullerene cage (see fig. 3). No differences are observed for the internal modes of the fullerene cage for $M@C_{74}$ (M = Sr, Ba, Eu). Exclusively one mode in the low frequency range ($\nu < 200$ cm^{-1}), which is attributed to the metal vs. cage mode, shows a dependence on the endohedrally enclosed metal atom. Molecular dynamic simulations on $Ba@C_{74}$ (CPMD) point to a metal vs. cage vibration at about 110 cm^{-1}. The experimental values are 134 cm^{-1} ($Sr@C_{74}$), 120 cm^{-1} ($Ba@C_{74}$) and 123 cm^{-1} ($Eu@C_{74}$), respectively (see fig. 3, inset). The plot of the frequency of the metal vs. cage mode below 200 cm^{-1} versus the square root of the reciprocal atomic mass of the corresponding metal atom of all endohedral fullerenes studies so far [4 - 7], exhibits a separation into two groups (see fig. 3). For each group, a linear function representing all data points has been found. This indicates a nearly identical force constant between the metal and the fullerene cage, which is dependent only to the (expected) charge state of the metal atom, but neither to the cage size nor to the cage isomer. Therefore, endohedral metal fullerenes are to be considered as inner salts $M^{x+}@C_n^{x-}$, without any covalent bonding character. Therefore, the frequency of the metal vs. cage mode is a first measure of the valence state of the endohedral metal.

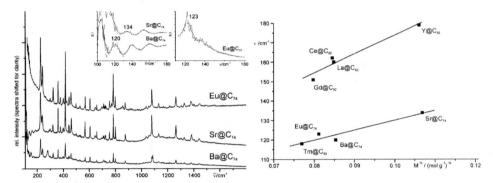

FIGURE 3. Raman spectra of $M@C_{74}$ (M = Ba, Sr, Eu; left; inset: zoom into the low frequency region) and plot of the frequency vs. the reciprocal square root of the atomic mass of the enclosed metal atom (right; see also ref. [4 - 7]).

EPR Spectroscopy

In agreement with literature reports [4], cw-X-band EPR measurements of $Eu@C_{74}$ at room temperature do not exhibit any signal. This finding is contraintuitive to the expectation of $Eu@C_{74}$ to be an inner salt of type $Eu^{2+}@C_{74}^{2-}$. Anyhow, upon cooling, cw-spectra exhibit signals at a temperature below 10 K. Interpretation of these spectra is far from trivial, hinting to a reduced symmetry. Besides the anisotropic zero field splitting present, $S = {}^7/_2$ systems exhibit components of higher exponents of S of order four and six, resulting in a complex dependence on the magnetic field. Pulse-EPR spectra of Eu@C_{74} exhibit signals at temperatures up to 20 K. Due to coupling of the electronic spin system to solvent protons (toluene), spin echo spectra exhibit field dependent echo modulations. Extraction of the fine structure parameters from spin echo signals of these powder spectra turned out to be unreliable. Therefore, cw-W-band EPR spectra were recorded (see fig. 4). Due to baseline drifts only the signal positions could be evaluated, but not the intensities. A consistent set of parameters for a $S = {}^7/_2$ spin system was derived from fitting simulated spectra, using the SimFonia program, (g = 2.08, D = 1950 G, E = 40 G). In first approximation these parameters fit even to the low field parts of the X-band spectra, while the simulation of the high field part suffers from the large value of the D-parameter. Further refinements are in progress. In any case, these preliminary results on the EPR spectroscopy investigations of Eu@C_{74} strongly point towards the presence of a $S = {}^7/_2$ spin system and a fast, temperature dependent relaxation caused by low lying thermally excited spin states.

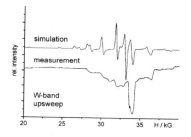 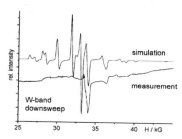

FIGURE 4. W-band spectra (left: upsweep, right: downsweep) and preliminary, simulated EPR spectra of Eu@C_{74} in toluene at 4K.

CONCLUSION

The VIS-NIR and Raman spectra of M@C_n (n = 74, 76; M = Ca, Sr, Ba, Eu) are dominated by the spectral features of the fullerene cage. Raman spectroscopic investigations on M@C_{74} as well as EPR spectroscopic investigations on Eu@C_{74} point out, that the chemical nature of endohedral metal fullerenes M@C_n is well described in form of inner salts, i.e. $M^{x+}@C_n^{x-}$.

REFERENCES

1. Haufe, O., Reich, A., Möschel, C., and Jansen, M., *Z. Anorg. Allg. Chem.*, 627, 23-27 (2001).
2. Haufe, O., Hecht, M., Grupp, A., Mehring, M., and Jansen, M., *PCCP*, submitted.
3. Okazaki, T., Lian, Y. F., Gu, Z. N., Suenaga, K., and Shinohara, H., *Chem. Phys. Lett.*, **320**, 435-440 (2000).
4. Kuran, P., Krause, M., Bartl, A. and Dunsch, L., *Chem. Phys. Lett.*, **292**, 580-586 (1998).
5. Krause, M., Hulman, M., Kuzmany, H., Kuran, P., Dunsch, L., Dennis, T. J. S., Inakuma, M., and Shinohara, H., *J. Mol. Struct.*, **521**, 325-340 (2000).
6. Krause, M., Kuran, P., Kirbach, U., and Dunsch, L., *Carbon*, **37**, 113-115 (1999).
7. Lebedkin, S., Renker, B., Heid, R., Schober, H., and Rietschel, H, *Appl. Phys. A-Mater. Sci. Process.*, **66**, 273-280 (1998).

CARBON NANOSTRUCTURE SYNTHESIS, PURIFICATION AND SEPARATION

Approaching the rational synthesis of carbon nanotubes

R. Herges, D. Ajami, S. Kammermeier

Institut für Organische Chemie, Universität Kiel, Otto-Hahn-Platz 4, D-24118 Kiel, Germany, e-mail: rherges@oc.uni-kiel.de

Abstract. The high temperature methods developed so far, yield heterogeneous mixtures of carbon nanotubes with different length, diameter, helicity and with defects alongside with amorphous carbon. Most applications, however, require samples with well defined physical properties and thus with uniform geometry. Template assisted growth methods and sophisticated purification methods not withstanding, many researchers in the field propose that a rational, wet-chemical synthesis should be developed to prepare bulk amounts of uniform nanotubes. We are aiming at this goal for more than 10 years and now present the status of our work. So far our tubes are short (8 A long) substructures of [4,4]armchair tubes which we therefore coined picotubes.

INTRODUCTION

Singled-wall carbon nanotubes (SWNT's) exhibit promising properties for their use in nanoelectronics. Field effect transistors and logic gates have been build using semiconducting SWNT's. The properties of these devices surpass their silicon counterparts in many aspects. However, there are a number of substantial problems to be solved before a carbon nanotube technology can be developed. One of the major drawbacks is the fact that the high temperature methods developed so far, yield heterogeneous mixtures of carbon nanotubes with different length, diameter, helicity and with defects alongside with amorphous carbon. Most applications, however, require pure samples with well defined physical properties and thus with uniform geometry. Template assisted growth methods and sophisticated purification methods not withstanding, many researchers in the field propose that a rational, wet-chemical synthesis should be developed to prepare bulk amounts of uniform nanotubes. We are aiming at this goal for more than 10 years and now present the status of our work. So far our tubes are short (8 A long) substructures of [4,4]armchair tubes which we therefore coined picotubes.

CP723, *Electronic Properties of Synthetic Nanostructures*, edited by H. Kuzmany et al.
© 2004 American Institute of Physics 0-7354-0204-3/04/$22.00

STRATEGY:

It is generally accepted among synthetic chemists that given "ample time money and skilled practioneers" any naturally occurring compound with even the most complicated structures can be synthesized. The rational synthesis of carbon nanotubes, however, causes particular problems. Because of the lack of functional groups and the symmetric all-carbon structure a distinct synthetic strategy and new synthetic methods have to be developed. In organic synthesis there are no tools for single molecule manipulation, so the strategy has to start with small building blocks with a well defined reactivity (rolling up a graphite sheet is not an option). Basically there are three different strategies to make tubes:[1]

1. Stacking rings
2. Assembling prefabricated, concave building blocks
3. Ring enlargement metathesis/dehydrocyclization

We pursue strategy 3.

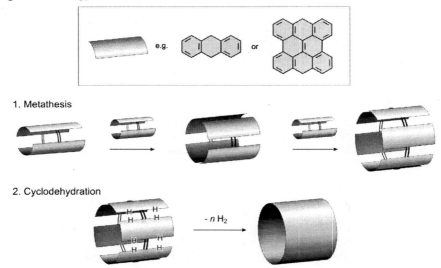

FIGURE 1. Ring enlargement metathesis and dehydrocyclization to synthesize carbon nanotubes.

What we need to start with, are aromatic plates like anthracene or naphthobianthrene which are connected by at least two double bonds to form a cycle. A ring enlargement metathesis reaction would then yield a dimer or higher cyclic oligomers and in a second consecutive step the tube walls are closed by dehydrocyclization.

EXPERIMENTS

Our starting material and "working horse" is tetradehydrodianthracene (TDDA) (two anthracene units as aromatic plates and two double bonds).[2] Because of its high strain energy the metathesis reactions are unequivocally driven towards ring enlargement. TDDA reacts with small annulenes like benzene to form belt-like conjugated systems. Unlike "normal" aromatic systems the inner lobes of the p orbitals all point towards the center of the system. The metathesis product of TDDA with benzene can be viewed as a small substructure of a [4,4]armchair nanotube. It is the first belt-like structure that was published in the literature.

28 %

belt-like conjugated

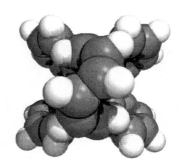

FIGURE 2. Metathesis of TDDA with benzene.[3]

Hence, the belt-like, in-plane conjugated bianthraquinodimethane building block enforces a belt-like conjugation also in the polyene bridge. A larger bridge should have a larger degree of conformational freedom and adopt a less strained trigonal planar configuration.

Reaction of TDDA with cyclooctatetraene, (the next higher homologue in the series of annulenes), according to our calculations, should give a twisted belt. In 1964 Heilbronner predicted that such Möbius annulenes should be aromatic with 4n electrons as opposed to the Hückel rule which predicts that "normal" aromatic systems are aromatic with 4n+2 electrons.[4]

This prediction roused much interest among theoretists as well as experimentalists. However, all attempts to synthesize Möbius annulenes so far failed.

In our case the in-plane, pyramidalized building block bianthraquinodimethane, should stabilize the 180° twist and thus make the Möbius compound more stable than all conceivable untwisted isomers. This effect can be easily rationalized using a cardboard model:

a) Roll up a rectangular strip of cardboard to form a half pipe and fix the bend by a stick connecting both ends. This is a model of the bianthraquinodimethane unit.
b) Then cut a disk with a large hole in the middle and cut the ring at any location you like. This broken ring represents the π system of the polyene bridge.
c) Try to connect both building blocks to form a ring. You immediately recognize that it is easier (involves less strain) to form a Möbius band with a 180° twist than a non-twisted ring.

How to translate this simple model into chemistry? To put this idea into (molecular) action we need a suitable starting material for the in-plane conjugated building block bianthraquinodimethane and the polyene bridge and a chemical reaction to connect both parts. We chose TDDA as the in-plane conjugated part, cyclooctatetraene (COT) to form the polyene bridge and ring enlargement metathesis to connect both parts. Figure 3 depicts this approach.

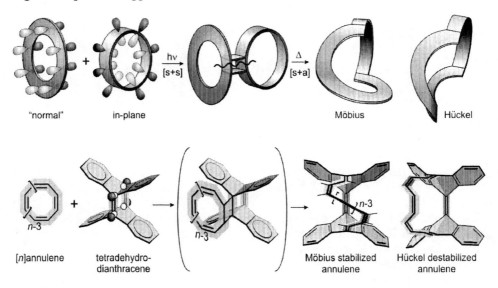

FIGURE 3. Ring enlargement metathesis to form a Möbius ring..

Upon irradiation a [2+2]cycloaddition should form a four-membered ring which immediately undergoes cycloreversion. Two conceivable ring-like products can form, an untwisted Hückel and a twisted Möbius type ring. According to our simple cardboard model the twisted Möbius ring should be more stable than the corresponding Hückel isomer. This is also confirmed by quantum chemical calculations (B3LYP/6-31G*). According to these DFT calculations the most stable Möbius isomer is 7 kcal/mol more stable than the most stable Hückel compound. The reaction in the laboratory, however, was not successful. No addition product was detected even after prolonged irradiation of both starting materials and applying a number of different reaction conditions. Cyclooctatetraene (COT) most probably acts as a triplet sensitizer, transfers triplet energy onto the TDDA which in turn undergoes an electrocyclic ring opening reaction to form bianthryl.

To circumvent these problems we had to find a synthetic equivalent for cyclooctatetraene which does not absorb light in the range we use for our metathesis reaction. Our choice was *syn*-tricyclooctadiene (TCO) which compared to cyclooctatetraene (COT) has two additional transannular single bonds. These bonds have to be cleaved at a later stage of the synthesis and one would end up with the same product we would have gotten if the metathesis with COT would have been successful.

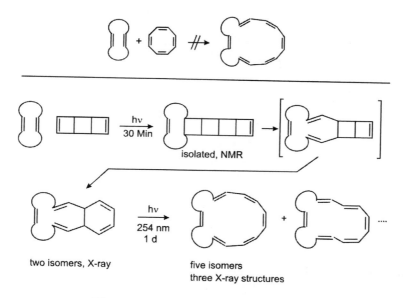

Figure 4. Metathesis with tricyclooctadiene.[5]

In our first attempt to react TCO with TDDA, irradiation with light of the wavelength >320 nm indeed yielded the metathesis product. However, only one of the two single bonds opened. Two isomers of this reaction product were isolated and characterized by X-ray analysis. Upon irradiation with a low pressure mercury lamp in a quartz apparatus we finally were able to open the second single bond and at the photostationary equilibrium obtained a mixture of several isomers. Five ring opened isomers were isolated and characterized, three of them by X-ray analysis. Three of the structures -as expected- exhibit a Möbius topology. We were lucky enough that one of the X-ray structures revealed a Hückel (non-twisted) topology. So we can now compare the properties of the Möbius (aromatic according to Heilbronners prediction) with the Hückel isomer (antiaromatic according to the Hückel rule).

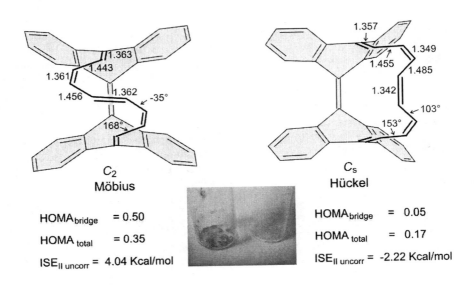

FIGURE 5. Properties of the Möbius and the Hückel isomer.

There are basically three different types of methods to determine the degree of aromaticity of a compound. They are based on the quantum chemical observables geometry, energy and magnetic properties. One of the most frequently used aromaticity probes based on geometry parameters is the bond length equalization method HOMA. According to this measure our Möbius annulene is about as aromatic as the central ring in phenanthrene, whereas the Hückel isomer is not aromatic at all. The ISE method of Schleyer et al. predicts a stabilization of the Möbius aromatic isomer of 4.04 kcal/mol (this is 20% of the value of benzene) and an antiaromatic destabilization of the Hückel isomer by 2.22 kcal/mol. Both methods (HOMA and ISE) finally confirm the prediction Heilbronner made 40 years ago.

As stated in the beginning TDDA should not only add to annules, but also react with itself to yield cyclic oligomers. We achieved the dimerization of TDDA by irradiation as a suspension in benzene in 35% yield. The resulting cyclic tetraanthracenylidene was termed picotube because with a length of 8.2 A and a diameter of 5.4 A it is smaller than the commercially available nanotubes.

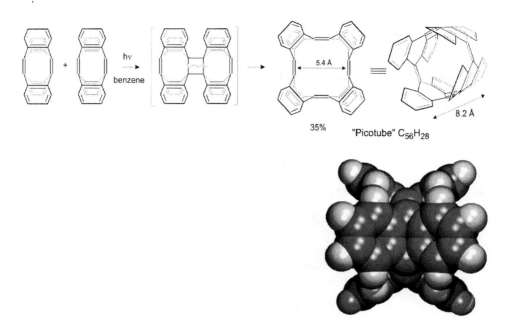

35%   "Picotube" $C_{56}H_{28}$

FIGURE 6. Dimerization of TDDA to form a pictube.[3]

To form a true short piece of a [4.4]armchair nanotube 8 hydrogen molecules have to be eliminated either thermally or by oxidation to close the tube walls completely. Unfortunately, all attempts so far failed. Semiempirical calculations predict the elimination of hydrogen to be extremely endothermic (8 steps each of which is endothermic with 25.8 kcal/mol). A [4.4]armchair tube with a diameter of 5.4 Å is extremely strained and therefore a straightforward solution to our problem is to increase the tube diameter. We recently achieved the dimerization of the tetraanthracenylidene (picotube) to form the octaanthracenylidene by irradiation of the picotube with a nitrogen laser. The yield is almost quantitative, the production rate, however, is still too small to do further experiments on a preparative scale. Theoretical calculations predict, that the dehydrocyclization (elimation of 16 hydrogen molecules) to form a short piece of a [8.8]nanotube is exothermic. Thus the reaction should even be feasible using the well established Kovacic conditions at room temperature.

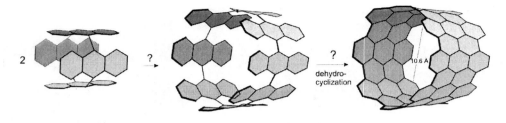

[8,8]nanotube

FIGURE 7. Dimerization of tetraanthracenylidene (picotube) to form an octaanthraceneylidene and the (still hypothetical) dehydrocyclization to yield a short piece of a [8,8]armchair nanotube.

The next step will be the "tube elongation". We pursue three different approaches:

a) a combination of a Diels-Alder reaction and oxidation (Clar's method)[6],[7]
b) chemical vapor deposition in the presence of $Fe(CO)_5$
c) fusing the picotubes inside a nanotube with a larger diameter[8]

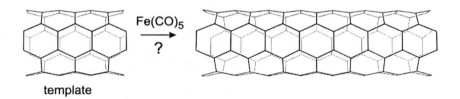

template

FIGURE 8. Tube elongation.

REFERENCES

[1] S. Kammermeier, P. G. Jones, R. Herges, *Angew. Chem. Int. Ed. Engl.* **1996**, *35*, 2669-71
[2] R. L. Viavatenne, F. D. Greene, L. D. Cheung, R. Majeste, L. M. Trefonas, *J. Am. Chem. Soc.* **1974**, *96*, 4342-4343.
[3] S. Kammermeier, R. Herges, *Angew. Chem. Int. Ed. Engl.* **1996**, *35*, 417-419.
[4] E. Heilbronner, *Tetrahedron Lett.* **1964**, 1923-1928.
[5] D. Ajami, O. Oeckler, A. Simon, R. Herges, *Nature* **2003**, 426, 819.
[6] E. Clar, *Chem. Ber.* **1948**, *81*, 52.
[7] E. Clar, Chem. Ber. **1932**, *65*, 850.
[8] In collaboration with Prof. Kuzmany (Wien)

Controlled growth of SWCNT on solid catalysts with narrow (n,m) distribution

Daniel E. Resasco*, Leandro Balzano[&], Jose E. Herrera*,
Olga Matarredona[&], and Liang Zheng*

* School of Chemical Engineering and Materials Science, University of Oklahoma
[&] SouthWest Nanotechnologies Inc. (SWeNT)
Norman, Oklahoma 73019 USA

Abstract. We have optimized a catalytic process to obtain high-quality SWNT over silica-supported Co-Mo catalysts in the temperature range 700-950°C. The excellent performance of this catalyst is based on the formation of a surface cobalt molybdate, in which, the Co-Mo interaction inhibits the Co metal agglomeration that typically occurs at high temperatures. This process is attractive because it can be readily scaled up, but at the same time is highly selective in terms of the diameter and chirality of the nanotubes that it produces.

INTRODUCTION

The production of single-walled carbon nanotubes (SWNT) with controlled diameter and chirality is a highly attractive target for scientists because the electronic properties of these unique materials are strongly related to their structure, which is typical described in terms of the chiral vector (n,m). In the last few years, we have concentrated on the tailoring of specific catalysts and operating conditions to grow SWNT by the Boudouart disproportionation of CO (decomposition into C and CO_2) [1-4].

Since the CO disproportionation is an exothermic reaction, at the high temperatures needed for the nanotube growth, the CO conversion can be limited by equilibrium. Therefore, since the reaction proceeds with a decrease in the number of moles, high CO pressures are required in order to counteract the effect of the temperature and drive the reaction in the forward direction. For example, at atmospheric pressure, the equilibrium conversion of a stream of pure CO is 83 % at 600°C, but only 15 % at 800°C and barely 1 % at 1000°C. For example, if the reaction is carried out isothermically at 700°C, the CO conversion at equilibrium shifts from 48% to 75 % when the pressure is increased from 1 atm to 10 atm. However, at higher temperatures, the conversions are very low and the effect of pressure is less pronounced. Therefore, the pressure needed to keep a moderate equilibrium conversion at temperatures above 800°C is indeed high. For example, to reach 30 % CO conversion at 900°C, 20 atm are needed. Therefore,

CP723, *Electronic Properties of Synthetic Nanostructures*, edited by H. Kuzmany et al.
© 2004 American Institute of Physics 0-7354-0204-3/04/$22.00

from a thermodynamic point of view, one would prefer to operate at low temperatures. However, it has been demonstrated that if the temperature is too low the selectivity towards SWNT drastically drops. Since the reaction is reversible, the buildup of CO_2 may adversely affect the SWNT production in a fixed bed reactor. We may expect a decrease in carbon deposition along the catalyst bed as the equilibrium condition is approached. Therefore, we have chosen to operate at low CO conversions in a fluidized bed reactor, at moderate pressures (1- 5 atm) and moderate temperatures (700-950°C).

Among a large number of catalysts investigated, we have found that bimetallic Co-Mo catalysts supported on silica exhibit the highest selectivity to SWNT. We have explained the excellent performance of the Co-Mo catalyst on the basis of the central role played by a surface cobalt molybdate. We have found that to be effective the catalyst requires both metals simultaneously present in a Co:Mo ratio significantly lower than unity to ensures that all the Co in the catalyst is forming a surface cobalt molybdate. This Co-Mo interaction inhibits the Co metal agglomeration that normally takes place at high temperatures. Cobalt agglomeration needs to be avoided to prevent the formation of less desirable carbon species such as graphitic nanofibers. The surface cobalt molybdate structure exhibits the right reducibility to resist decomposition unless it is in the presence of CO at high temperatures. These studies have been the basis for the development of a commercial process called CoMoCAT™ that can produce single-walled carbon nanotubes in large scale.

For many applications in nanoelectronics and nanosensors it is essential to have a nanotube material with specific electronic properties. The I-V characteristics of nanotubes are directly related to their diameter and chirality. Therefore, a process that allows controlling in a reproducible way the structure of nanotubes has a remarkable edge over non-selective processes. The nanotubes produced by the CoMoCAT™ process exhibit a uniquely narrow distribution of diameters, which can be controlled by adjusting the process parameters. The narrowest distribution is obtained at the lowest synthesis temperature. This characteristic of the CoMoCAT™ product has been confirmed by photoluminescence analysis performed in collaboration with the group of Prof. Weisman at Rice University [5]. It was observed that when dispersed in an appropriate solvent (i.e. Na-DDBS [6]) only two types of nanotubes represent the majority of the semiconducting nanotubes present in the sample synthesized at 750°C. By contrast, a similar analysis of the competing HIPCO material displays a much broader distribution of both, diameter and chirality. The two types of nanotubes observed in CoMoCAT sample are the (6,5) and (7,5), whose diameters are 0.75 nm and 0.82 nm, respectively. This result is in perfect agreement with the 0.8 nm average diameter measured by Raman spectroscopy, TEM, and STM. Interestingly, the distribution of chiralities is also very narrow. Both, the (6,5) and (7,5) nanotubes have a chiral angle near 27 degrees. By contrast, the HIPCO material exhibits a much broader distribution of chiralities.

We have further characterized this material by optical absorption in aqueous suspension using the DNA-stabilization method described by Zheng et al.[7]. The single-stranded DNA-assisted dispersion results in isolated nanotubes in aqueous suspension, which allows for a good resolution in the optical absorption spectrum. As shown in Fig. 1, the spectrum in the 500-1100 nm range clearly exhibits the first and second van Hove optical transitions (S_{11} and S_{22}) for the (6,5) and (7,5) nanotubes, the most abundant (n,m) semiconducting structures previously observed on this material by spectrofluorimetry [5]. It appears that in this particular sample the (6,5) is more abundant than the (7,5). Although the band at around 650 nm can contain contributions of both (7,6) and (7,5) nanotubes, which have S_{22} transitions at 647 and 644 nm, respectively [8], the band at 1025 nm is most certainly due to (7,5).

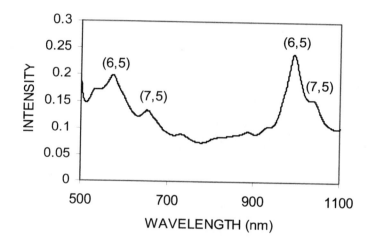

FIGURE 1. Absorption spectrum of the SWNT sample produced at 750°C suspended by sonicating in ssDNA aqueous solution.

The Raman spectra of all the samples synthesized in the temperature range 700-900°C exhibit a low D band, indicative of good quality nanotubes, with strong bands in the breathing mode region. Figure 2 shows the portion of the spectrum below 500 nm, corresponding to the breathing mode region using a 633 nm excitation laser. The main band at 282 nm is consistent with the frequency expected for the (7,5) nanotube [8]. We expect to be able to see a strong BM signal for this nanotube since the excitation wavelength (633 nm) is close to the resonance of the S_{11} band gap (644 nm). On the other hand, the wavelength for the most abundant nanotube in this sample (6,5) would be in resonance at 567 nm. Therefore, that Raman band is not observed in this spectrum. The second intense band at 260 nm is most probably due to the (7,6) nanotube, for which the resonance occurs at a wavelength of 647 nm, very close to that of the excitation laser. The low intensity of this band is in agreement with the low concentration of this nanotube as seen in the fluorimetric study.

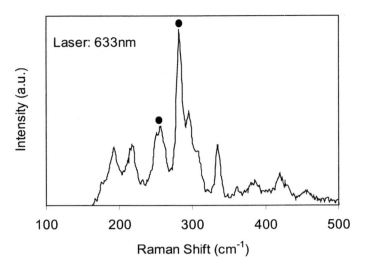

FIGURE 2. Raman spectrum of the SWNT sample produced at 750°C, purified in HF to remove the silica support. The two dots indicate the bands assigned to (7,5) and (7,6) nanotubes.

Due to the presence of the solid silica substrate that separates the growing nanotubes during the synthesis, the resulting bundles of SWNT are significantly thinner than what are typically obtained with methods in which the catalyst is in the vapor phase (i.e., arc discharge, laser ablation, HIPCO). While each of the bundles produced by any of the vapor phase methods contain 50 to 100 nanotubes, those obtained in the CoMoCAT process only contain 10-20 nanotubes. A sample with thinner bundles has several important advantages over one with thicker bundles. For example, for applications in flat panel displays (field emission) thinner bundles result in much lower voltages requirements for a given operating emission current. Lower onset voltages in field emission have a great impact on the cost and viability of flat panel displays. Similarly, in the area of polymer composites, thinner bundles can produce conducting composites with lower nanotube loadings, increasing the transparency of the material and reducing the cost.

Figure 3 illustrates the field emission characteristics of a SWNT sample produced in the CoMoCAT process at 850°C, compared to a sample produced by a different synthesis method. The I-V characteristics were obtained in a vacuum chamber at about 10^{-7} Torr on a 1 cm-diameter cylindrical electrode-counterelectrode system operating at a 0.25 mm spacing. A Keithley 237 source-measure unit was used for linearly ramping the voltage up to 1100 V and down to zero while measuring the current with pA sensibility. In all cases, the samples obtained in the CoMoCAT process showed good stability, meaning little deterioration in the sample after reaching a current density of almost 5 mA/cm^2, as reflected by the low hysteresis of the I vs. V curve. By contrast, a sample prepared by arc discharge (MER sample) displayed a much lower current and lower stability under the same operating conditions. Most interestingly, it has been found that the I-V characteristics can be systematically and reproducibly varied by varying the nanotube synthesis parameters.

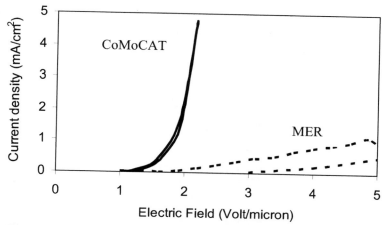

FIGURE 3. Field emission at 10^{-7} Torr. I-V characteristics of a SWNT sample (prepared at 850C by CoMoCAT) compared to a sample prepared by arc discharge (MER).

In summary, the CoMoCAT process exhibits the dual benefit of potential scalability and excellent control of nanotube structure. A very narrow distribution of nanotubes approaching monodispersity can be obtained.

ACKNOWLEDGEMENTS

Financial support from DoE Basic Energy Sciences (grant No. DE-FG03-02ER15345) and NSF Grant No. CTS-0308619 is gratefully acknowledged. SWeNT Inc. thanks OCAST, NSF and NASA for SBIR support.

REFERENCES

1. D. E. Resasco, W. E. Alvarez, F. Pompeo, L. Balzano, J. E. Herrera, B. Kitiyanan, and A. Borgna, *J. Nanopart. Res.* 2002, *4* 131
2. B. Kitiyanan, W.E. Alvarez, J. H. Harwell, and D. E. Resasco, *Chem.Phys. Lett.* 2000, *317*, 497
3. W. E. Alvarez, B. Kitiyanan, A. Borgna, and D. E. Resasco, *Carbon*, 2001, *39* 547
4. J. E. Herrera, L. Balzano, A. Borgna, W. E. Alvarez, D. E. Resasco, *J. Catal.* 2001, *204*, 129
5. S. M. Bachilo, L. Balzano, J. E. Herrera, F. Pompeo, D. E. Resasco, and R. B. Weisman, *J. Amer. Chem. Soc.* (2003), 125, 11186
6. O. Matarredona, H. Rhoads, Z. Li, J. H. Harwell, L. Balzano, and D. E. Resasco *J. Phys. Chem. B* (2003), 107, 13357
7. M. Zheng, A. Jagota, E. D. Semke, B. A. Diner, R. S. Mclean, S. R. Lustig, R. E. Richardson, N. G. Tassi, *Nature Materials* (2003), 2, 338
8. S. M. Bachilo, M. S. Strano, C. Kittrell, R. H. Hauge, R. E. Smalley, R. B. Weisman, Science, 298 (2002) 2361

Blowing Carbon Nanotubes to Carbon Nanobulbs

D. S. Su[a], Z.P. Zhu[a,b], Y. Lu[b], R. Schlögl[a], G. Weinberg[a], Z.Y. Liu[a]

a: Department of Inorganic Chemistry, Fritz Haber Institute of the MPG, Faradayweg 4-6,
D-14195 Berlin, Germany
b: State Key Laboratory of Coal Conversion, Institute of Coal Chemistry, Chinese Academy of Sciences,
030001 Taiyuan, China

Abstract. We report the blowing of multi-walled carbon nanotubes into carbon nanobulbs. This is realized in a unique tube growth environment generated by explosive decomposition of picric acid mixed with nickel formate. The carbon spherical bulbs are characterized by large dimensions (up to 900 nm), thin walls (around 10 nm), and fully hollow cores. The walls are in graphitic structure of sp^2 hybridized carbons. Bulb-tube assemblies are found as intermediate derivatives of blowing. A joint action of the filled high-pressure gases and the structural defects in the carbon nanotubes is responsible to the formation of the carbon nanobulbs. Our finding may indicate the possibility to engineer the carbon nanotubes to the designed nanostructures.

INTRODUCTION

Since the discovery of fullerene molecules of carbon [1] and the report on carbon nanotube [2], carbon in various novel forms such as buckles, rings, ribbons and multiple-graphene onions has been discovered or reported [3-9]. The nanocarbons in well-shaped form are considered to be possible due to the flexibility of carbon to form pentagonal and heptagonal carbon-rings. The combination of these two basic structural units with the hexagonal carbon-rings opens the world of nanocarbon with a variety of geometrical configurations. Exotic carbon nanostructures such as carbon calabashes [10-11], or carbon double helice and braids [12] are reported. Some of the non-planar carbon nanostructure can be explained by nucleation from a pentagonal atom ring or by substitution of hexagon in aromatic carbon rings by pentagons leading to curving and closed carbon structures. However, the growth mechanism for calabashes or double helices and braids still needs to be studied.

Recently, we reported on the spherical hollow carbon nanobulbs produced by an explosive decomposition of picric acid in the presence of nickel catalysts [13]. The structure of the nanobulbs is unique, characterized by large dimensions (up to 900 nm), thin walls (around 10 nm), and fully hollow cores. Unlike carbon spheres or carbon calabashes, the formation mechanism of carbon nanobulbs is tightly related to the thermal and mechanic properties of carbon nanotubes under special conditions. The nanobulbs are not formed directly from carbon species, rather they are derivated from carborn nanotubs by a mechanism like blowing.

CP723, *Electronic Properties of Synthetic Nanostructures*, edited by H. Kuzmany et al.
© 2004 American Institute of Physics 0-7354-0204-3/04/$22.00

EXPERIMENTAL

Picric acid and nickel formate (weight ratio of 3:1 ~ 5:1) were mixed physically and loaded into a sealed stainless steal pressure vessel (10.8 cm^3) with a loading density of 0.25 g/cm^3. The decomposition reaction was induced by heating to 310 °C at a rate of 20 °C/min. The reaction, lasting on a microsecond scale, generated pressure of about 40 MPa (shock wave, after which the equilibrium pressure is about 15 MPa) and temperature of about 930 °C. After the reaction, the vessel is cooled to room temperature in air and emptied of gaseous products (dominant with CO, CO_2, N_2, and H_2O, as revealed by gas chromatography), and the solid products are collected. For the acid treatment, 50 mg sample was introduced into 100 ml 6 M HCl aqueous solution and refluxed for 16 h with magnetic stirring. After the refluxing, the suspension was filtrated, washed with deionized water and ethanol, and dried at 110 °C overnight. SEM images were taken on a Hitachi S-4000 scanning electron microscope with a field-emission gun operated at 15-kV accelerating voltage. TEM images were taken using a Phillips CM200 FEG electron microscope operated at 200 kV. This machine is equipped with a GATAN imaging filter GIF100 for electron energy loss spectroscopy (EELS) measurement.

RESUTLS AND DISCUSSION

Fig. 1a shows a SEM micrograph of the solid products from the decomposition of the picric acid and nickel formate mixture. Spherical structures are found along with fibre-like material. TEM investigation reveals that the fibre-like structures are multi-walled carbon nanotubes, while the spherical structures are carbon hollow nanobulbs (rather than solid carbon spheres). Most of the nanobulbs contain nickel particles formed during the decomposition of nickel formate (Fig. 1b). But few empty nanobulbs are also found in the product. The diameters of the nanobulbs vary from 100 to 900 nm. The yield of carbon nanobulbs is about 30% of the deposits.

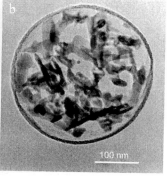

FIGURE 1. a) SEM image showing that the product contains large amount of bulbs along with carbon nanotubes. b) TEM image of a bulb containing nickel particles.

The bulbs exhibit a wall thickness in the range of 6- 25 nm with most of around 10 nm. As it can be seen, the walls are uniform in thickness over the entire sphere and consist of only carbon, as revealed by EELS and EDX. Carbon K-edge EELS spectrum also confirms that carbon is in sp^2 hybridization state and consequently a graphitic short-range ordering. High-resolution TEM image (Fig. 2a) reveals that the walls are also well ordered in long range, constructed by an assembly of concentric graphene shells with a layer distance of 0.34 nm. The nickel particles in the bulbs can be simply removed by treatment with 6 M HCl. Fig. 2b presents a typical bulb after the treatment, which clearly shows that the bulb has been completely emptied.

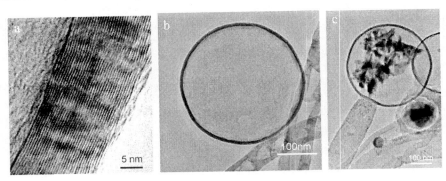

FIGURE 2. a) High-resolution TEM of a nanobulb wall; b) TEM image of an empty bulb; c) An intermediate derivative showing a bulb-tube assembly.

The obtained bulbs, characterized by large dimensions, thin walls (diameter to wall-thickness ratio 100:1) and fully hollow cores are an unique carbon structure. They differ from onion-like multi-graphene shell carbon or solid carbon spheres in the hollow core. The carbon nanobulbs are also different from the graphitic hollow carbon calabashes found among the solid carbon spheres synthesized by a mixed-valent oxide-catalytic carbonization process [10], or from the carbon calabashes in a shock-produced carbon melt [11]. However, they are common in the large size and hollow-shell graphitic structure. Unlike fullerenes and solid carbon spheres, the formation and growth of nanobulbs or calabashes cannot be explained by the nucleation from a pentagonal atom ring producing an inward surface with positive curvature and therefore a spiral growth.

To understand how the bulbs are formed, we have to study the electron micrograph in Fig.1 again. Although freestanding bulbs are found, some of them connect chemically with carbon nanotubes in a flask-like structure. Bulbs at the middle of bamboo-like tubes are also found. These bulb-tube assemblies are symmetrical in their morphology along the tube axis, and the thickness of bulb walls is always thinner than the thickness of the tube wall (Fig. 2c). They can be considered as intermediate derivatives and reflect the "frozen" stages during the formation, i.e., the bulbs are formed from tubes by their volume expansion, accompanied by a decrease for the thickness of tube (or bulb) wall as the surface increases. The driving force of the expansion of the carbon nanotube could be the positive pressure difference between the inside and outside of the tube. As mentioned above, high pressure and high temperature are generated by decomposition. During the tube growth, hot and high-pressure gases fill the tube. Once the temperature outside the tubes drops after the decomposition, the temperature difference between inside and outside of the tube causes the pressure difference [13]. This induces the tube expansion and leaves behind flask-like assemblies and nanobulbs. This process is very similar to the situation in conventional glass blowing.

Beside the positive pressure difference, tubes in closed form and defects are additional conditions for the blowing. The fact that the most blowing starts on the dome of a tube could be due to the high density of defects on the cap of the tubes (for instance, pentagon rings or Stone-Walses defects). However, the blowing of tubes to the large bulbs reported here is only possible if carbon nanotubes exhibit, at least during their generation, excellent thermoplasticity and expansibility, in opposition to the usual opinion that the thermal expansion coefficient of carbon nanotubes is near zero. Our finding suggests that it is possible to engineer tubular structures at nano-scale into various shaped devices by adjusting and controlling reaction environments.

SUMMARY

In summary, large carbon nanobulbs are produced by an explosive decomposition of picric acid at presence of nickel formate. The blowing of carbon nanotubes, a joint action of the filled high-pressure gases and the structural defects in the tubes, leads to the formation of the unique new carbon nanostructure. This finding exhibits the possibility to engineer tubular structures at nano-scale to designed forms.

ACKNOWLEDGMENTS

The work is partially supported by NSFC (59872047) and by the Deutsche Forschungsgemeinschaft (Project SCHL 332), performed in frame of ELCASS. Z. P. Zhu thanks the A.v.H. foundation for fellowship support.

REFERENCES

1. Kroto, H.W., Heath, J.R., O'Brien, S.C., Curl, R.F., Smalley, R.E., Nature **318**, 162 (1985).
2. Iijima, S., Nature **354**, 56 (1991).
3. Liu, J., Dai, H., Hafner, J. H., Colbert, D. T., Smalley, R. E. Nature **385**, 780 (1997).
4. Sano, M., Kamino, A., Okamura, J., Shinkai, S. Science **293**, 1299 (2001).
5. Martel, R., Shea, H. R., Avouris, Ph. Nature **398**, 299 (1999).
6. Falvo, M. R., Clary, G. J., Taylor, R. M., Brooks, F. P., Washburn, Jr. S., Superfine, R. Nature, **389**, 582 (1997).
7. Lu, J. P. *Phys. Rev. Lett.* **1997**, *79*, 1297.
8. Chopra, N. G., Benedict, L. X., Crespi, V. H., Cohen, M. L., Louie, S. G., Zettl, A. Nature **377**, 135 (1995).
9. Ugarte, D. Nature **359**, 707 (1992).
10 Wang. Z.L., Yin, J.S., *Chem. Phys. Lett.* **289**, 189 (1998).
11. Rietmeijer, F.J.M.,, Schultz, P. H., Bunch, Th. E., *Chem. Phys. Lett.* **374**, 464 (2003).
12. Lin, J., Zhang, X., Zhang, Y., Chen, X., Zhu, J., *Mater. Resear. Bul.* **38**, 261 (2003).
13. Zhu, Z.P., Su, D.S., Lu, Y., Schlögl, R., Weinberg, G., Liu, Z.Y., *Adv. Mater.* **16**, 443 (2004)

On the temporary growth of multi-walled carbon nanotubes

K. Bartsch and A. Leonhardt

*Leibniz Institute of Solid State and Materials Research Dresden, Helmholtzstr. 20,
D-01069 Dresden, Germany*

Abstract. The growth of aligned multi-walled carbon nanotubes (MWNT) on thin Co layers using microwave CVD has been investigated. The growth rate of the MWNT with tubular structure was found to be higher than those of the bamboo-like MWNT. The estimated activation energies for the tubular and the bamboo-like growth point to a different state of the catalyst particles during the growth. A molten particle state is assumed in the case of the tubular structure. The tube growth was time limited, which could be related to the meta stable catalyst particle state.

INTRODUCTION

Aligned multiwalled carbon nanotubes have been prepared by various chemical vapor deposition processes for some years, and especially, by bias assisted plasma CVD using substrates coated with thin catalyst layers (e. g. from Fe, Ni, Co) . Due to the width of the size distribution of the catalyst particles, formed from the layers during the process, dimensional and structural changes appear in CVD MWNT layers. For example, the two basic structures known as tubular and bamboo-like structure, respectively, are very often simultaneously present in the obtained MWNT layers and there are also apparent differences in length and diameter of the tubes. The growth of aligned MWNT on oxidized silicon substrates coated with thin cobalt layers by microwave assisted CVD was investigated with respect to the particle size and state as well as the tube structure.

EXPERIMENTAL

The deposition was performed in a simple vertical quartz tube reactor mounted in a microwave cavity. Oxidized silicon (thickness of SiO_2: 1 µm) coated with 2 nm thick Co layers by magnetron sputtering was used as substrate (size: 9 mm x 9 mm). The gas feedstock was a H_2-CH_4 mixture (H_2: 100 sccm, CH_4: 0.6 sccm in case of the fig. 5 and 0.8 or 2.0 sccm in case of the fig. 6). Heating of the substrates occurred by the microwave plasma, which was additionally assisted by a dc-bias voltage of 400 V (current: 100 – 150 mA). The deposition temperature amounted to the range 1123 – 1373 K (measured by a two colour pyrometer). The substrates were subjected to a short thermal pre-treatment in a pure hydrogen plasma (1173 K, 10 min). SEM, TEM and HRTEM were used to characterize the samples.

CP723, *Electronic Properties of Synthetic Nanostructures*, edited by H. Kuzmany et al.
© 2004 American Institute of Physics 0-7354-0204-3/04/$22.00

RESULTS AND DISCUSSION

The obtained MWNT layers consisted of bunches of individual tubes, aligned perpendicular to the substrate surface. According to the SEM and TEM investigations, the tip growth mechanism occurred. HRTEM investigations demonstrated the tubular [1, 2] and the bamboo-like [3-5] structure to have been formed simultaneously in the entire range of deposition temperature, but with different abundance in dependence on the deposition time and temperature. After short deposition time the tubular structure was prevailing, whereas long deposition times dominantly yielded the bamboo-like structure (deposition temperature 1273 K). Fig. 1 and 2 show the tips of MWNT bunches after different deposition time. It is visible from fig. 1 that smaller spherical and droplet-like catalyst particles are positioned at the MWNT tips (dark particles), which are characteristic of the tubular structure. After the longer deposition time larger cone-like particles are revealed, that are typical for the bamboo-like structure (fig. 2). The insets of the figures show the structures in HRTEM resolution.

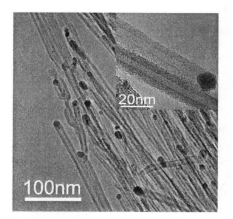

Figure 1. Tubular MWNT, 10 min

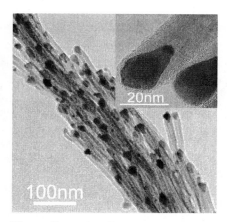

Figure 2. Bamboo-like MWNT, 120 min

Further investigations were devoted to the particle size distribution in dependence on the process time. Fig. 3 shows the TEM micrograph of a lamella, prepared by the focussed ion beam method, from a sample after thermal treatment in a pure bias assisted hydrogen microwave plasma. It is obvious that Co particles (dark) of different size are formed. Furthermore, the SiO_2 interlayer appears to be attacked. Particle size distribution curves for a thermally treated substrate (T = 1073 K) and for the particles in the tips of tubes after different deposition times are given in the diagram shown in fig. 4. No doubt, the curves are shifted to larger particle diameters with time. The curve for the thermally treated substrate and that for the longest deposition time reveal their maximum at the same position. The findings discussed above indicate at least the tubular growth to be faster than the bamboo-like one and the length growth of the tubes (in both cases) to be limited by time. The dominance of the bamboo-like structure in the bunches after long deposition time, the size distribution of the particles after thermal pre-treatment and long deposition time imply that the particles promoting the growth of the bamboo-like structure are formed in excess during the process.

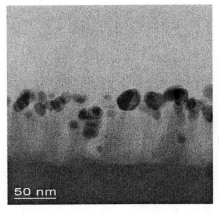

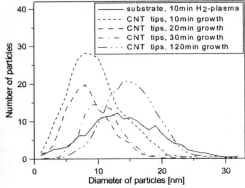

Figure 3. Co particles on substrate surface after thermal treatment, SiO₂ interlayer partially etched (TEM bright field micrograph)

Figure 4. Size distribution of Co particles on substrate after thermal treatment and in the tips of the MWNT (estimated by TEM)

Fig. 5 demonstrates the tube length in dependence on the deposition time. Indeed, the growth rate is higher for short deposition time which the tubular structure is observed for. Furthermore, the assumed time limitation of the growth appears at about 20 min for the tubular structure and at about 100 min for the bamboo-like structure. The limitation of the MWNT growth on catalyst layers which was also found for other kinds of CVD processes [5, 6], is controversially discussed in literature. We assume, that sudden change of the meta stable particle state into a stabile state stops the growth by precipitation of closed carbon shells surrounding the catalyst particles [7].

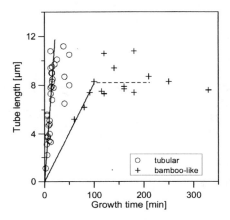

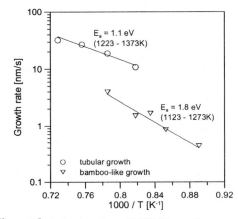

Figure 5. Tube length vs. growth time

Figure 6. Arrhenius plots of MWNT growth

38

Assuming the growth to be controlled by the carbon diffusion through the catalyst particles [8], the growth rate must be related to the particle size, which is frequently demonstrated in the literature [9, 10]. We assume furthermore that the particle state governs the tube structure and acts on the growth rate. Other than in the case of the bamboo-like structure, we assume the catalyst particles to be in molten state for the tubular tubes. The spherical shape of the particles might support this assumption. Different particles states should be related to different activation energies for the tubular and bamboo-like growth. Fig. 6 shows the dependence of the growth rate on the inverse deposition temperature. The deposition time was 3 – 10 min for the tubular MWNT and 50 min for the bamboo-like MWNT. From the slope of the linear curves the activation energies were calculated, which are given in the diagram. The value obtained for the bamboo-like structure is in agreement with data given by Smith [11] (E_a = 1.8 eV, 1125 – 1370 K). The activation energy for the tubular growth is significantly lower and does not contradict the assumption of a molten particle state.

CONCLUSIONS

The particle size and state as well as the deposition time control at a fixed temperature the length growth and structure of aligned MWNT deposited on thin Co catalyst layers. Short deposition times favour the tubular growth at adequate temperature. The different growth rates of the tubular and the bamboo-like tubes are not only caused by the catalyst particle size but also by the particle state. The estimated activation energies for the growth do not contradict the assumption of a liquid particle state in the case of the tubular growth and a solid particle state for the bamboo-like growth. A reason for the limited growth of the tubes is seen in the meta stable catalyst particle state.

References

1.   Zhang, W.D., Wen, Y., Liu, S.M., Tjiu, W.C., Xu, G.Q., Gan, L.M., Carbon **40**,1981-1989 (2002)
2.   Li, W.Z., Xie, S.S., Qian, L.X., Chang, B.H., Zou, B.S., et al., Science **274**,1701-1703 (1996)
3.   Fan, S.S., Chapline, M.G., Franklin, N.R., Tombler, T.W., et al., Science **283**,512-514 (1999)
4.   Lee, C.J., Park, J., Kang, S.Y., Lee, J.H., Chem Phys Lett **323**,554-559 (2000)
5.   Jeong, H.J., Shin, Y.M., Jeong, S.Y., Choi, Y.C., et al. Chem Vapor Deposition **8**,11-15 (2002)
6.   Bower, C., Zhou, O., Zhu, W., Werder, O.J., Jin, S., Appl Phys Lett **77**,2767-2769 (2000)
7.   Bartsch, K., Leonhardt, A., Carbon xxx(2004), in press
8.   Baker, R.T.K., Carbon **27**,315-323 (1989)
9.   Chhowalla, M., Teo, K.B.K., Ducati, C., et al., J Appl Phys **90**,5308-5317 (2001)
10. Choi, Y.C., Shin, Y.M., Lee, Y.H., Lee, B.S., et al., Appl Phys Lett **76**,2367-2369 (2000)
11. Smith, R.P., Trans Metall SocAIME **476**,230 (1964)

Synthesis, properties and possible applications of helical carbon nanotubes

A. Szabó[1], A. Fonseca[1], A. Volodin[2], C. Van Haesendonck[2],
L. P. Biró[3], J. B.Nagy[1]

[1] *Facultés Universitaires Notre Dame de la Paix, 61 Rue de Bruxelles, B-5000 Namur, Belgium.*
[2] *Laboratorium voor Vaste-Stoffysica en Magnetisme, Katholieke Universiteit Leuven, B-3001 Leuven, Belgium.*
[3] *Research Institute for Technical Physics and Materials Science, H-1525 Budapest, P.O.Box 49, Hungary.*

Abstract. Helical carbon nanotubes were produced on silica supported Co catalysts by chemical vapour decomposition of acetylene. The catalysts were prepared applying two methods: ion-adsorption precipitation and sol-gel technique. We compared the quality of the products obtained on these catalysts, the amount and characteristics of coiled carbon nanotubes. The characterization of products was based on transmission electron microscopy (TEM), scanning electron microscopy (SEM) observations and on thermogravimetric measurements.

Coiled nanotubes have been found to be of various shapes, diameter and pitch. Some of them are extremely long, 2.5 μm, with regular helices. The average outer diameter of coils are about 16 nm, the pitch is in the range 50-400 nm and the length about 1-1.4 μm. The helix-shaped windings of the helical carbon nanotubes reveal characteristic mechanical resonances, which are determined by the elastic modulus, mass, shape and dimensions.

INTRODUCTION

Since the first observation of carbon nanotubes (CNTs) by Iijima [1], several papers have been published concerning various techniques of their synthesis, purification and properties [2-3]. Helical carbon nanotubes were predicted by Dunlap [4] and Ihara et al. [5]. The coiling of the proposed structures originated from the regular insertion of pentagons and heptagons in the hexagonal honeycomb network.

Relatively large amount of curved and coiled nanotubes among the straight ones produced by catalytic vapour decomposition stimulated several studies on the modelling of the coiling mechanism [6, 7]. Recently Biró et al. [7] applied the idea of Laszlo and Rassat by assembling more complex patterns of azulenes and hexagons and showed that specific wrappings of these structures lead to a new variety of toroidal, coiled, screwlike and pearl-necklace-like formations. The coiling appears naturally by rolling up a Haeckelite-like stripe and does not require the insertion of additional polygons. The pentagons and heptagons are important parts of these structures and their very specific arrangement is the basis for the formation of regular helical coils.

The helical carbon nanotubes are described by the coil diameter, where observations range from 10 nm to 1 μm and the pitch- distance between adjacent corresponding points along the axis of the helix- which has been observed to take

CP723, *Electronic Properties of Synthetic Nanostructures*, edited by H. Kuzmany et al.
© 2004 American Institute of Physics 0-7354-0204-3/04/$22.00

values from 10 nm up to 5 μm. They could be used in composites because they would be better anchored in their embedding matrix than straight nanotubes. Their shape would favour a better load transfer to the matrix than in the case of straight tubes, and possibly easier infiltration. They could be attractive objects for nanoelectro-mechanical systems, too [8].

In this present work we report the synthesis of coiled carbon nanotubes produced on silica supported cobalt catalysts by decomposition of acetylene. We were interested of the influence of the catalyst preparation methods and of the influence of metal loading on the quality of carbon deposit concerning the formation of coiled nanotubes.

EXPERIMENTAL

Preparation methods of the catalysts:

a) Ion-adsorption precipitation method

The appropriate amount of $Co(CH_3COO)_2.4H_2O$ (Aldrich) was dissolved in distilled water. 3 g of silicagel (Merck part. diam. 15-40 μm) were added to the solution under stirring. The mixture was stirred about 20 min. The pH was controlled and adjusted to pH 9 using ammonia solution (Vel). After 48 h the pH of the as-prepared mixture was controlled and set again to 9, just before the filtration. Then the precipitate was washed with distilled water, dried at 120 °C overnight and grind.

b) Sol-gel method

This technique is based on the hydrolysis of tetraethoxysilane (TEOS-Acros) in an acidic or basic environment. Preparation process: 10 ml of ethanol (Fluka) and 10 ml of TEOS was mixed, then 15 ml of 1.5 M solution of $Co(CH_3COO)_2.4H_2O$ (Aldrich) was added to the mixture and stirred for 20 min. A few drops of conc. HF (Vel) were added to promote the reaction and the stirring was continued for another 20 min. The as-prepared gel was dried at 60 °C overnight, then grind and calcined for 1 h at 450 °C in air.

The reactions were carried out in a fixed bed flow reactor using acetylene as carbon source and N_2 as inert gas. Every time 1 g of catalyst was introduced in the reactor. Reaction temperature was kept at 700 °C, reaction time was usually 30 min., gas flow rate was 30 ml/min of C_2H_2 in 300 ml/min of N_2. The quality of carbon deposit was observed by microscopic techniques: TEM (Tecnai, Philips), SEM (XL Series, Philips); the purity was controlled by thermogravimetric measurement. The carbon deposit was calculated as follows:

$$\text{Carbon deposit (\%)} = 100 \ (m_{tot}-m_{cat}/m_{cat}),$$

where m_{tot} is the mass of total product and m_{cat} is the mass of dried catalyst.

RESULTS AND DISCUSSION

The first part of our results is summarized in Table 1. Here we compare the products obtained on the catalyst prepared by method a) with different Co loadings (2.5-12.5 wt.%). The comparison is based on the microscopic observations.

TABLE 1. Carbon deposit and quality, for catalysts prepared by method a).

Wt.% of Co	2.5	5	10	12.5
Coiled tubes	+	+++	+++	+++
Other tubes	++	++++	++++	++++
Carbon deposit (%)	40	50	60	60

In the case of 2.5 wt.% metal content, the product was not rich in carbon nanotubes and coils. There were a lot of amorphous carbon and catalyst particles. In the other cases, the catalysts had a good activity and their surfaces were totally covered by CNTs. We observed several morphologies of nanotubes such as tubes with different curvatures, helical tubes, spirals. The coiled tubes were examined and generally, the pitch varied from 50 to 200 nm and the coil diameter was in the range of 50-150 nm. We observed several coils with length up to 5 μm. The coils usually screwed themselves and in some cases they were forming interesting structures. The total carbon deposit was about 40-60 % and the amount of coiled nanotubes was approximately 5 % (Fig.1).

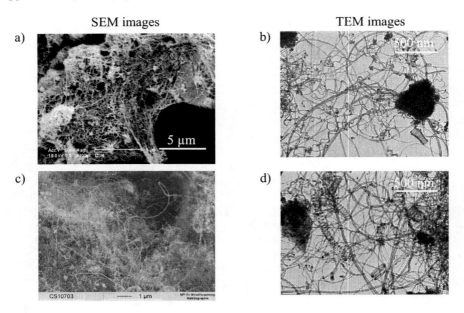

FIGURE 1. CNTs produced on Co/SiO$_2$ catalysts prepared by method a) (30 min reaction time at 700 °C, 30 ml/min of C$_2$H$_2$, 300 ml/min of N$_2$). a) and b) 5 wt.% Co loading; c) and d) 10 wt.% Co loading.

Most of the tubes do not contain metal particles and their tips are open. In some cases we observed coils that after some coiling continue as straight tubes. Probably the activity of the catalyst changed during the reaction and this is the reason why the form of the tube changed. The pitch and coil diameter generally do not change noticeably. In case of long tubes that screwed themselves, the orientation of coiling could also change.

On the SEM pictures of the samples synthesized on the catalysts prepared by method b) (sol-gel method; Fig. 2a) there are high amounts of CNTs, among which some are coiled nanotubes. Based on TEM pictures we made an approximate estimation of the amount of coils in the samples to be about 15 %. The carbon deposit was about 40 % and the diameter of the tubes was in the range 15-30 nm. The pitch values were found in the range of 30-130 nm and the coil diameters were 35-100 nm (Fig. 2). There were some exceptions, i.e. pitch up to 1100 nm, coil diameter up to 300 nm. The coiled carbon nanotubes found in these samples were very regular and there are more tubes like 'telephone wires' or springs in the product. Most of the helical tubes are long coils.

SEM image   TEM image

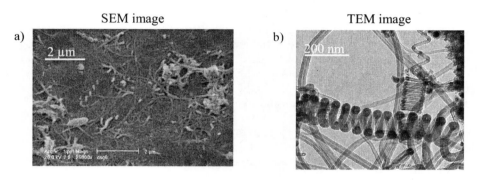

FIGURE 2. Coiled nanotubes produced on Co/SiO$_2$ catalyst prepared by sol-gel method (30 min reaction at 700 °C, 30 ml/min of C$_2$H$_2$, 300 ml/min of N$_2$).

Some of the coils were suspended in ethanol and transferred to silicon plates for the measurements of their mechanical properties [9]. After adsorption on an oxidized silicon substrate, the coiled CNTs form three-dimensional structures with freely suspended windings (Fig 3). The helix-shaped windings reveal characteristic mechanical resonances, which are determined by the elastic modulus, mass, shape, and dimensions. The suspended windings are resonantly excited in situ at the fundamental frequency by an ultrasonic transducer connected to the substrate. When the tip of the atomic force microscope (AFM) is positioned above the winding, the cantilever is unable to follow its fast oscillations. Nevertheless, an oscillation-amplitude dependent signal is generated due to the nonlinear force-to-distance dependence [10]. The helix-shaped CNTs can be used as convenient mechanical resonant sensors.

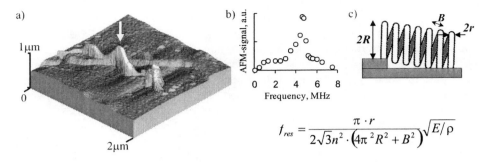

FIGURE 3. AFM image of a suspended coiled nanotube bridging a step formed by the edge of a gold pad on a silicon substrate (a). Vertical displacement of the AFM tip, which is positioned above the coiled nanotube (see white arrow in (a)), as a function of the excitation frequency (b). Model of a suspended helical carbon nanotube (c), used for calculations $(B \sim 0.5\ \mu m;\ 2r \sim 20\ nm;\ 2R \sim 0.5\ \mu m;\ n \sim 4$ is the number of windings; E is the elastic modulus ~ 0.2 TPa; ρ is the density $\sim 1300\ kg/m^3$).

As a conclusion, helical carbon nanotubes were produced on silica supported Co catalyst by chemical vapour decomposition of acetylene. Catalysts were prepared by sol-gel and ion-adsorption precipitation method. The carbon products grown on them were examined. The coiled nanotubes produced possess various morphologies: slightly curved, springs or 'telephone wire' like tubes. Products grown on the first type of the catalysts were rich on springs with regular helicity. The proportion of helical tubes was about 15%, in the sample. In the second type, the amount of coils was about 5% and they presented more a sinusoid-like shape, with large helicity.

ACKNOWLEDGEMENTS

The authors acknowledge the E.C. (RTN contract NANOCOMP HPRN-CT-2000-00037), the Belgian program PAI (P5/1), Region of Wallonia (SYNATEC convention n° 0014526) and OTKA (grant T043685) for financial support.

REFERENCES

1. S. Iijima, Nature **354**, 56 (1991).
2. T.W. Ebbsen, Annuel Rev. Mat. Sci. **24**, 235 (1994).
3. C. Journet, P. Bernier, Appl. Phys. A **67**, 1 (1998).
4. B. J. Dunlap, Phys. Rev. B **46**, 1933 (1992).
5. S. Ihara, S. Itoh, J. Kitakami, Phys. Rev. B **48**, 5643 (1993).
6. A. Fonseca, K. Hernádi, J. B.Nagy, Ph. Lambin, A. A. Lucas, Synthetic Metals **77**, 235 (1996).
7. L. P. Biró, G. I. Márk, A. A. Koós, J. B.Nagy, Ph. Lambin, Phys. Rev. B **66**, 165405 (2002).
8. M. Ahlskog, E. Seynaeve, R. J. M. Vullers, C. Van Haesendonck, A. Fonseca, K. Hernádi, J. B.Nagy, Chem. Phys. Lett. **300**, 202 (1999).
9. A. Volodin, M. Ahlskog, E. Seynaeve, C. Van Haesendonck, A. Fonseca and J. B.Nagy, Phys. Rev. Lett. **84**, 3342 (2000).
10. A. Volodin, C. Van Haesendonck, R. Tarkiainen, M. Ahlskog, A. Fonseca, J. B. Nagy, Appl. Phys. A **72**, S75 (2001)

A low cost method for the synthesis of carbon nanotubes and highly Y-branched nanotubes

O.T. Heyning[a,b], L. Kouwenhoven[b], P. Bernier[a], M. Glerup[a,c]*

[a]GDPC(UMR5581), Université Montpellier II, Pl. E. Bataillon, 34095 Montpellier, Cedex 5, France
[b]Department of Applied Physics and DIMES, Delft University of Technology, Lorentzweg 1, 2628 CJ Delft, The Netherlands
[c]Grenoble High Magnetic Field Laboratory, MPI-FKF/CNRS, 25, Avenue des Martyrs, 38042 Grenoble, Cedex 9, France

Abstract. Using a novel simplified low cost set-up in a CVD type of process using the aerosol technique, carbon nanotubes and branched nanotube trees have been synthesized. In this set-up a catalyst precursor solution of metal-salts dissolved in water is sprayed into a furnace as aerosols. In the furnace a mixture of carbon reactant gas and hydrogen forms a heated atmosphere. In this atmosphere the active catalyst particles are formed and reduced *in-situ* and the growth of the nanotubes occurs. Controlled by a number of parameters, the active catalyst particles induce the growth of carbon nanotubes and highly Y-branched nanotubes.

INTRODUCTION

Since the discovery of fullerenes [1] and carbon nanotubes [2], research in the field of carbon nanostructures has gained a lot of attention.

The main problem for the application of nanotubes is the absence of well-controlled, low-cost and high-yield methods for production. Chemical vapour deposition (CVD), is a well-known method for the synthesis of carbon nano-fibers, multi-walled nanotubes (MWNTs) [3,4,5], and single-walled nanotubes (SWNTs) [6,7]. The CVD process requires lower reaction temperature and gives the possibility to grow nano-structures directly on wafers or devices. The only disadvantage of this method is the lack of possibilities for making it a continues method. Equipping a classical CVD set-up with an aerosol injector can be a way to change the batch type process into a continuous synthesis method.

Aerosol-CVD syntheses are generally performed with a complex catalyst precursor, which is mixed with a carbon solvent serving as the carbon source for the growth of the CNTs. We recently demonstrated that metal-salts/complexes are suitable catalyst precursors for the synthesis of nanotubes [8]. This work presents a further extension of the approach of lowering the costs of the synthesis of carbon nanotubes. We here demonstrate that it is possible to use water as the solvent for the catalyst precursors, and cheap carbon gasses as the carbon source. The results given in this communication

CP723, *Electronic Properties of Synthetic Nanostructures*, edited by H. Kuzmany et al.
© 2004 American Institute of Physics 0-7354-0204-3/04/$22.00

shows that with this set-up it is possible to produce carbon nanotubes. Using this low-cost synthesis approach we were able to form high quality and very pure samples of branched nanotubes.

EXPERIMENTAL

In Fig. 1 a sketch of the experimental set-up is given, which is also presented in detail in [15]. A commercial aerosol injector is connected to a round bottom flask containing the solution of the catalyst precursor, dissolved in distilled water. The aerosol injector is run by a mix of hydrogen and argon gases and has a re-circulation system leading the excess liquid away from the injector and back to the flask. The outlet of the aerosol injector is fitted to the entrance of a quartz tube (7 cm in diameter, 120 cm long) which is placed in a horizontal high-temperature furnace. The generated aerosols flow into the heated quartz tube. The catalyst particles are formed *in-situ* as the water evaporates and as the metal-ions are reduced by the hydrogen gas. Between the aerosol injector and the entrance of the quartz tube an additional gas inlet is mounted. Through this inlet a controlled mix of hydrogen and a carbon gas enters the oven where the aerosols are injected. Reactions have been carried out in the temperature range from 800 to 1150°C. As metal catalyst precursors iron(II)acetate ($[Fe(CH_3CO_2)_2]$, Aldrich 95%) was used. The salt was dissolved in distilled water to give solutions with iron concentrations on 32 mM predicted to generating catalyst particles with a diameter on around 20 nm [8]. As the carbon source methane has been used, in flows from 0.125 nl/min to 0.8 nl/min. The rate of hydrogen to the carbon source is varied from 0.12:1 to 4:1.

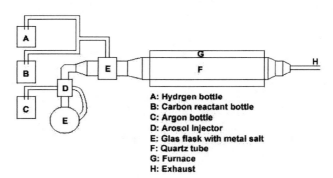

A: Hydrgen bottle
B: Carbon reactant bottle
C: Argon bottle
D: Arosol Injector
E: Glas flask with metal salt
F: Quartz tube
G: Furnace
H: Exhaust

FIGURE 1: Sketch of the experimental set-up. The aerosol injector is connected to a glass flask containing the metal salts/complexes dissolved in water.

Scanning electron microscopy (SEM) images were recorded without prior preparation of the samples (Jeol 6300F at 15 kV, field emission gun). The nanotubes were dispersed in absolute ethanol before preparing the grids for transmission electron microscopy (TEM) imaging (Philips CM20, 200kV).

RESULTS

Highly Y-branched MWCNTs were synthesized using the conditions described above. In Fig. 2 are shown the tree-like nanotubes which were synthesized at 975°C with methane as a carbon reactant. The bamboo-like tubes seem to have a fairly constant outer diameter of approximately 20 nm, and wall thickness on around 7 nm. This corresponds well to the predicted diameter of the active catalyst.

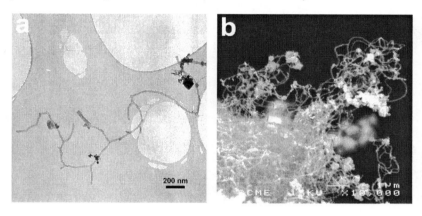

FIGURE 2: TEM (a) and SEM (b) images of branched carbon nanotube trees. This synthesise were performed at 975°C with iron-acetate dissolved in water as the catalyst precursor, and methane as the carbon reactant gas. Image (b) shows the homogeneity of the sample and the recurrence of branches in the nanotubes.

It has been suggested by several groups that the Y-branched of nanotubes arises from 6 heptagon defects which will provoke the three 120° angles [9,10]. In Fig. 3 is given a sketch of a calculated Y-junction.

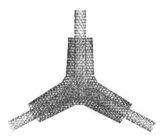

FIGURE 3: A sketch of a Y-junction nanotube, based on the model suggested by G.E. Scuseria [9,10]. The regular angles on 120° arises from six heptagon defects.

A thorough investigating of the TEM images showed that the highly Y-branched nanotubes always consist of a main trunk with branches grown only in one direction. This implies that it is a single catalyst particle that gives rise to the growth of the Y-branched tubes. Further investigation will have to be carried out for an elucidation of the mechanism behind the growth of these particular nanostructures.

CONCLUSION

It is demonstrated that nanotubes and branched nanotubes can be synthesized using water as the solvent for the catalyst precursor in an aerosol experiment. This can be seen as a demonstration that it is possible to lower the synthesis costs of nanotubes since this method significantly limits preparation time and the prices both for the catalyst precursors and for the catalyst precursor solvent. It was also shown for the first time that highly Y-junction nanotubes can be synthesised in high purity.

Acknowledgements: MG acknowledges the contract HPRN-CT-2000-00157 for financial support. Dr. M. Holzinger is acknowledged for valuable discussions and beautiful illustrations. J.-P. Selzner, and Prof. D. Laplaze are thanked for valuable discussions.

REFERENCES

1. H.W. Kroto, J.R. Heath, S.C. O'Brien, R.F. Curl, R.E. Smalley, *Nature 318* 162 (1985)
2. S. Iijima, *Nature 35* 456-58 (1991)
3. A. Oberlin, M. Endo, T. Koyama, *J Cryst. Growth. 32* 335 (1976)
4. G.G. Tibbetts, *J. Cryst. Growth.66* 632 (1984)
5. N.M. Rodriguez, *J. Mater. Res.8* 3233 (1993)
6. J.H. Hafner, M.J. Bronikowski, B.R. Azamiam, P. Nikolaev, A.G. Rinzler, D.T. Colbert, R.E. Smalley, *Chem. Phys. Lett. 296* 195 (1998)
7. H.M. Cheng, F. Li, G. Su, H.Y, Pan, L.L. He, X. Sun, M.S. Dresselhaus, *Appl. Phys. Lett. 72* 3282 (1998)
8. M. Glerup, H. Kanzow, R. Almairac, M. Castignolles, P. Bernier, *Chem. Phys. Lett. 337* 293 (2003)
9. G.E. Scuseria, *Chem. Phys. Lett. 195* 534 (1992)
10. T. Rueckes, K. Kim, E. Joseluich, G.Y. Tsang, C.L. Cheung, C.M. Leiber, *Science, 289* 94 (2000)

Raman-Spectroscopy on Metallic SWNTs Separated from Semiconducting SWNTs with Dielectrophoresis

Frank Hennrich[1,†], Ralph Krupke[1,†], Hilbert v. Löhneysen[2,3], Manfred M. Kappes[1,4]

[1]*Forschungszentrum Karlsruhe, Institut für Nanotechnologie, D-76021 Karlsruhe*
[2]*Physikalisches Institut, Universität Karlsruhe, D-76128 Karlsruhe*
[3]*Forschungszentrum Karlsruhe, Institut für Festkörperphysik, D-76021 Karlsruhe*
[4]*Institut für Physikalische Chemie, Universität Karlsruhe, D-76128 Karlsruhe*

Abstract. Alternating current (ac) dielectrophoresis was used to deposit single-walled carbon nanotubes (SWNTs) made by high pressure decomposition of CO (HiPco) between gold microelectrodes. The SWNTs were suspended as individuals in D_2O and 1% of the surfactant sodium dodecyl sulfate (SDS). We performed Raman-spectroscopy studies on the deposited tubes with a confocal Raman microscope showing that we achieved a separation of metallic from semiconducting SWNTs.

INTRODUCTION

Carbon nanotubes are unique 1-D macromolecules having considerable potential as building blocks in future nanoscale electronics. Metallic SWNTs could function as leads in a nanoscale circuit, whereas semiconducting SWNTs could perform as nanoscale field effect transistors. However, for the realization of a nanotube based electronics it is necessary to manipulate metallic and semiconducting SWNTs separately. Site-selective deposition of aligned bundles of SWNTs has been demonstrated using applied electric fields[1]. We present a method for the separation of metallic from semiconducting SWNTs, which is based on dielectrophoresis. SWNTs develop an induced dipole moment when subjected to an ac electric field[2]. A Raman spectroscopy study on dielectrophoretically prepared samples demonstrates the functionality and efficiency of the separation method.

EXPERIMENTAL

For the preparation of an individual SWNTs suspension, 50 mg of raw HiPco-SWNTs soot was suspended in 100 ml D_2O containing 1 weight% of the surfactant sodium dodecylsulfate (SDS) under sonication[3]. After sonication the suspension was centrifuged at 154,000 g for 4 h and the upper 90 % of the supernatant was than

CP723, *Electronic Properties of Synthetic Nanostructures*, edited by H. Kuzmany et al.
© 2004 American Institute of Physics 0-7354-0204-3/04/$22.00

carefully decanted. The suspension is characterized by sharp features in its electronic absorption spectrum associated with optical allowed interband transitions and shows strong fluorescence which indicates that the suspension contains mainly micelle-coated individual nanotubes rather than bundles.

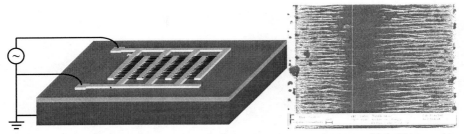

FIGURE 1. Electrode wiring scheme (not to scale). The electrodes are ~30 nm thick, 200 nm wide and have a gap of 1 μm on a p-type silicon substrate with 600 nm thermally oxidized SiO_2. Gold was used as the top electrode material (~30 nm) with titanium as an adhesion layer (~2-3 nm). Black lines represent tubes after deposition. SEM image of tubes trapped between electrodes via dielectrophoresis.

Microelectrodes were prepared with standard e-beam lithography and wired to a function generator as schematically depicted in Figure 1. For the dielectrophoresis the generator was operated at a frequency $f = 10$ MHz and a peak-to-peak voltage $V_{p-p} = 10$ V. After switching on the generator, a drop of suspension of individual SWNTs (≈ 10 μl) was applied onto the chip. After a delay of ten minutes, the drop was gently blown of the surface by a stream of nitrogen gas. Finally the generator was switched off. The resulting sample was characterized with a confocal Raman Microscope (Witec, CRM-200) excited with a Ar^+ ion laser (514.5 nm, Spectra Physics). The laser spot had a diameter of ~1.25 μm at the sample with a power density of ~ 10^5 W/cm². When taking Raman spectra of the dielectrophoretically deposited aligned SWNTs the polarisation of incident light was generally chosen in parallel to the aligned tubes. All Raman spectra were recorded using a grating of 600 grooves/mm and a corresponding spectral resolution of 3.75 cm⁻¹. We also studied the sample with Scanning Electron Microscopy (LEO 1530 SEM).

A reference sample was prepared by depositing SWNTs onto a Si substrate by simply applying a drop of suspension of individual SWNTs (~ 10 μl) and letting it dry.

RESULTS AND DISCUSSION

Figure 1 shows a SEM micrograph of the sample after deposition of SWNTs via dielectrophoresis. It indicates, that the individually suspended tubes form bundles again when trapped on the surface. They are highly aligned along the electric field lines, present during deposition.

Figure 2 shows Raman spectra obtained by accumulating respectively ten single Raman spectra measured on different spots of the aligned SWNTs and on the reference sample. In the RBM region of the reference sample it shows two bands with one band dominated by a feature at 187 cm[-1] and the other band comprising of three peaks at 247, 263 and 271 cm[-1]. The frequency of the radial breathing mode w_{RBM} is proportional to the inverse diameter: $w_{RBM} = c_1 / d + c_2$, where w_{RBM} is the RBM frequency and $c_1 = 223.5$ nm·cm[-1] and $c_2 = 12.5$ cm[-1] are empirically derived parameters[4], with d being the diameter of the tubes.

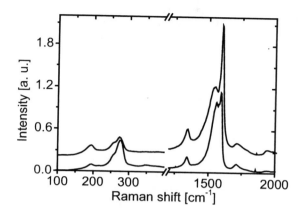

FIGURE 2. Raman spectra of SWNTs deposited via dielectrophoresis compared to a reference sample deposited on Si without application of an electric field.

For a given excitation wavelength only those SWNTs contribute to a Raman signal that can be resonantly excited via optical allowed interband transitions[5]. Three types of SWNTs can possibly be resonantly excited with the laser energy $E_{exc} = 2.41$ eV used in our experiment: i) semiconducting SWNTs having diameters > 1.1 nm, ii) quasi-metallic SWNTs with (n-m) mod 3 = 0 and iii) metallic SWNTs with n = m, both having diameters < 1.1 nm. Therefore we expect to measure metallic SWNTs RBM frequencies in the range between 238-280 cm[-1] and semiconducting SWNTs RBM frequencies ranging between 175-213 cm[-1]. We conclude from our measurements on the reference sample that the lower RBM band originates from at least one type of semiconducting SWNT and that the higher RBM band originates from metallic SWNTs. On the SWNTs deposited via dielectrophoresis again we observe the two bands associated with metallic and semiconducting SWNTs. However the intensity ratio is different, indicating that we achieved an enrichment of metallic nanotubes.

Comparison of the G-mode regions of both spectra in Fig. 2 provides further support for the above conclusion of the enrichment of metallic SWNTs in the dielectrophoretically deposited sample. In semiconducting nanotubes both components of the G-band, ω_{G-} and ω_{G+}, show a Lorenzian lineshape, with ω_{G+} being stronger in intensity than ω_{G-}. In metallic nanotubes both components are of equal intensity and ω_{G-} is much broader, exhibiting an asymmetric Breit-Wigner-Fano lineshape[5]. In this context the data in Fig. 2 is in agreement with the enrichment of metallic tubes in our sample prepared by dielectrophoresis.

Dielectrophoresis is operative in separating metallic from semiconducting SWNTs in aqueous suspension due to a simple electromechanical model. Using the frequency $f = 10$ MHz, electrophoretic forces vanish because of time averaging and only the induced dipole moment interacting with the inhomogeneous electric field gives rise to a translational motion along the electric field gradient[6,7]. For the simplest case, the time-averaged dielectrophoretic force for a dielectric sphere is expressed as

$$\vec{F}_{DEP} \propto \varepsilon_m \frac{\varepsilon_p - \varepsilon_m}{\varepsilon_p + 2\varepsilon_m} \nabla E_{rms}^2 \qquad (1)$$

where ε_p and ε_m are the dielectric constants of the particle and the solvent medium respectively and E_{rms} is the average field strength[7]. Positive dielelectrophoresis, the attraction towards the higher field strength, occurs for $\varepsilon_p > \varepsilon_m$. The static dielectric constant for semiconducting SWNTs, ε_S, has been calculated in ref. 2 and was found to be inversely proportional to the square of the band gap E_G. For tubes with $d = 0.8 - 1.4$ nm the smallest possible band gap values E_G are between 0.5 and 1.0 eV for the semiconducting tubes in our samples. Translating band gaps into dielectric constants, we obtain finite dielectric constants for semiconducting SWNTs with $\varepsilon_S < 5$. For metallic SWNTs we expect a very large absolute value of the dielectric constant. In fact it has been suggested that the polarisability of metallic SWNTs is effectively infinite[2]. With eq. 1 we calculate the sign of the dielectrophoretic force for individual SWNTs in D_2O ($\varepsilon_{D2O} = 80$) and assuming that $\varepsilon (10$ MHz$) \approx \varepsilon (0)$. For semiconducting SWNTs we derive a *negative* dielectrophoresis, whereas for the metallic and quasi-metallic SWNTs the dielectrophoresis is *positive*, explaining why metallic SWNTs are deposited on our electrodes while the semiconducting SWNTs remain in suspension.

REFERENCES

[1] Krupke, R. et al. *Appl. Phys. A* **76**, 397 (2003).
[2] Benedict, L. X., Louie, S. G., Cohen, M. L., *Phys. Rev. B* **52**, 8541 (1995).
[3] O'Connell, M. J. et al., *Science* **297**, 593 (2002).
[4] Bachilo, S. M. et al, *Science* **298**, 2361 (2002).
[5] Dresselhaus, M. S., Dresselhaus, G., Jorio, A., Filho, A. G. S., Saito, R., *Carbon* **40**, 2043 (2002).
[6] Pohl, H. A., *Dielectrophoresis* (Cambridge University Press, 1978).
[7] Jones, T. B., *Electromechanics of Particles* (Cambridge University Press, Cambridge, 1995).

Production of Aligned Carbon Nanotube Films and Nitrogen Doped Carbon Nanotube Films from the Pyrolysis of Styrene

Yi Zheng Jin[1*], Wen Kuang Hsu[2], Yan Qiu Zhu[3], Paul C P Watts[1], Yu Lun Chueh[2], Li Jen Chou[2], Harold W Kroto[1], David R M Walton[1]

[1]*School of Life Sciences, University of Sussex, Brighton BN1 9QJ, UK*

[2]*Department of Material Science and Material Engineering, National Tsinghua University, Hsinchu 300, Taiwan*

[3]*School of Mechanical, Materials, Manufacturing Engineering & Management, University of Nottingham, Nottingham NG7 2RD,UK*

Abstract. Styrene is used as a carbon source in a CVD process to obtain aligned carbon nanotube films. Changing the carrier gas from argon to ammonia introduces nitrogen into the tubes. SEM, TEM and HRTEM show the well-aligned structures, which appear to exist as macrobundles. EELS analyses have verified the existence of 3.3 wt.% nitrogen in the tube. Irradiation experiments show that this technique can be used to manipulate NCNTs.

INTRODUCTION

Aligned carbon nanotubes have aroused interest due to their attractive anisotropic and mechanical properties [1]. Doping with various elements is a promising approach to controlling the electronic structure of carbon nanotubes (CNTs) [2]. Nitrogen-doped carbon nanotubes (NCNTs) are metallic and exhibit strong electron donor states near the Fermi level [2]. Different solvents including toluene, xylene, benzene *etc.* have been employed as the carbon feedstock [3-5]. Styrene, cheap and commercially available, is a basic building block for numerous polymers such as polystyrene (PS), acrylonitrile-butadiene-styrene (ABS), styrene-acrylonitrile (SAN), styrene-butadiene rubber (SBR) *etc.* In this paper, we describe CNT and NCNT films produced by the pyrolysis of styrene. Experiments on NCNTs show that irradiation is a promising approach to mechanical manipulation.

Experimental

A typical two-stage furnace was utilized for chemical vapour deposition (CVD) of styrene. Aligned CNT films were produced using a ferrocene-styrene solution (3.5 wt.%) as the carbon feedstock. A quartz substrate was located inside a quartz tube in a high temperature furnace (1000 °C), with an argon (Ar) gas flow (*ca.*

CP723, *Electronic Properties of Synthetic Nanostructures*, edited by H. Kuzmany et al.

30 sccm). The ferrocene-styrene solution was injected into the low temperature furnace (250 °C), and subsequently vaporised. The vapour was carried by Ar into the high temperature furnace (1000 °C) where the material grew. After the system was cooled, aligned CNT films were observed on both sides of the quartz substrate and the inner wall of the reaction tube. When ammonia (NH$_3$) was employed as the carrier gas instead of Ar, NCNT films were produced.

Results and discussion

Figure 1a shows a SEM image of an as-grown CNT film. Figure 1b is an enlarged image of Figure 1a, demonstrating that the film consists of high density aligned CNTs. The CNTs within arrays are free of carbon particles with uniform diameters. The thickness of the films can easily attain several hundred microns by extension of a ferrocene-styrene solution feed (Figure 1c). The catalytic iron particles are detached from the substrate and reside at the CNT tips. Aligned CNT ropes are occasionally observed on top of the film. Iron particles are often encapsulated in the tips of the CNTs and occasionally form long Fe-crystals filling the hollow section of the tubes. Further examination shows that the encapsulated iron is a γ phase crystal, in accord with our previous results [6]. The ferrocene concentration is strongly linked to the CNT yield. A 1.0 wt.% ferrocene-styrene solution gives a lower CNT yield by comparison with the 3.5 wt.% solution. In the absence of ferrocene, pyrolysis of neat styrene produces uniform carbon spheres (Figure 1d). HRTEM observations indicate that the synthesised CNTs are highly graphitized, only carbon profile, but no nitrogen can be detected from the EELS mapping.

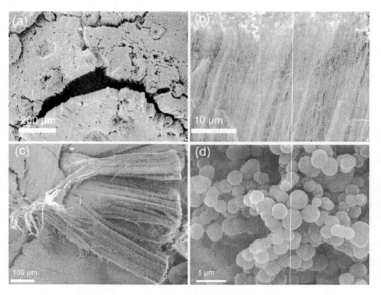

FIGURE 1. (a) SEM image of as-grown CNT films. (b) higher magnification of (a). (c) SEM image of a CNT array from a film (d) SEM image of carbon spheres.

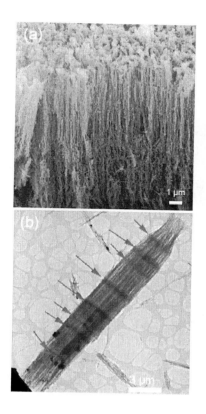

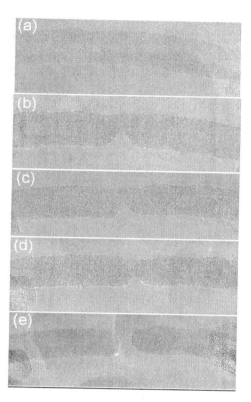

FIGURE 2. (a) SEM image of as-grown NCNT films. (b) TEM image of a NCNT bundle. The moiré-fringe of the NCNT bundle is marked.

FIGURE 3. A series of HRTEM images of one individual NCNT exposed under a 300kV electron beam. The time interval between two images is 15 seconds

Changing the carrier gas from Ar to NH_3 incorporates nitrogen into the nanotubes. Figure 2a is an SEM image showing the highly orientated NCNT bundles. After high-energy sonication, most of the NCNTs preserve their alignment, indicating the high stability of bundles (Figure 2b). Furthermore, the formation of moiré-fringe (arrows, Figure 2b) suggests ultra-alignment, as compared to other CNT bundles reported. EELS mapping indicates *ca.* 3.3 wt.% of nitrogen between the inner carbon walls and the hollows of the NCNTs. The nitrogen introduces extra defects into the carbon lattice. Thus, a 300 keV electron beam can easily irradiate the NCNTs. Figure 3 demonstrates the irradiation and subsequent breakage of an individual NCNT when exposed under the electron beam for one minute. In contrast, CNTs without nitrogen doping are stable during exposure of a 300 KeV electron beam for 5 min. This confirms that the tube irradiation is a potential approach to NCNTs manipulation.

ACKNOWLEDGMENTS

This work was supported by the EPSRC (UK) and National Science Council (Taiwan). We thank J. Thorpe and D. Randall (Sussex) for assistance with the electron microscopy.

REFERENCES

1. Huczko, A., *Appl. Phys. A-Mater.* **74**, 617-638 (2002).
2. Fischer, J. E., *Accounts Chem Res* **35**, 1079-1086 (2002).
3. Singh, C., Shaffer, M. S., Windle, A, H., *Carbon* **41**, 359-368 (2003).
4. Andrews, R., Jacques, D., Rao, A. M., Derbyshire, F., Qian, D., Fan, X., Dickey, E. C., Chen, J., *Chem Phys Lett* **303**, 467-474 (1999).
5. Mayne, M., Grobert, N., Terrones, M., Kamalakaran, R., Ruhle, M., Kroto, H. W., Walton, D. R. M., *Chem Phys Lett* **338**, 101-107 (2001).
6. Watts, P. C. P., Hsu, W. K., Kotzeva, V., Chen, G. Z., *Chem Phys Lett* **366**, 42-50 (2002)

Limited Number of Possible Mean Diameters in the Evaporation Synthesis of Single-Walled Carbon Nanotubes

O. Jost[1], W. Pompe[1], X. Liu[2], and J. Fink[2]

[1] Institute of Materials Science (IfWW), Dresden University of Technology, D-01062 Dresden
[2] Institute for Solid State Research (IFF), IFW Dresden, P.O. Box 270016, D-01171 Dresden

Abstract. Diameter dependencies for the pulsed laser evaporation synthesis of single-walled carbon nanotubes (SWCNT's) have been studied by optical absorption spectroscopy. Nanotube diameter changes found in our experiments are always very small and depend on a limited number of physical parameters of the synthesis environment and on the catalyst composition. This is in strong contrast to SWCNT synthesized by the decomposition of organic precursors (CVD). From extensive synthesis experiments we find a window for the *mean* nanotube diameter for the evaporation synthesis that opens at roughly 0.95 nm and closes at roughly 1.55 nm. Possibilities to increase this nanotube diameter window are discussed.

INTRODUCTION

Very recently we could show that the growth mechanisms of single-wall carbon nanotubes (SWCNT's) in evaporation methods and in methods based on the decomposition of organic precursors (CVD) are different [1]. This results in different mean SWCNT diameters and diameter distributions [1]. The diameter of SWCNT synthesized by CVD methods depends only on the size of the templating catalyst particle leading to mean nanotube diameters between fractions of a nanometer and ten nanometers and more [2]. On the other hand it appears to be quite difficult to synthesize nanotubes in evaporation methods having mean diameters far away from the commonly found 1.2-1.3 nm. Here we show, that the conditions of the synthesis environment indeed determine - and limit- the possible mean nanotube diameters obtainable from evaporation-based methods.

EXPERIMENTAL DETAILS

The SWCNT synthesis procedure is the same as given in detail in Ref. [1]. A pressed target pellet consisting of charcoal and metal catalyst at a total of 2 at.-% was placed on the axis of a quartz tube close to the center of a resistance-heated tube furnace. The temperature of the tube wall in the center of the furnace was kept constant at selected temperatures between 900-1150°C. One Q-switched Nd:YAG laser (wavelength 1064 nm, pulse duration 8 ns, pulse energy 1700 mJ, pulse frequency 10

CP723, *Electronic Properties of Synthetic Nanostructures*, edited by H. Kuzmany et al.
© 2004 American Institute of Physics 0-7354-0204-3/04/$22.00

Hz) was used to evaporate the target surface. A water-cooled copper collector was positioned in the hot gas flow 30 mm behind the target to quench the soot down to temperatures well below 100°C. The furnace tube was filled with flowing Ar, He or N_2 gas ($v = 2$cm/s in the hot zone) at a constant pressure of 90 kPa. For the characterization of the SWCNT diameter distribution of bulk samples we used an optimized express method based upon optical absorption spectroscopy (OAS). Details can be found in Ref. 1. After a background correction, a peak analysis (Gaussians) can be used to determine the mean diameter of the nanotube diameter distribution [3]. Five subsequent SWCNT synthesis runs at constant experimental conditions resulted in a standard deviation of the determined mean diameters of not more than 0.003 nm (meaning an excellent nanotube diameter reproducibility).

RESULTS AND DISCUSSION

Before we start to discuss our results we want to highlight that we discuss *mean* nanotube diameters only and not the width of the nanotube diameter distribution. The width of the diameter distribution did not change very much in our experiments.

Following Ref. 1, the by far most important synthesis parameters influencing the nanotube diameter in evaporation-based approaches are temperature, gas type, gas pressure and the catalyst parameters (type, composition). Representative data for these synthesis parameters are shown in Fig. 1. The temperature dependence of the nanotube diameter for a CoNi catalyst does not remarkably depend on the gas and gas pressure (Fig. 1a). A temperature difference of 250 K results always in a nanotube diameter change of roughly 0.1 nm, a quite small value. With other catalysts (Figs. 1d and 1e) the situation is quite similar - a nanotube diameter change of more than 0.1 nm for a temperature difference of 250 K is not observable.

The laser evaporation synthesis of nanotubes is temperature-limited and proceeds between 800°C and the temperature of the carbon-metal catalyst eutectics (around 1300°C for the commonly used CoNi catalysts) [4,5]. These 500 K between the upper and lower temperature limits the possible diameter change to about 0.2 nm. Nevertheless, special carbon precursors [6] or high-melting catalysts [5] such as binary RhPt catalysts might make it possible to extend the nanotube formation temperature range to 500-1750°C. In those cases, additional changes of the mean nanotube diameter of 0.3 nm might be possible. With other words, a maximum diameter change of 0.5 nm by temperature variations might be possible in special cases.

Diameter changes by variations of the gas type and gas pressure (Figs. 1a and 1b) appear to be even smaller than that found for temperature variations. For pressure changes from 30-120 kPa, the accompanied diameter change amounts to only 0.05 nm. It depends in a logarithmic manner (Fig. 1b) on the gas pressure. For the laser evaporation synthesis, we have nanotube formation limits at around 10 kPa because of unfavorable gas phase processes at small pressures [4] and at around 200 -300 kPa due to experimental limits. This should not allow to change the *mean* nanotube diameter by more than 0.1 nm through gas pressure variations (Fig. 1b). A different situation might

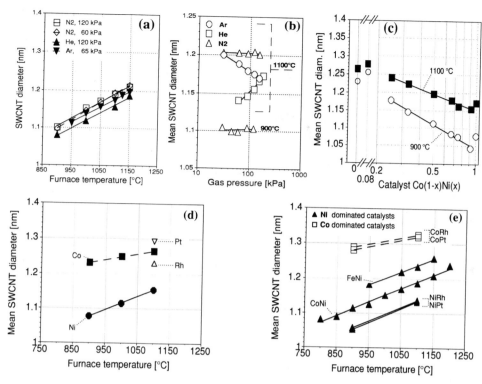

FIGURE 1. SWCNT diameter dependencies: For different temperatures, gases and gas pressures using a $Co_{0.5}Ni_{0.5}$ binary catalyst (a, b); For different catalyst compositions and temperatures in N_2 at 90 kPa (c, d, e); For different compositions of a CoNi binary catalyst (c); For different single metal catalysts (d) and different binary metal catalysts (e) at different temperatures.

arise from the use of very heavy gases in the synthesis environment. According to Ref. 1, diameter changes by pressure variations are probably due to the collision frequency (gas pressure) of gas atoms with carbon atoms at the root of the growing nanotube, with the mass difference between the collision partners determining the slope (and direction, see Fig. 1b) of the SWCNT diameter change with increasing gas pressure. A very heavy noble gas might therefore increase the sensitivity of the nanotube diameter on pressure variations. Xe has roughly 4 times the Ar mass. This might mean a 4 times higher diameter sensitivity on Xe pressure variations compared to Ar pressure variations, i.e. possibly a total change of the nanotube diameter of 0.2 nm in Xe between 30-120 kPa compared to 0.05 nm for Ar (Fig. 1b). However, we cannot provide supporting experimental data at the moment.

Finally, the catalyst composition at constant temperatures is known to influence the nanotube diameter (Figs. 1c-1e). The data show that it is possible to change the nanotube diameter by 0.2 nm by choosing different catalysts (single metals or binary catalysts). Since the diameters found for binary catalysts (Figs. 1c and 1e) are almost "in between" those of the only single elements that are suitable to grow SWCNT's by laser evaporation (Ni, Co, Rh, and Pt, see Fig. 1d) we are almost limited to the above mentioned 0.2 nm diameter change by different catalysts. However, there is an exception. From an almost complete survey in different transition metal elements as a component in catalyst mixtures we found that lanthanides and also chromium are suitable to result in an additional and general increase of the nanotube diameter by up to 0.1 nm (not shown in Fig. 1). This finally gives us a total of roughly 0.3 nm for possible changes of the *mean* nanotube diameter by catalyst composition variations.

To summarize, for the laser evaporation synthesis of SWCNT we can easily change the mean diameter of nanotubes by 0.2 / 0.1 / 0.3 nm for temperature / pressure / catalyst composition changes, respectively. Because these parameters seem to influence the diameter independent from each other one can directly combine the synthesis parameters to get additive effects and maximized nanotube diameter changes (up to about 0.6 nm). This is a very small value compared to the CVD synthesis of nanotubes for which *mean* diameters can differ by more than 10 nm [2].

In the laser evaporation synthesis, the nanotube abundance depends on the same synthesis parameters which are important for the nanotube diameter. Usually, conditions which lead to mean nanotube diameters of roughly 1.2 nm yield an optimum abundance. Changing the nanotube diameter then necessarily means lowering the nanotube abundance to some extent. By using an optimum diameter of 1.2 nm and the above derived range of 0.6 nm one can vary the *mean* nanotube diameter between 0.95 nm and 1.55 nm (1.45 nm without lanthanides) around the optimum value at 1.2 nm. However, the achievable nanotube abundances for these two SWCNT diameter limits might be quite low.

ACKNOWLEDGEMENTS

This work was supported by the BMBF. We thank H. Zöller for experimental assistance and L. Dunsch for his support regards the optical measurements.

REFERENCES

[1] O. Jost *et.al.*: J. Nanosci. Nanotech. **4** (2004) [in press].

[2] Y. Li *et al.*: J. Phys. Chem. B **105**, 11424 (2001); C.L. Cheung *et al.*: J. Phys. Chem. B **106**, 2429 (2002).

[3] X. Liu *et al.*: Phys. Rev. B **66**, 045411 (2002); O. Jost *et.al.*: Appl. Phys. Lett. **75**, 2217 (1999).

[4] O. Jost *et.al.*: J. Phys. Chem. **B 106**, 2875 (2002).

[5] H. Kataura *et al.*, *Carbon* **38**, 1691 (2000).

[6] S. Zhang, S. IIjima, Appl. Phys. Lett. **75**, 3087 (1999).

Mass Production of Multiwalled Carbon Nanotubes by Chemical Vapor Deposition

A. Magrez, Cs. Mikó, J.W. Seo, R. Gaál and L. Forró

Institut of Physics of Complex Matter, Ecole Polytechnique Fédérale de Lausanne
1015 Lausanne, Switzerland

Abstract. We elaborated a continuous production method for multiwalled carbon nanotubes (MWCNTs) based on a rotary tube furnace. MWCNTs were synthesized by chemical vapor deposition of C_2H_2 over $Fe_{1-x}Co_x$ catalyst supported by carbonate. Growth parameters were optimized in a fixed bed furnace in order to increase the yield and the product quality. We have found an optimum for x=0.25, at 700°C by using $CaCO_3$ as support. Co content does not only influence the quantity but also the quality of the product. The increase of the growth temperature as well as the use of other alkaline earth carbonates led to a decrease of the MWCNT yield. Based on these results, we determined the optimum conditions for our rotary tube furnace (rotation speed, gas flux, inclination) and achieved production of 500g/day of purified MWCNTs

INTRODUCTION

Since the discovery of multiwalled carbon nanotubes (MWCNTs) [1], numerous attempts have been made to synthesize such products on a large-scale and at low-cost. Currently, Chemical Vapor Deposition (CVD) over supported catalyst has become the most popular technique to scale up the production.

Very recently, we have shown that by using a rotary tube furnace an up scaling of the production is possible, preliminarily we obtained about 100g/day of purified MWCNTs [2]. In this paper, we systematically study the influence of the catalyst composition, nature of the support, growth temperature and retention time of the catalyst in the furnace on the production yield and the product quality. By optimizing all growth parameters, we obtain a yield of 500g/day.

RESULTS AND DISCUSSION

MWCNTs were synthesized by CVD of acetylene (C_2H_2) over $Fe_{1-x}Co_x$ catalyst supported by alkaline earth carbonates (MCO_3 with M= Mg, Ca, Sr, Ba). For the catalyst preparation, cobalt acetate tetrahydrate and iron nitrate nonahydrate corresponding to $Fe_{1-x}Co_x$ composition are dissolved in water, and carbonate support is dispersed subsequently. The solution is dried overnight at 120°C to obtain the catalyst in the form of a brownish powder. The total concentration of the metal relative to carbonate is about 5wt% [3].

CP723, *Electronic Properties of Synthetic Nanostructures*, edited by H. Kuzmany et al.
© 2004 American Institute of Physics 0-7354-0204-3/04/$22.00

Parametric Study

We performed at first a parametric study in a fixed bed furnace in which a small quantity of catalyst was exposed to C_2H_2 and N_2 fluxed at 1 and 70L/hour respectively. Catalyst composition, support material and retention time were optimized.

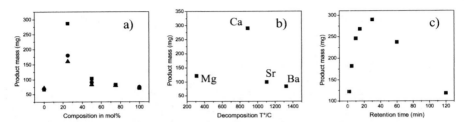

FIGURE 1. Parametric study: Evolution of the product mass versus a) the catalyst composition (x in $Fe_{1-x}Co_x$) after 30min at 700°C(■), 720°C(●) and 740°C(▲) (using $CaCO_3$ as support); b) the decomposition temperature of the support material (cation M of the carbonate MCO_3 is indicated) of $Fe_{0.75}Co_{0.25}$ catalyst after 30min at 700°C; c) the retention time at 700°C using $CaCO_3$ as support for 100mg of catalyst.

In figure 1a), we plotted the evolution of the product mass versus the catalyst composition (x) after 30 minutes at different temperature. Clearly, a maximum is visible at x=25mol% for all three temperatures applied; the yield is higher for all binary alloys compared to pure metals. Applying higher temperature than 700°C leads to a decrease of the yield. Typically, 250mg of MWCNTs was obtained after purification from 100mg of x=25mol% catalyst supported by $CaCO_3$ when the growth was performed at 700°C.

In figure 1b, a clear dependence of the decomposition ($MCO_3 \rightarrow MO + CO_2$) temperature of the support on the yield is observed. One can notice that the highest yield is obtained when the growth temperature is close to the decomposition temperature of the support. Consequently, for a growth temperature of 700°C, $CaCO_3$ is the most adequate support.

The cobalt content (x) has also a significant influence on the quantity as well as the quality of the product. From SEM observations (figure 2a), we cannot detect catalytic particles for purified samples with x ≤ 25mol%. The outer diameter of the CNTs is narrowly distributed around 25nm. TEM observations confirm the SEM results: for x≤25mol%, carbon is present only as CNTs (figure 2b) while for higher cobalt content, carbon nanoparticles are present as side-products.

For the valuation of the CNTs, characteristics, such as partial filling with catalytic particles, straightness related to walls defects and/or graphitisation, warrant the quality of the CNTs.

For CVD production, we selected catalyst and support to be easily removed during the purification step. Nevertheless, CNTs can be partially filled by catalytic particles because of the growth process [4]. In such conditions, acidic purification is not effective for embedded catalyst particles since they are not accessible to acid. In our purified samples, we observed partial filling of CNTs with catalytic particles. The number of encapsulated particles increases with the cobalt content. Up to 80% of CNTs are partially filled when Co/CaCO$_3$ is used.

On other hand, iron enrichment of the catalyst is leading to an improvement of the wall structure, in agreement with [5], but seems not to influence the CNTs diameters.

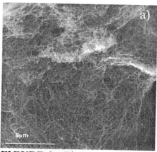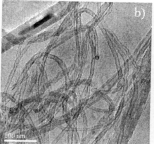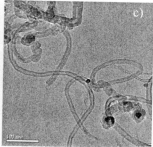

FIGURE 2. Electron microscopy studies. SEM (a) and TEM (b) pictures of the product obtained at 700°C with Fe$_{0.75}$Co$_{0.25}$/CaCO$_3$ catalyst. TEM picture (c) for Fe$_{0.25}$Co$_{0.75}$/CaCO$_3$ catalyst.

Finally, we studied the carbon yield of the catalyst powder in the furnace versus the retention time t. Figure 1c shows maximum yield at 30min of reaction time for 100mg of catalyst. Two different processes are needed to explain the evolution of the carbon yield. The first part of the reaction (2 to 30 minutes) is dominated by the growth of the CNTs. During this time, there was no clear variation in the length or in the outer diameter of CNTs indicating that the growth kinetic of tubes is much faster than our shortest reaction time (2 min). For t > 30min, the oxidation of already precipitated carbon dominates. We assume that by this time enough C$_2$H$_2$ was fluxed into the reactor so that all available catalytic particles were completely consumed, and the only possible reaction is the oxidation of carbon by oxygenated species, which can be present in the reactor either after the decomposition of the salts (precursors of the catalyst), or due to oxygen traces introduced as impurity with N$_2$. This phenomenon is presumably correlated with CNT quality degradation.

Mass Production

Obviously, some of the optimum parameters are strongly correlated: the gas flux and the retention time to the quantity of the catalyst. In our continuous production system, larger quantity of powder is passing at the same time through the heated zone of the rotary tube. Therefore we had to enrich the gas with acetylene and increase the retention time of the powder. The latter is controlled by the rotation speed and the inclination of the tube. After several assays, they were set to 1rpm and 15°, respectively. Parameters, such as temperature, catalytic composition, support and (Fe$_{1-x}$Co$_x$)/MCO$_3$ ratio, were taken from the parametric study.

In our standard fixed bed reactor, the semi continuous production can yield up to 6.5g/day of purified MWCNTs. With the rotary tube furnace technique, we achieved a production rate of about 500g/day of high quality purified MWCNTs.

CONCLUSION

This paper reports on our achievement of large-scale production (500g/day) of purified and high quality MWCNTs using $Fe_{1-x}Co_x$ catalyst supported by alkaline earth carbonate in a rotary tube furnace. The first step was the optimization of the growth parameters in a fixed bed furnace. This study led to the following optimum parameters: temperature 700°C, retention time for 100mg of catalyst 30 min, support material $CaCO_3$, catalyst composition $Fe_{0.75}Co_{0.25}$.

The composition of the catalyst controls the quantity of the product and its quality, whereas raising the growth temperature leads to a decrease of the yield. Cobalt-rich catalyst produces mixture of nanoparticles and nanotubes, partially filled with catalyst. CNTs produced with $x \leq 25mol\%$ catalyst showed the best wall structure.

ACKNOWLEDGMENTS

Authors thank the European Commission (RTN Program, NANOCOMP network, RTN 1-1999-00013), NCCR "Nanoscale Science" of the Swiss National Science Foundation, and the TopNano21 program 5933.1 for financing this investigation. Cs. M. thanks for the grant of Swiss National Science Foundation (NSF 61534). We are also grateful to Centre Interdisciplinaire de Microscopie Electronique (CIME) at EPFL for access to electron microscopes as well as for technical support.

REFERENCES

1. Iijima, S., *Nature* **354**, 56 (1991).
2. Couteau, E., Hernadi, K., Seo, J.W., Thien-Nga, L., Gaal, R., and Forro, L., *Chem. Phys. Lett.* **378**, 9 (2003).
3. Seo, J.W., Couteau, E., Umek, P., Hernadi, K., Marcoux, P., Lukic, B., Miko, Cs., Milas, M., Gaal, R., and Forro, L., *New Journal of Physics,* **5**, 120.1 (2003).
4. Helveg, S., Lopez-Cartes, C., Sehested, J., Hansen, P.L., Clausen, B.S., Rostrup-Nielsen, J.R., Abild-Pedersen, F., Norskov, J.K., *Nature,* **427**, 426 (2004).
5. Huang Z.P., Wang, D.Z., Wen, J.G., Sennett, M., Gibson, H., Ren, Z.F., *Appl. Phys. A*, **74**, 387 (2002).

Electrochemical Synthesis Of An Array Of Aligned Polypyrrole Nanotubes

Pierre R. Marcoux, Laszló Forró

Institute of Physics of Complex Matter
Ecole Polytechnique Fédérale de Lausanne, 1015 Lausanne, Switzerland.

Abstract. Polypyrrole nanotubes were electrochemically synthesized in commercial polycarbonate filtration membranes with different pore sizes. After dissolving the template membrane, polymer nanotubes are arranged in a high-density array in which the individual tubes protrude from the surface like the bristles of a brush. The synthesized arrays are characterized through scanning electron microscopy and infrared spectroscopy. The possibility to make the same kind of arrays from functionalized monomers is investigated. As a conclusion, different methods to make arrays of functionalized polypyrrole nanotubes are discussed.

INTRODUCTION

Among the different strategies to synthesize nanoscopic materials reported in the literature, template synthesis is an elegant approach. This technique consists of including metallic or organic constituents inside the void spaces of nanoporous host materials. Though there now exist a huge range of hosts, track-etched membranes present a significant advantage, because they lead to the production of different kinds of nanotubes and nanowires with monodisperse diameters. Different kinds of polymers can be synthesized inside these membranes, giving rise to some hollow nanotubes made of polymers. [1,2] We chose to study polypyrrole since it shows several interesting features : it can be designed into different shapes (films, [3] nanotubes, [4] nanowires [5]) and can be chemically derivatized. Furthermore, it is conductive and has sensing properties. [6] Finally, this polymer is porous and therefore can be used for encapsulation, [7] an application in which we are interested. That is why we made some arrays of aligned nanotubes. This article deals with the electrochemical synthesis of arrays of polypyrrole hollow nanotubes, using different polycarbonate filtration membranes. The second part of the article focuses on the chemical derivatization of these nanotubes.

1. SYNTHESIS OF POLYPYRROLE NANOTUBES

The polymerization of pyrrole is based on the oxidation of the monomer. This initiation reaction yields a radical cation that reacts with a growing polymer chain (Figure 1). The polymerization can be chemically initiated (by $FeCl_3$ for example), or electrochemically initiated, depending on the way monomers are oxidized. In the case

CP723, *Electronic Properties of Synthetic Nanostructures*, edited by H. Kuzmany et al.

of electrochemical initiation, polymerization stops as soon as you stop applying the difference of potential. This allows for a more precise control on the length of synthesized nanotubes.

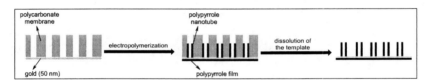

FIGURE 1. Radical cation formation of pyrrole monomer and the chain propagation during the synthesis of polypyrrole.

We performed the electrochemical polymerization of pyrrole in a three electrodes cell, at a constant potential (+0.8 V). A polycarbonate filtration membrane (Whatman) was covered with gold on one side and used as the work electrode. 50 nm of gold was deposited on membranes by plasma sputtering. The thickness of gold layers was measured by a profilometer (Tecnor). Since the oxidation of monomer occurs at the surface of the work electrode, polypyrrole grow inside pores, starting from the gold layer and yielding nanotubes. It also grows on the other side of the gold layer, yielding thus a film of polypyrrole (Figure 2). Our experiments were done in water with distilled pyrrole (0.1 M), using polystyrene sulfonate (Aldrich, 70 000 g.mol^{-1}) as an electrolyte and a commercial Ag/AgCl electrode (Metrohm) as a reference electrode. We paid a particular attention to the homogeneity of lengths of nanotubes, which depends on the distribution of current densities that flow from the counter-electrode to the work electrode. In order to get a better homogeneity we designed an electrochemical cell in which the work electrode and the counter-electrode (made of platinum) have the same size, are parallel and face each other.

polycarbonate membrane		polypyrrole nanotube	

[Figure 2 scheme: polycarbonate membrane with gold (50 nm) → electropolymerization → polypyrrole nanotube / polypyrrole film → dissolution of the template → nanotube array]

FIGURE 2. Scheme illustrating the template-assisted process: a polycarbonate membrane is covered with gold and then used as a work electrode. Polypyrrole growth starts from the gold layer.

After the polymerization step, the membrane is washed in a big amount of hot distilled water, then washed with ethanol and dried in air. The polycarbonate part, which is actually the template, is dissolved into chloroform to finally provide an array of polymer nanotubes, standing on a layer of gold and polypyrrole (Figure 2). These arrays are then observed through scanning electronic microscopy (SEM). We noticed that every pore of the membrane is filled with a nanotube. As a consequence, the surface density of arrays (nanotubes per mm^2) depends only on the density of pores of the initial membranes. We worked with 0.2 µm (diameter of pores) membranes that had 3.4 10^6 pores/mm^2, and 0.4 µm membranes with 1.0 10^6 pores/mm^2 (measured by SEM). As shown on the Figures 3a and 3b, 500 secondes are enough to grow at room temperature nanotubes that are 3.3 µm long. At high magnifications, striations on the walls of nanotubes can be observed. They are explained in literature by the preferential orientation of polypyrrole chains perpendicularly to the axis of nanotubes. [8] Figure 3c is the image of an array made in a 0.2 μm pores membrane: the surface density is in this case much higher than the one obtained with 0.4 μm membranes. For some applications such as encapsulation or sensing properties, a higher surface density should make an array more efficient. Infrared absorption spectra of our synthesized nanotubes showed peaks characteristic of polypyrrole. [9,10]

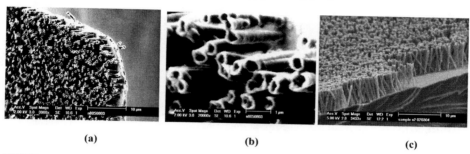

| (a) | (b) | (c) |

FIGURE 3. (a) and (b): Scanning electron micrographs of polypyrrole nanotubes grown 500 secondes in a template with 0.4 μm pores. The image (b) shows striations on the walls of nanotubes. Figure (c) shows 5.4 μm long nanotubes grown 400 secondes in a template with 0.2 μm pores.

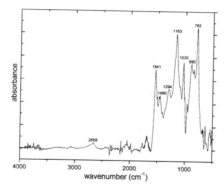

FIGURE 4. FTIR spectrum of polypyrrole nanotubes synthesized in a template with 0.4 μm pores.

2. SYNTHESIS OF FUNCTIONALIZED POLYPYRROLE NANOTUBES

To provide polypyrrole with a specific function, it is of importance to derivatize polymer chains (for example, to insure a high selectivity in the case of a biosensor based on polypyrrole). [11] As far as we are concerned, we are currently investigating the synthesis of nanotubes made from a functionalised monomer, the 1-(2-cyanoethyl)pyrrole, in order to make some polypyrrole nanotubes derivatized with cyanide function. The −C≡N group is interesting since it can easily yield two other fundamental chemical functionalities: carboxylic group through oxidation of cyanide in an aqueous solution of KOH, and amine group through reduction of cyanide with aluminum hydride. These both reactions are currently under investigation. Figure 5 shows the infrared absorption spectrum of nanotubes made from the cyanoethylpyrrole monomer. The peak at 2249 cm^{-1} is characteristic for the –C≡N group.

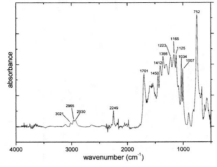

FIGURE 5. FTIR spectrum of nanotubes made from 1-(2-cyanoethyl)pyrrole grown 600 secondes (at 1.0 V potential) in a template membrane with 0.4 µm pores.

PERSPECTIVES AND CONCLUSION

We have made some arrays of aligned polypyrrole hollow nanotubes, that may find applications in the field of encapsulation. We are currently investigating the synthesis of polypyrrole nanotubes functionalised with amine or carboxylic groups. These groups should allow for covalent bonds between nanotubes and proteins.

ACKNOWLEDGMENTS

The authors are very thankful to Guillaume Camarda for building the electrochemical cell and to Henri Jotterand for sputtering gold and helpful discussions. We also thank Nicolas Xanthopoulos for performing XPS measurements. This work is supported by the NANOTEMP European Research Network.

REFERENCES

1. Martin, C. R., *Chem. Mater.* **8**, 1739-1746 (1996).
2. Martin, C. R., *Acc. Chem. Res.* **28**, 61-68 (1995).
3. Jérôme et al, *Angew. Chem. Int. Ed.* **37**, 2488-2490 (1998).
4. Dumoustier-Champagne, S. and Stavaux, P.-Y., *Chem. Mater.* **11**, 829-834 (1999).
5. Cosnier et al, *Chem. Commun.*, 414-415 (2003).
6. Pernaut, J. M. and Reynolds, J. R., *J. Phys. Chem.* **104**, 4080-4090 (2000).
7. Parthasarathy, R. V. and Martin, C. R., *Nature* **369**, 298-301 (1994).
8. Menon, V. P., Lei, J. and Martin, C. R., *Chem. Mater.* **8**, 2382-2390 (1996).
9. Jang, J. and Hak Oh, J., *Chem. Comm.* 2200-2201 (2002) in Electronic Supplementary Information.
10. Jang, J. and Yoon, H., *Chem. Commun.* 720-721 (2003) in Electronic Supplementary Information.
11. Deore, B., Chen, Z. and Nagaoka, T., *Anal. Chem.* **72**, 3989-3994 (2000).

Metallic/Semiconducting Nanotube Separation and Ultra-thin, Transparent Nanotube Films

Zhihong Chen*, Zhuangchun Wu, Jennifer Sippel and Andrew G. Rinzler

Department of Physics, University of Florida, Gainesville FL, USA
Present address: IBM T.J. Watson Research Center Yorktown Heights, NY 10598

Abstract: Perhaps the most wonderful feature of carbon nanotubes is that they are synthesized in both metallic and semiconducting variants, and perhaps the most problematic feature is that they are synthesized in both metallic and semiconducting variants. The intimate mixture hampers numerous envisioned applications and separation of the nanotubes into their respective electronic transport classes has emerged (after high yield synthesis and purification) as the next great materials challenge. Recently, several groups (us among them) have shown progress in attacking the problem. We will elaborate on our bromine exposure/centrifugation based method and the evidence leading to our conclusion that the separation results from an interplay between the nanotubes, bromine and the surfactant. We will also elaborate on our method for production of ultra-thin (and thereby transparent), optically homogeneous, nanotube films used in the spectroscopic absorbance based assay of the metallic/semiconducting nanotube content. Such films, by virtue of their electrical conductivity constitute a new class of electrically conducting, transparent electrode that may one day rival ITO in the ubiquity of its applications. The potential significance of the nanotubes in solving contact barrier problems between metals and semiconductors is also discussed.

INTRODUCTION

The separation of semiconducting from metallic single wall nanotubes (SWNTs) constitutes the next logical step in providing uniform, nanotube based, building blocks for use in rational, large-scale supramolecular assembly of functional devices. We have accordingly pursued such efforts and recently reported a method for carrying out the seperative enrichment of samples in semiconducting or metallic nanotubes [1]. The method relies on the exposure of surfactant suspended nanotubes to a controlled dose of bromine followed by centrifugation. Assay of the separately collected sedimented and supernatant fractions show the nanotubes remaining in suspension to be enriched in semiconducting nanotubes while those in the sediment are correspondingly enriched in metallic nanotubes. The original idea for the method involved two components: 1) that band structure differences between the metallic and semiconducting nanotubes would yield enhanced charge transfer doping of Br with the metallic nanotubes (supported by density functional calculations); and 2) that the more highly doped Br/metallic nanotube complex would acquire a sufficient increase in density over the less doped Br/semiconducting tubes to permit the former to sediment out under centrifugation. While the method works to yield seperative enrichment, experiments and modeling have shown that the increased density of the Br/metallic tubes is not likely the dominant mechanism for the separation. Instead we infer that the mechanism relies on a disruption of the surfactants ability to suspend the more highly doped metallic nanotubes. Here we report our evidence for this conclusion.

CP723, *Electronic Properties of Synthetic Nanostructures*, edited by H. Kuzmany et al.
© 2004 American Institute of Physics 0-7354-0204-3/04/$22.00

We begin by reviewing the evidence for separation. Figure 1 shows the absorbance spectrum for a thin film of SWNTs (laser vaporization grown, HNO₃/cross-flow purified) deposited on quartz following a bake cycle in flowing argon at 600°C (thick gray curve). The peaks labeled SC1 and SC2 are associated with absorption by semiconducting nanotubes, while the peak labeled M is associated with metallic nanotubes in the sample. Our assay of the relative concentration of metallic or semiconducting nanotubes relies on forming ratios between the integrated intensities of one of the semiconducting peaks and the metallic peak. To facilitate extraction of these integrated intensities the spectrum is fitted to 5 Lorentzians (3 dotted curves and 2 filled curves). The quality of the 5 Lorentzian fit to the spectrum is also shown (solid black line). The spectrum in Fig. 1 is labeled sediment because the SWNT film in this case was formed from the sedimented fraction of a centrifuged, bromine exposed sample. Br exposure is discussed below. Figure 2 shows the absorbance spectrum, and 5 Lorentzian fit for the corresponding, separately collected, supernatant from that centrifugation.

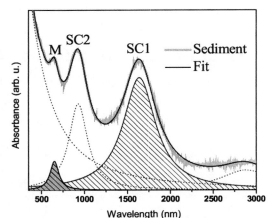

FIGURE 1. Absorbance spectrum of baked SWNT film and 5 Lorentzian fit to spectrum. Sediment, see text for details.

Labeling the semiconducting nanotube integrated intensity S, and the metallic tube integrated intensity M, the relative concentrations of semiconducting and metallic nanotubes are obtained from,

$$C_M = \frac{\beta M}{S + \beta M} \quad \text{and} \quad C_s = \frac{S}{S + \beta M}$$

respectively, where β is a factor accounting for differing absorption and scattering cross-sections for the two classes of nanotubes. Taking the ratio of these concentrations yields,

$R = \frac{S}{\beta M}$. For the supernatant we

have then, $R_{sup} = \frac{S_{sup}}{\beta M_{sup}}$, while

for the sediment, $R_{sed} = \frac{S_{sed}}{\beta M_{sed}}$.

For a given centrifugation, the ratio of these ratios between the supernatant and sediment

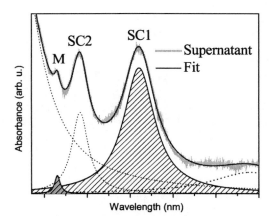

FIGURE 2. Absorbance spectrum of baked SWNT film and 5 Lorentzian fit to spectrum. Supernatant, see text for details.

fractions, $\dfrac{R_{sup}}{R_{sed}} = \dfrac{S_{sup}M_{sed}}{S_{sed}M_{sup}}$,

eliminates the unknown β, and provides a quantitative measure of the separation efficiency.

Table 1 lists the result of such analysis for several experiments in which the type of pretreatment ultrasonication (cup horn: C, or bath: B) time of ultrasonication (h: hours, m: minutes) and Br concentration was varied. Two controls are also shown in which no Br was added to the solution prior to the 12 hr, 24,000g centrifugation used in all experiments. The

TABLE 1. The separation efficiency for experiments in which the pretreatment ultrasonication and Br concentrations were varied. See text for details.

Sample	Br conc. (mg/ml)	S2 R_{sup}/R_{sed}	S1 R_{sup}/R_{sed}
batch 1; B11h	0 (control)	0.69	0.72
batch 1; B12h	0 (control)	0.83	0.80
batch 1; B14h	8.0	1.66	1.58
batch 1; B12h	5.3	1.93	2.14
batch 1; C2h,B6h	8.0	1.38	1.41
batch 1; C2h,B6h	10.9	1.15	1.05
batch 2; B18h	8.0	2.02	2.84
batch 2; B11h,C5m	1.6	1.57	1.89
batch 2; B6h,C10m	1.3	1.32	1.31

surfactant used in these cases was Triton X-100 (Union Carbide). In the column labeled S1, the SC1 absorbance peak was used for the analysis while in the column labeled S2, the SC2 peak was used. Ratio's greater than one indicate preferential sedimentation of metallic nanotubes. Interestingly, the controls indicate a small reverse effect from the Br exposed samples, i.e. for the controls the semiconducting nanotubes settle preferentially. The variability in the separation efficiency among similar Br concentrations was traced to the efficiency with which the pretreatment ultrasonication unbundled the nanotubes to individuals.

Since charge transfer doping the nanotubes dramatically affects their absorption bands [2, 3], and the nanotubes were purposely doped, the use of absorbance to assay the relative metallic/semiconducting nanotube concentration required that the nanotubes be dedoped prior to spectral recording. This was facilitated by our method of forming uniform, optically transparent, thin films of the nanotubes for the spectral recording. Simply heating the Br doped nanotubes to 600°C in flowing argon fully dedopes them. This is shown in Figure 3. The thick gray line there is the spectrum of a nanotube film, baked in this manner, prior to Br exposure. The solid black line is the spectrum of this film following exposure to Br vapors and the dot-dashed line is the spectrum of this film following another bake cycle. As is evident from the complete recovery of the original spectrum the bake cycle fully dedopes the nanotubes.

To establish the initial degree of Br doping in this set of experiments, diluted Br in DI water was added, drop wise, to a 3% Triton X-100 surfactant suspension of nanotubes in DI water with solution spectral recording between Br additions. Water absorption bands obscured the SC1 band so the SC2 band was monitored. It was found that at 8 mg/ml Br concentration the SC2 absorbance first began to show a slight decrease in intensity, indicating that nanotubes were being doped.

Modeling, along with the spectra can provide useful limits regarding the idea that density differences between the more highly Br doped metallic nanotubes and the Br doped semiconducting nanotubes could be responsible for the separation. Figures 4A and 4B show the density of electronic states (DOS) for a representative semiconducting (12,8), and metallic (10,10) nanotube, respectively, within a simple tight binding model [4]. Only states below the Fermi level (0 eV) are shown and the DOS is normalized here in electrons/eV/nm of nanotube length. The integrated

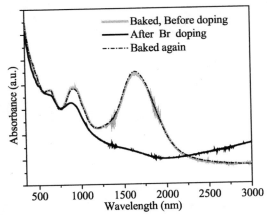

FIGURE 3. Demonstration that 600°C baking in flowing, ultra high purity argon fully dedopes Br doped nanotubes. See text for details.

areas under the density curves represent then the number of holes on the nanotube. Recall that the Br doping criteria were that SC2 barely change. Assuming one charge transfer per Br, this implies that the doping, at most, involved the gray area under the curve of Fig. 4A, resulting in 1.05 Br ions/nm of tube length. Turning now to the metallic nanotube, since M was not observed to change, the doping for the metallic nanotubes at most involved the gray area under the curve of Fig. 4B, resulting in 1.85 Br ions/nm of tube length. Hence the maximum difference in Br associated with the metallic over the semiconducting nanotubes (within this rigid band model) is 0.8 Br/nm of tube length. A nanometer of nanotube in this diameter range has ~160 C atoms so that the greatest mass increase of such a "maximally" doped metallic nanotube over a "maximally" doped semiconducting nanotube is only 3%. It seems unlikely that such a small change in mass could be responsible for the separation seen, leaving open the question of the separation mechanism.

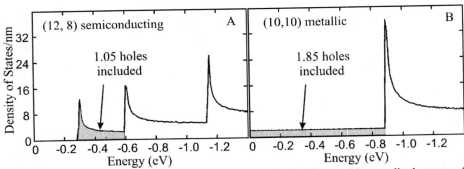

FIGURE 4. Density of electronic states, within a simple tight binding model, normalized per nm of nanotube length, for a representative (A) semiconducting (12,8) and (B) metallic (10,10) nanotube, respectively. See text for description of shaded regions.

An important indication that Triton X-100 played a critical role in the separation came when sodium duodecyl sulfate (SDS) was substituted as the surfactant. An experiment using 3 wt% SDS and 8 mg/ml Br yielded far less sediment, following centrifugation, than the experiments using Triton. This was nevertheless sufficient to record spectra and perform the analysis to find that there was no metallic-semiconducting nanotube separation. To determine if the SDS somehow prevented the Br doping of the nanotubes, a solution based spectrum with these concentrations was recorded. Under these conditions however, SC2 was completely gone. In fact it was found by drop-wise addition of highly diluted Br to the nanotubes in SDS that it (remarkably) took ~10,000 times *less* Br in SDS to dope the nanotubes to the same degree as occurred with Triton X-100 (indicated by the point at which SC2 first decreased). Hence, while the nanotubes were far more heavily doped in the case of SDS, few nanotubes sedimented out during the centrifugation (and no separation was observed). While it is possible that the SDS surfactant shell imparts far greater buoyancy to the nanotubes than Triton X-100, this seems unlikely.

A further clue regarding the separation mechanism is provided by the poor separation observed in the highest Br concentration experiment (10.9 mg/ml) using Triton X-100. In that experiment flocculation was visible in the supernatant following the centrifugation. We also note that the Br concentrations in the more successful separations are not far below this clearly excessive concentration. Triton X-100 like most non-ionic surfactants has limited ionic strength tolerance (as evidenced by the cloud point reduction with added salt), after which it losses its activity as a surfactant. We therefore hypothesize that the near instability in the Triton X-100, induced by the high Br concentrations, plays a dominant role in the separation mechanism; the higher Br concentration associated with the metallic nanotubes is just sufficient to push their surfactant stabilization over the edge, causing the metallic nanotubes to lose their surfactant shells. As shown by O'Connell, *et al.* [5] it is only because those shells possess a hydrated density substantially smaller than 1 g/cc that the ~1.3 g/cc density nanotubes can, once surrounded by a surfactant shell (giving the net object a density near one), remain in suspension under centrifugation. Once that shell is lost the nanotubes sediment out. Clearly further experiments are required to elucidate the mechanism with greater certainty. A telling experiment would be to use Triton X-100 surfactant at reduced Br concentrations but bringing the surfactant closer to its cloud point by elevating the temperature during the centrifugation.

We also discussed our method for fabricating controlled thickness, homogeneous films of pure SWNTs sufficiently thin to be transparent over technologically relevant regions of the electromagnetic spectrum. Optical properties of these films appear as another contribution in this volume [6]. The simultaneous transparency and electrical conductivity of such films is likely to lead to a number of applications. Aspects of the fabrication method have already appeared [1,7] and will be elaborated elsewhere [8], here we briefly summarize a feature makes these films particularly exciting. Contact barriers between metals and semiconductors in many existing and proposed electronic and photonic devices remain a problem. These barriers require application of increased voltages to inject current into the devices. This leads to large local electric fields and increased power dissipation both of which exacerbate electro-migration of the contact materials. This impacts both the efficiency and lifetime of the devices (the latter because the electro-migrated species usually degrade device performance). We have shown that the transparent nanotube films permit Ohmic electrical contact to be made between various metal electrodes and p-GaN [7], a material that has historically proved difficult to contact electrically. Transparency of the film was useful because the p-GaN in that case was part of a GaN/InGaN light emitting diode and the blue light was emitted through the transparent nanotube film. The low barrier, Ohmic contact implies a favorable band structure line-up between the nanotubes and the GaN. This is already useful, however the real significance emerges upon recalling that the nanotube Fermi level can readily be shifted large amounts by simple chemical charge transfer doping. Hence by controlling the degree of doping the nanotubes may solve the contact barrier problem for a broad range of materials. A possible concern is that the loosely bound charge transfer dopants could themselves electro-migrate to create problems. If the barrier to transport across the junction is small however, then the potential drop across the junction, which is the driving force for electro-migration, can also be kept small. Experiments to test these ideas are presently underway.

1. Z. Chen, X. Du, M-H. Du, C. D. Rancken, H-P. Cheng, A. G. Rinzler, Nano Letters, **3**, 1245 (2003)
2. P. Petit, C. Mathis, C. Journet, and P. Bernier, Chem. Phys. Lett. **305**, 370 (1999).
3. S. Kazaoui, N. Minami, R. Jacquemin, H. Kataura, and Y. Achiba, Physical Review B 60, 13339 (1999).
4. R. Saito, G. Dresselhaus, M. S. Dresselhaus, *Physical Properties of Carbon Nanotubes*, Imperial College Press, 1998, London.
5. M. J. O'Connell, S. M. Bachilo, C. B. Huffman, V. C. Moore, M. S. Strano, E. H. Haroz, K. L. Rialon, P. J. Boul, W. H. Noon, J. Ma, R. H. Hauge, R. B. Weisman and R. E. Smalley. *Science,* **297**, 593-596 (2002)
6. F. Borondics_, K. Kamarás_, Z. Chen, A.G. Rinzler, M. Nikolou and D.B. Tanner, In these AIP conference proceedings.
7. K. Lee, Z. Wu, Z. Chen, F. Ren, S. J. Pearton, A. G. Rinzler, Nano Lett. **4**, 911 (2004).
8. Z. Wu, Z. Chen, X. Du, J. Logan, J. Sippel, M. Nikolou, K. Kamaras, J. R. Reynolds, D. B. Tanner, A. F. Hebard and A. G. Rinzler, in preparation (2004).

Manipulating Carbon Nanotubes with Nucleic Acids

Ming Zheng

DuPont Central Research and Development
Wilmington, DE 19880, USA

Abstract. Single-stranded DNA (ssDNA) forms stable complex with CNT and effectively disperses CNT into aqueous solution. We found that a particular ssDNA sequence (d(GT)n, n = 10 to 45) self-assembles into an ordered supramolecular structure around individual CNT, in such a way that the electrostatic properties of the DNA-CNT hybrid depend on tube type, enabling CNT separation by anion-exchange chromatography. Optical absorption and Raman spectroscopy showed that the separation is bimodal based on both electronic properties and the diameters of CNTs: early fractions are enriched in the smaller diameters and metallic tubes, whereas late fractions are enriched in the larger diameters and semiconducting tubes. In this presentation, I will also provide an update on our effort in single (n, m) type carbon nanotube separation.

DNA is a natural-occurring polymer that plays a central role in biology. Many unique properties of DNA have inspired a search for its non-biological applications. Molecular recognition between complementary strands of a double-stranded DNA has been used to construct various geometric objects at the nanometer scale[1], and to organize the assembly of colloidal particles [2,3]. The π-stacking interaction between bases in DNA has prompted the exploration of its electronic properties for possible use in molecular electronics[4]. Less utilized are the potential inorganic substrate-binding properties of DNA, in contrast to recently demonstrated efficacy of oligopeptides for this purpose[5-7].

A large molecular library can be formed by ssDNA, which offers an intriguing possibility for carbon nanotube binding: depending on its sequence and structure, aromatic nucleotide bases in ssDNA may be exposed to form π-stacking interactions with the side-wall of carbon nanotubes. To search for DNA sequences that bind carbon nanotubes, we followed a well-established *in vitro* evolution procedure[8]. We discovered that DNA binding to carbon nanotubes is extremely effective and facile. Specific conditions for dispersion depend on the source of carbon nanotubes. For purified HiPco nanotubes, we found that practically any ssDNA would work in the presence of a denaturant, with mild sonication. The primary role of denaturant appears to disrupt base-pairing; which can be prevented by avoiding dG(guanine):dC(cytosine) and dA(adenine):dT(thymine) base-paring interactions in the sequence design. For as-produced HiPco nanotubes, in addition to ssDNA, short double-stranded DNA and total RNA extracted from *Saccharomyces cerevisiae* and *Escherichia coli* can also disperse carbon nanotubes. In this case, however, vigorous sonication is needed for

CP723, *Electronic Properties of Synthetic Nanostructures*, edited by H. Kuzmany et al.

effective dispersion. DNA-CNT solutions are stable for months at room temperature. Removal of free DNA by either anion exchange column chromatography or nuclease digestion does not cause nanotube flocculation, indicating that DNA binding to carbon nanotubes is very strong. Even though there is no reliable way to determine the percentage of individually dispersed nanotubes in solution, all of our evidence suggests that DNA converts bundled CNT into individual tubes. The electronic absorption spectra of DNA-CNT solutions show well-resolved structures and systematic dependence on pH. We have also observed strong near IR fluorescence from DNA-CNT solutions. These features are characteristic of individually dispersed nanotube solutions, obtained with surfactants after intense sonication treatment [9]. Atomic force microscopy measurements show that DNA-CNTs have diameters ranging from 1 to 2 nm. The diameters are larger than the 0. 7 to 1.1 nm range expected for HiPco tubes, but are consistent with DNA coating of nanotubes (Fig.1).

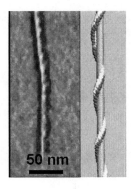

FIGURE 1. AFM image of random polyd(CT) wrapped around a carbon nanotube in good agreement with the proposed molecular wrapping model.

To explore the influence of the DNA sequence and length on dispersion efficiency, we tested some simple patterns of sequence with defined lengths. The dispersion efficiency was measured by the optical absorption intensity of dispersion solution at 730 nm, which is relatively invariant with respect to pH changes. Among fixed length (60–mer) homo-polymers that can be made by solid-phase synthesis, we found that poly d(T) has the highest dispersion efficiency. Among the four different lengths (60-, 30-, 21- and 15-mer) of poly d(T) examined, we found that d(T)30 gave the highest yield.

In comparison with other polymers that also disperse carbon nanotubes [10], DNA appears to be much more efficient. In a typical experiment, 1 mg of DNA can disperse equal amount of as-produced HiPCO CNT in 1 mL volume, yielding 0.2 to 0.4 mg/mL CNT solution after removal of non-soluble material by centrifugation. Such obtained CNT solution can be further concentrated by ten-fold to give a soluble CNT concentration as high as 4 mg/mL. DNA chain flexibility and backbone charge may all contribute to such high dispersion efficiency. In addition, DNA offers the advantage of defined length and sequence, and well-developed chemistries for functionalization. The latter provides a way to functionalize the carbon nanotube surface without modifying it covalently. As a demonstration, we made DNA-CNT using an oligonucleotide that was modified at one end with biotin. We observed biotin-dependent deposition of nanotubes onto streptavidin-coated agarose beads. This demonstration also illustrates a way to control the placement of carbon nanotubes on a solid substrate.

Carbon nanotube (CNT) separation is an enabling step for many potential applications and fundamental studies that require defined nanotube structures and properties. CNTs can be classified into two categories based on their electronic structures: metallic and semiconducting tubes. The latter can be further classified by tube diameters, since the band gap of a semiconducting tube, a critical parameter that needs to be controlled for nanoelectronic applications, is inversely proportional to its diameter. Reports in the literature indicate that it is possible to separate metallic from semiconducting tubes by taking advantage of differences in their physical or chemical properties [11-14]. Diameter-based separation is more difficult, because differences in the physical/chemical properties caused by diameter changes are smaller, and variations in tube length could be a dominant factor in physical based separation methods. We have identified an oligonucleotide sequence that self-assembles into a highly ordered structure on CNT, allowing not only metal/semiconducting separation, but also diameter-dependent separation. We found that anion-exchange chromatography provides a macroscopic means to assay for electrostatic properties of nanoscale DNA-CNT hybrids. More specifically, we found that the outcome of anion-exchange based DNA-CNT separation, as measured by optical absorption spectral changes from fraction to fraction, is strongly dependent on the DNA sequence. To explore this dependence, we conducted a systematic but limited search of the huge ssDNA library under identical chromatographic conditions. We found that sonication effectively cuts CNTs in the presence of DNA, and that short CNTs increase sample recovery from an anion exchange column. In a typical dispersion experiment, a DNA/CNT mixture was kept in an ice-water bath and sonicated (Sonics, VC130 PB) for 120 min at a power level of 8 W. The average length of CNTs after this level of sonication is ~ 140 nm as measured by AFM. We tested simple homo-polymers of dA, dC and dT, and sub-libraries composed of random combinations of two of the four nucleotides (dG, dA, dT, dC): poly d(A/C), poly d(A/G), poly d(A/T), poly d(C/T), poly d(C/G) and poly d(G/T). Among these, poly d(G/T), poly d(G/C) showed the largest variation in the optical absorption spectra from fraction to fraction. To narrow down the choice of sequence, we then tested representative sequences in the poly d(G/T) and poly d(G/C) sub-library. We found that the best separation was obtained with a sequence of repeats of alternating G and T, d(GT)n, with total length ranging from 20 to 90 bases (n = 10 to 45).

In a report published earlier [15], we provided evidence showing diameter as well as electronic property based separation of HiPco carbon nanotubes dispersed by d(GT)n sequences. Using carbon nanotubes synthesized by the "CoMoCat" process as the starting material [16], we found that individual (n,m) type enriched carbon nanotubes can be obtained by our separation process. In this experiment, unpurified single wall carbon nanotubes from Southwest Nanotechnologies (SWeNT, Norman, OK) and single-stranded DNA d(GT)30 were used. Anion exchange chromatography separation was done as described. The SWeNT tubes have two major type of semiconducting tubes (6,5) and (7, 5), as was characterized in the literature [16]. In Fig.2, optical absorption profiles of eluted fractions 35, 39 ,40 along with that of the starting material are shown. The absorption spectrum of f35 has two dominant peaks at 990 nm and 574 nm. These are close to the literature assignment of 975 nm (E11) and 567 nm (E22) for the (6, 5) tubes, but are red-shifted [16]. Similarly, peaks at 928 nm and 702 nm can be assigned to (9, 1) tubes (same diameter as (6, 5)); peaks at ~ 970 nm (overlapping with 990 nm peak) and 674 nm can be assigned to (8, 3). By intensity comparison, it is clear that f35 is largely enriched with (6, 5) tubes.

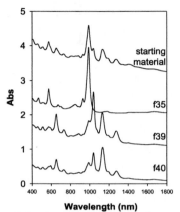

FIGURE 2. Absorption spectra of fractions from anion exchange chromatography separation of d(GT)30-dispersed CoMoCat tubes.

To better understand the role of the d(GT)n sequence we used for CNT separation, we used atomic force microscopy (AFM) to study the DNA assembly on CNT. Molecular modelling suggests that ssDNA can adopt many different modes of binding to CNT, with little difference in binding free energies. These modes include helical wrapping with different pitches, consistent with our observations. In contrast, d(GT)n exhibits a qualitatively different binding to CNT than most of the other ssDNA sequences. AFM measurements show that d(GT)n-CNT hybrids have a much more uniform periodic structure with a regular pitch of ~18 nm. Whatever structure d(GT)n has on CNT, it appears to be very sensitive to minute changes in the structure of the nucleotide bases (Fig.3 mutation data). Replacement of G or T by their homologues, inosine or uridine, respectively, decreases the degree of separation on the anion exchange column. Replacing d(GT)n with d(GGTT)n/2 that has identical chemical composition also decreases the extent of the separation.

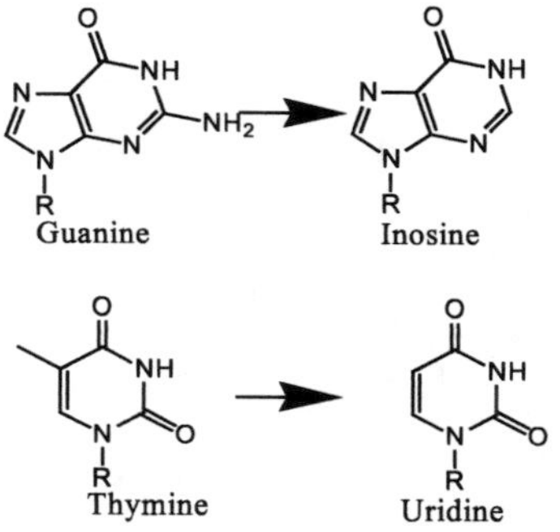

FIGURE 3. Structures of G, T and their homologues.

It is known that GT-rich sequences can self-assemble into supramolecular structures that play important roles in telomere replication[17]. More relevantly, GT-rich sequences have been observed to form nano-wires involving hydrogen-bonding interactions among different strands[18]. As a model for the d(GT)n-CNT structure, we propose that two anti-parallel d(GT)n strands interact with each other through hydrogen bonds to form a double-stranded strip, which then wraps around the CNT with close-packed bases resembling molecular tiles lying on the side-wall of the nanotube [15]. Such a double-helical structure is built on the unique hydrogen bonding network between two d(GT)n strands, and is expected to be more rigid and to have fewer allowed conformations than a single-helical structure. A rigid DNA structure that creates an identical charge distribution within a given type of nanotube is necessary for successful separation. This is probably why most sequences disperse CNTs but do not give good separation.

ACKNOWLEDGMENTS

This work comes from the Molecular Electronics group at DuPont Central Research and Development. I thank my colleague Anand Jagota, Scott Mclean and Ellen Semke for their collaboration.

REFERENCES

1. Seeman, N. C. DNA engineering and its application to nanotechnology. *Trends Biotechnol* **17**, 437-43 (1999).
2. Alivisatos, A. P. et al. Organization of 'nanocrystal molecules' using DNA. *Nature* **382**, 609-11 (1996).
3. Mirkin, C. A., Letsinger, R. L., Mucic, R. C. & Storhoff, J. J. A DNA-based method for rationally assembling nanoparticles into macroscopic materials. *Nature* **382**, 607-9 (1996).
4. Arkin, M. R. et al. Rates of DNA-mediated electron transfer between metallointercalators. *Science* **273**, 475-80 (1996).
5. Wang, S. et al. Peptides with selective affinity for carbon nanotubes. *Nat Mater* **2**, 196-200 (2003).
6. Whaley, S. R., English, D. S., Hu, E. L., Barbara, P. F. & Belcher, A. M. Selection of peptides with semiconductor binding specificity for directed nanocrystal assembly. *Nature* **405**, 665-8 (2000).

7. Dieckmann, G. R. et al. Controlled assembly of carbon nanotubes by designed amphiphilic Peptide helices. *J Am Chem Soc* **125**, 1770-7 (2003).

8. Wilson, D. S. & Szostak, J. W. In vitro selection of functional nucleic acids. *Annu Rev Biochem* **68**, 611-47 (1999).

9. O'Connell, M. J. et al. Band gap fluorescence from individual single-walled carbon nanotubes. *Science* **297**, 593-6 (2002).

10. O'Connell, M. J. Reversible water-solubilization of single-walled carbon nanotubes by polymer wrapping. *Chem. Phys. Lett.* **342**, 265-271 (2001).

11. Chattopadhyay, D., Galeska, I. & Papadimitrakopoulos, F. A route for bulk separation of semiconducting from metallic single-wall carbon nanotubes. *J Am Chem Soc* **125**, 3370-5 (2003).

12. Krupke, R., Hennrich, F., Lohneysen, H. & Kappes, M. M. Separation of metallic from semiconducting single-walled carbon nanotubes. *Science* **301**, 344-7 (2003).

13. Weisman, R. B. Carbon nanotubes: Four degrees of separation. *Nat Mater* **2**, 569-70 (2003).

14. Zheng, M. et al. DNA-assisted dispersion and separation of carbon nanotubes. *Nat Mater* **2**, 338-42 (2003).

15. Zheng, M. et al. Structure-based carbon nanotube sorting by sequence-dependent DNA assembly. *Science* **302**, 1545-8 (2003).

16. Bachilo, S. M. et al. Narrow (n, m)-Distribution of Single-Walled Carbon Nanotubes Grown Using a Solid Supported Catalyst. *J. Am. Chem. Soc.* **125**, 11186-11187 (2003).

17. Williamson, J. R. G-quartet structures in telomeric DNA. *Annu Rev Biophys Biomol Struct* **23**, 703-30 (1994).

18. Marsh, T. C., Vesenka, J. & Henderson, E. A new DNA nanostructure, the G-wire, imaged by scanning probe microscopy. *Nucleic Acids Res* **23**, 696-700 (1995).

Direct Growth of Carbon Nanofibers on Plastic Substrates

M. Cantoro[1], S. Hofmann[1], B. Kleinsorge[1], G. Csanyi[2], M. C. Payne[2], A. C. Ferrari[1], J. Robertson[1]

[1]University of Cambridge, Engineering Department, Cambridge CB2 1PZ, UK
[2] University of Cambridge, Cavendish Laboratory, Cambridge CB3 0HE, UK

Abstract. Carbon nanofibers were successfully grown onto Cr covered thin polyimide plastic substrates by plasma-enhanced chemical vapour deposition. Growth is visible at a temperature as low as 200 °C, without detectable substrate degradation. Patterned 100 nm-size Ni catalyst features allow selective deposition. We propose that surface carbon transport on the metal catalyst is the low activation energy pathway in low temperature nanofiber growth, as shown by activation energy measurements and first principles density functional calculations.

INTRODUCTION

Carbon nanotubes (CNTs) are of great interest to the scientific and industrial communities due to their remarkable electronic and mechanical properties. They have been synthesized by various techniques, such as arc discharge, laser ablation and chemical vapour deposition (CVD). The latter has the unique advantage of allowing selective growth directly on a substrate with nanoscale accuracy, avoiding the need for expensive post-deposition purification and manipulation. In particular, plasma-enhanced CVD (PECVD) [1-5] enable access to a lower temperature range respect to thermal CVD, therefore substrates not suitable for high temperature applications, such as plastics, can be used. Polyimide plastic substrates are common in microelectronics as interlayer dielectrics or passivation layers and can be structured by plasma etching or laser ablation. They have also been used as substrate material for flexible thin-film transistors, being compatible with established microelectronics fabrication processes. They also have the advantage of low cost, shock resistance, flexibility and light-weight.

RESULTS AND DISCUSSION

A 177 μm-thick commercial Kapton® polyimide foil was used as substrate. A 70 nm-thick conductive Cr layer was sputtered onto it by means of rf magnetron sputtering. 100 nm dots of 6 nm-thick Ni catalyst were patterned onto its surface by e-beam lithography. Vertically aligned carbon nanofibers (CNFs) were grown using a dc-PECVD system at a temperature of 200 °C [4], striking a dc discharge between the gas inlet (anode) and a resistively heated sample holder (cathode). The temperature was measured with a thermocouple mounted on a polyimide substrate of equivalent

CP723, *Electronic Properties of Synthetic Nanostructures*, edited by H. Kuzmany et al.
© 2004 American Institute of Physics 0-7354-0204-3/04/$22.00

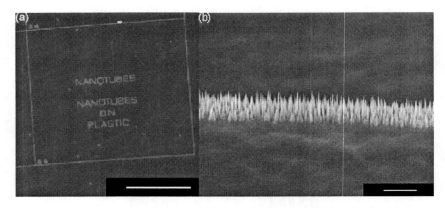

FIGURE 1. SEM micrographs of vertically aligned CNFs grown from e-beam patterned Ni catalyst on a Cr covered polyimide plastic foil. Scale bars: (a) 100 μm, (b) 1 μm.

original sample thickness. A 1.5 mbar, 200:30 sccm NH_3:C_2H_2 gas mixture was used, and the dc discharge was maintained for 1 hour. Teo *et al.* [6] argued that high power plasmas can result in very high temperatures at the substrate holder in a dc-PECVD system. They reported a substrate temperature of 700 °C for 200 W plasma power at 5-12 mbar pressure, even in the absence of additional heating. In this condition they observe a very interesting intense glow pattern in the cathode region [6]. In our experiment we have a discharge current of typically 30 mA. This implies that our plasma power is below 20 W, ensuring the minimization of plasma heating effects. Indeed, we do not observe any thermal glowing in the cathode region. The polyimide substrate exhibits no bending after the CNF deposition and maintains its flexibility, further confirming the low temperature.

Fig. 1(b) shows short and slightly tapered structures. The dilution of the carbon feedstock with ammonia minimizes the deposition of amorphous carbon, as shown by results of high-resolution transmission electron microscopy (HRTEM) analysis [5]. The Ni catalyst particles are seen at the top of the CNFs, suggesting a tip-growth mechanism.

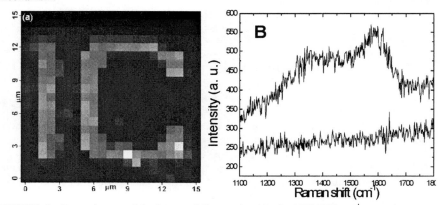

FIGURE 2. Grayscale map of the integrated Raman signal in the 1200-1700 cm^{-1} spectral range, acquired scanning the "IC" letters of the "plastic" CNF pattern of fig 1(a). The mapped area is 15 x 15 μm. The scanning step is 750 nm. The top spectrum reported in (b) is representative of the CNF spectra, whilst the bottom spectrum in (b) shows no carbon signal outside the pattered region.

The selectivity of the deposition is confirmed by Raman spectroscopy. Fig. 2(a) shows a map of the Raman signal, integrated in the 1200-1700 cm^{-1} spectral region, acquired by scanning a 15x15 μm area of the sample in Fig. 1(a). The map was taken by coupling a Physik Instruments piezoelectric scanning stage to a Renishaw Raman spectrometer at 514.5 nm excitation energy. The background spectra in Fig. 2(b) shows that no carbon signal is detected outside the CNFs patterned region. The CNFs patterns show the characteristic D and G peaks [3,7].

Figure 3 plots the growth rate as a function of the deposition temperature for CNFs previously deposited on Si substrates by thermal CVD and by high and low temperature PECVD [4,8]. Whilst the thermal growth shows an activation energy of ~1.2 eV, the plasma assisted growth gives a much lower value (0.23 eV). This is very similar to the experimentally measured activation energy for carbon surface diffusion on polycristalline Ni (0.3 eV) [9]. We find a similar low activation energy of 0.2-0.4 eV when using Co as catalysts. In order to explain our results, the carbon diffusion on the catalyst surface and the stability of the gas used for nanofibers growth (C_2H_2) were investigated by first principles plane wave density functional calculations using the CASTEP code [10] (ultrasoft pseudopotentials, 270 eV cutoff, 3x3x1 k-point mesh, 2x2 supercell). According to the Wulff construction for an fcc catalyst cluster, small metal particles reconstruct to show mainly (111) planes and, to a lower extent, (100) planes, Fig. 4(a). Figure 4 also shows ball and stick models of

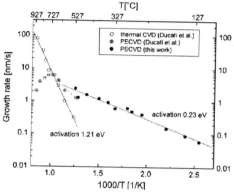

FIGURE 3. CNFs growth rate vs. temperature for thermal CVD and PECVD [4,8].

possible carbon diffusion paths. We find a barrier for surface diffusion on Ni(111) of ~0.5 eV, Fig. 4(b), and ~2 eV for Ni(100), Fig. 4(c). For the Ni(100) surface we can identify a subsurface (but not bulk) diffusion path with 0.7 eV activation energy, Fig. 4(d). Similar results are found for Co. Our calculations suggest that, once C atoms are available on the catalyst surface, they can flow with very low activation energy on the (111) surface. This agrees with recent in-situ TEM growth studies, which experimentally detected carbon flow on Ni(111) during thermal CNF growth and found that carbon surface diffusion over the catalyst particle seems to play a significant role [11].

We then calculated the activation energy for C_2H_2 hydrogen abstraction. This is significantly higher than 0.5 eV on the Ni(111) surface. This explains why, to allow for the hydrogen abstraction on the catalyst surface, an extra barrier arises in pure thermal growth. Indeed, in the present case, the plasma creates C and C-C radicals, eliminating the need for surface dissociation. We thus propose that bulk diffusion can contribute at high temperature, but surface diffusion is always active.

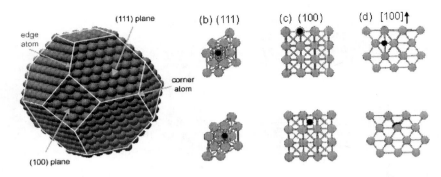

FIGURE 4. (a) Wulff construction for a fcc catalyst cluster. (b,c,d) possible carbon diffusion paths on (111) and (100) planes.

In conclusion, CNFs were successfully grown on a thin polyimide substrate at a temperature as low as 200 °C. SEM and Raman analysis confirmed the selectivity of the PECVD process. Surface diffusion of carbon on the catalyst particle can play a significant role in low temperature CNF growth.

ACKNOWLEDGEMENTS

This work was supported by the EU project CARDECOM GRD1-2001-41830. A.C.F. acknowledges funding from The Royal Society. The authors thank F. Festy for help with the PI stage set-up. CASTEP calculations were performed at CCHPCF, Cambridge.

REFERENCES

1. Z. F. Ren et al. *Science* **282**, 1105 (1998).
2. V. I. Merkulov et al. *Appl. Phys. Lett.* **76**, 3555 (2000).
3. M. Chhowalla et al. *J. Appl. Phys* **90**, 5308 (2001).
4. S. Hofmann, C. Ducati, B. Kleinsorge, and J. Robertson, *Appl. Phys. Lett.* **83**, 135 (2003).
5. S. Hofmann, C. Ducati, B. Kleinsorge, and J. Robertson, *Appl. Phys. Lett.* **83**, 4661 (2003).
6. K .B. K. Teo et al. Nano Lett. **4**, 921 (2004)
7. A. C. Ferrari, J. Robertson, Phys. Rev. B **61**, 14095 (2000)
8. C. Ducati et al., *J. Appl. Phys.* **92**, 3299 (2002).
9. J. F. Mojica, and L. L. Levenson, *Surf. Sci.* **59**, 447 (1976).
10. V. Milman et al., *Int. J. Quantum Chem.* **77**, 895 (2000).
11. S. Helveg et al., *Nature* **427**, 6973 (2004).

PROPERTIES OF SINGLE-WALL CARBON NANOTUBES

Changes in carbon nanotube electronic properties by collisions with inert gases

Kim Bolton[*,1], Arne Rosén[1], Hugo Romero[2] and Peter Eklund[2,3]

[1]*Experimental Physics, School of Physics and Engineering Physics, Göteborg University and Chalmers University of Technology, SE-412 96, Göteborg, Sweden*
[2]*Department of Physics and [3] Department of Materials Science and Engineering, The Pennsylvania State University, University Park, Pennsylvania 16802, USA*

Abstract: Collisions with inert gases increase the thermoelectric power and resistivity of single-walled carbon nanotube (SWNT) thin films as $M^{-1/3}$, where M is the mass of the colliding gas. Molecular dynamics simulations show that there is a similar mass dependence for the collision-induced deformation and phonon excitation of SWNTs, which suggests that the electron scattering is due to the induced dents and/or phonons. The simulations also show that the $M^{-1/3}$ dependence is obtained under a wide range of collision conditions (*e.g.*, SWNT temperature, incident scattering angle and colliding gas energy).

INTRODUCTION

Since all atoms of a SWNT form part of the nanotube surface, the chemical environment in contact with the nanotube is expected to affect electronic transport in the nanotube wall. For example, transconductance of SWNT field effect transistors is sensitive to 2-200 ppm ambient NO_2 and NH_3 [1], and the electrical resistance and thermoelectric power (TEP) of SWNT films is affected by the physisorption of aromatic hydrocarbons [2]. In these examples the change in resistivity and TEP is believed to be dominated by either electron transfer or weak interactions between the adsorbed gas and the nanotubes that induce changes in the electron and hole free carrier lifetimes. However, due to their weak radial forces, SWNTs are easily deformed and collisions by ambient gases deform the nanotube and excite phonons [3]. This may provide a new scattering channel for electron transport that may dominate for inert gases (where gas-SWNT electron transfer is negligible). Indeed, as discussed below, TEP and resistance of SWNTs (measured experimentally) as well as collision-induced deformation and phonon excitation (simulated) increase as $M^{-1/3}$, where M is the mass of the inert gas.

METHODS

The TEP and four-probe resistance measurements were carried out on thin films of purified, bundled SWNTs deposited on a quartz substrate (CarboLex,

---

*Corresponding author: kim@fy.chalmers.se
Tel- +46-31-7723294

CP723, *Electronic Properties of Synthetic Nanostructures*, edited by H. Kuzmany et al.
© 2004 American Institute of Physics 0-7354-0204-3/04/$22.00

Inc.; arc-discharge method). The films were vacuum annealed at ~1000°C for 12 hours before attaching thermocouples (chromel-Au/7 at% Fe) and electrical (copper) leads with silver epoxy to four corners of the sample. The specimens, that contained *ropes* of SWNTs of 1.0-1.6 nm in diameter, were placed in a turbo-pumped vacuum chamber (~1x10^{-7} Torr) where transport measurements were made in the presence of the gases. The samples were vacuum-degassed *in situ* at 200 °C to remove oxygen and water [4].

The potential energy surfaces and simulation methods used here have been detailed elsewhere [3]. The Brenner potential was used for the SWNT intramolecular potential, and interactions between the gas and SWNT were described by Lennard-Jones (12-6) potentials fit to experimental equilibrium distances and binding energies. The simulations focused on the (10,0) SWNT, which is sufficiently narrow (~0.8 nm) to allow for computationally tractable simulations, while also being sufficiently broad to obtain data representative of experiment, and were performed for a range of SWNT temperatures, initial scattering angles and gas kinetic energies. To evaluate the collision-induced SWNT deformation and phonon excitation, simulations were also performed with SWNTs at 0 K, where there is no thermal background noise.

RESULTS AND DISCUSSION

As discussed elsewhere [5], the linear relationship between the collision-induced changes in TEP and resistivity is consistent with the creation of a new scattering channel for the conduction electrons in the metallic tubes. The slopes of these linear relationships are found to be independent of the gas pressure [5]. Hence, the $M^{-1/3}$ dependence shown in Figure 1 for the maximum TEP change, ΔS_{max}, and also found for the change in resistivity, does not depend on the frequency of the gas-SWNT collisions (the data in the figure was obtained at 1 atm). Instead, it reflects a change in the SWNT properties due to individual collisions with the gas molecules.

Figure 1 also shows the change in collisional energy transfer obtained from simulations of SWNTs at 1300 K (ΔE_{1300}) and 0 K (ΔE_0). The results at 1300 K were obtained for an initial relative translational energy E_i=1.4 kcal mol^{-1} and incident angle θ_i=45°, and those at 0 K for E_i=4.0 kcal mol^{-1} and θ_i=0°. It is clear that the $M^{-1/3}$ dependence seen for ΔS_{max} is also observed for the collisional energy transfer (or phonon excitation), and that it is insensitive to the collision conditions. It is also evident that the collision-induced deformation, ΔD_0, determined by the maximum radial displacement of any atom in the SWNT during the collision, has an $M^{-1/3}$ dependence (results are for T=0 K, E_i=2.0 kcal mol^{-1}, θ_i=45°). This indicates a strong link between ΔS_{max} and the collision-induced SWNT phonon excitation and/or deformation. Simulations for other collision conditions yielded similar results.

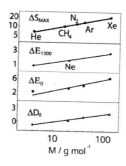

FIGURE 1. Maximum TEP change ΔS_{max} ($\mu V\ K^{-1}$), energy transfer (kcal mol^{-1}) for a SWNT at 1300 K (ΔE_{1300}) and 0 K (ΔE_0) and SWNT deformation ΔD_0 (Å) due to collisions with inert gases. All three properties show an $M^{-1/3}$ dependence, where M is the gas mass. The linear fits are $\Delta S_{max}=3.99M^{0.33}$, $\Delta E_{1300}=0.55M^{0.35}$, $\Delta E_0=0.91M^{0.39}$ and $\Delta D_0=0.04M^{0.35}$.

Details of the phonon excitation were studied by monitoring the time-dependent SWNT power spectra during gas collisions. The power spectra shown in Figure 2 are weighted so that the area under the spectrum equals the average SWNT total energy, induced by the collision (Xe, T=0 K, E_i =13 kcal mol^{-1}, θ_i=45°). The spectrum in Figure 2a was obtained from the first 5 ps of the collision, which includes the gas-SWNT impact, and the spectrum in Figure 2b was obtained from the second 5 ps, and shows the SWNT phonons a short time after impact. It is clear that, in accordance with simple kinetic arguments, the heavy Xe atom imparts more energy to the nanotube than the lighter Ne and He atoms, and that a large amount of collision-induced energy remains in the SWNT after 10 ps. It is also evident that there are fewer peaks in the spectrum obtained between 5 and 10 ps, which shows that the energy in the very low

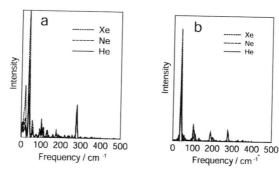

FIGURE 2. Power spectra showing the SWNT phonons that are induced by collisions with Xe, Ne and He gases. Panel a shows the phonons during the first 5 ps of collision (and includes the gas-SWNT impact) and panel b is obtained from the second 5 ps.

frequency phonons (about 20 cm^{-1} in Figure 2a) either i) flows into the other phonons or ii) flows axially away from the collision center of the SWNT.

To identify the character of the SWNT deformation, the displacement of a ring of SWNT carbon atoms was followed during a collision. The 'ring' of carbon atoms that is followed includes the carbon atom closest to the point of impact by the gas. It is clear from Figure 3a that the impact flattens one side of the tube (*i.e.*, causes a dent in the tube), and at later times other modes are dominant, especially the squash mode shown in Figure 3b.

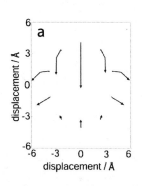 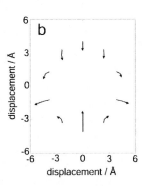

FIGURE 3. Cross-section of the SWNT showing the displacement of a ring of carbon atoms at the point of impact by Xe (a) and after the collision (b). Atomic displacements are scaled by 10 for clarity.

In summary, the similar $M^{-1/3}$ dependence of the collision-induced changes in TEP and SWNT deformation/phonon excitation indicates that there is a direct link between these properties. Since electron scattering from low frequency phonons is expected to be weak, we suggest that the important scattering channel is due to SWNT deformation. Also, the simulations suggest that this scattering channel should be important for a wide range of collision conditions.

REFERENCES AND NOTES

1. J. Kong, N.R. Franklin, C. Zhou, M.G. Chapline, S. Peng, K. Cho, H. Dai, *Science* **287**, 622 (2000).
2. G.U. Sumanasekera, B.K. Pradhan, H.E. Romero, K.W. Adu, P.C. Eklund, *Phys. Rev. Lett.* **89**, 166801/1 (2002).
3. K. Bolton, A. Rosén, *Phys. Chem. Chem. Phys.* **4**, 4481 (2002).
4. H.E. Romero, G.U. Sumanasekera, G.D. Mahan, P.C. Eklund, *Phys. Rev. B* **65**, 205410 (2002).
5. H.E. Romero, K. Bolton, A. Rosén, P.C. Eklund, submitted.
6. This work was supported by the National Science Foundation (PCE), Penn State University (PCE, HR) and the Swedish Foundation for Strategic Research (AR).

Thermal Measurements on Multi-wall Nanotubes

E. Brown, L. Hao, J. C. Gallop, and J. C. Macfarlane*

National Physical Laboratory, Queens Road, Teddington, Middlesex, TW11 0LW, UK
*University of Strathclyde, Glasgow, G4 0NG, UK

Abstract. The electrical and mechanical properties of carbon nanotubes (CNTs) have been studied in depth at a single nanotube level; the same though cannot be said for their thermal properties.

We have measured the thermal properties of multi-wall carbon nanotubes (MWNTs) using a temperature sensing scanned microscope probe. An arc-grown MWNT bundle is attached to a thermal probe with several individual MWNTs protruding from the end of the bundle. The system is operated in high vacuum and the temperature of the thermal probe may be controlled anywhere between 300 and 600 K. Then, under piezo displacement, the MWNT is brought into contact with a substrate surface (HOPG, highly oriented pyrolitic graphite). The substrate can be cooled from room temperature to ~100 K. The heat flow down the MWNTs in contact with the substrate can be recorded as a function of the temperature difference across their ends. Simultaneous measurements of the electrical conductance and force applied to the nanotube are taken.

The thermal and electrical conductance curves of different MWNTs are presented; the size of the conductance steps observed and the correlation between electrical and thermal conductance steps are discussed together with the effect of oxygen adsorption on the properties of the MWNTs.

INTRODUCTION: PROPERTIES OF CARBON NANOTUBES

CNTs have been predicted to show extremely high thermal conductivity [1] and are also expected to display a 1-dimensional (1D) phonon density of states.[2] However measurements on individual nanotubes have been very few.[3]

We suggested that carbon nanotubes should also be the ideal test beds for the study of ballistic phonon transport and of thermal conductance quantisation.[4] This has been theoretically confirmed in a recent publication by T Yamamoto et al.[5] Thermal conductance quantisation occurs in a dielectric quantum wire at low enough temperatures when its dimensions are smaller than its phonon mean free path.[6,7] The quantum of thermal conductance is G_{th}. The total number of phonon channels available in a CNT is three times the number of atoms in its unit cell: 3N. N can be readily calculated from the (n, m) indices of the CNT.[8] A (10,10) CNT (diameter 1.4 nm) has 120 phonon channels while a (200,200) CNT (diameter 27.5 nm) has 2400 phonon channels. Thus if at a given temperature T this CNT is a ballistic phonon conductor its thermal conductance will be $2400 \times G_{th}(T)$.

MWNTs display electrical conductance quantisation even at room temperature, as observed in several experiments.[9,10] The quantum of electrical conductance is G_o.

CP723, *Electronic Properties of Synthetic Nanostructures*, edited by H. Kuzmany et al.
2004 American Institute of Physics 0-7354-0204-3/04/$22.00

CNTs are expected to contribute $2G_o$ per conducting carbon shell,[8,10] however, contributions of $1G_o$ and $1/2G_o$ have also been reported.[9] The expressions for G_{th} and G_o are shown in equation (1): h: Planck's constant; k_B: Boltzmann constant, T: temperature, e: electron charge.

$$G_{th} = \frac{\pi^2 k_B^2 T}{3h} = 9.456 \times 10^{-13} \frac{W}{K^2} \times T, \; G_o = \frac{2e^2}{h} = \frac{1}{12.9k\Omega}. \tag{1}$$

THERMAL MEASUREMENTS ON MULTI-WALL NANOTUBES

We aim to measure the thermal properties of individual CNTs and to study evidence for ballistic phonon transport and thermal conductance quantisation in CNTs.

Experimental Method and Set-up

A diagram of the experiment is shown in Fig. 1. An arc-produced MWNT fibre is fixed on a Veeco thermal probe. The tip of the Veeco probe can be heated to ~600 K.

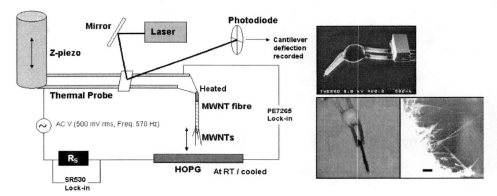

FIGURE 1. The experimental set-up (left). On the right are shown an SEM picture of the Veeco Thermal probe (top), a photograph of an MWNT fibre mounted on the thermal probe and an SEM picture of the individual MWNTs protruding from the fibre's tip (the scale bar is 1μm).

The probe is lowered on a graphite substrate (which can be cooled down to ~100 K) until a preset force is applied by the MWNTs protruding from the fibre's tip onto the substrate. The probe is then moved towards and away from the substrate under piezo control at a constant speed (typically 20 nm/s) while thermal and electrical conductance and the force of the CNTs on the substrate are recorded. The electrical and thermal conductance of the MWNTs are recorded at the same time to ensure that the thermal conductance through a single CNT is being measured. When the signature of quantum electrical conductance is seen it is clear that only one CNT is involved. All measurements are done in high vacuum (~10^{-5} mbar).

Conductance Curves

Electrical and thermal conductance measurements are compared expressed in units of G_o and G_{th} respectively (Fig. 2). The electrical conductance step is due to the longest protruding MWNT losing contact with the substrate and its height of $\sim 2G_o$ is that expected for ballistic conductance in a CNT.

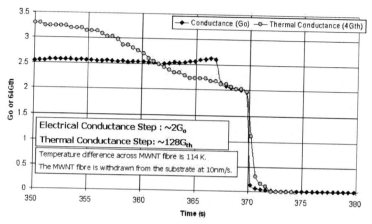

FIGURE 2. Graph showing thermal and electrical conductance variations as the MWNT fibre is moved away from the substrate.

The thermal conductance has a corresponding step height of ~ 128 G_{th} which is an order of magnitude lower than expected given the average size of the MWNTs in the fibre (typically 15-40 nm). The presence of ballistic electron transport suggests that the MWNT has few defects therefore we attribute the low thermal conductance values to phonon-phonon scattering which determines the phonon mean free path, quenches the ballistic phonon transport and reduces the conductance of the CNT.

Thermal conductance steps in the range 10^{-8}-10^{-6} W/K were observed when the temperature difference across the MWNT fibre was 110-150 K (one end of the fibre was at room temperature and the other was heated). These values well compare with thermal conductance measurements on individual MWNTs,[3] which report a thermal conductance of 10^{-7} W/K at 300K. This shows the high purity of the MWNTs in the fibre and the lack of scattering within the MWNT fibre.

On some occasions steps in the electrical conductance with no corresponding thermal conductance steps were observed (and vice versa). The missing electrical conductance steps can be due to semiconducting CNTs in the MWNT fibre. We have attributed the missing thermal conductance steps to MWNT carbon shells sliding past one another as the MWNT is moved away or towards the substrate. This because, while the electrical conductance of a MWNT is changes significantly with the removal of a conducting shell [11] the thermal conductance is a 2D process at these temperatures [3] and thus will be less affected by it.

In Fig. 3 thermal and electrical conductance steps are compared for different DTs (the temperature difference across the MWNT fibre, with one end at room temperature and the other heated). There is a large reduction in the size of the electrical conductance steps at higher DTs. We attribute this to O_2 desorption. O_2 is readily absorbed by CNTs and transforms semiconducting CNTs in metallic ones.[12] At higher temperatures and in vacuum O_2 desorbs more readily causing a higher reduction in the number of metallic CNTs contributing to the electrical conductance.

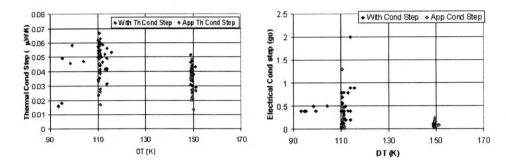

FIGURE 3. Comparison of the size of thermal and electrical conductance steps at different DTs.

CONCLUSION

In our work we have investigated the electrical and thermal conductance of MWNTs. We observed ballistic electron transport but no ballistic phonon transport, however, lowering the operating temperature below the onset of phonon-phonon scattering predominance could solve this problem. The effect of oxygen adsorption on the transport properties of the MWNTs has also been discussed, and an explanation for the sometimes-missing electrical or thermal conductance steps has been presented.

REFERENCES

1. Berber, S., Kwon, Y-K., and Tománek, D., *Phys. Rev. Letters* **84**, 4613-4616 (2000).
2. Hone, J., Llaguno, M. C., Biercuk, M. J., Johnson, A. T., Batlogg, B., Benes, Z., and Fischer, J. E., *Appl. Phys. A* **74**, 339-343 (2002).
3. Kim, P., Shi, L., Majumdar, A., and McEuen, P. L., *Phys. Rev. Letters* **87**, 215502 (2001).
4. DTI Quantum Programme 2004-2007, http://www.dti.gov.uk/nms/consult/quantum_draft_for_comment.pdf, p. 83-89.
5. Yamamoto, T., Watanabe, S., and Watanabe, K., *Phys. Rev. Letters* **92**, 075502 (2004).
6. Rego, L. G. C., and Kirczenow, G., *Phys. Rev. Letters* **81**, 232-235 (1998).
7. Schwab, K., Henriksen, E. A., Worlock, J. M., and Roukes, M. L., *Nature* **404**, 974-976 (2000).
8. Saito, R., Dresselhaus, G., and Dresselhaus, M. S., *Physical Properties of Carbon Nanotubes*, London: Imperial College Press, 1998.
9. Frank, S., Poncharal, P., Wang, Z. L., and de Heer, W. A., *Science* **280**, 1744-1746 (1998).
10. Urbina, A., Echeverría, I., Pérez-Garrido, A., Díaz-Sánchez, A., and Abellán, J., *Phys. Rev. Letters* **90**, 106603 (2003).
11. Collins, P. G., and Avouris, Ph., *Appl. Phys. A* **74**, 329-332 (2002).
12. Collins, P. G., Bradley, K., Ishigami, M., and Zettl, A., *Science* **287**, 1801-1804 (2000).

Pathogenic Activity of 1D Nanocarbons: *In Vivo* Studies on Guinea Pigs

A. Huczko*[1], H. Lange[1], M.Sioda[1], M.Bystrzejewski[1], P.Nejman[2], H. Grubek-Jaworska[2], K.Czumińska[3] and J. Glapiński[4]

[1]*Department of Chemistry, Warsaw University, 1 Pasteur, 02-093, Warsaw, Poland*
[2]*Department of Pneumology, Medical University of Warsaw, 61 Żwirki i Wigury, 02-091 Warsaw, Poland*
[3]*Faculty of Veterinary Medicine, Warsaw Agriculture University, 159c Nowoursynowska, 02-787 Warsaw, Poland*
[4]*Institute of Biocybernetics and Biomedical Engineering PAN, 4 Trojdena, 02-109 Warsaw, Poland*

Abstract. Pathogenic activity of different 1D nanocarbons was studied in vivo on guinea pigs. Long exposure of airways to specific 1D nanomaterial can affect the respiration and induce pathological processes in lung tissue.

INTRODUCTION

The possible health and safety implications should be addressed within any project funded on nanomaterials. Among others, genuine concern has been raised about the inhalation of nanoparticles [1]. One study showed recently that carbon nanotubes, as asbestos-like material, can damage lung tissue in mice [2]. Earlier we reported the preliminary results of physiological testing of fullerenes [3] and carbon nanotubes [4, 5]. In this paper, the results of *in vivo* physiological testing of different 1D nanocarbons are presented. Comparing to our former exploratory study [5], the number of tested materials and animals was increased, the exposure period was longer and more elaborated analytical techniques were used to study an inflammatory response.

EXPERIMENTAL

The lungs of anesthetized guinea pigs were intratracheally instilled with 12.5 mg of the different 1D nanocarbons (Fig. 1). The material was suspended in 0.5 ml of sterile saline with the addition of SDS surfactant. The control group of animals was administered with a nanocarbon-free fluid. Following 90 d post-instillation exposure guinea pigs were anesthetized, tracheae were cannulated and spontaneous breathing was suppressed. Lung resistance (LR) to inflation was measured by constant flow respirator with pressure transducer located between the trachea and respiratory pump. LRs were continuously recorded breath by breath during 7 min Then the bronchoalveolar lavage (BAL) was performed using standard procedure and BAL fluid cells and supernatants were analysed for estimation of inflammatory processes. Finally, the animals were euthanized for histopathological study of the lunges.

CP723, *Electronic Properties of Synthetic Nanostructures*, edited by H. Kuzmany et al.
© 2004 American Institute of Physics 0-7354-0204-3/04/$22.00

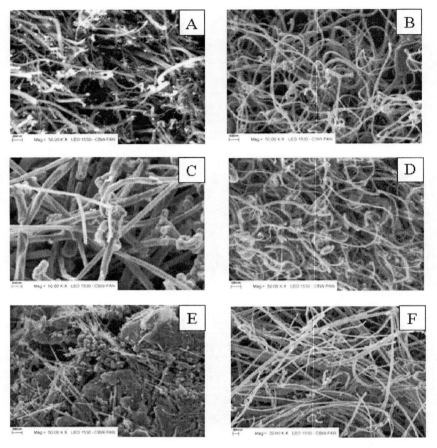

FIGURE 1. HR SEM images of tested 1D materials: A – multiwalled carbon nanotubes (MWCNTs) produced by CCVD technique; B – MWCNTs, 95% purity (by NanoLab, GB); C – carbon nanofibers (by Showa Denko, Japan); D – MWCNTs, 80% purity (by NanoLab, GB); E – MWCNTs produced by carbon arc technique (material from the core of cathode deposit); F – carbon nanofibers (by Pyrograf, USA)

RESULTS

For all experimental animals, excluding the control group, organizing pneumonitis (*bronchiolitis obliterans* organizing pneumonia) with focal nonspecific desquamative interstital pneumonia-like ("DIP-like") reaction without fibrosis or with mild peribronchiolar fibrosis were observed (Fig. 2). Simultaneously an increase of lung resistance was observed only in guinea pigs instilled with carbon arc nanotubes. The microscopic analysis of BAL-fluid cells revealed the infiltration of bronchoalveolar space with inflammatory cells in all the experimental animals excluding the control and animals instilled with carbon nanofibers by Pyrograf. In some BAL-fluid cells the nanofibers were located inside of alveolar macrophages (in cases of guinea pigs instilled with Showa Denko or Pyrograf). The spherical or irregular shape of carbonaceous materials were observed inside of some phagocytes of animals instilled with CVD or carbon arc nanotubes (Fig. 3). Extracellular nanofibers were detected in BAL-fluid of guinea pigs treated with nanofibers by Pyrograf. The protein concentration in BAL-f of experimental animals did not differ significantly as compared to the control although exudate to alveoli was observed in all the experimental groups histopathologically.

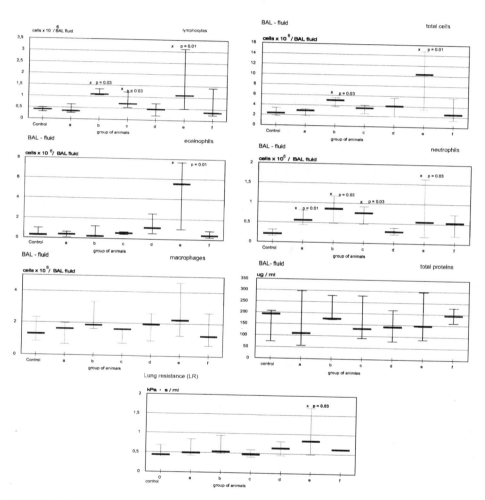

FIGURE 2. Results of BAL analyses for different 1D nanocarbons comparing to control group

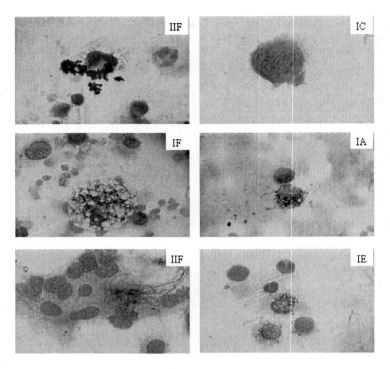

FIGURE 3. Photomicrographs of the BAL-fluid cells of guinea pigs intratracheally instilled with different 1D nanocarbons: IA, IC, IE, IF – intracellular and IIE, IIF – extracellular deposits of respective materials

Our present results in connection with previously performed study [5] indicate that both the time of the exposure and the characteristics of the intratracheally introduced 1D nanocarbon are critical for induction of measurable pathology of lungs.

ACKNOWLEDGMENTS

The work was supported by the Committee for Scientific Research (KBN) under Grant No. 3 PO5A 126 22.

REFERENCES

1. C. Sealy, *Materials Today* **6**, 1 (2003).
2. R.F. Service, *Science*, **300**, 243 (2003).
3. A. Huczko, H. Lange, E. Całko, *Full. Sci. Techn.*, **7**, 935 - 938 (1999).
4. A. Huczko, H. Lange, *Full. Sci. Techn.*, **9**, 247-250 (2001).
5. A. Huczko, H. Lange, E. Całko, H. Grubek-Jaworska, P. Droszcz, *Full. Sci. Techn.*, **9**, 252-254 (2001).

Comparison of Current-Voltage Characteristics of Nanofibres and Nanotubes

A. B. Kaiser[*], B. Chapman[*], U. Schlecht[†] and M. Burghard[†]

[*] MacDiarmid Institute for Advanced Materials and Nanotechnology, SCPS,
Victoria University of Wellington, P O Box 600, Wellington, New Zealand
[†] Max-Planck-Institut für Festkörperforschung, Heisenbergstr. 1, 70569 Stuttgart, Germany

Abstract. The current-voltage characteristics of a number of nanoscale materials show some striking similarities despite the differing conduction mechanisms expected in different materials. We make a comparison of experimental I-V characteristics of Ag-V_2O_5 nanofibre networks with those of carbon nanotube networks and polyacetylene nanofibres, using as a base the generic expression for the nonlinear conductance given by our numerical calculations for fluctuation-assisted tunnelling and thermal activation. We find a remarkably similar change from linear behaviour at high temperatures to nonlinear conduction at lower temperatures.

INTRODUCTION

The advent of nanofibres and nanotubes has opened up new possibilities for conducting materials that show novel behaviour due to their limited size and the quasi-one-dimensional nature of conduction [1,2]. For example, polyacetylene nanofibres show temperature-independent Zener-type tunnelling at very low temperatures [3] that we have shown [4] could arise from tunnelling of the conjugated bond pattern along single polyacetylene chains.

In this paper, we compare our new data on the evolution of the current-voltage (I-V) characteristics with changing temperature for Ag-V_2O_5 nanofibre networks with earlier data for other nanoscale materials: single-wall carbon nanotube (SWCNT) networks and individual polyacetylene nanofibres, as measured by Kim et al. [5] and Park et al. [6], respectively. We also compare the I-V characteristics to the generic expression for nonlinear conductance suggested by our numerical calculations for fluctuation-assisted tunnelling and thermal activation.

Ag-V_2O_5 NANOFIBRE NETWORKS

Vanadium pentoxide (V_2O_5) can be made in the form of fibres of molecular dimensions (with cross-section approximately 1.5 nm by 10 nm and length several μm) that can (for example) be used as chemiresistors in sensor applications [7]. The structure of the fibres is one-dimensional and conduction is attributed to hopping (thermally activated tunnelling) between V^{5+} and V^{4+} sites.

CP723, *Electronic Properties of Synthetic Nanostructures*, edited by H. Kuzmany et al.
© 2004 American Institute of Physics 0-7354-0204-3/04/$22.00

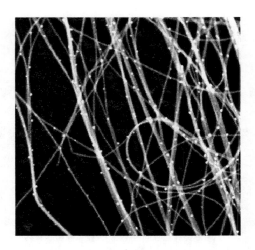

FIGURE 1. Image of the network of Ag-V$_2$O$_5$ nanofibres with attached Ag clusters obtained by scanning transmission electron microscopy (S-TEM) operated in the High-Angle Annular Dark-Field (HAADF) mode; the width of the image is approximately 720 nm.

In the present case, Ag-V$_2$O$_5$ nanofibres were grown in an aqueous solution via polycondensation of vanadyl acid [8]. Ag ions were added to the solution to speed up the growth of the fibres. Note that Ag cluster are attached to the fibres, as can be seen in the scanning transmission electron microscope image of Figure 1. Transport measurements were made on networks of fibres deposited on an interdigitated electrode array with 10 μm separation between electrodes.

CURRENT-VOLTAGE CHARACTERISTICS

Figure 2 shows the I-V characteristics of an Ag-V$_2$O$_5$ nanofibre network for a range of temperatures. The I-V characteristics were symmetric to within experimental error with respect to reversal of the voltage direction, so only the positive quadrant is shown. Figures 3(a) and 3(b) show the positive quadrant of the I-V characteristics of an individual polyacetylene nanofibre [6] and of a SWCNT network [5], that also show symmetric behaviour.

It is clear that there is a remarkable similarity between the I-V characteristics for all three materials, which show a similar dependence on applied voltage V as well as a similar qualitative evolution with temperature. As shown by the fitted lines, all the data in these figures are well fitted by the expression:

$$I = G_0 V \exp(V/V_0), \tag{1}$$

i.e. conductance

$$G = I/V = G_0 \exp(V/V_0). \tag{2}$$

The parameter G_0 is the low-field conductance, i.e. the ratio I/V as $V \to 0$. G_0 shows a strong increase with temperature in each material - approximately activated behaviour in the case of Ag-V_2O_5 and the polyacetylene nanofibre, but consistent with fluctuation-induced tunnelling behaviour [10] for the SWCNT network [5,9]. The parameter V_0 is the scale parameter for the exponential increase in conductance in Eq. (2). V_0 shows only a relatively small increase with temperature for all three materials.

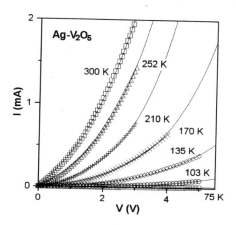

FIGURE 2. *I-V* characteristics of a network of Ag-V_2O_5 nanofibres at temperatures from 300 K down to 75 K; the lines are fits of the data to Eq. (1).

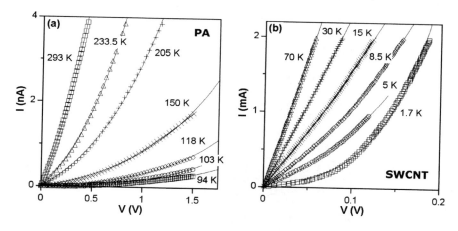

FIGURE 3. *I-V* characteristics for a range of temperatures of (a) a polyacetylene nanofibre (PA) of diameter 20 nm doped with iodine as measured by Park et al. [6], and (b) a single-wall carbon nanotube network (SWCNT) measured by Kim et al. [5]. In each case the lines show fits to Eq. (1).

The origin of the fitting expression is our full numerical calculations of fluctuation-induced tunnelling and thermal activation (extending the fluctuation-induced tunnelling model of Sheng [10] to cases of higher conductivity). These calculations give the generic expression [9]:

$$I = G_0 V \frac{\exp(V/V_0)}{1 + h[\exp(V/V_0) - 1]},$$

(3)

of which Eq. (1) is the $h \to 0$ limit.

For $h > 0$, the exponential increase of conductance slows down at higher voltages V (ultimately yielding ohmic behaviour with conductance G_0/h as $V \to \infty$) - there are small indications of such a slowing down in some of the data sets in the figures above.

CONCLUSIONS

The data shown demonstrate a remarkable similarity in the development of nonlinearities in the I-V characteristics as temperature is lowered in Ag-V$_2$O$_5$ nanofibre networks, SWCNT networks and polyacetylene nanofibres.

The overall shape of these nonlinearities is well described by Eq. (1) given by our numerical calculations for fluctuation-assisted tunnelling and thermal activation. This generic result appears to be applicable more widely for conduction involving tunnelling and thermal activation, not just for the specific model used in the calculations.

ACKNOWLEDGMENTS

ABK and BC acknowledge support from the Marsden Fund administered by the Royal Society of New Zealand.

REFERENCES

1. Roth, S., *One-Dimensional Metals*, Weinheim: VCH, 1995.
2. Saito, R., Dresselhaus, G., and Dresselhaus, M. S., *Physical Properties of Carbon Nanotubes*, London: Imperial College Press, 1998.
3. Park, J. G., Kim, B., Lee, S. H., Kaiser, A. B., Roth, S., and Park, Y. W., *Synth. Met.* **135-136**, 299-300 (2003).
4. Kaiser, A. B., and Park, Y. W., *Synth. Met.* **135-136**, 245-247 (2003).
5. Kim, G. T., Jhang, S. H., Park, J. G., Park, Y. W., and Roth, S., *Synth. Met.* **117**, 123-126 (2001).
6. Park, J. G., Kim, G. T., Krstic, V., Lee, S. H., Kim, B., Roth, S., Burghard, M., and Park, Y. W., *Synth. Met.* **119**, 469-470 (2001).
7. Schlecht, U., Besnard I., Vossmeyer T., and Burghard, M., in *Molecular Nanostructures*, edited by H. Kuzmany et al., New York: American Institute of Physics, 2003, pp. 491-494.
8. Schlecht U., in preparation
9. Kaiser, A. B., Rogers, S. A., and Park, Y. W., *Mol. Cryst. Liq. Cryst.* (2004) in press.
10. Sheng, P., *Phys. Rev. B* **21**, 2180-2195 (1908).

Superconductivity in Long and Short Molecules

A.Yu. Kasumov[1,2], K.Tsukagoshi[1,3], M. Kawamura[1], T. Kobayashi[1],
Y.Aoyagi[1,4], T. Kodama[5], V.T. Volkov[2], Yu.A. Kasumov[2], I.I. Khodos[2],
D.V. Klinov[6], M. Kociak[7], R. Deblock[7], S. Guéron[7], H. Bouchiat[7]

[1]RIKEN, Hirosawa 2-1, Wako, Saitama 351-0198, Japan.
[2]Institute of Microelectronics Technology RAS, Chernogolovka, Moscow district, 142432, Russia.
[3]PRESTO, JST, Honcho 4-1-8, Kawaguchi, Saitama, Japan.
[4]Department of Information Processing, Tokyo Institute of Technology, Nagatsuda 4259, Midori,
Yokohama, Kanagawa 226-8502, Japan.
[5]Department of Chemistry, Tokyo Metropolitan University, Minami-Ohsawa 1-1, Hachioji, Tokyo 192-
39, Japan.
[6]Shemyakin-Ovchinnikov Institute of Bioorganic Chemistry, RAS, Miklukho-Maklaya 16/10, Moscow
117871, Russia.
[7]Laboratoire de Physique des Solides UMR 8502 - Universite Paris-Sud, Bat. 510, 91405 Orsay cedex,
France.

Abstract. We present the results of experimental study of superconductivity in individual molecules of carbon nanotubes, DNAs and metallofullerenes. Critical currents of supeconductor-molecule-superconductor junctions were extensively studied as a function of temperature and magnetic field. The mechanism of current induced superconductor-normal state transition for a long molecule (carbon nanotubes and DNAs) is the creation of phase slip centres and for a short molecule (metallofullerens) - multiple Andreev reflections. We observe an influence of spin state of encapsulated atom on the induced superconductvity in a metallofullerene molecule.

Molecular electronics is considered one of the alternatives to traditional semiconducting electronics, capable of increasing the density of devices in integrated circuits to up to a trillion per centimeter square [1]. However given that the heat removal possibility is less than 50 W/cm^2 [2], each device mounted with such high density should dissipate at most 50 pW. This is 3 orders of magnitude less than the power dissipated by the known molecular switches [1,3].

Two fundamental restrictions (von Neuman's and Heizenberg principle) and two technical restrictions (heat removal and switching time of a modern transistor) define the region of integrated (fast!) molecular electronics (Fig. 1). The restrictions remain very small region for room temperature molecular electronics, but the region is unreachable for semiconducting molecular devices because of another technical restriction - high operating voltage, order of 1 V [1,3]. A possible solution to this problem is use of superconducting molecular devices with operating voltage order of 1 mV. So, study of superconductivity in individual molecules is very important for future molecular electronics.

Here we report the study of proximity induced and intrinsic superconductivity in long molecules (carbon nanotubes and DNAs) and proximity induced superconductivity in short molecules (Gd@C$_{82}$ metallofullerens). The main result for

CP723, *Electronic Properties of Synthetic Nanostructures*, edited by H. Kuzmany et al.

carbon nanotubes is observation of suppression of intrinsic superconductivity by quantum phase slips (QPS). A QPS is a topological vortex-like excitation of the superconducting phase field, which only exists in 1D superconductors. In addition to the Tc depression QPSs produce a finite sub-Tc linear resistance apart from the usual temperature-independent contact resistance. This effect is indeed observed experimentally [4], and can be compared in a quantitative way to theory [5] (Fig. 2). The rather good agreement found there supports the notion that ropes represent 1D superconductors in the few-channel limit where QPSs can be experimentally observed in a clear manner from the temperature-dependent resistance below Tc.

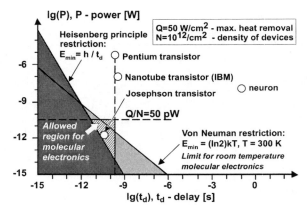

FIGURE 1. Limits for molecular electronics.

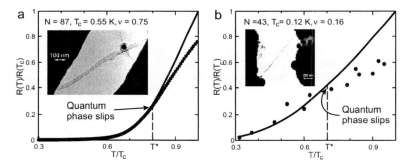

FIGURE 2. Temperature dependence of the linear resistance below T_c for 2 individual ropes of carbon nanotubes (N – number of metallic tubes in a rope; v – dimensionless parameter from the theory [5]). Open circles denote experimental data (with substracted residual resistance). Insets: TEM micrographs of corresponding samples.

We emphasize the importance of the interaction of interactions of DNA molecules with the underlying substrate. For most commonly used substrates like mica or silicon oxide the interaction between the molecule and the surface is very strong and induces a very large compression deformation of deposited DNAs. Thickness of such compressed DNAs is 2-4 times less than the diameter (about 2nm) of native Watson-Crick B-DNA. We confirm the insulating character of DNA on such substrates. On the other hand when the substrate is treated (functionalized) in order that deposited molecules keep their original thickness we observe their conducting behavior, both from conducting AFM and transport measurements on molecules connected to platinum electrodes. This conductivity persists down to very low temperature (0.1K) where it exhibits a non-ohmic behavior with a power low singularity in the bias dependence of the differential resistance typical of one-dimensional conductors with Coulomb interactions [6] (Fig. 3a,b).

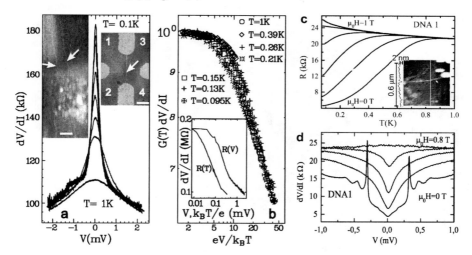

FIGURE 3. (a) Bias dependence of the differential resistance for different temperatures between 0.1 and 1K. Note the asymmetry of the curves above 0.4 mV. The excitation current is 0.1 nA. Right inset: optical picture of the slits etched in the 3nm thick Pt film on mica. The wide part of the slits were etched using a focused laser beam, the narrow submicron part was etched using a focused ion beam. Scale bar is 20 μm. Left inset: AFM picture of DNA molecules combed across the slit. 2 molecules are clearly visible (shown by arrows) on the Pt contacts (in dark) but not on the etched mica across the slit which rough surface state impedes good visualization of the molecules. Scale bar is 250 nm. (b) Scaling behavior of the differential resistance in the temperature range between 0.1 K and 1 K and voltage range between 0.02 and 0.4 mV. Inset: Temperature dependence of the resistance of sample depicted on the insets compared to the bias dependence measured at 0.1K on a log-log scale indicating a power law increase of the resistance at low temperature and low bias. (c) dc resistance of DNA as a function of temperature, below 1 K, for different values of the magnetic field: $\mu_0 H$ = 0, 0.2, 0.4, 0.6, 0.8, and 1 T. Inset: AFM profile of DNA. (d) Voltage dependence of the differential resistance of DNA measured at 50 mK in magnetic fields of 0, 0.2, 0.4, 0.6, 0.8, and 1 T (ac excitation current was 1 nA at 30 Hz).

We have also performed conductivity measurements on double-stranded DNA molecules deposited by a combing process across a submicron slit between rhenium/carbon metallic contacts. Conduction is ohmic between room temperature and 1 K. The resistance per molecule is less than 100 kOhm and varies very slowly with temperature. Below 1K, which is the superconducting transition temperature of the contacts, we observe proximity induced superconductivity (Fig. 3c,d). This implies in particular that DNA molecules can be metallic down to mK temperature, and furthermore that phase coherence is achieved over several hundred nanometers [7].

We also report the first study of transport through a nanometer size molecule in contact with superconducting electrodes and direct observation of the molecule by high resolution transmission electron microscopy (HRTEM) (Fig. 4a). As a molecule we used a metallofullerene molecule, $Gd@C_{82}$, which has a diameter of about 1 nm. Metallofullerene molecules have the same mechanical stability as fullerenes, but their doping by a metallic atom (in our case Gd) acting as a donor favors charge transfer through the molecule [8]. The possibility of doping organic molecules with one or two metal atoms recalls silicon electronics, where doping determines the transport properties of a device. The search for a molecule that would play the same role in molecular electronics is a crucial issue.

FIGURE 4. (a) Actual HRTEM image of $Gd@C_{82}$ molecular dimer between electrodes. (b) Schematic picture of the molecular dimer between superconducting electrodes. The black dots symbolize the Gd atoms inside the fullerene cage. (c) Voltage dependence of the differential resistance; differential resistance peaks in accordance with theory are indicated by arrows. Suppression of MAR and peaks on dV/dI at voltage $V_n = 2\Delta/ne$ [9]. Additional peaks are due to Quantum Dot effect [10].

We find that a junction containing a single metallofullerene dimer between superconducting electrodes (Fig. 4b) displays signs of proximity-induced superconductivity. In contrast, no superconductivity remains in a junction containing a cluster of dimers. These results can be understood, taking into account multiple Andreev reflections (MAR) [9], and the spin states of Gd atoms [11] (Fig. 4c).

In conclusion we have discovered and studied superconductivity in individual molecules. The discovery opens the door for superconducting molecular electronics. It is the only way to overcome fundamental and technical restrictions for integrated molecular electronics.

We thank the Russian Foundation for Basic Research, Solid State Nanostructures, and International Science and Technology Center for financial support.

REFERENCES

1. J.R. Heath, M.A. Ratner, *Physics Today* **56(5)**, 43 (2003).
2. J.D. Meindl, Q. Chen, J.A. Davis, *Science* **293**, 2044 (2001).
3. J. Appenzeller, D.J. Frank, *Appl.Phys.Lett.* **84**, 1771-1773 (2004).
4. A.Yu. Kasumov et al., *Phys.Rev. B* **68**, 214521 (2003).
5. A.De Martino & R. Egger, Phys.Rev. B, (2004), will be published.
6. A Yu. Kasumov et al., *Appl.Phys.Lett.* **84**, 1007 (2004).
7. A.Yu. Kasumov et al., *Science* **291**, 280 (2001).
8. H. Shinohara, *Rep.Prog.Phys.* **63**, 843 (2000).
9. M. Octavio, M. Tinkham, G.E. Blonder and T.M. Klapwijk, *Phys.Rev.* **B 27**, 6739 (1983).
10. G. Johansson, E.N. Bratus, V.S. Shumeiko and G. Wendin, *Phys. Rev.* **B 60**, 1382 (1999).
11. A.Yu. Kasumov et al., cond-mat/0402312.

Effect of Irradiation on Aligned Carbon Nanotube Fibers

Cs. Mikó, M. Milas, J.W.Seo, E. Couteau, N. Barišić, R. Gaál, L. Forró

Institute of Physics of Complex Matter, Ecole Polytechnique Fédérale de Lausanne, Lausanne, Switzerland

Abstract. We carried out in-situ resistivity measurements on macroscopic oriented ropes of single wall carbon nanotubes in a transmission electron microscope. We have found a minimum in the resistivity as a function of irradiation dose. This minimum is interpreted as a result of a twofold effect of the irradiation: the domination of covalent bond formation between tubes in a bundle due to broken bonds in the tube walls and the amorphization of the sample at high dose.

INTRODUCTION

Carbon nanotubes have allured much attention because of their unique physical properties. In the last decade, literature has reported considerable progress in the large-scale synthesis [1-3] as well as the purification of CNTs using different techniques. In order to exploit their feasible application, one of the major challenges is the processing of CNTs on macroscopic scale, for instance into a macroscopic fiber.

For an individual single-walled carbon nanotubes an exceptionally high Young's modulus has been demonstrated both in theory and in experiments; therefore, the CNT-based composite materials are expected to have extraordinary mechanical properties as well. Unfortunately the single-walled nanotubes (SWNTs) preferentially form bundles, where tubes interact via van der Waals force, and therefore easily slide along their axis. Hence, the mechanical strength of a macroscopic SWNT fiber is dominated by their shear modulus and not by their Young's modulus, accordingly not as high as expected.

Recently, it has been demonstrated that electron beam irradiation can create covalent bonds between surfaces of SWNTs [4] just as cross-linked CNTs within SWNTs rope resulting in a considerable increase of Young's modulus.

Kis et al. [5] pointed out that both 80 and 200 kV beam energy could displace carbon atoms, and induce cross-links between the single CNTs, enhancing the shear modulus for low electron irradiation dose.

In this contribution, we report on effect of electron irradiation on electrical properties of macroscopic fibers consisting of SWNTs [6]. From electrical point of view, CNT can show either semiconducting or metallic behavior depending on their chirality's. Ropes of SWNTs assembled to a macroscopic fiber show semiconducting properties due to the predominant presence of semiconducting nanotubes (~70%). As

CP723, Electronic Properties of Synthetic Nanostructures, edited by H. Kuzmany et al.
© 2004 American Institute of Physics 0-7354-0204-3/04/$22.00

single CNTs are much shorter than the contact electrodes on the fiber it can be considered as random resistor network of many weakly connected nanotubes.

This is the paragraph spacing that occurs when you use the Enter key.

EXPERIMENTAL DETAILS

SWNT bundles were assembled to a macroscopic fiber according to Gommans et al. [7] For in-situ transport measurements, a sample of 3 mm length was cut from a long rope and mounted on a TEM Cu ring with four gold electrodes, which were glued by conducting silver paint and electrically isolated from the supporting Cu ring by mica. This configuration enabled the in situ measurement of the resistance in a TEM (Hitachi TE 700 operating at 200 kV) during irradiation as well as ex situ measurements by interrupting the irradiation and placing the sample in a He cryostat for resistivity vs temperature measurements in the 4.2–300 K temperature range.

The irradiation was carried out for approximately 29 h at a constant electron flux of about $2.4*10^{13}$ electrons cm^{-2} s^{-1}. In order to avoid long exposure time, the incident beam was focused to a diameter of 15 μm, increasing the flux to $2*10^{16}$ electrons cm^{-2} s^{-1}, and this spot was moved step by step along the fiber between the two center electric contacts.

RESULTS AND DISCUSSION

Our result of the in-situ measurement is shown in fig.1, presenting the resistivity of the SWNT fiber vs irradiation time.

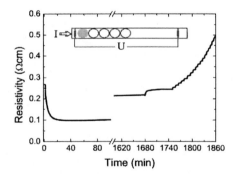

FIGURE 1. The resistivity of a SWNT fiber as function of electron irradiation time. The schematic drawing in the inset shows the contact arrangement and the irradiated spot size with respect to sample dimensions. The step-like variation towards the end of the irradiation is the result of the stepwise moving of the irradiated zone along the sample.

Before the irradiation the ropes had a typical 4-probe resistivity of 0.2 Ωcm ± 0.08 Ωcm, which agrees with Vigolo et al [8]. At the beginning, the resistivity fell approximately to 0.07 Ωcm ± 0.02 Ωcm and reached a minimum after about 35 minutes of irradiation time. Past this minimum, the resistivity increased linearly with the irradiation time with a slope of about 1.15 Ω/min. This dependence remained the same for many hours. We focused then the beam to a 15-micron diameter spot and irradiated each part until the resistivity saturated, then we moved to a neighboring section along the fiber axis. This gives rise to the step-like time dependence shown in fig.1.

Besides measuring the resistivity during irradiation, we measured the resistance as function of temperature before the irradiation, at the minimum resistivity point and after the irradiation. Figure 2 shows these three curves. As it can be seen, except for the magnitude of the resistance, there is no significant change in their behavior.

The temperature dependence of the resistivity is well described by $\rho \propto \exp(T_0/T)^{1/2}$ indicating Coulomb-repulsion limited hopping conduction like in many granular carbon materials [9].

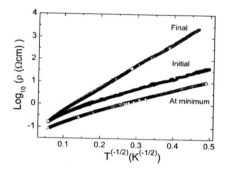

FIGURE 2. The resistivity of pristine and irradiated nanotubes as a function of temperature. The temperature dependence indicates hopping like conduction.

The results of the measurements can be explained as follows. The electron irradiation creates displacements of the carbon atoms in nanotubes, leading to very reactive broken chemical bonds. This has a two-fold effect on the resistivity: one process is the cross-linking of nanotubes in a bundle, improving its electric conductivity by decreasing the individual tube-tube distance. The other effect however is the damage of the graphite layer by these point defects, increasing the resistivity with increasing incident electron as well as defect concentration. This increase is linear at the beginning, but it gets saturated when the system is completely transformed into amorphous carbon. Consequently, for low damage level, the creation of covalent bonds between two nanotubes dominates and leads to an improvement in the conductivity while the point defects have little effect on the on-tube conduction.

SUMMARY

We carried out *in-situ* resistivity measurements on macroscopic oriented ropes of single wall carbon nanotubes in a transmission electron microscope.

Our results suggest that the cross-links created by electron irradiation also change the electronic conductivity of the fiber significantly yielding a dramatic decrease of the resistance. As electron irradiation continues, the irradiation is destructive to the CNT wall structure and leads to an increase of the resistance.

ACKNOWLEDGMENTS

The authors thank the European Commission (RTN Program, NANOCOMP network, RTN1-1999-00013), NCCR Nanoscale Science of the Swiss National Science Foundation, and the Swiss National Science Foundation (grant number 61534) for financial support. We are also grateful to Centre Interdisciplinaire de Microscopie Electronique (CIME) at EPFL for access to electron microscopes as well as for technical support.

REFERENCES

1. S. Iijima, Mater. Sci. Engin. B 1993, **19**, pp 172.
2. C. Journet, W. K. Maser, P. Bernier, A. Loiseau, M. L. Delachapelle, S. Lefrant, P. Deniard, R. Lee, J. E. Fischer, Nature 1997, **388**, pp 756.
3. K. Hernadi, A. Fonseca, P. Piedigrosso, M. Delvaux, J. B. Nagy, D. Bernaerts, J. Riga, Cat. Lett. 1997, **48**, pp 229.
4. B. Smith and D. E. Luzzi, J. Appl. Phys. 2001, **90**, pp3509.
5. A. Kis, J.-P. Salvetat, Thien-Nga Lee, E. Ljubovic, A. Kulik, W. Benoit, J. Brugger, L. Forro, Nature Materials 2004, **3**, pp 153.
6. Cs. Miko, M. Milas, J.W. Seo, E. Couteau, N. Barisic, R. Gaal, L. Forro, Appl. Phys. Lett. 2003, **83**, pp 4622.
7. H. H. Gommans, J. W. Alledredge, H. Tashiro, J. Park, J. Magnuson, G. A. Rinzler, J. Appl. Phys 2000, **88**, pp 2509.
8. B. Vigolo, A. Penicaud, C. Coulon, C. Sauder, R. Pailler, C. Journet, P. Bernier, P. Poulin, Science, 2000, **290**, pp 1331.
9. A. W. P. Fung, Z. H. Wang, M. S. Dresselhaus, G. Dresselhaus, R. W. Pekala, M. Endo, Phys. Rev. B 1994, **49**, pp 17325.

Luminescence from pillar suspended single-walled carbon nanotubes

J. Lefebvre[*], P. Finnie[*], and Y. Homma[§]

[*]Institute for Microstructural Sciences, National Research Council, Montreal Road, Ottawa, Ontario, K1A OR6, Canada.

[§]NTT Basic Research Laboratories, Nippon Telegraph and Telephone Corporation, 3-1 Morinosato-Wakamiya, Atsugi, Kanagawa 243-0198, Japan with Author's Affiliation

Abstract. This paper reviews our recent effort in the study of photoluminescence (PL) from single-walled carbon nanotubes (SWNTs) suspended between pillars. Luminescence is obtained from both ensembles and individual SWNTs in air, at room temperature without any post-growth processing. PL spectra reveal some of the one-dimensional characteristics expected for this material system, such as narrow asymmetric linewidths, and polarized emission and absorption.

INTRODUCTION

Single-walled carbon nanotubes (SWNTs) emit bandgap photoluminescence (PL) when isolated in a surfactant solution [1-3], or when isolated on substrates with nanotubes suspended between pillars [4-6]. In the latter case, PL emission in air at room temperature is readily obtained from as-grown samples. Luminescence can also be obtained from isolated SWNTs directly in contact with insulating substrates [7-8]. For all methods of sample preparation, interaction with other nanotubes and with the environment is minimized, a key factor for radiative recombination of electron-hole pairs to occur at the band edge of semiconducting nanotubes. Since the discovery of PL from SWNTs, progress has been rapid. This paper reviews our most recent work on both ensembles of many nanotubes and on single, individual nanotubes. For ensembles, photoluminescence excitation (PLE) mapping shows remarkable similarity between micelle-encapsulated nanotubes and pillar suspended nanotubes. The PL and PLE spectra are greatly simplified by scaling down to the single nanotube level. Single nanotube spectra reveal clear signatures of the one-dimensional nature of SWNTs, including strongly polarized absorption and emission, asymmetric lineshape, and sub-$k_B T$ linewidths.

CP723, *Electronic Properties of Synthetic Nanostructures*, edited by H. Kuzmany et al.
© 2004 American Institute of Physics 0-7354-0204-3/04/$22.00

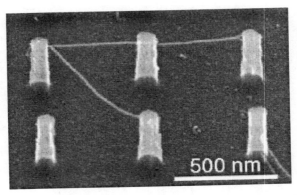

FIGURE 1. Scanning electron microscopy image showing individual single-walled carbon nanotubes suspended between pillars (side view with sample tilted approximately 45 degree). SWNTs are grown by chemical vapor deposition directly on Si/SiO_2. Photoluminescence is obtained in air, at room temperature without any special handling or post-growth processing of these samples.

RESULTS AND DISCUSSION

Recently, we have shown that SWNTs suspended on pillars are luminescent [4]. An example of such a sample is presented in Fig. 1, where chemical vapor deposition with methane as carbon source and Fe or Co as catalyst are used to grow SWNTs directly on pillars [9]. Photoluminescence (PL) from such samples is readily observed without any special handling or post-growth processing. Measurements are performed in air, at room temperature using different lasers with energy resonant with E_{22}, the second lowest optically allowed transition in semiconducting nanotubes. PL can also be observed off resonance, but with greatly reduced intensity. The dispersed PL is focussed on a liquid nitrogen cooled InGaAs photodetector array with sensitivity from visible to 1.65 μm (0.75 eV).

Figure 2 shows a PL excitation (PLE) map obtained from an ensemble of SWNTs. In this type of measurement, a series of PL spectra are accumulated for different excitation energies, in this case using a tunable Ti:sapphire laser. As demonstrated in Ref. [2], a PLE map probes the various singularities in the JDOS (joint density of states) in a non-interacting picture, or what are likely excitonic levels in an interacting electron-hole pair picture. Based on the work from the Rice University group [2,10], each PLE peak in Fig. 2 can be assigned to a SWNT species with a given set of (n,m) indices. The peak position along the emission axis is a measure of E_{11}, while its position along the excitation axis is a measure of E_{22}, which relates to the first and second van Hove singularity, respectively. In Fig. 2, two SWNT species dominate the spectrum, namely (9,8) and (9,7). Five other SWNT species (three are labeled) can be identified in Fig. 2, and a systematic comparison with SWNTs in surfactant solution can be made. We find that both E_{11} and E_{22} peaks are redshifted in the surfactant solution. The average difference is 28 meV for E_{11} and 16 meV for E_{22}, with a slight

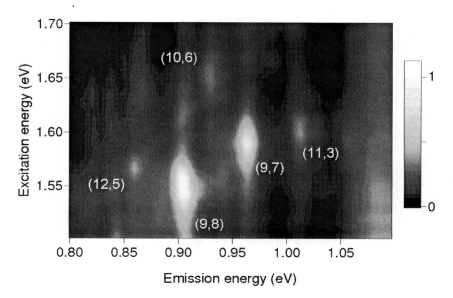

FIGURE 2. Photoluminescence excitation map of an ensemble of suspended SWNTs measured in air, at 300 K. The spectrum is dominated by two SWNT species, the (9,8) with $E_{11} = 0.904$ eV and $E_{22} = 1.550$ eV, and the (9,7) with $E_{11} = 0.964$ eV and $E_{22} = 1.588$ eV. Other SWNT species identified in the spectrum are labeled with the assigned (n,m) values.

dependence on (n,m) [5]. This difference can be attributed to the effects of environment.

In addition to ensemble measurements, PL from individual nanotubes on isolated pillars can be obtained as easily [6]. Compared to PL from an ensemble of SWNTs, PL spectra from individual SWNTs are greatly simplified, with a single emission peak free from the background emission from other nanotube species. Examples of such spectra from an individual (9,8) SWNT are presented in Fig. 3 and 4. The PL from suspended SWNTs is generally quite bright. For typical excitation intensities $\approx 10^{16}$ photons/s (1 mW focussed to a 10 μm diameter spotsize, or 10^{14} photons/μm²•s), the PL peak intensity for the brightest nanotubes is $\approx 10^{6}$ photons/s (with $\approx 0.4\%$ collection efficiency). The linewidth is found to be as narrow as 9 meV, a value well under the thermal energy of 25 meV at 300 K. The lineshape is asymmetric with a longer tail on the high energy side of the peak. In Fig. 3 b), a Lorentzian functional form is plotted and clearly reveals this asymmetry. These two characteristics, along with the strong polarization of light emission and absorption, are clear signatures of the reduced dimensionality of SWNTs.

Figure 2b) presents the power dependence. We find that the PL intensity increases linearly with excitation intensity up to power densities as large as 1 MW/cm². As shown in Fig. 2a), the lineshape remains unchanged over at least three orders of magnitude in excitation power.

In many material systems, ensemble measurements can lead to significant broadening of spectral features. For a given species of SWNTs suspended between

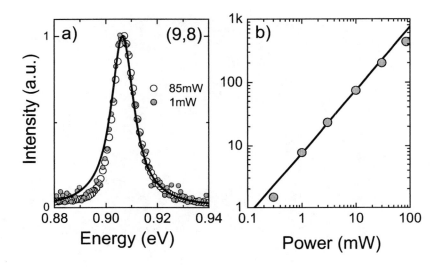

FIGURE 3. Photoluminescence spectrum from an individual (9,8) SWNT. a) PL spectra for two excitation intensities (E_{LASER} = 1.58 eV). The continuous line is the Lorentzian functional form (FWHM = 11 meV). b) Integrated PL intensity as a function of excitation power (2 μm diameter spotsize). The continuous line is the functional form I = 7.5 P, where I is the intensity and P is the power.

pillars, the PL from an ensemble is very similar to that of an individual SWNT. Figure 4 presents this result for the (9,8) SWNT species. In both PL emission (Fig. 4 a)) and excitation (Fig. 4 b)), the spectra overlap very closely showing the absence of inhomogeneous broadening within a given SWNT species. In contrast, significant inhomogeneous broadening has been reported for SWNTs in micelles on glass [8]. Fig. 4 b) also shows that within the scatter of the data, the absorption peak at E_{22} can be well fit with a Lorentzian functional form. The FWHM (full width at half maximum) for E_{22} is typically 35 meV, approximately three to four times larger than for E_{11}.

Photoluminescence and photoluminescence excitation spectra have been obtained from isolated SWNTs suspended between pillars on substrate. The spectral characteristics reveal the one-dimensionality of the SWNT materials system, the limited inhomogeneous broadening within a given (n,m) species, and the effect of different environmental conditions.

ACKNOWLEDGMENTS

We thank R. L. Williams and J. M. Fraser for numerous discussions and everyday assistance in the laboratory. Partial financial support was provided by the NEDO International Joint Research Grant Program.

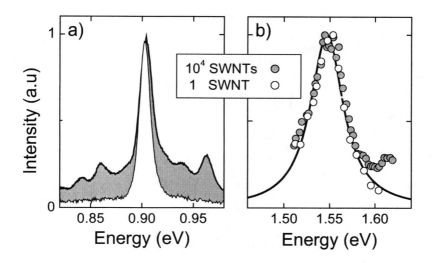

FIGURE 4. Comparison between the photoluminescence spectrum of an individual nanotube with a nanotube ensemble. a) PL emission around E_{11} for resonant excitation of the (9,8) SWNT species at E_{22}=1.550 eV. For ensemble measurements (filled gray), other SWNT species contribute to the emission. b) Optical absorption around E_{22} of the (9,8) SWNT species for PL detection at E_{11}=0.904 eV. The Lorentzian functional form (continuous line) provides a good fit of the absorption profile (FWHM = 44 meV).

Correspondence should be addressed to J.L. (email: jacques.lefebvre@nrc.ca)

REFERENCES

1. O'Connell, M. J., Bachilo, S. M., Huffman, C. B., Moore, V., Strano, M. S., Haroz, E., Rialon, K., Boul, P. J., Noon, W. H., Kittrell, C., Ma, J., Hauge, R. H., Smalley, R. E., and Weisman, R. B., *Science* **297,** 593-596 (2002).
2. Bachilo, S. M., Strano, M. S., Kittrell, C., Hauge, R. H., Smalley, R. E., and Weisman, R. B., *Science* **298,** 2361-2366 (2002).
3. Lebedkin, S., Hennrich, F., Skipa, T., and Kappes, M. M., *J. Phys. Chem.* B **107,** 1949-1956 (2003).
4. Lefebvre, J., Homma, Y., and Finnie, P., *Phys. Rev. Letters* **90,** 217401 (2003).
5. Lefebvre, J., Fraser, J. M., Homma, Y., and Finnie, P., *Appl. Phys. A: Mat. Sci. Proc.* **78,** 1107-1110 (2004).
6. Lefebvre, J., Finnie, P., and Homma, Y., *Phys. Rev. B* **69,** 075403 (2004).
7. Misewich, J. A., Martel, R., Avouris, Ph., Tsang, J. C., Heinze, S., and Tersoff, J., *Science* **300,** 783-786 (2003).
8. Hartschuh. A., Pedrosa, H. N., Novotny, L., and Krauss, T. D., *Science* **301,** 1354-1356 (2003).
9. Homma, Y., Kobayashi, Y., Ogino, T., and Yamashita, T., *Appl. Phys. Letters* **81,** 2261-2263 (2002).
10. Weisman, R. B., and Bachilo, S. M., *Nanoletters* **3,** 1235-1238 (2003).

Strain-Induced Shifts of the Photoluminescence of Single-Walled Carbon Nanotubes in Frozen Aqueous Dispersions

Katharina Arnold [a,b], Sergei Lebedkin [a], Frank Hennrich [a] and Manfred M. Kappes *[,a,b]

[a] *Forschungszentrum Karlsruhe, Institut für Nanotechnologie, D-76021 Karlsruhe, Germany*
[b] *Institut für Physikalische Chemie, Universität Karlsruhe, D-76128 Karlsruhe, Germany*

Abstract. Significant shifts of photoluminescence (PL) emission-excitation resonances have been observed by freezing and cooling of water-surfactant dispersions of single-walled carbon nanotubes (SWNTs) down to 16 K. The PL resonances correspond to E_{11}^S, E_{22}^S electronic energies of specific (n,m) nanotubes. The shifts occur mainly in the interval of ~150-200 K, are reversible and similar for SWNT dispersions with different surfactants and viscosity-increasing additives. The sign of the shifts is determined by the $(n-m)$ *mod* 3 rule, whereas the shift magnitude depends on a chiral angle, being the smallest for the large angles. These results are in agreement with tight-binding model calculations of Yang et al. for SWNTs under uniaxial compression (apparently caused by thermal contraction of the ice matrix in our case). This indicates a high sensitivity of electronic properties of SWNTs to mechanical strain and suggests an extended, 'rod'-like configuration of nanotubes in frozen dispersions.

INTRODUCTION

The near-infrared photoluminescence spectroscopy of dispersed (individual) semiconducting SWNTs can provide rich information about electronic properties of carbon nanotubes and their dependence on the nanotubes structure, treatment, surroundings, etc. [1-6]. The PL emission-excitation resonances correspond to electronic transition energies E_{11}^S (band gap emission) and E_{ii}^S, i = 1, 2, 3…(excitation) between van Hove singularities in the valence and conductance zones of semiconducting nanotubes. E_{11}^S / E_{22}^S emission-excitation resonances are particularly characteristic and can be readily assigned to nanotubes with specific diameters and chiralities described by the chiral vector indices (n,m) [2].

The photoluminescence of SWNTs has so far been studied at ambient temperature [1-6]. In this work we investigated the PL properties of frozen water-surfactant dispersions of SWNTs and found significant shifts of E_{11}^S and E_{22}^S energies by decreasing the temperature down to 16 K. Two types of shifts, with nearly opposite directions, are observed for different (n,m) nanotubes classified by the $[(n-m)$ *mod* 3] value (1 or 2 for semiconducting SWNTs). The shift magnitude depends on chiral angle, being the largest for small angle values. These results are in agreement with tight-binding model calculations of Yang et al. for the band gap energies of SWNTs under axial compression [7]. A high sensitivity of electronic properties of SWNTs to mechanical strain is not only of fundamental interest, but might also be of importance for possible practical applications of carbon nanotubes.

CP723, *Electronic Properties of Synthetic Nanostructures*, edited by H. Kuzmany et al.
© 2004 American Institute of Physics 0-7354-0204-3/04/$22.00

EXPERIMENTAL

The raw SWNTs produced by high-pressure catalytic decomposition of carbon monoxide (HiPco method) [1] were dispersed with an ultrasonic dispergator in D_2O containing either 1 wt.% of sodium dodecylbenzene sulfonate (SDBS) or 1 wt.% of sodium dodecylsulfate (SDS) surfactants. The dispersion was centrifuged for 2 hours at 150.000 g to remove the majority of denser nanotube bundles and metal catalytic particles. A supernatant solution was mixed with water-soluble polymers, typically with polyvinylpyrrolidone, in order to increase viscosity and to reduce aggregation of dispersed nanotubes upon freezing.

Near-infrared PL spectra were measured down to 16 K using a closed-cycle optical cryostat (Leybold) and a Bruker FTIR 66v/S spectrometer equipped with a liquid nitrogen cooled germanium detector and a monochromatic light source, as described elsewhere [6]. PL was excited in the range of 500-900 nm covering the E_{22}^S energies of nanotubes in the HiPco material. Dispersions of SWNTs were placed on a cold finger of the cryostat in a 'cuvette' formed by two quartz windows sealed with an O-ring. Presented PL maps were corrected for the wavelength-dependent excitation intensity (in relative photon flux units), but not for instrumental response of the FTIR spectrometer and absorption of SWNTs.

RESULTS AND DISCUSSION

Cooling untreated dispersions of SWNTs down to a freezing temperature results in a decrease of the PL intensity (in particular for nanotubes with higher emission energies) and a broadening of the PL. We attribute these effects to aggregation of individually dispersed nanotubes, initially into small bundles. A weaker, broader and on average red-shifted PL is expected from bundles with a few semiconducting nanotubes as compared to ensemble of individually dispersed ones due to intertube interactions and energy transfer to nanotubes having smaller band gap energies (longer emission wavelengths).

We have found that the aggregation can be reduced by fast freezing of dispersions (technically difficult to realize in an optical cryostat) as well as by viscosity-increasing additives. We have tested several water-soluble polymers and gelating compounds and obtained the best results by addition of ~1-5 wt.% of polyvinylpyrrolidone (PVP). Only a moderate degradation of the PL was observed for dispersions containing PVP even after several freeze-thaw cycles.

Fig. 1 shows PL contour maps (PL intensity vs. excitation and emission wavelengths) for a dispersion of HiPco nanotubes at room temperature and after cooling down to 35 K. Significant shifts of the PL emission-excitation resonances and changes of their shapes are clearly seen in Fig. 1. The resonances shifted simultaneously and mainly in the temperature interval of ~150-200 K. The PL map changed only slightly below ~100 K. All these effects were reversible by increasing the temperature. Similar shifts as in Fig. 1 were also observed for dispersions of HiPco nanotubes with other surfactant (SDS) and viscosity-increasing additives (e.g., sodium carboxymethylcellulose).

The shifts of the PL emission-excitation resonances (corresponding to E_{11}^S and E_{22}^S energies) between room temperature and 16 K together with (n,m) indices of the emitting nanotubes are shown in Fig. 2. It is easy to see that there are two types of tubes with

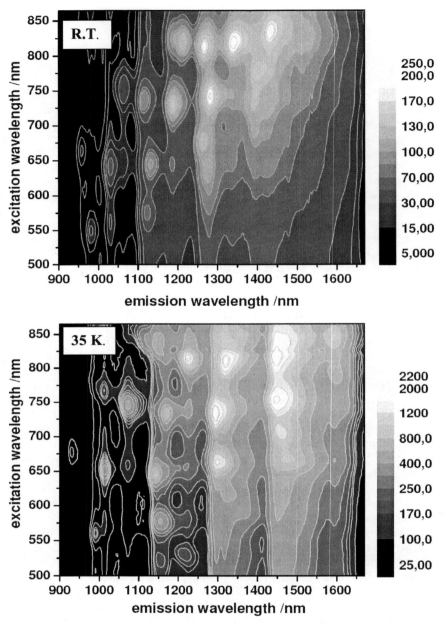

FIGURE 1. Photoluminescence contour maps of a SWNT dispersion in D$_2$O/ 1 wt.% SDBS / 3 wt.% PVP at room temperature and after cooling down to 35 K. The PL relative intensity scale (right bar, a.u.) is the same for the both maps.

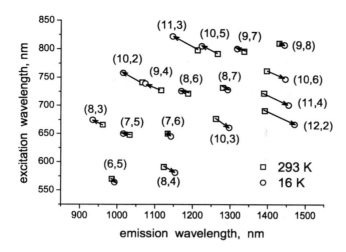

FIGURE 2. Shifts of photoluminescence excitation-emission resonances for different (n,m) nanotubes dispersed in D_2O/ 1 wt.% SDBS / 3 wt.% PVP. The (n,m) structural indices are assigned from the photoluminescence data at room temperature according to Ref. 2.

nearly opposite shifts classified by the value of $(n-m)\,mod\,3$. The nanotubes with $(n-m)\,mod\,3 = 1$ demonstrate a decrease of E_{11}^S and an increase of E_{22}^S by cooling, whereas the nanotubes with $(n-m)\,mod\,3 = 2$ demonstrate the opposite trends. The smallest shifts of E_{11}^S and E_{22}^S are observed for nanotubes with a large chiral angle, α, i.e. with $n - m = 1$ (Fig. 2). These are, for instance, (7, 6) and (6, 5) nanotubes having $\alpha = 27.4°$ and $27°$, respectively. In contrast, (12, 2) nanotubes ($\alpha = 7.6°$) demonstrate remarkably large shifts of E_{11}^S (-5.6 %) and E_{22}^S (+3.5 %).

The $(n-m)\,mod\,3$ rule and dependence on chiral angle are fully consistent with tight-binding model calculations of Yang et al. for the band gap energies (E_{11}^S) of SWNTs under uniaxial compression [7]. This agreement strongly suggests mechanical strain effects as the reason for the PL changes in frozen SWNT dispersions. The uniaxial compression may be a result of a preferentially extended, rod-like geometry and efficient 'trapping' of SWNTs in the ice matrix.

The temperature dependence of PL shifts may be caused by complicated thermomechanical properties of frozen SWNT dispersions containing additionally large amounts of a surfactant and a viscosity-increasing additive. It is known that ice has a relatively large thermal expansion coefficient between ~150 K and 273 K, which strongly decreases below ~150 K. This seems to be consistent with a small PL shift below ~100 K. A similar behavior (small PL shift) at temperatures down to ~200 K might be due to the presence of free water molecules (interface water) around SWNTs down to this temperature.

There are also other issues of the strain-induced changes of the photoluiminescence of SWNTs, which require further investigation. According to the caculations of Yang et al.,

119

the observed -5.6% shift of E_{11}^S (see above) would develop under the strain of about -0.5 %. This estimate is in agreement with modified tight-binding model caclulations of Ding et al. [8] as well as with direct electronic transport measurements on individual SWNTs stressed with an AFM tip [9]. For an axial Young's modulus of 1 TPa, this would correspond to an axial compression stress as high as ~5 GPa. It remains to be understood, how such huge values of the strain and stress can be developed and maintained in frozen SWNT dispersions.

ACKNOWLEDGEMENTS

The support of this work by the Deutsche Forschungsgemeinschaft and by the BMBF is gratefully acknowledged. The authors thank Prof. R. E. Smalley for a sample of HiPco carbon nanotubes and Dr. Ralph Krupke for helpful discussions.

REFERENCES

1. O'Connell, M. J., Bachilo, S. M., Huffman, C. B., et al., *Science* **297**, 593 (2002).
2. Bachilo, S. M., Strano, M. S., Kittrell, C., et al., *Science* **298**, 2361 (2002).
3. Lebedkin, S., Hennrich, F, Skipa, T., Kappes, M. M., *J. Phys. Chem. B***107**, 1949 (2003).
4. Hartschuh, A., Pedrosa, H. N., Novotny, L., Krauss, T. D., *Science* **301**, 1354 (2003).
5. Lefebvre, J., Homma, Y., Finnie, P. *Phys. Rev. Lett.* **90**, 217401 (2003).
6. Arnold, K., Lebedkin, S., Hennrich, F, Krupke, R., Renker, B., Kappes, M. M., *New J. Phys.* **5**, 140 (2003).
7. Yang, L., Anantram, M. P., Han, J., Lu, J. P., *Phys. Rev. B* **60**, 13874 (1999).
8. Ding, J.W., Yan, X.H., Liu, C.P., Tang, N.S., *Chin. Phys. Lett.* **21**, 704 (2004).
9. Minot, E. D., Yaish, Y., Sazonova, V., et al. *Phys. Rev. Lett.* **90**, 156401 (2003).

Mechanical Dynamics of Chiral Carbon Nanotubes: Magnetochyrodynamic Effects

V. Krstić[1], G. Wagnière[2], G.L.J.A. Rikken[3]

[1]*Grenoble High Magnetic Field laboratory, Max-Planck-Institut für Festkörperforschung/CNRS, B.P. 166, F-38042 Grenoble, France*
[2]*Institute of Physical Chemistry, University of Zürich, Winterthurerstr. 190, Zürich, Switzerland*
[3]*Laboratoire Nationale des Champs Magnétiques Pulsés, CNRS/INSA/UPS, B.P. 14245, F-31432 Toulouse France*

Abstract. Carbon nanotubes exist in three configurations named zigzag, armchair and chiral. Out of these only the so-called chiral nanotubes are real chiral molecular structures whereas the other two are achiral. Due to their chiral character, the mechanical dynamics of chiral nanotubes in a magnetic field are differing from their achiral counterparts. It will be shown by fundamental symmetry considerations that the movement of the chiral nanotubes depends bi-linearly on the magnitude and the direction of the magnetic field and its linear momentum. For an estimation of the magnitude of these effects, the quantum mechanical model of a free electron on a helix is considered. The model can be solved analytically and will be quantitatively evaluated for chiral carbon nanotubes.

INTRODUCTION

A system is chiral (from the Greek χ ε ῍ρ hand) when it exists in two non-superimposable forms that can only be interconverted by parity reversal, and not by any combination of time-reversal and spatial rotations.[1] Chirality is a fundamental symmetry property, and at the same time an essential aspect of biochemistry as many biologically relevant molecules are chiral. Carbon nanotubes can exist in achiral (zigzag, armchair) and chiral forms. Each of the chiral forms exist in two different versions, called enantiomers, each of which is the mirror image of the other and which are named right (D) or left (L) handed.

Recently, a new magneto-optical optical effect was observed in chiral systems - the magneto-chiral anisotropy (MChA).[2-5] This effect corresponds to a polarization independent term in the dielectric constant of a chiral\optical medium which is proportional to $\kappa^{D/L} k_{ph} \cdot B$ where k_{ph} is the wavevector of the light, B is the external magnetic field and $\kappa^D = -\kappa^L$ a material parameter accounting for the chiral nature of the system.[6,7] This represents a magnetically induced spatial anisotropy, the sign of which depends on the handedness of the medium. MChA has also been observed recently in photochemistry.[8]

However, MChA is not restricted to optical phenomena but exists also in transport phenomena in chiral objects.[9] Indeed, for carbon nanotubes it has been shown that a magnetochiral contribution of the form $\kappa^{D/L} k \cdot B$ exists in the magneto-resistance, where k is the average wavevector of the electrons carrying the current and thus directly

CP723, *Electronic Properties of Synthetic Nanostructures*, edited by H. Kuzmany et al.
© 2004 American Institute of Physics 0-7354-0204-3/04/$22.00

proportional to the currrent.[10] In particular the observation of the MChA in the magnetotransport in carbon nanotubes shows that the current must have a chiral component, the current follows a helical path. In this Letter we will examine the MChA of mechanical effects for the specific example of chiral carbon nanotubes.

There are several well-known mechanical effects due to a magnetic field. Anisotropic objects, or objects having an anisotropic magnetic susceptibility can be aligned by a magnetic field, proportional to B^2. Also proportional to B^2 is the magnetostrictive effect.[11] All objects experience a force in a magnetic field gradient, proportional to $M \cdot \nabla B$, M being the magnetization, that can lead to magnetic levitation.[12] The Einstein-deHaas effect is the only known linear magneto-mechanical effect, and corresponds to an angular momentum that any object acquires if a magnetic field is applied.[13] The semi-classical explanation of this effect is that applying a magnetic field causes Larmor precession of the electrons of the object and that angular momentum conservation requires the body to rotate in the opposite sense in order to compensate for the angular momentum of the electrons. The inverse of the Einstein-deHaas effect is called the Barnett effect [14], and describes the magnetization induced in an object which is set into rotation. Classically speaking, the object's electrons lag behind when the object is set in rotation, and this corresponds to a circular current, and therefore to an axial magnetization. For a detailed discussion of these two so-called gyromagnetic effects, see Ref. [15].

When considering magneto-dynamical effects in the case of a chiral object, imposing time-reversal and parity-reversal symmetry, one readily finds that any chiral object possessing a magnetic moment m is allowed to have a linear momentum contribution of the form

$$p = \kappa^{D/L} m \qquad (1)$$

where $\kappa^L = -\kappa^D$ and $\kappa^{D/L}$ can be called the magnetochirodynamical coupling parameter. Similarly, the inverse effect corresponds to the circumstance that any chiral object in motion is allowed to posses a magnetization, which in turn generates a magnetic field given by

$$B = \widetilde{\kappa}^{D/L} v \qquad (2)$$

where $\widetilde{\kappa}^L = -\widetilde{\kappa}^D$, that is, any chiral object with velocity v generates a magnetic field B. The only additional symmetry restriction on the parameters $\kappa^{D/L}$ and $\widetilde{\kappa}^{D/L}$ is that they must be odd under charge conjugation, i.e., be proportional to an odd power of the object's electric charge.

THEORETICAL MODEL

The microscopic origin of these new effects, described by the last two equations, lies in the coupling of angular and linear momentum in chiral systems (like e.g. for the motion of a particle on a helix) on one hand, and the inherent coupling between magnetic and mechanical angular moments on the other hand (like for the

122

gyromagnetic effects). The combination of these two effects results in a coupling between linear moments and magnetic moments in chiral systems. This, as will be shown in the following, leads to a dynamical magneto-chiral anisotropy (DMChA), described by Eq. (1) and its inverse effect (IDMChA), described by Eq. (2). The symmetry arguments above are very general, and apply to any chiral system.

In order to get an estimate of the magnitude of the DMChA, i.e., the magneto-chirodynamical parameter $\kappa^{D/L}$, for a carbon nanotube, the tube is approximated by the free electron on a helix model [16], a model that has explained the observation of MChA in the magneto-transport of chiral carbon nanotubes.[10,17] This observation implies a helical component in the motion of the electrons within the tube. The shape of the helix with radius a and period $d \equiv 2\pi|b|$ resembles the possible motion of an electron in a chiral nanotube, ($b > 0$ corresponds to a right-handed and $b < 0$ to a left-handed helix/tube). Now, the quantum-mechanical problem of the free electron on a helix of N turns in an axial magnetic field, using cyclic boundary conditions [16,17] and neglecting spin, has to be solved. The Hamiltonian of the system reads

$$\hat{H} = -\frac{\hbar^2}{2m_e(a^2+b^2)}\frac{\partial^2}{\partial\varphi^2} + i\hbar\frac{e}{2m_e}\frac{a^2}{(a^2+b^2)}B_{ext}\frac{\partial}{\partial\varphi} + \frac{e^2a^2}{8m_e}B_{ext}^2 \qquad (3)$$

where $\boldsymbol{B}_{ext} = B_{ext} \cdot \hat{z}$ with $\hat{z}$ being the unit vector in direction of the longitudinal axis of the carbon nanotube/helix. Using the normalized eigenstates $\Psi_n(\varphi) = (2\pi|b|N)^{-1/2}\exp(in\varphi/N)$ of the Hamiltonian allows for determining the axial component of the linear momentum p_z [16] and thus its expectation value for the state n (n an integer),

$$\langle p_{z,n} \rangle = \frac{\hbar n}{N(a^2+b^2)}\frac{b}{|b|}. \qquad (4)$$

Similarly, the expectation value of the axial component of the angular momentum l_z of the electron (neglecting its spin) for the state n [16] is found to be

$$\langle l_{z,n} \rangle = \frac{\hbar n}{N(a^2+b^2)}\frac{a^2}{|b|}. \qquad (5)$$

From the expectation value of the axial angular momentum the corresponding axial magnetic moment can now be derived:

$$\langle m_{z,n} \rangle = -\frac{e}{2m_e}\langle l_{z,n} \rangle = -\frac{e\hbar n}{2m_e N(a^2+b^2)}\frac{a^2}{|b|}. \qquad (6)$$

Combining of Eq. (4) and Eq. (6) finally yields for the magnetochirodynamical coupling parameter

$$\kappa^{D/L} \equiv \frac{\langle p_{z,n} \rangle}{\langle m_{z,n} \rangle} = -\frac{2m_e b}{ea^2}. \tag{7}$$

It is noteworthy to stress at this point that the coupling parameter turns is independent of the eigenstate n. Thus, the coupling between angular and linear momentum solely depends on the helix geometry, or more general on the chiral character of the object under investigation. This implies that even for many electrons on the helix, Eq. (7) still holds, as long as electron-electron interaction can be neglected.

RESULTS

The results of the above calculations allow now to estimate the magnitude of DMChA for a carbon nanotube. Assuming a carbon nanotube with a diameter of $2a = 1$ nm and $d = 0.1$ nm, from Eq. (7) a magnetochirodynamical coupling parameter $\kappa^{D/L} \approx 4.5 \cdot 10^{-3}$ kg/Asm is found.

To evaluate Eq. (1) for a chiral nanotube, the nanotube mass density $\rho_{tube} = m_{tube}/V$ has to be introduced, which can be estimated to be about 2 g/cm^3 (density of graphite). m_{tube} is the mass of the nanotube and V the corresponding volume. Then, Eq. (1) yields a nanotube velocity

$$v_{tube} = \rho_{tube}^{-1} \left| \frac{\kappa^{D/L}}{\mu_0} \frac{\chi}{1+\chi} \right| B_{ext}. \tag{8}$$

where μ_0 is the permeability in vacuum and χ the magnetic susceptibility of the nanotube. Assuming the chiral nanotube to be metallic, the nanotube is paramagnetic and χ has a value of about $3 \cdot 10^{-4}$ for a tube diameter of 1 nm. [18,19] In an external magnetic field of 1 T aligned parallel to the axis of the nanotube, Eq. (8) predicts a nanotube velocity of approximately $0.5 \cdot 10^{-3}$ m/s.

It has to be noted at this point that the theoretical considerations presented apply only for perfect, infinitely long carbon nanotubes and the resulting estimates give an upper limit for the effect. Reduction of the effect arises from the finite length of a nanotube which will lead to charge accumulation at the tube ends and the corresponding electric field should be incorporated self-consistently into the Hamiltonian. Experimentally, this problem can be circumvented by short circuiting the tube at its ends. Similarly, imperfections in the tube will lead to inelastic scattering, which will reduce the effect. This reduction is determined by the ratio between the tube transit time T and the inelastic scattering time τ. Ballistic transport has been observed in high quality carbon nanotubes at room temperature [20,21], implying $\tau > T$. Under such conditions, one can assume that the estimates above are close to the real dynamical response of the nanotube in an external magnetic field.

In summary, on the basis of symmetry arguments, the MChA could be generalized to magneto-dynamical properties of a chiral object and this was illustrated by microscopic considerations for a chiral nanotube. By using the quantum-mechanical model of a free electron on a helix as an approximation for a real chiral nanotube, the magnitude of the DMChA for such a system was estimated.

REFERENCES

1.   Barron, L. D., *J. Am. Chem. Soc.* **108**, 5539 (1986).
2.   Rikken, G. L.J.A., and Raupach, E., *Nature* **390**, 493 (1997).
3.   Kleindienst, P., and Wagnière, G., *Chem. Phys. Lett.* **288**, 89 (1998).
4.   Rikken, G. L.J.A., and Raupach, E., *Phys. Rev. E* **58**, 5081 (1998).
5.   Vallet, M., et al., *Phys. Rev. Lett.* **87**, 183003 (2001).
6.   Portigal, D.L., and Burstein, E., *J. Phys. Chem. Solids* **32**, 603 (1971).
7.   Baranova, N. B., Bogdanov, Yu.V., and Zeldovich, B.Ya., *Sov. Phys. Usp.* **20**, 870 (1977).
8.   Rikken, G. L.J.A., and Raupach, E., *Nature* **405**, 932 (2000).
9.   Rikken, G. L.J.A., Fölling, J., and Wyder, P., *Phys. Rev. Lett.* **87**, 236602 (2001).
10.   Krstić, V., Roth, S., Burghard, M., Kern, K., and Rikken, G.L.J.A., *J. Chem. Phys.* **117**, 11315 (2002).
11.   de Lacheisserie du Trémolet, E., *Magnetostriction*, CRC Press, Boca Raton, 1993.
12.   Jayawant, B.V., *Electromagnetic levitation and suspension techniques*, Edward Arnold, London, 1981.
13.   Einstein, A., and deHaas, W. J., *Verh. Deut. Phys. Ges.* **17**, 152 (1915).
14.   Barnett, S. J., *Phys. Rev.* **6**, 171 (1915).
15.   Vonsovskii, S.V., *Magnetism*, Wiley, New York (1976).
16.   Tinoco, I.., and Woody, R.W., *J. Phys. Chem.* **40**, 160 (1964).
17.   Krstić, V., and Rikken, G.L.J.A., *Chem. Phys. Lett.* **364**, 51 (2002).
18.   Lu, J.P., *Phys. Rev. Lett.* **74**, 1123 (1995).
19.   London, F., *J. Phys. Rad.* **8**, 397 (1937). Latil, S., PhD Thesis, University of Montpellier II (2001).
20.   Hertel, T., and Moos, G., *Phys. Rev. Lett.* **84**, 5002 (2000).
21.   Bachtold, A., et al., *Phys. Rev. Lett.* **84**, 6082 (2000).

CHARACTERIZATION OF
CARBON NANOTUBES

Scanning Conductance Microscopy of Carbon Nanotubes and Polyethylene Oxide Nanofibers

Cristian Staii[1], Nicholas J. Pinto[2], and Alan T. Johnson, Jr.[1]

[1]*Department of Physics and Astronomy and Laboratory for Research on the Structure of Matter, University of Pennsylvania, Philadelphia, Pennsylvania 19104*
[2]*Department of Physics and Electronics, University of Puerto Rico, Humacao, Puerto Rico 00791*

Abstract. We have developed a quantitative model that explains the phase shifts observed in Scanning Conductance Microscopy, by considering the change in the total capacitance of the tip-sample-substrate system. We show excellent agreement with data on samples of (conducting) single wall carbon nanotubes and insulating polyethylene oxide (PEO) nanofibers. Data for large diameter, conducting doped polyaniline/PEO nanofibers are qualitatively explained. This quantitative approach is used to determine the dielectric constant of PEO nanofiber $\varepsilon_f = 2.88 \pm 0.12$, a general method that can be extended to other dielectric nanowires.

INTRODUCTION

Scanning Conductance Microscopy (SCM) is a scanning probe technique that can probe the conductivity of nanoscale structures without electrical contacts.[1] An SCM image records the oscillation phase of a driven voltage-biased AFM cantilever as a function of the tip position (Fig. 1a). Earlier results[2] show that in SCM, λ-DNA has zero phase shift,[1] SWNTs (both semiconducting and metallic) always show a negative phase shift (Fig. 1b) while insulating polyethylene oxide (PEO) nanofibers show a positive phase shift that increases with fiber diameter (Fig. 1c). We also reported[2] that conducting doped polyaniline/poly(ethylene oxide) (PAn/PEO) fibers with diameter larger than 30nm show a negative-positive-negative ("double dark line") contrast in SCM (Fig. 1d). Although a model for SCM has been proposed,[1] it does not account for all these observations.

In the following sections we present an improved model for SCM that explains these observations quantitatively and significantly enhances the analytic power of the technique. Moreover, we then use SCM to measure the dielectric constant of the insulating PEO nanofibers. Finally, we provide a semi-quantitative explanation for the negative- positive-negative phase shift observed for conducting PAn/PEO nanofibers.

EXPERIMENTAL

For the SCM measurements we use CVD grown SWNTs (likely a combination of single tubes and small bundles),[3] insulating poly(ethylene oxide) (PEO) nanofibers and conducting nanofibers made from a blend of polyaniline doped with

CP723, *Electronic Properties of Synthetic Nanostructures*, edited by H. Kuzmany et al.
© 2004 American Institute of Physics 0-7354-0204-3/04/$22.00

camphorsulfonic acid and insulating polyethylene oxide (PAn.HCSA/PEO).[2] The substrate for all experiments is a 200nm SiO_2 layer on top of a p-type degenerately doped Si wafer. SCM is a dual-pass technique. In the first line scan, the tip acquires a topography profile in tapping mode. In the second (interleave) line scan (Fig. 1a), the tip travels at a defined height above the surface. A DC voltage is applied to the tip, and the cantilever is mechanically driven at its free oscillation resonant frequency. The SCM image records the phase of the cantilever oscillation as a function of tip position. SCM images were taken using W_2C – coated tips with curvature radius R=30-60nm, quality factor Q=150 and spring constant k=0.65-1 N/m.

RESULTS

SCM images for SWNTs, an insulating PEO fiber, and a conducting PAn.HCSA/PEO fiber, together with the corresponding line scans are presented in Fig. 1b–d. SWNTs show a negative phase shift (Fig 1b) whereas the phase shift for insulating PEO fibers is always positive (Fig. 1c). Finally, conducting nanofibers with diameter larger than 30nm show a negative-positive-negative phase shift (Fig. 1d).

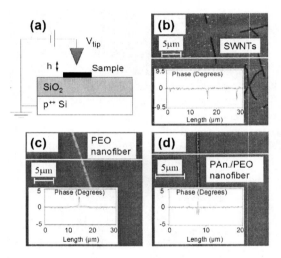

FIGURE 1. (a) Schematic of SCM. (b)-(d) SCM images of SWNTs (b), insulating PEO nanofiber (diameter: 10-100nm) (c), and conducting PAn.HCSA/PEO nanofiber (diameter 100nm) (d). Insets show line scans along the black lines. Differences in phase shift are explained in the text.

We explain these images quantitatively by modeling the cantilever as a driven harmonic oscillator with resonant frequency ω_0 and spring constant k. When the tip is scanned at height h above the bare SiO_2 substrate, the electrostatic force between metallic tip and surface leads to a decrease of the resonant frequency of the cantilever[4] and therefore to a negative value Φ_0 for the *background* phase lag:

$$\tan(\Phi_0) = -\frac{Q}{2k}C_1''(h)V_{tip}^2 \qquad (1)$$

where $Q = \gamma\omega_0$ is the quality factor of the cantilever and $C_1(h)$ is the capacitance of the tip-substrate system. The value of Φ_0 is independent of the tip horizontal position and is used as the reference zero in Fig. 1b-d. When the tip is at height h above the sample (SWNT, nanofiber) the total capacitance of the system changes to $C_2(h)$. Again, assuming that the electrostatic forces are small, the phase shift relative to that over the bare substrate is:

$$\tan(\Phi - \Phi_0) \approx \frac{Q}{2k}\left(C_1''(h) - C_2''(h)\right)V_{tip}^2 \qquad (2)$$

Equations 1 and 2 predict that the tangents of both the phase background value and the phase shift above the nanowire vary linearly with V_{tip}^2. This is seen in data (not shown) taken at different scan heights above the bare SiO_2 substrate, SWNTs, and PEO fibers[4]. Equation 2 predicts that the sign of the phase shift in SCM is determined by the change in the second derivative of the total capacitance of the system.

We calculate the capacitances called for in eq. 1, 2 using a simplified model for the geometry of the bare substrate, a thin SWNT, and a large diameter polymer nanofiber[4] (Fig. 2a,b). Since the PEO fiber diameter D is comparable to the tip radius R_{tip}, we model the PEO fiber as an insulating plate of thickness D and dielectric constant ε_f. The geometrical models presented in Fig. 2 give simple analytical solutions for the total capacitance of the system. For example, for h=30nm over the bare substrate the model predicts $C_1'' = 102\mu F/m^2$; the experimental value[4] (from the slope of the line of $\tan(-\Phi_0)$ vs. V_{tip}^2) is: $C_1'' = 95\mu F/m^2$, in excellent agreement with the predictions. Using the model of Figure 2b, the capacitance of the SWNT to each metal plate separately is found analytically,[4] and then combined in series to give C_2 (h). We find $C_2''(h) > C_1''(h)$ for all scan heights h=10-100nm. Thus equation 2 predicts a *negative* phase shift, in agreement with the measurements. The predictions are in excellent quantitative agreement with the data (Fig. 2c) for intermediate scan heights (h=30-50nm) where the model geometry is appropriate[4]. The observed phase shifts depend on the nanotube length and are almost independent of the tube diameter. The minimum detectable length (limited by the noise in the phase channel[1]) is about 0.4μm. Also, at least when the measurements are done in atmosphere there is no detectable difference between semconducting and metallic SWNT's.

For insulating PEO fibers the analytical solutions of the geometrical models[4] predict that $C_2''(h) < C_1''(h)$ for all scan heights, so *positive* phase shifts (as observed in Fig. 1c).

For conducting PAn.HCSA/PEO nanofiber we note that as the tip approaches the fiber from the side at height h above the substrate, two forces act on the cantilever: the capacitive force from the tip-substrate interaction F_1 and an additional *attractive* force F_{tf} due to the tip-fiber interaction. This additional force leads to a *decrease*[4] in the phase Φ and thus to a negative phase shift $\Delta\Phi$. When the tip is directly above the fiber, the phase shift is due to the capacitive coupling between tip and nanofiber.

A simple geometrical model[4] predicts a positive phase shift above the conducting fiber. This is the origin of the "negative-positive-negative" phase shift observed experimentally (Fig.1d).

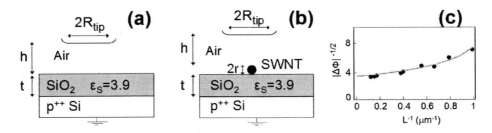

FIGURE 2. (a),(b) Models for the tip-sample geometry. The AFM tip (radius R_{tip}=30-50nm) scans at a constant height h above the sample. (b) The SWNT is approximated with a cylinder of radius r=1nm and length L=1-10μm at distances h, t from the tip and Si substrate respectively, both modeled as conducting plates. (c) $|\Delta\Phi|^{-1/2}$ versus L^{-1} for SWNTs. The solid curve represents the theoretical prediction based on the model of Fig. 2b and eq 2, and the dots are data for h=30nm and V_{tip} =+5V.

As an important practical application, we use the model to determine the dielectric constant ε_f of PEO [$(CH_2- CH_2 -O)_n$] nanofibers. From the slope of the plot of tan($\Delta\Phi$) vs V_{tip}^2, the predictions for the total capacitance and eq. 2, we find the fiber dielectric constant (at the cantilever oscillation frequency of 48 KHz) to be $\varepsilon_f = 2.88 \pm 0.12$.

This measured value is between the tabulated dielectric constants for polyethylene $(CH_2-CH_2)_n$, ε_{PE}= 2.28 to 2.32, and that for polyoxymethylene $(CH_2-O)_n$, ε_{POM}= 3.6 to 4 at 1 KHz and room temperature.[5] This is consistent with the higher polarizability of the C-O bond compared to that of the C-C bond. As expected, the measured value is also higher than ε_{PEO}=2.24 found for PEO at optical frequencies.[6] This method is general and can be used to determine the dielectric constant of other insulating nanowires.

ACKNOWLEDGMENTS

This work was supported by the Laboratory for Research on the Structure of Matter (NSF DMR00-79909) and the National Science Foundation NIRT (PHY-0103552).

REFERENCES

1. Bockrath, M., Markovic, N., Shepard, A., Tinkham, M., Gurevich, L., Kouwenhoven, L.P., Wu, M.W., Sohn, L.L., *Nano Letters* **2**, 187 (2002).
2. Zhou, Y., Freitag, M., Hone, J., Staii, C., Johnson, A.T., Pinto, N.J., MacDiarmid, A.G., *Appl. Phys. Lett.* **83**, 3800 (2003).
3. Radosavljević, M., Freitag, M., Thadani, K.V., Johnson, A.T., *Nano Letters*, **2**, 761 (2002).
4. Staii, C., Johnson, A.T., Pinto, N.J., *Nano Letters,* **4**, 859 (2004).
5. *Polymer Handbook*, 4th ed.; John Wiley and Sons: New York, 1999, pp. V/13, V/100.
6. Takahagi, T., Saiki, A., Sakaue, H., Shingubara S., *Jpn. J. Appl. Phys.* **42**, 157 (2003).

Characterization and Gas Adsorption on Multi-Walled Carbon Nanotubes Before and After Controlled Chemical Opening

M. Rachid Babaa*, Edward McRae*, Cedric Gommes¶, Sandrine Delpeux‡, Ghouti Medjahdi*, Silvia Blacher¶, François Beguin‡

* *Laboratoire de Chimie du Solide Minéral, UMR-CNRS 7555, Université Henri Poincaré-Nancy I, BP 239, 54506 Vandoeuvre-lès-Nancy, France*
¶ *Laboratoire de Génie Chimique, Université de Liège, B6a, B-4000 Liège, Belgium*
‡ *Centre de Recherche sur la Matière Divisée, CNRS-Université d'Orléans, 1B rue de la Férollerie, 45071 Orléans cedex 02, France*

Abstract. We report on the surface characterization of multi-walled carbon nanotubes (MWNTs) before and after chemical opening using an HNO_3 oxidation followed by a CO_2 treatment. Transmission electron microscopy (TEM) and image analysis were performed to characterize the structural evolution. Adsorption isotherms were measured for Kr at 77.3 K to study the effect of this treatment on the surface crystallinity and to give evidence of the accessibility of the inner channels of the tubes. The specific surface area was determined for each sample as was the presence or absence of a hysteresis loop in the adsorption–desorption cycle. The results lead to concluding that this treatment is very efficient for purification and selective opening but also results in significant decrease of the surface crystallinity.

INTRODUCTION

Carbon nanotubes have been the focus of recent interest due to of their very attractive properties and large applications potential. Their inner channels present great potential for gas storage and for building nano-composites by filling the tubes; they also allow studying confinement effects on chemical reactions and the consequences of such confinement on the physical properties of a variety of species. As-produced nanotubes are in general capped at both ends. Many chemical methods have been used to open them, typically with HNO_3 [1] or by O_2 [2] or CO_2 [3] oxidation. Electro-chemical means have also been called upon [4].

CP723, *Electronic Properties of Synthetic Nanostructures*, edited by H. Kuzmany et al.
© 2004 American Institute of Physics 0-7354-0204-3/04/$22.00

This work shows that opening of MWNTs can be performed by oxidative reaction that occurs preferentially at the ends of the tubes. In addition to the impact on structure and diameters, the effect of such a treatment on the surface homogeneity is also studied.

EXPERIMENTAL DETAILS

MWNTs were prepared by decomposition of acetylene at 600 °C on Co particles from solid state solution ($Co_xMg_{(1-x)}O$) as described elsewhere [5]. The opening technique [6] comprises two steps: first, nanotubes are treated by HNO_3 69 % weight at 130°C for 35 min under agitation. This step leads to open tubes and to elimination of catalytic particles. After washing in distilled water, the nanotubes are submitted to a complementary step using CO_2 oxidation at 525°C for about one hour. The aim of this second step is to eliminate carbon nanoparticules resulting from the first step of the treatment.

TEM studies were performed on a CM 20, 200 KV Philips apparatus with an unsaturated LaB6 cathode. Image analysis was done on TEM photos following the methodology described in reference [7]. Adsorption isotherms were performed from the very first stages of adsorption (P/Po $\approx 10^{-6}$). The experimental set-up is described elsewhere [8].

RESULTS AND DISCUSSION

Figure 1 displays TEM images of the sample before (a) and after 60% weight loss (b). It is seen that the tubular morphology is preserved but most ends became uncapped after oxidation. Images also indicate that tubes became shorter after treatment. The image analysis method was used in order to understand the scenario of burning of MWNTs during the oxidative process: it allows determining the outer and inner radii distributions. Two series of 10 images taken from a pristine and an oxidized MWNT after 60% weight loss were analysed. After the oxidative treatment, the outer radii slightly decrease and the inner radii increase. Consequently, the wall thickness distribution decreases (c.f., figure 2). These results suggest that after removing the extremities of the tubes which are more reactive because of the curvature, both the outermost and innermost walls were then burned but the kinetics are slow compared to that at the tube extremities.

An adsorption isotherm of Kr on pristine MWNTs at 77 K is presented in Fig. 3. The curve exhibit two distinct steps separated by a horizontal plateau. The steps were attributed to two successive monolayer condensations on the uniform or quasi-uniform patches of the surface, namely the external walls of the nanotubes [9]. The step pressures are determined by the decrease in the substrate attraction forces with increasing distance from the surface.

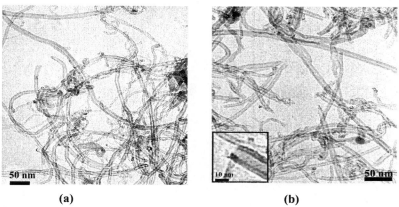

(a)   (b)

FIGURE 1. TEM micrograph of MWNTs: (a) pristine sample; (b) oxidized sample 60% weight loss; inset shows opened end.

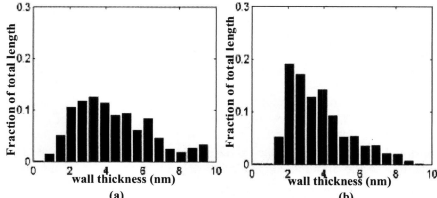

(a)   (b)

Figure 2. Statistical distributions of number of nanotubes as a function of wall thickness. (a) pristine MWNTs, (b) oxidized MWNTs 60% weight loss.

No hysteresis being observed between adsorption and desorption processes does not confirm that capillary condensation does not occur inside the tubes. Nevertheless, it is consistent with TEM observations which show that the majority of tubes are closed. In the case of oxidized MWNTs, the step becomes significantly wider (Fig. 3, upper curve). Compared with the pristine MWNTs, the step height which is proportional to the area of the uniform patches of the surface (i.e. amorphous carbon does not contribute to the step-like adsorption), decreases after oxidation. However, a second step cannot be distinguished. This demonstrates that the oxidation generates defects on the surface and thus an increase in superficial heterogeneity. In fact, the increase of Kr adsorption at very low pressures might be assigned to the creation of microporosity in addition to other attractive sites. The plateau between the two steps became inclined, and a small hysteresis is observed. This reveals the opening of a number of tubes as observed by TEM. The hysteresis is

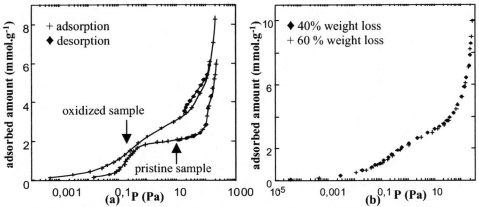

Figure 3. (a) Adsorption isotherm of Kr at 77.3 K on pristine MWNTs and on oxidized MWNTs after 60% weight loss.
(b) Adsorption isotherms on oxidized MWNTs for 40% and 60% weight loss.

small and thus cannot be taken as confirmation that the totality of the inner surface is accessible to Kr molecules. The BET specific surface area has increased after the oxidative treatment. By taking 14.7 Å^2 as the molecular cross-section of Kr, we obtain S_1 = 161 m^2/g for the pristine sample, and S_2 = 296 m^2/g after oxidation.

Figure 3b shows the adsorption isotherms of Kr at 77.3 K on oxidized MWNTs after 40% and 60% weight loss. The isotherms are seen to be identical. We thus conclude, based on the ensemble of the results provided by the various techniques used, that the HNO_3-CO_2 oxidation efficiently opens a number of tubes, and simultaneously leads to burning the tubes along their tube axes and removal of the innermost and outermost graphene sheets of the MWNTs.

REFERENCES

1. S.C. Tsang, Y.K. Chen, P.J.F. Harris, M.L.H. Green, Nature **372**, 159-162 (1994).
2. P.M Ajayan, T. W. Ebbesen, T. Ichihachi, S. Iijima, K. Tanigaki, h. Hiura, Nature **362**, 523-525, (1993).
3. S. C. Tsang, P. J. F. Harris and M. L. H. Green, Nature **362**, 520-522 (1993).
4. J.M. Skowroński, P. Scharff, N. Pfänder and S. Cui, Adv. Mat **15**, 56-57 (2003).
5. S. Delpeux, K. Szostak, E. Frackowiak, S. Bonnamy; F. Beguin, J. Nanosci. Nanotechnol. **2**, 481-484 (2002).
6. F. Béguin, S. Delpeux-Ouldriane, K. Szostak, French Patent fr 0210115 (8 August 2002).
7. C. Gommes, S. Blacher, K. Masenelli-Varlot, Ch. Bossuot, E. McRae, A. Fonseca, J.-B. Nagy, J.-P. Pirard, Carbon **41**, 2561-2572 (2003).
8. M.-R. Babaa, N. Dupont-Pavlovsky, E. McRae, K. Masanelli-Varlot, Carbon, in press.
9. K. Masenelli-Varlot, E. McRae, N. Dupont-Pavlovsky, Appl. Surf. Sci. **196**, 209-215 (2002).

Wide Range Optical Studies on Transparent SWNT Films

F. Borondics[*], K. Kamarás[*], Z. Chen[†], A.G. Rinzler[†], M. Nikolou[†] and D.B. Tanner[†]

[*]*Research Institute for Solid State Physics and Optics, Hungarian Academy of Sciences, P. O. Box 49, Budapest, H-1525, Hungary*
[†]*Department of Physics, University of Florida, Gainesville, FL 32611, U.S.A.*

Abstract. We present transmission spectra from the far infrared through the ultraviolet region on freestanding SWNT films at temperatures between 40 and 300 K. Several interesting features are observed in the low-frequency part of the spectrum: the Drude-like frequency dependence of the metallic tubes as well as a (sample-dependent) peak in the conductivity around 0.01 eV. We also studied the accidental nitrate doping of the SWNT samples during purification by nitric acid. As-prepared purified samples exhibit increased metallic absorption and decreased interband transitions; these features disappear on heating in vacuum.

INTRODUCTION

Frequency-dependent optical measurements on carbon nanotubes can render a wealth of information about their electronic structure. Early optical evidence for metallic (or semimetallic) carriers in single-wall carbon nanotubes came from reflectance spectra obtained in a wide temperature and frequency range [1]. Similar transmission studies on airbrushed nanotubes [2] and freestanding films [3] have not included temperature dependence. Here we present temperature-dependent spectra and the effect of accidental nitrate doping during purification, which influences the far-infrared part of the spectrum as well as the interband transitions.

EXPERIMENTAL

Self-standing SWNT films were prepared by filtration of a dilute solution [3] of laser-produced material [4]. The sample used for optical measurements was transferred to a graphite template with an approximately 4 mm hole. This sample geometry and the high transparency of the film allowed us to measure transmission with high reproducibility. We used two different spectrometers with liquid helium cryostats: in the FIR through MIR range a Bruker 113v spectrometer, in the NIR, visible and UV a modified Perkin-Elmer grating spectrometer. Spectra were obtained before and after heat treatment in vacuum at 1000° C(to remove HNO_3 that p-dopes the material; the procedure is hereafter referred to as "baking"). We examined three different samples. Samples 1 and 2 were of the same batch, one baked and one unbaked; sample 3 was from a subsequent batch,

CP723, *Electronic Properties of Synthetic Nanostructures*, edited by H. Kuzmany et al.
© 2004 American Institute of Physics 0-7354-0204-3/04/$22.00

here data were taken on the same specimen before and after baking.

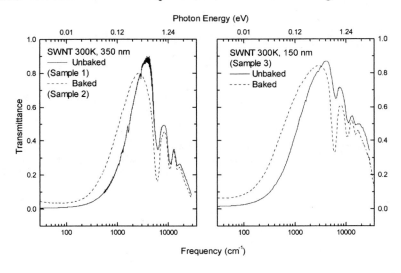

FIGURE 1. Transmission of SWNT films before and after baking.

RESULTS AND DISCUSSION

Figure 1 shows the transmission of the samples at room temperature. The films reach transmission values of up to 80 per cent around 3000 cm^{-1} and are highly transparent (above 30 per cent) throughout the whole midinfrared range. Baking increases transparency in the far infrared and decreases it somewhat at higher frequencies. For application purposes, obviously the featureless midinfrared region is of importance, but the information about fundamental materials properties is carried by the highly absorbing structures in the far and near infrared.

The transmission must be consistent with Kramers-Kronig relations, just as is the reflection, and the phase shift on transmission may be estimated from a Kramers-Kronig integral on $\ln T$, just as with the reflectance. After computing the phase, one may extract the complex refractive index (and all other optical constants) using equation (1).

$$\sqrt{T}e^{i\Theta} = \frac{4N}{(N+1)^2 e^{i\delta} - (N-1)^2 e^{i\delta}},\tag{1}$$

where $\delta = \frac{\omega N d}{c}$. The sample thickness d can be a fitting parameter as well, but in our case we used 350 nm for Sample 1 and 2 and 150 nm for Sample 3, obtained from AFM measurements.

We fitted the transmission data by a Drude-Lorentz model. In the fits to the optical conductivity σ_1 of Sample 2 (totally removed HNO$_3$) one can clearly see a low frequency oscillator (Figure 2.). This phenomenon has been predicted, attributed to a secondary gap caused by the curvature of certain semimetallic tubes [5] or the symmetry breaking by the neighboring tubes in metallic nanotube bundles [6]. Ouyang et al. [7] have measured both gaps on individual nanotubes by tunneling spectroscopy; based on their results, we attribute the 8 meV peak to a curvature-induced gap in low-gap semiconducting tubes. Consequently, we assign the broader Drude contribution to metallic (armchair) tubes. The sample dependence of the low-frequency behavior is probably caused by the fact that the small gaps are easily filled by extrinsic carriers even at minimal doping.

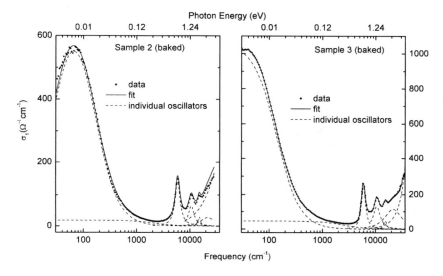

FIGURE 2.   Fits to room-temperature optical conductivity in Sample 2 and 3.

Figure 3. summarizes the oscillator strength values for all measured temperatures in Sample 3. It is apparent that observable changes on dedoping occur almost exclusively in the narrow Drude part. This change is accompanied by a sharpening and strengthening of the high-frequency van Hove transitions, already shown by Hennrich et al. [8], as the valence bands of high-gap semiconducting tubes are filled by electrons. The oscillator strength shows little temperature dependence. Such behavior is expected in normal metals, however, we see also very weak dependence of the scattering rates on temperature. This is consistent with previous optical [1] and dc measurements and points to the possibility of exotic transport mechanisms in nanotubes.

CONCLUSIONS

We measured the transmission spectra of ultrathin SWNT films in a wide temperature and frequency range. The low-frequency part of the spectrum can be fitted with two components, which we attribute to armchair and low-gap semiconducting tubes, respectively. Optical signatures of hole doping can be observed only in the semiconducting

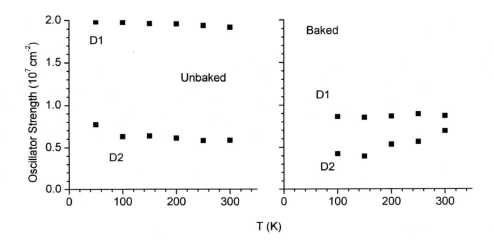

FIGURE 3. Temperature dependence of low-frequency oscillator strength in Sample 3 before and after baking.

tubes (both low- and high-gap). The temperature dependence of the optical functions is much weaker than what would be expected for conventional metals.

ACKNOWLEDGMENTS

This work was supported by the following grants: OTKA 034198, ARO grant DAAD19-99-1-0316, NSF (ECS-0210574), US NSF-INT 9902050.

REFERENCES

1. Ugawa, A., Rinzler, A. G., and Tanner, D. B., *Phys. Rev. B*, **60**, R11305 (1999).
2. Itkis, M. E., Niyogi, S., Meng, M. E., Hamon, M. A., Hu, H., and Haddon, R. C., *Nano Lett.*, **2**, 155 (2002).
3. Hennrich, F., Lebedkin, S., Malik, S., Tracy, J., Barczewski, M., H. Rösner, and Kappes, M., *Phys. Chem. Chem. Phys.*, **4**, 2273 (2002).
4. Rinzler, A. G., Liu, J., Dai, H., Nikolaev, P., Huffmann, C. B., F. J. Rodriguez-Macías, Boul, P. J., Lu, A. H., Heymann, D., Colbert, D. T., Lee, R. S., Fischer, J. E., Rao, A. M., Eklund, P. C., and Smalley, R. E., *Appl. Phys. A*, **67**, 29 (1998).
5. Kane, C. L., and Mele, E. J., *Phys. Rev. Lett.*, **78**, 1932 (1997).
6. Delaney, P., Choi, H. J., Ihm, J., Louie, S. G., and Cohen, M. L., *Nature*, **391**, 466 (1998).
7. Ouyang, M., Huang, J.-L., Cheung, C. L., and Lieber, C. M., *Science*, **292**, 702 (2001).
8. Hennrich, F., Wellmann, R., Malik, S., Lebedkin, S., and Kappes, M. M., *Phys. Chem. Chem. Phys.*, **5**, 178 (2003).

Bulk synthesis and characteristic properties of boron nitride nanostructures: nanocapsules and nanotubes

E. Borowiak-Palen[1,2*], M. H. Rümmeli[1], M. Knupfer[1], G. Behr[1], T. Gemming[1], R. J.

Kalenczuk[2], T. Pichler[1]

1) Leibniz-Institute for Solid State and Materials Research Dresden,P. O. Box 270016, Dresden D-01171, Germany

2) Institute of Chemical and Environment Engineering, Technical University of Szczecin, Poland

Abstract: A new synthesis route for the formation of boron nitride (BN) nanocapsules by means of a substitution process using single wall carbon nanotubes as templates, with yields of > 95% is presented. It is also shown that these BN nanocapsules can act as ideal reference samples for the determination of the relative sp^2 to sp^3 configuration in BN species; a value that is crucial for the physical properties of these nanostructures. The data are compared to multiwall boron nitride nanotubes (MWBNNT) which were also produced using a substitution procedure, and commercial h-BN.

INTRODUCTION

Concentric shell carbon clusters, also referred to as carbon onions, were first discovered by Ugarte [1] in 1992. Due to the structural similarity of h-BN and graphite, the formation of concentric shell BN nanostructures (spherical clusters with incomplete closed shells, encapsulated polyhedral BN nanoparticles or BN onions and nested BN fullerenes) came as no surprise [2-8]. The ever-increasing interest in BN nanostructures is based on their large band gap leading to potential applications in photoluminescence devices [9] and nano-electronic-magnetic devices [10]. However, thus far, with all of these methods, BN nanocapsules are only present as one or a few among many BN species, for example h-BN sheets and MWBNNT. The substitution reaction of carbon nanotubes used for the production of diverse heteronanotubes [11-15] is also a suitable method to obtain fullerene like BN nanostructures here referred to as "nanocapsules".

In this work we present an adapted substitution process that has been optimised for the preparation of BN nanocapsules in high yields, c.f. high yield carbon onion production [16]. To the best of our knowledge this is the first time such nanocapsules have been obtained with a volume yield of 95% and by means of a substitution process. The presented results will be compared to the MWBNNT produced also by a similar substitution process [13] and commercial h-BN (Merck).

CP723, *Electronic Properties of Synthetic Nanostructures*, edited by H. Kuzmany et al.
© 2004 American Institute of Physics 0-7354-0204-3/04/$22.00

EXPERIMENTAL

The starting material was a black powder containing 40% of SWCNT produced by a laser ablation [17]. The nitrogen source was nitrogen gas, and that of boron, boron oxide. The SWCNT and B_2O_3 were mixed to a ratio of 1:5 respectively and placed in an Al_2O_3 crucible, which was then placed in a graphite holder. The graphite holder plus contents was then introduced to the centre of a radial high frequency chamber furnace. The detailed description of the set-up is given in Ref. [13]. Once the system had been fully evacuated the vacuum pump was shutdown and a static nitrogen gas pressure of 950 mbar was established within the oven chamber. The temperature was then raised to 1300°C (80°C/min.) and kept for a time of 4 h in order to activate the substitution process. The key to the formation of the concentric BN nanocapsules lies in the rapid cooling down to room temperature (within 5 min.) upon completion of the substitution process (slow cooling yields BN lamella like structure-see below). The reaction results in a grey substrate. MWBNNT were produced in the same set-up.

The EELS measurements were performed in a purpose built high-resolution spectrometer described elsewhere [18]. The energy resolution was set to 200 meV and the momentum resolution was 0.1 $Å^{-1}$. The IR spectra were measured with the spectral resolution set to 2 cm^{-1} (0.25 meV) using a Bruker IFS 113V/88 spectrometer. Fourier transform Raman spectroscopy was performed with a Bruker RFS 100/s. with a resolution of 2 cm^{-1}.

RESULTS AND DISCUSSION

Fullerene-like BN nanostructures are observed under rapid cooling of the reaction product (see Fig.1 a-b). This can be explained by a fragmentation or curling of the tube walls under the quenching process. The same process parameters but under gentle cooling results in the lamella-like BN structures, presented in Fig.1 c. Typical high resolution electron microscopic images of BN nanocapsules are depicted in Fig.1 on different scales ranging from 2 nm up to 100 nm. It can be seen that while the size of the capsules vary, the general shape of each structure is similar. The diameters of the nanocapsules range between 5 and 120 nm with 3 to 30 walls. A statistical analysis of several TEM micrographs shows that most of the nanocapsules have a mean diameter peaking at 40 nm. This is similar to the diameter and number of walls of the MWBNNT produced by the same technique without the rapid quenching [13]. The low magnification TEM image shown in Fig. 1. b highlights that the sample is comprised almost entirely of BN nanocapsules which is in contrast with other production methods for BN nanocapsules that do not achieve such high yields. In addition, the local electronic structure, bonding and chemical environment was studied using core level excitations. The site selective excitations are probed by core level EELS to check sample composition. The analysis of the EELS C1s edge showed that less than 5% carbon is present in the sample.

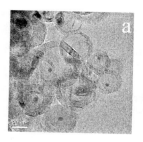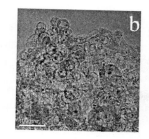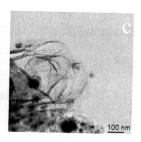

FIGURE 1 (a-b) TEM micrographs of BN-nanocapsules and **(c)** micrographs of lamella-like BN nanostructures.

This highlights the efficiency of the substitution process. To determine the actual sample composition, the EELS spectra of the B1s and N1s excitation edges were compared with those obtained from commercial polycrystalline h-BN and MWBNNT and are depicted in Fig 2. Taking into account the different form factors for the scattered electrons a boron to nitrogen ratio of 1:1 is obtained, as expected for BN. The most pronounced peak in the B1s excitation edge is the so-called π^* resonance at 192 eV (fig. 4a –solid line). This position is characteristic for hexagonal BN in an sp^2 environment. For other hexagonal structures like BC_3, or B_2O_3 the peak is shifted to higher energy of 193 eV or 194 eV, respectively. Both the B1s and N1s π^* resonances from the BN nanocapsules are lower than those for h-BN and MWBNNT.

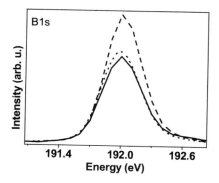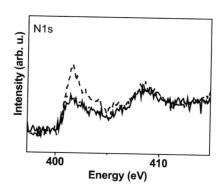

FIGURE 2 B1s and N1s core edges of BN-nanocapsules (solid line), h-BN (dashed line) and MWBNNT (dotted line).

This is due to a *texturing* effect, that is, since the π orbital is sensitive to excitations perpendicular to the BN sheets [19] there exists a preferential orientation between the BN nanocapsules film and that of h-BN leading to differences in the π^* resonances. This is easily understood in terms of the h-BN crystals being more planar in form and preferentially orientated on the c-axis whilst the nanocapsules are more spherical in form and so have a random orientation. The intensity of π^* resonances of MWBNNT is in between those of the above mentioned BN nanostructure indicating that the MWBNNT is more preferentially orientated in comparison to the BN-nanocapsules but less so than h-BN due to their curvature.

Electron diffraction studies show that the spacing between BN sheets in the BN nanocapsules is enlarged by ca. 4.5% as compared to the reference h-BN sample. The same trend is observed for the in-plane lattice parameter, a, and it is very close in to that of MWBNNT [13].

To summarize, we have presented a fast and highly efficient synthesis route to produce BN nanocapsules with a narrow distribution of diameters and shells. This is achieved by an adapted substitution process using SWCNT as templates followed by a rapid cooling treatment. The morphology was studied by TEM, the chemical composition studied by core level EELS showed the 1:1 B:N composition with remaining carbon impurities from partly reacted raw material below 5%. The EELS and ED results evidenced close similarity in the structural and electronic properties between BN-nanocapsules, h-BN and MWBNNT. The advantage of BN-nanocapsules lays in their random orientation what derivates sp^2 to sp^3 ratio in BN species.

ACKNOWLEDGEMENTS

We thank the DFG PI 440/1. M. H. R. thanks the EU for support through a MC fellowship. The authors are grateful to K. Müller, R. Hübel and S. Leger for technical assistance and O. Jost for delivering the sample of SWCNT.

REFERENCES

1. D. Ugarte, Nature (London),358 (1992) 707.
2. T. S. Bartniskaya, et. al, J. Less-Common Met. 117 (1986) 253.
3. L. Boulanger, et. al, Chem. Phys. Lett., 234 (1995) 227.
4. F. Banhart, et. al, Chem. Phys. Lett., 231 (1994) 98.
5. M. Terrones, et. al, Chem. Phys. Lett., 259 (1996) 568.
6. O. Stephan, et. al, Appl. Phys. A: Mater. Sci. Process., 67 (1998) 107.
7. V. V. Pokropivny, et. al, J. of Sol. St. Chem., 154 (2000) 214.
8. T. Oku, et. al, Mat. Sci. and Eng. B, B74 (2000) 206.
9. D. Golberg, et. al, Chem. Phys. Lett., 323 (2000) 185.
10. D. Golberg, et. al, Carbon, 38 (2000) 2017.
11. W. Han, et. al, Chem. Phys. Lett., 299 (1999) 368.
12. W. Han, et. al, Appl. Phys. Lett., 73 (1998) 3085.
13. E. Borowiak-Palen, et. al, ChemComm, 1 (2003) 82.
14. E. Borowiak-Palen, et. al, Chem. Phys. Lett., 378 (2003) 516.
15. X. D. Bai, et. al, Appl. Phys. Lett., 77 (2000) 67.
16. T. Pichler, et. al, Phys. Rev. B, 63 (2001) 155415.
17. O. Jost, et. al, Chem. Phys. Lett., 339 (2001) 297.
18. J. Fink, Adv. Electron. Electron Phys. 75 (1989) 121.
19. G. G. Fuentes, et. al, Phys. Rev. B, 67 (2003) 35429.

Structure and Electronic Properties of Suspended Single Wall Carbon Nanotubes

A. Hassanien[1], M. Tokumoto

Nanotechnology Research Institute, AIST, 1-1-1 Umezono, Tsukuba, Ibaraki 305-8568, Japan

Abstract. We report on scanning tunneling microscopy and spectroscopy measurements of single wall carbon nanotubes that are suspended over Au(111) substrate. In comparison with supported nanotubes, we find two remarkable differences: first there is less shift in the energy band structure, second is the occurrence of persistent charge density oscillations. The undamped behavior of these oscillations on suspended nanotubes is a manifestation of long coherence length and reflects the fact that the dephasing mechanism is mainly due to interactions between nanotubes and substrates. The scattering of electron waves, indicates less perfection of the nanotube lattice, and originates from defects on the nanotube. We present high-resolution image of a unique type of such defects in pristine nanotubes where vacancy on nanotube wall is observed for the first time.

INTRODUCTION

The early STM experiments on single wall carbon nanotubes (SWNTs) provided very valuable information on their structures and electronic properties [1]. In these experiments one needs to deposit tubes over a metallic substrate to allow electrons to tunnel to/from a STM tip. However, due to substrate-nanotube interaction many of their properties could not be explored. A band asymmetry around the zero bias and broadening in van Hove singularities were attributed to a signature of nanotube-substrate interaction [1]. Another signature of substrate-nanotube interactions is the short-range modulation of electron waves close to nanotube edges [2] and defects [3, 4]. In this work we examine the role of substrate-nanotube interactions using room temperature STM experiments on suspended nanotubes. The results show there is clearly less band asymmetry on suspended nanotubes compared to supported ones. Moreover, we have observed interference pattern that is only visible on the suspended part and missing on the supported tubes. In addition, we didn't see any damping of interference oscillation on suspended nanotubes, which indicates, that the coherence length is much larger than what was deduced on supported tubes [4].

EXPERIMENTAL

The SWNTs of this study is synthesized using laser ablation method at Rice University. The tubes were sonicated in dichloroethane for 20 minutes prior to being

[1] Author to whom correspondence should be addressed. Electronic mail: Abdou.Hassanien@aist.go.jp

CP723, *Electronic Properties of Synthetic Nanostructures*, edited by H. Kuzmany et al.
© 2004 American Institute of Physics 0-7354-0204-3/04/$22.00

cast on atomically flat gold substrates. The gold substrate is specially fabricated to yield small crystalline voids into Au(111) surface. Atomically resolved topographic images of carbon nanotube were obtained by recording the tip height at constant current. Typical bias parameters are 500 mV and 300 pA for bias voltage and tunneling current respectively. The STS measurements were carried out by interrupting the feedback loop while recording the current at different bias voltage.

RESULTS AND DISCUSSION

Fig. 1 shows a true atomic resolution on supported SWNTs. The lattice structure is clearly seen where carbon atoms are forming hexagonal rings on a cylindrical surface i.e nanotube wall. The distance between any two vortices, the c-c bond length, is 1.4±0.1 Å. The dark areas are the centers of carbon hexagons; their orientation with respect to the nanotube axis gives a chiral angle of 10±1°. I/V STS spectra is shown in fig.1(b). The current is nearly zero up to a threshold of 0.5 V, then increases gradually, which is typical behavior for a semiconducting SWNTs. Inset of Fig. 1(b) displays the differential conductance, which is a measure of the density of states, versus bias voltage. A gap of about 0.8 eV is not symmetrically positioned around the zero bias voltage. This has been attributed to doping effect from Au(111) substrate which tends to shift the Fermi energy toward the valence band of the SWNTs. The shift is quite dramatic that the Fermi level is aligned with the valence band edge. This means that the DOS inside the band gap is zero and gives another evidence that the nanotube is a semicoducting one. This is consistent with previous STM study at 5 K [1].

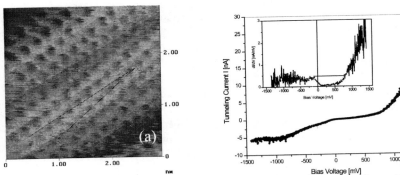

Figure 1. (a) Atomic resolution STM image in topographic mode. Dark areas are the centers of the carbon hexagons which form a chiral angle of 10° with the tube axis. (b) I-V spectroscopy of nanotube in (a) displays a semiconducting behavior, the inset is the dI/dV as a function of bias voltage. A large shift of the Fermi level toward band edge is clearly visible and is attributed to substrate-nanotube interactions.

Now we examine the structure and the electronic properties of nanotubes suspended over voids in the Au(111). In Fig. 2(a) we show an STM image of suspended SWCNTs. The tubes are bridging a void without any bending at the edges. Fig. 2(b) is atomic resolution image of suspended SWNTs. A slow oscillation with a period of 1.1

nm is superimposed on the lattice structure. The line profile is shown in fig.2(c). This oscillation results from interference of excited electron waves within 0.05 eV energy window. The oscillation persists along the whole length of the suspended part.

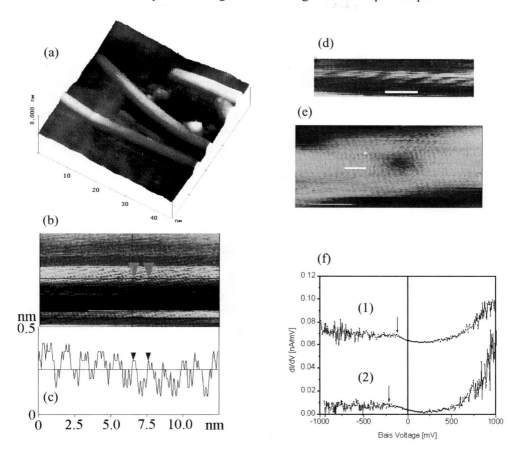

Figure 2. (a) STM topographic image of suspended SWCNTs. (b) atomic resolution image shows the lattice structure of the nanotube. The image shows also a slower oscillation, which results from interference of electron waves on the nanotube. The line profile of (b) is shown in (c). Imaging parameters are 50 mV and 300 pA for bias voltage and tunneling current respectively. The oscillation period depends on the bias voltage as shown in (d); bias parameters 30 mV and 300 pA; scale bar is 2 nm. (e) Topographic image of a vacancy on the nanotube wall. Scattered charge density are clearly visible in the vicinity of the defect, scale bar is 2 nm. (f) dI/dV curves at two different positions (1) and (2); 5 and 10 nm away from the substrate edge. Curves are off-set for clarity. As we move closer to the substrate the Fermi level gets closer to the valence band edge. This shift demonstrates clearly the effect of substrate interaction on electronic properties of SWNTs.

In our previous STM study on supported tubes, we observed similar oscillation close to the nanotube ends [2], however there is a fundamental difference between this and oscillations on unsupported nanotubes. In the first type there is damping factor which washes the interference pattern within 6 nm away from the nanotube edge while in the second type we see no damping behavior. Similar decay behavior of the interference pattern was observed in the vicinity of defects on SWNTs [3,4]. The decay behavior of oscillations in supported SWCNTs is a manifestation of interaction process that led to decoherence of electron waves. This could be due to electron-electron (e-e) interactions and electron-phonon interaction processes. Hertel and G. Moos [5] have studied the electron dynamics in SWNTs bucky paper and concluded that e-e interaction is the most dominant mechanism of the phase relaxation even at room temperature (if we consider energy widow of few meV). In principal, e-e interactions can be due to substrate/nanotube or nanotube/nanotube interactions as their measurements were carried out on SWNTs ropes. Our observation of undamped oscillations on individual suspended SWNT indicates that the electron decoherence process is relaxed as nanotube/substrate interaction is minimized. The fact that these oscillations exist at room temperature is quite remarkable. However, it is not unusual for them to occur as we consider interference of electron waves just few meV above the Fermi energy (30-50 meV). The scattering of electron waves is caused by defects in the nanotube lattic. In fig. 2 (f) we show one example of defects on the nanotub wall where a vacacy can be clearly seen. We have carried out STS studies on suspended SWNTs and found remarkable difference compared with the supported tubes. Results are shown in fig. 2(f), where STS were taken at two different positions on the suspended nanotube. (1) and (2) curves are taken 5 nm and 10 nm away from the edge respectively. The Fermi level in (2) moves closer toward zero bias indicating less doping from substrate. The shift in the Fermi energy between the two curves is ~0.1 eV. This, indeed, provides another evidence that e-e interactions between nanotube and substrate influence the electronic properties.

In summary, we have presented room temperature STM investigation on suspended nanotubes in order to understand the influence of substrate on their electronic properties. High-resolution STM images show interference of electron waves that are persistent on the suspended nanotubes. The existence of interference pattern is correlated with defects on SWNTs. I-V spectroscopy on suspended SWNTs shows a shift in Fermi energy toward zero bias, which indicates less doping from the substrate.

ACKNOWLEDGMENTS

The authors would like to acknowledge the financial support from AIST and NEDO (nanocarbon project).

REFERENCES

1. J. W. G. et al, Nature **391**, 59 (1998). T. W. Odom et al , Nature **391**, 62 (1998); A. Hassanien et al, Appl. Phy. Lett. **73**, 3839 (1998).
2. A. Hassanien, et al, Appl. Lett. Phys. **78,** 808-810 (2001).
3. A. Hassanien, et al, AIP Conference Proceeding Vol. 633, pp.271-274 (2002
4. M. Ouyang et al Phy. Rev. Lett. **88,** 066804 (2002).
5. T. Hertel and G. Moos, Chem. Phys. Lett. **320,** 359 (2000)

STM investigation of irradiated carbon nanotubes

Z. Osváth, G. Vértesy, G. Pető, I. Szabó, J. Gyulai, W. Maser* and L. P. Biró

Research Institute for Technical Physics and Materials Sciences, H-1525 Budapest, P.O. Box 49, Hungary
*Instituto de Carboquimica, CSIC, C/Maria de Luna, 12, E-50015 Zaragoza, Spain

Abstract. Multi-wall and single wall carbon nanotubes (MWCNTs, SWCNTs) dispersed on graphite (HOPG) and gold substrates were irradiated with two different doses of Ar^+ ions. The irradiated samples were investigated by scanning tunneling microscopy (STM) and spectroscopy (STS). The investigation of SWCNTs irradiated with low-dose Ar^+ ions revealed isolated defects in the nanotube-walls, similar to the hillocks observed earlier on ion irradiated HOPG. These results are in agreement with recent predictions, which attribute the STM features produced by ion irradiation to local modifications of the local electronic structure. The high-dose irradiation of MWCNTs produced numerous defects in the nanotube-walls, and the surface of the nanotubes appeared rugged in the STM images. The rugged surface can be regarded as a summation of the individual hillocks, measured at the defect sites.

INTRODUCTION

Irradiation of carbon nanotubes with charged particles can induce interesting phenomena like welding [1], cross-linking [2] or coalescence [3] of nanotubes. Recent experiments show that the structure and the dimensions of carbon nanotubes can be modified by electron beam irradiation [4]. In-situ XPS and AES analysis [5] also showed that irradiation with Ar^+ ions produces dangling bonds (vacancies) on the surface of nanotubes and may cause shrinkage of the nanotube-diameters. Atomic vacancies and other topological defects may be present in the as-grown nanotubes also, and they can influence the electrical transport properties of the nanotubes [6]. Therefore, the properties of nanotubes with defects have to be investigated.

The present work confirms that during STM investigation, the local defects produced by irradiation appear as hillock-like features, which is in agreement with recent theoretical studies [7]. For irradiation, we used MWCNTs produced by the arc-discharge method, and commercial SWCNTs. The nanotubes were ultrasonicated in toluene, and dispersed on HOPG and gold substrates. The dispersed MWCNTs were irradiated with Ar^+ ions of 800 eV, using a high dose of $D=1 \times 10^{15}$ cm^{-2} (irradiation time: t=1 min, normal incidence). In the case of SWCNTs the energy of 30 keV and a low dose of $D=5 \times 10^{11}$ cm^{-2} was used (t=15 min, normal incidence). Without further manipulation, the irradiated samples were investigated with STM and STS under ambient conditions.

CP723, *Electronic Properties of Synthetic Nanostructures*, edited by H. Kuzmany et al.
© 2004 American Institute of Physics 0-7354-0204-3/04/$22.00

RESULTS AND DISCUSSION

The low-dose Ar^+ irradiation of SWCNTs produced point defects on the nanotube-walls. Figure 1 shows that these defects appear as protrusions (hillock-like features) in the STM images, similar to the hillocks observed during STM investigation of irradiated HOPG surface [8].

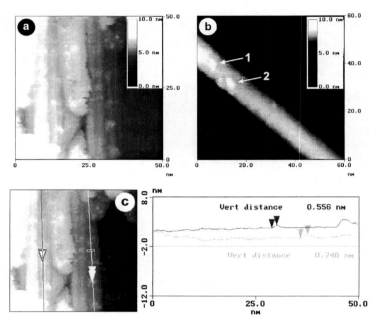

FIGURE 1. (a) Irradiated bundle of SWCNTs. The hillock-like features (larger tunneling currents) correspond to ion-induced defect sites. (b) Individual SWCNT with two defects labeled *1* and *2*. (c) Cross-sectional lines along the nanotubes presented in (a).

Tight-binding calculations predicted [7] that a vacancy in the nanotube-wall should result a maximum in the tunneling current, when imaged with the STM, and this maximum is due to the modified density of states (DOS) near the defect site. Our results show that this protrusions in the topographic images are in the range of 0.2 – 0.9 nm. The width of the hillocks is in the range of 10 nm, which is larger than the theoretical value. This is due to the geometry of the STM tip, which may differ substantially from the ideal tip assumed in the calculations. From our measurements one cannot decide which kind of defects were produced. Most probably they are vacancies, interstitial carbon atoms and clusters [9, 10]. It has been shown that the large hillocks observed on irradiated graphite surface can only be caused by interstitial clusters [11]. Taking into account that carbon interstitials have high mobility at room temperature and they can form clusters of C_2- or C_3-carbon molecules [10], we expect that these clusters can form easily inside the nanotube, resulting larger hillocks in the STM images. Furthermore, the dangling bonds produced by irradiation were probably saturated by adsorbed atoms, before we investigated the sample with STM, therefore we cannot exclude the influence of these adsorbates on the height of the hillocks.

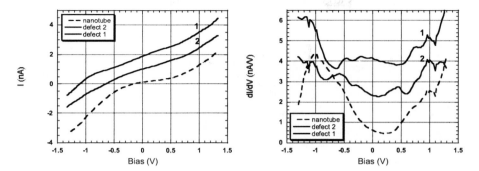

FIGURE 2. I-V curves and the corresponding dI/dV plots measured on the SWCNT presented in Fig. 1 (b), above the defects *1* and *2* (continuous lines), and above a location distanced from the defects (dashed line).

We performed STS measurements above the two defects labeled *1* and *2* on the nanotube seen in Figure 1 (b). We recorded STS curves above the two hillocks and also between the hillocks, where a defect-free region is assumed. The I-V characteristics and the corresponding differential conductance plots are presented in Figure 2. One can see that both I-V curves measured at the defect sites (continuous lines) are almost straight, compared with the curve recorded between the defects (dashed line). This suggests a metallic character at the defect sites. The corresponding dI/dV curves also reveal a plateau at the Fermi energy and small peaks in the plots corresponding to the defects, located at up to 0.5 eV below and above the Fermi energy, which are missing from the dI/dV plot belonging to the defect-free region. These peaks represent additional electronic states created during defect formation and they are similar to theoretical curves, which show the variation of the local DOS near a vacancy and near a "pentagon-one dangling bond" defect (5-1db) [7]. The "5-1db" defect can form by atomic rearrangement of the carbon atoms near a vacancy. This defect is more stable than a vacancy and the height of the hillock produced in the STM images is roughly the same as for a vacancy [7].

The MWCNTs were irradiated with high-dose Ar^+ ions of 800 eV. Figure 3 shows that the surface of the nanotubes and the surface the gold substrate are more ragged after irradiation. Previous experiments on ion-induced defect creation on graphite surface [8, 9] showed that the irradiation-induced hillocks form independently of the incoming ion energy. Thus it is reasonable to assume that hillocks formed also in our irradiated MWCNTs, but the overall amorphiazation induced by the high dose of ions ($D=1x10^{15}$ cm^{-2}) made impossible to observe them individually. We interpret the ragged surfaces of the MWCNTs as the summation of these individual hillocks.

In conclusion, we presented direct observation of small-scale defects produced on the nanotube-walls by Ar^+ irradiation. In agreement with theory, the defects appear as hillocks in the STM images, which is attributed to the modified electronic properties at the defect sites. Adsorbates may also influence the height of these protrusions.

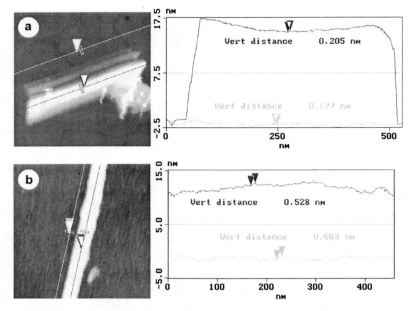

FIGURE 3. MWCNTs before (a) and after (b) irradiation with high-dose Ar⁺ ions. The cross-sections show that after irradiation the MWCNTs and the substrate have ragged surfaces.

ACKNOWLEDGMENTS

This work was supported by the EU, in the frame of the RTN contract NANOCOMP, HPRN-CT-2000-00037, and by OTKA grants T 43685 and T 43704 in Hungary.

REFERENCES

1. M. Terrones, F. Banhart, N. Grobert, J.-C. Charlier, H. Terrones, and P. M. Ajayan, Phys. Rev. Lett. **89**, 075505-1 (2002)
2. A. Kis, G. Csányi, J.-P. Salvetat, Thien-Nga Lee, E. Couteau, A. J. Kulik, W. Benoit, J. Brugger, and L. Forró, Nature Materials **3**, 153 (2004)
3. M. Terrones, H. Terrones, F. Banhart, J.-C. Charlier, and P. M. Ajayan, Science **288**, 1226 (2000)
4. P. M. Ajayan, V. Ravikumar, J.-C. Charlier: Phys. Rev. Lett. **81**, 1437 (1998)
5. Y. Zhu, T. Yi, B. Zheng, L. Cao, Appl. Surf. Sci. **137**, 83 (1999)
6. J. W. Park, J. Kim, J.-O. Lee, K. C. Kang, J.-J. Kim, K.-H. Yoo, App. Phys. Lett. **80**, 133 (2002)
7. A. V. Krasheninnikov, K. Nordlund, Phys. Solid State **44**, 470 (2002)
8. L. Porte, M. Phaner, C. H. de Villeneuve, N. Moncoffre, and J. Tousset, Nucl. Instrum. Meth. B **44**, 116 (1989)
9. J. R. Hahn and H. Kang, Phys. Rev. B **60**, 6007 (1999)
10. K. P. Reimann, W. Bolse, U. Geyer, and K. P. Lieb, Europhys. Lett. **30**, 463 (1995)
11. A. V. Krasheninnikov, and V. F. Elesin, Surf. Sci. **454-456**, 519 (2000)

A Resonant Raman Study of SWNTs under Electrochemical Doping

P. M. Rafailov, M. Stoll, J. Maultzsch and C. Thomsen

Institut für Festkörperphysik, Technische Universität Berlin
Hardenbergstr. 36, 10623 Berlin, Germany

Abstract. The first and second-order high-energy Raman modes of single-walled carbon nanotubes were investigated as a function of electrochemical doping for excitation energies in the range 477 – 780 nm. In the metallic resonance window (560 – 740 nm) a new peak was detected in the HEM region and found to exhibit spectacular shifts upon doping. For excitation energies above and below the metallic resonance range this new peak is absent. The second-order D mode also shifts with doping; the shift, however, strongly depends on excitation energy. In double resonance this corresponds to a wave-vector dependence of the doping-induced shift of the phonon branches.

INTRODUCTION

Electrochemical doping via double layer charging in a electrolyte solution can provide a fine tuning of the mechanical and electronic properties of carbon nanotubes[1]. This tuning can be monitored with Raman spectroscopy by means of a standard three-electrode cell equipped with a quartz window. The shift of the high-energy SWNT mode (HEM) upon double-layer charging in an electrolytic solution has been intensively studied[1, 2, 3] at the most common laser excitation of 515 nm and was found to be ≈ 250 cm^{-1}/hole/C-atom (1.5 cm^{-1}/V). The second-order D mode also exhibits a considerable shift upon doping[4]. From doping-induced frequency shifts useful information on the bond length change can be obtained[2, 5] as in the double-layer model the transferred charge can be quantified. However, as the HEM and the D mode are double-resonant in nature, it should be taken into account that different laser excitations will probe different regions of the bond-stretching dispersion branches whose shift upon doping may be wave-vector dependent. Here we present a Raman investigation of the whole HEM dispersion branch by excitation-energy dependent measurements of the HEM and the second order of the D mode of a SWNT bundle sample[6].

EXPERIMENTAL

A stripe of SWNT paper with surface density of $\approx 5 \cdot 10^{-5}$ g[6] was prepared as a working electrode in a three-electrode cell equipped with quartz windows. A Metrohm - Potentiostat was employed for charging at constant potentials. A platinum wire and Ag/AgCl/3 M KCl served as auxiliary and reference electrode, respectively. The work-

CP723, *Electronic Properties of Synthetic Nanostructures*, edited by H. Kuzmany et al.
© 2004 American Institute of Physics 0-7354-0204-3/04/$22.00

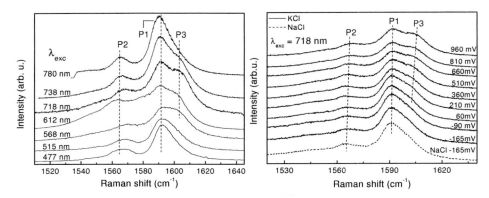

FIGURE 1. **Left**: The HEM band as excited with various laser wavelengths at a constant doping level of ≈1 V (0.005 holes/C-atom).
Right: The HEM band at $\lambda_{exc} = 718$ nm. At this wavelength the developtment of the new peak P3 on increasing the doping level can be followed best.

ing electrode was only partly dived into the solution and was electrically contacted at his dry end.

Several different aqueous solutions (concentration 1 M) were applied in the voltammetric measurements: LiCl, NaCl and KCl. The Raman spectra were recorded with a DILOR triple grating spectrometer equipped with a CCD detector. The 477, 488, 514.5 and 568 nm lines of an Ar^+/Kr^+ laser, as well as a dye laser and a Titanium-Sapphire laser were used for excitation and the spectral resolution was 2 - 6 cm^{-1}.

RESULTS AND DISCUSSION

As can be appreciated from Fig. 1, outside the metallic resonance window (560 – 740 nm) the HEM band in the doped state preserves its well known shape [main peak at 1593 cm^{-1} (P1) and a second-strongest peak at 1567 cm^{-1} (P2)]. In contrast, upon red excitation of the doped SWNTs we observe a qualitative change of what is known as the resonant HEM band of metallic tubes: a new peak (P3) emerges and separates from high-frequency side of P2. The appearance of this peak can either be attributed to an onset of an intercalation process, or it can be assumed that phonons in metallic tubes shift more strongly upon doping. The former suggestion is based on the similarity to the Raman spectrum of p-type graphitic intercalation compounds, where an additional peak appears above 1600 cm^{-1}[7]. However, intercalation may be ruled out as in no Raman experiment on intercalated nanotubes thus far the influence of the charge transfer showed up only upon red excitation and was completely absent outside the metallic resonance window. We therefore tentatively attribute P3 to phonons from the high-energy dispersion branch of metallic tubes (note that because of the double resonance[8] a whole section of the dispersion branch can contribute to the HEM when varying the doping level). From fitting the spectra to Voigt profiles a shift between 3 and 4 cm^{-1}/V was established for P3. Although the error in determination of the P3 peak positions is especially high, it is obvious that its doping-induced shift is about twice as high as that of P1 and P2 ($\approx$1.5 cm^{-1}/V for all wavelengths). The stronger shift of P3 may arise from a peculiar redistribution of phonon states upon shifting the Fermi level from its special position at the band-crossing point in metallic SWNTs[9].

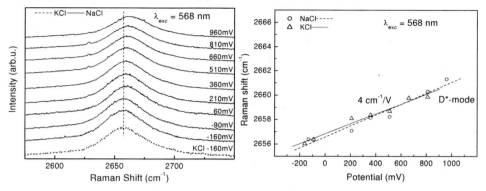

FIGURE 2. **Left**: Raman spectra of the D^* mode at several doping levels for $\lambda_{exc} = 568$ nm. **Right**: The frequency of the D^* mode as a function of the applied potential for two different solutions: NaCl (circles) and KCl (triangles).

We further checked the doping behavior of the second-order D mode: D^* (the D mode itself was too weak to be reliably fitted). With appropriately chosen laser excitation the D^* mode exhibits a considerable shift upon doping as is seen from Fig. 2 for $\lambda_{exc} = 568$ nm. We observe a surprisingly strong dependence of the doping-induced shift on the excitation energy: from no shift to above 4 cm^{-1}/V in going from blue to red excitation. The following table summarizes the D^* shift slopes (in cm^{-1}/V) found in the examined spectral range:

Electrolyte / λ_{exc}	477 nm	488 nm	514.5 nm	568 nm	601 nm	612 nm
NaCl	0	0	2	4.1	4.5	4.4
KCl	0	0	1.9	4.4	4.8	4.8

Table 1: Doping-induced D^* shift (in cm^{-1}/V) at several excitation wavelengths.

To explain the D^* mode behavior we recall that both D and D^* mode originate from phonons in the vicinity of the graphite K-point and appear in the Raman spectra due to a defect-induced double-resonant process[11]. Due to the steep dispersion of the high-energy TO phonon branch[10], increasing the excitation energy leads to smaller wave vectors and higher frequency of the doubly-resonant phonon[11]. This is illustrated in Fig. 3a.

Doping shifts the phonon branch. This alters the double-resonance condition as only a unique pair of electronic-transition energy and phonon energy can fulfill it. Different frequency slopes of the D^* mode at different excitation energies thus imply that the doping-induced shift varies with wave-vector. This is schematically shown in Fig. 3b.

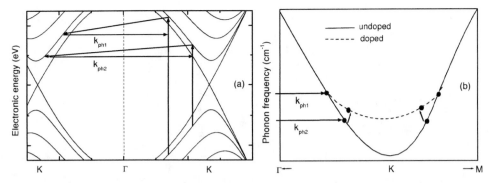

FIGURE 3. **a**: Double-resonant Raman scattering process that reveals the D and D^* modes for two different excitations.
b: Phonon dispersion relation without (solid line) and with doping (dashed line) and the corresponding double-resonantly excited phonon states. Arrows denote the same phonon wave-vectors as in **a**.

CONCLUSIONS

The main peak P1 of the HEM contains a contribution from modes of metallic SWNT (P3) that are revealed in the Raman spectra via electrochemical doping. Furthermore, the high-energy bond-stretching phonon branches of SWNTs shift non-uniformly upon electrochemical doping depending on the k-vector. One important implication from these results is that estimates of doping-induced strain from frequency shifts should take into account the different shift slopes of the HEM constituents and of the D^* mode itself at different excitation energies. Especially the D^* mode shift is unsuitable for such estimates and the HEM shift should be used at higher energy excitations where, the metallic tubes are not resonant.

REFERENCES

1. L. Kavan, P. Rapta, L. Dunsch, M. Bronikowski, P. Willis and R. Smalley, *J. Phys. Chem. B*, **105**, 10764 (2001).
2. M. Stoll, P. M. Rafailov, W. Frenzel and C. Thomsen, *Chem. Phys. Lett.*, **375**, 625 (2003).
3. C. P. An, Z. V. Zardeny, Z. Iqbal, G. Spinks, R. H. Baughman and A. Zakhidov, *Synth. Met.*, **116**, 411 (2001).
4. O. Lourie and H. D. Wagner, *J. Mater. Res.*, **13**, 2418 (1998).
5. S. Gupta, M. Hughes, A. H. Windle and J. Robertson, *J. Appl. Phys.*, **95**, 2038 (2004).
6. free-standing nanotube paper (mean diameter 1.3 nm) prepared by S. Roth and U. Dettlaff, MPI - Stuttgart
7. J. J. Song and D. D. L. Chung, P. C. Eklund and M. S. Dresselhaus, *Solid State Comm.* **20**, 1111 (1976).
8. J. Maultzsch, S. Reich and C. Thomsen, *Phys. Rev. B*, **65**, 233402 (2002).
9. J. Maultzsch, S. Reich and C. Thomsen, *Phys. Rev. Lett.*, **91**, 087402 (2003).
10. J. Maultzsch, S. Reich and C. Thomsen, *Phys. Rev. Lett.*, **92**, 075501 (2004).
11. J. Maultzsch, S. Reich and C. Thomsen, *Phys. Rev. B*, **64**, 121407(R) (2001).

Raman on Carbon Nanotubes Using a Tunable Laser and Comparison with Photoluminescence

A. Jorio[1], C. Fantini[1], M. de Souza[1], R. Saito[2], Ge. G. Samsonidze[4], G. Dresselhaus[5], M. S. Dresselhaus[3,4] and M. A. Pimenta[1]

[1]Departamento de Física, Universidade Federal de Minas Gerais and Instituto de Nanociências, Belo Horizonte, MG, 30.123-970 BRAZIL
[2]Department of Physics, Tohoku University and CREST JST, Sendai 980-8578, JAPAN
[3]Department of Physics, [4]Department of Electrical Engineering and Computer Science, [5]Francis Bitter Magnet Laboratory, Massachusetts Institute of Technology, Cambridge, MA 02139-4307 USA

Abstract. Stokes and anti-Stokes Resonance Raman experiments on carbon nanotubes using a tunable laser system are reported. Both arc discharge and HiPco samples are investigated with laser excitation energies varying continuously from 1.6eV up to 2.7eV. The usual features, i.e., the RBM, D-band and G-band are analyzed, as well as the intermediate frequency modes (IFMs) appearing between the RBM and the D/G band spectral region. The results are analyzed based on electron and phonon structure. Step-like dispersive behavior is observed for the IFMs and this anomalous result is analyzed based on Raman selection rules plus quantum confinement of electrons and phonons. An experimental determination of the van Hove singularity energies E_{ii} is compared with photoluminescence results. The single versus double resonance nature of the G band SWNT spectra is briefly addressed.

INTRODUCTION

In recent years, the development of spectroscopic techniques as a tool for the study and characterization of single-wall carbon nanotubes (SWNTs) has been intense. Resonance Raman [1] and joint optical absorption-photoluminescence [2] spectroscopies can be cited as the two most important techniques for SWNT study, with Raman studies dating back to 1997 [3] and photoluminescence (PL) appearing very recently, late in 2002 [2]. What makes these techniques important for the development of carbon nanotube science is the observation of experimental results at the single nanotube level. For these spectroscopic techniques, therefore, the science is not developed for "general nanotubes", but for "a specific (n,m) nanotube".

Some contradictions, however, seems to appear when systematically comparing the experimental results recently obtained by using joint optical absorption-photoluminescence techniques with previous studies using resonance Raman spectra. The fact that different kinds of samples have been used, exhibiting different diameter distributions and different nanotube environments and using different measurement techniques, makes it complicated to analyze the photophysics of carbon nanotubes quantitatively with a unified approach. This paper addresses some efforts toward merging the physics of resonance Raman and optical absorption-photoluminescence

CP723, *Electronic Properties of Synthetic Nanostructures*, edited by H. Kuzmany et al.
© 2004 American Institute of Physics 0-7354-0204-3/04/$22.00

spectroscopy, as well as gaining new insights into the resonance nature of the Raman scattering processes originating from different Raman features.

EXPERIMENTAL

Stokes and anti-Stokes Resonance micro-Raman experiments on SWNTs using a Dilor XY triple-monochromator equipped with a CCD detector are reported. Ar-Kr, HeNe, Ti:Sapphire and Dye lasers have been used to provide the excitation laser energies (E_{laser}) varying from 1.52eV up to 2.71eV. Arc discharge and HiPco samples (in bundles and wrapped with SDS in aqueous solution) have been investigated.

RESULTS AND DISCUSSIONS

Radial Breathing Modes

Figure 1 shows measurements of the radial breathing mode (RBM) features for SWNTs wrapped in SDS in aqueous solution obtained using different E_{laser} values. The profiles of SWNT families are obtained [4], similarly to optical absorption-photoluminescence results [2]. For semiconducting SWNTs, the electronic transition energies between van Hove singularities (E_{22}^S) show agreement with previous PL measurements [2]. Metallic SWNTs can be measured by resonance Raman spectroscopy, so that their E_{11}^M values can be determined. The RBM frequencies for every observed SWNT can be analyzed on the basis of the (n,m) assignment previously published [2].

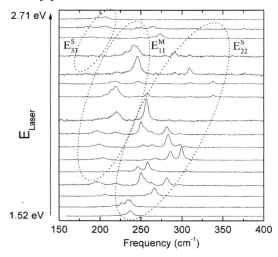

FIGURE 1. Stokes and anti-Stokes RBM spectra for SWNTs wrapped in SDS in aqueous solution obtained in the region of the Ti:Sapphire laser excitation. RBM profiles for families of SWNTs with ($2n+m$) constant are observed.

Different (n,m) assignments can be made without changing the profile of SWNT families, but by shifting entire families to fit the new (n,m) choice. The three sets of points in Figure 2(a) shows three possible (n,m) assignments for the RBM frequencies observed for HiPco SWNT samples. This figure plots how the observed values deviate from the $\omega_{RBM} = 223.5/d_t+12.5$ (dotted line), that is related to the (n,m) assignment proposed by Bachilo et al. [2]. The set of points with constant zero value corresponds to this (n,m) assignment. Open and filled curves are, respectively, for metallic and semiconducting SWNTs. Metallic SWNTs do not exhibit the same $\omega_{RBM} = 223.5/d_t+12.5$ dependence that is observed for semiconducting SWNTs.

The other two sets of points above and below the zero line correspond to a shift of the (n,m) family assignments for higher and lower diameters, respectively. The functions inside the boxes represent possible fits for these sets of points. Also plot in Fig.2 are the curves for isolated SWNTs ($\omega_{RBM} = 248/d_t$, solid line) [5] and for DWNTs ($\omega_{RBM} = 234/d_t+14$, dashed line) [6].

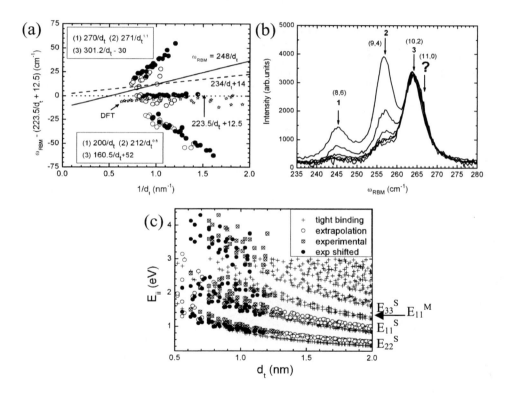

FIGURE 2. (a) Deviation from different models (lines) and (n,m) assignment choices from the $\omega_{RBM} = 223.5/d_t+12.5$ choice proposed by Bachilo et al. [2] (see text). (b) Six RBM spectra taken with different laser lines between 1.65eV and 1.74eV. The spectra were normalized to the peak at 264cm^{-1} to clearly show the absence of the expected $(11,0)$ SWNT, according to the (n,m) assignments from Ref.[2]. (c) Kataura plot based on tight binding ($\gamma_0 = 2.9$eV, $a_{C-C} = 0.144$nm) and experimental results (see text).

The lower set of points, corresponding to a shift of the (n,m) families for lower diameters has some advantages: (i) metallic and semiconducting SWNTs exhibit the same diameter dependence; (ii) there is no longer an absence of zigzag SWNTs, as proposed in Ref. [2]; Figure 2(b) shows the RBM spectra for one family of $(2n+m)$ constant. According to the assignment from Ref.[2], this family should have $2n+m = 22$, that has four (n,m) nanotubes, as displayed in Fig.2(b). Considering a family shift for $2n+m = 19$, only three SWNTs are predicted $((9,1)$, $(8,3)$, and $(7,5))$, in agreement with the observation of only three RBM peaks; (iii) the assigned electronic transition energies (E_{ii}) get closer to the previously determined tight binding values, parameterized by different optical experiments [7]. Figure 2(b) shows the change in E_{ii} considering this different (n,m) assignment. The crosses come from tight binding, the open circles marked with crosses from Raman scattering [4] and photoluminescence [2] measurements, and the open circles are the extrapolation of photoluminescence results considering a polynomial fit of the data. The filled circles correspond to the new set of E_{ii} values considering the shifted assignment to lower d_t SWNTs.

Intermediate Frequency Modes

Figure 3 shows the measurement of intermediate frequency mode (IFM) features for arc discharge SWNT bundles obtained with different E_{lasers} [8]. The dispersive behavior of these IFM features has been previously published [9]. However, it is clear from Fig.3 that the dispersion is not a continuous function of E_{laser}, but it is rather a *step-like* dispersive behavior. Raman peaks increase and decrease in intensity, while keeping a constant mode frequency. This result cannot be explained by just considering the resonance with electronic van Hove singularities, since for this diameter/E_{laser} range, the E_{ii} values for different (n,m) SWNTs lie close together. To understand this result, the presence of van Hove singularities in the *phonon* density of states must be considered. When the phonon vHSs are taken into account, a specific selective process is obtained and only a few (n,m) SWNTs contribute strongly to the Raman spectra.

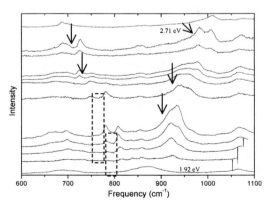

FIGURE 3. IFM features for arc discharge SWNT bundles obtained with different E_{lasers} from 1.92eV up to 2.71eV [8].

Disorder Induced (D) and tangential (G) bands

Figure 4 shows the measurements of the D and G band features for a fiber of SWNTs obtained with $E_{lasers} = 2.71eV$ [10]. The spectra are obtained at two different locations on the sample, spectrum (a) at the center of the fiber, and spectrum (b) at the edge of the fiber, where nanotube defects and impurities are expected to be abundant.

Spectrum (a) only exhibits two strong G modes, assigned as A symmetry modes, since both the incident and scattered beams are polarized along the nanotube axis [11]. The D band is two orders of magnitude lower in intensity than the G band peaks, indicating that the spectrum is dominated by a first-order single resonance Raman process.

Spectrum (b), obtained with the same sample but in a location where defects are expected to be abundant, shows the presence of many peaks within the G band lineshape, as well as a strong D band feature. This spectrum therefore exhibits D and G band disorder-induced double resonance features, as proposed by Maultzsch et al. [12]. Comparison of spectra (a) and (b) makes it possible to separate contributions from single- and double-resonance processes for the G band in carbon nanotubes, and makes it clear that the G band spectra from defect-free samples is stronger and is composed of first-order features.

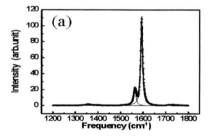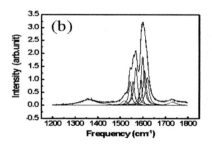

FIGURE 4. D and G band features for a fiber of SWNTs obtained at two different locations: (a) within the body of the aligned fiber; (b) at the edge of the fiber.

CONCLUSIONS

Resonance Raman measurements and comparisons with photoluminescence spectra show that there is good agreement between the electronic transition energies E_{ii} for SWNTs obtained by the two techniques on the same samples. The (n,m) assignment must obey geometrical family patterns, which help with the development of a unique (n,m) assignment. Arguments are given suggesting that the definition of assignments of specific experimental points to specific families is still open.

Observation of a *step-like* dispersive behavior for the intermediate frequency modes reveals the importance of phonon quantum confinement in turning on and off specific resonance Raman scattering processes in carbon nanotubes.

Experimental results on the G band for nondefective and defective materials shows that the G band Raman spectra from nondefective SWNTs originates from a single resonance Raman scattering process.

ACKNOWLEDGMENTS

The authors acknowledge financial support from the Instituto de Nanociênceias, Brazil, and under the NSF DMR 01-16042 grant. The authors acknowledge Prof. A. G. Rinzler, K. I. Winey and R. Haggenmueller for SWNT/poly(methyl methacrylate) composite fibers, L. O. Ladeira for arc discharge SWNTs, and M. Strano for SWNTs wrapped with SDS in aqueous solution.

REFERENCES

1. A.Jorio, M.A. Pimenta, A.G. Souza Filho *et al.*, New Journal of Physics **5**, 1.1-1.17 (2003); M.S. Dresselhaus, G.Dresselhaus, A.Jorio *et al.*, Carbon **40**, 2043-2061 (2002); And references therein.
2. S.M.Bachilo, M.S.Strano, C.Kittrell *et al.*, Science **289**, 2361 (2002); M.J.O'Connel, S.M.Bachilo, X.B.Huffman *et al.*, Science **297**, 593 (2002).
3. A.M.Rao, E.Richter, S.Bandow *et al.*, Science **275**, 187 (1997).
4. C. Fantini, A. Jorio, M. Souza *et al.* Umpublished.
5. A. Jorio et al. Phys. Rev. Letters **86,** 1118 (2001)
6. Pfeiffer et al. (this proceedings)
7. M. S. Dresselhaus and P. Eklund, Adv. in Phys. **40**, 705 (2000).
8. C. Fantini, A. Jorio, M. Souza et al. Umpublished.
9. L.Alvarez, A. Righi, S.Rols *et al.*, Chem.Phys.Lett. **320**, 441 (2000).
10. M. Souza, A.Jorio, C. Fantini *et al*. Umpublished.
11 A.Jorio, M.A. Pimenta, A.G. Souza Filho *et al.*, Phys. Rev. Lett. **90**, 107403 (2003).
12. J.Maultzsch, S.Reich, and C.Thomsen, Phys. Rev. B **65**, 233402 (2002).

Local phonon modes of single-walled carbon nanotubes observed by near-field Raman spectroscopy

A. Hartschuh*, A. J. Meixner* and L. Novotny†

*Physikalische Chemie, Universtät Siegen, 57068 Siegen, Germany
†The Institute of Optics, University of Rochester, Rochester, New York 14627, USA

Abstract. Near-field Raman spectroscopy with high spatial resolution is applied to study single-walled carbon nanotubes (SWNT) on glass. A sharp, laser illuminated metal tip acts as near-field source causing an enhanced Raman signal within close proximity of the tip. We present optical images of different Raman modes with a resolution below 20 nm. Using tip-enhanced Raman spectroscopy, different tubes structures are distinguished on the nanoscale and highly localized phonon modes are revealed.

INTRODUCTION

Raman spectroscopy has been demonstrated to provide both chemical specificity and sensitivity allowing for the investigation of the structural properties of an individual tube. By probing the radial breathing mode (RBM) frequency and making use of the resonance Raman effect a unique identification of the structural parameters of a SWNT can be achieved (see e. g. [1, 2, 3, 4]). All of this work has been done using standard confocal microscopy where the diffraction barrier limits the achievable resolution to about 300 nm. In this contribution, we use tip-enhanced Raman imaging and spectroscopy with sub 20 nm spatial resolution to perform local spectroscopy of nanometer-scale features of SWNT. The technique allows to combine very detailed spatial information with the spectral resolution required for structural selective probing of the RBM of tubes.

EXPERIMENTAL

Our experimental setup is based on a confocal microscope in combination with a probe head for shear-force detection. The laser excitation is provided by a HeNe laser operating at 632.8 nm. The laser output is converted into a radially polarized beam by passing through a mode converter [5] to provide a longitudinal field component required for the field enhancement at a metal tip [6, 7]. A more detailed description of the experimental setup can be found in [8, 9].

CP723, *Electronic Properties of Synthetic Nanostructures*, edited by H. Kuzmany et al.
© 2004 American Institute of Physics 0-7354-0204-3/04/$22.00

HIGH-RESOLUTION RAMAN IMAGING OF SWNT

Figure 1(a) shows the near-field Raman image of SWNT on glass in a 1 x 1 μm^2 scan area established by detecting the intensity of the G-band upon laser excitation at 632.8 nm. In figure 1(b), the simultaneously acquired topographic image of the sample is presented. Cross sections taken along the dashed lines in figure 1(a) and (b) are shown in (c) and (d) for the optical and the topographic image, respectively. The minimal width of the observed optical features is about 30 nm, far below the diffraction limit of light at this wavelength. Depending on the tip diameter a spatial resolution below 20 nm can be reached [8, 9, 10].

The optical and the topographic image are closely correlated and the SWNT can be easily identified in both images. However, the intensity of the detected Raman signal does not scale with the topographical height of the observed tube features as can be seen by comparing the cross sections in figure 1(c) and (d). While the strongest Raman signal occurs at about 300 nm, the corresponding height of the tube at this position is less than 1 nm. In contrast, the 5 nm heigh topographical feature at 800 nm caused presumably by a tube bundle renders a much weaker Raman signal.

The observed differences between the optical and the topographic image can be understood in terms of resonance Raman scattering [1, 11]. The Raman scattering strength of SWNT is known to be strongly dependent on the resonance Raman effect that reflects the difference between photon energies and the electronic energies of a particular tube. The resonance Raman effect enhances the signal essentially and is the reason why even single tubes can be observed. Some of the tubes that are observed topgraphically in figure 1(b) will be non-resonant at 632.8 nm and therefore do not appear in the optical image in figure 1(a).

NANOSCALE SPECTROSCOPY OF THE RADIAL BREATHING MODE

The high spatial resolution of 20 nm achieved by the tip-enhanced technique implies that the detected Raman signal originates from a sample area of 20 nm in diameter. Besides providing images of very high spatial resolution, this technique opens up the possibility to perform local spectroscopy of SWNT on the nanoscale. In the following we will show that the high spectral resolution required for Raman spectroscopy is maintained.

In figure 2(a) the topographic image of SWNT on glass is shown for a 1 x 1 μm^2 scan area. Two very thin tubes of about 0.8 nm in height are observed, one in vertical direction starting from the top and a second in horizontal direction within the lower third of the image. In the center of the image, two nearly cirular features with a height of about 5 nm are detected that appear to be particles.

The near-field Raman image of the same sample area formed by integrating the intensity detected within the range of the G-band is presented in figure 2(g). The two thin tubes observed in the topographic image do not render a detectable Raman signal and are therefore non-resonant. In the center of the scan area however, a strong G-band signal appears that is nearly uniform on a length of about 250 nm indicating the presence

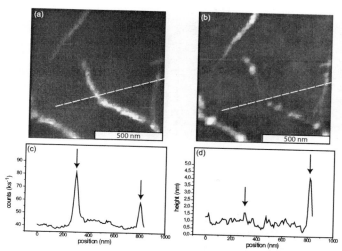

FIGURE 1. Simultaneous near-field Raman image (a) and topographic image (b) of SWNT on glass. Scan area 1 x 1 μm^2 . The Raman image is acquired by detecting the intensity of the G-band upon laser excitation at 632.8 nm. (c) Cross section taken along the indicated dashed line in the Raman image. (d) Cross section taken along the indicated dashed line in the topographic image. The arrows in (c) and (d) mark the positions of the maxima observed in the optical cross section in (c).

of an additional resonant tube.

The conventional Raman spectrum detected by confocal microscopy in the center of the scan area is shown in figure 2(b) for the spectral range of the RBM. The dashed circle in figure 2(a) indicates the confocal detection area. The Raman spectrum shows a single peak at 195 cm^{-1} with a width of about 17 cm^{-1} (FWHM) and Lorentzian lineshape indicating a single tube structure (n,m). Near-field Raman images of the same sample area are presented in figure 2(c) and (e) covering the low energy region of the RBM peak from 180 to 190 cm^{-1} for (c) and the high energy region from 190 to 200 cm^{-1} for (e). While in (c) the brightest spot occurs in the center of the scan area, in (e) it occurs in the lower part. In figure 2(d) and (f) the corresponding Raman spectra are shown detected on top of the bright spots marked in (c) and (e). Both spectra show a single RBM peak with a width of about 11 cm^{-1} but different center frequencies of 188 cm^{-1} for (d) and 197 cm^{-1} for (f). From the near-field studies we see that the RBM peak observed in the conventional spectrum in figures 2(b) results in fact from a superposition of different RBM originating from different sample regions within the confocal detection area.

Another remarkable result of the near-field measurement is that the sample regions with strong RBM signals are very small, about 20 nm for both contributions (see fig. 2(c) and (e)). Inbetween, the RBM signals disappears as can be seen from the spectrum in figure 2(h). In contrast to the highly localized RBM, the G-band intensity is nearly uniform (fig. 2(g)). A possible explanation for the observed spectral and spatial variations of the RBM would be the presence of two very short tubes of about 20 nm in length and slightly different diameters. In this case, an additional tube needs to be present inbetween the two short tubes to generate the uniform G-band signal. On the other hand,

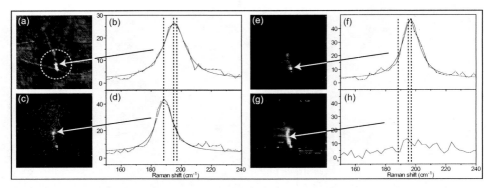

FIGURE 2. (a) Topographic image of a 1 x 1 μm^2 scan area of SWNT on glass. (b) Confocal Raman spectrum detected at the marked position in (a). The dashed circle represents the confocal detection area. (c), (e) and (g) near-field Raman images measured simultaneously with the topography in (a) established by integrating the observed Raman signal within different spectral ranges (see text). (d), (f) and (h) near-field Raman spectra observed at the positions marked in (c), (e) and (g) respectively. The dashed lines mark the spectral positions of the different maxima in (b), (d) and (f).

the strongest signal for the higher RBM frequency occurs at the position where the particles and the thin horizontal tube are observed in the topographic image (see fig. 2(a) and (e)). The presence of the particles and the tube could increase the RBM frequency of a single tube [12] and might also cause local variations in the electronic energies resulting in different resonance Raman enhancement. While the spectral information is extensive, the sample properties seem to be very complex and a clear distinction is not possible.

HIGH-RESOLUTION IMAGES OF THE END OF A SWNT

In figure 3 near-field Raman images detected at the end of a SWNT are presented. The images are formed by integrating the intensity within four selected spectral windows within a scan range of 80 nm by 80 nm. The image of the high-frequency range of the G-band from 1570 to 1640 cm^{-1} (fig. 3(c)) shows a nearly uniform intensity and the spatial extension of the SWNT can be clearly seen. For the other spectral windows, a significantly different contrast is observed. The signal in the low-frequency range of the G-band from 1500 to 1570 cm^{-1} is strongest at the very end of the SWNT (fig. 3(b)) whereas the G'-band from 2565 to 2670 cm^{-1} vanishes towards the end (fig. 3(c)). The D-band from 1260 to 1380 cm^{-1} (fig. 3(a)) shows a weak increase at the end.

A similar image contrast regarding the low-frequency range of the G-band was observed for some other SWNT, but not in all cases. It is unclear if in all cases the respective tube ends are opened or capped. In addition to the existing models describing the electronic density of states used to qualify results achieved by STS and STM (see e.g. [13]), theoretical models for the local phonon modes for differently terminated tubes need to be developed.

166

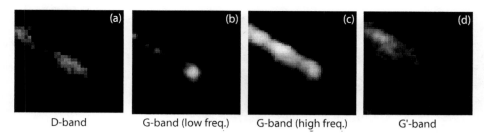

| D-band | G-band (low freq.) | G-band (high freq.) | G'-band |

FIGURE 3. Raman images detected for selected spectral windows at the end of a SWNT. Scan range 80 x 80 nm^2. (a) D-band intensity from 1260 to 1380 cm^{-1}. (b) low-frequency region of the G-band (1500 to 1570 cm^{-1}). (c) high-frequency region of the G-band (1570 to 1640 cm^{-1}). (d) G'-band intensity from 2565 to 2670 cm^{-1}.

SUMMARY

In this contribution, we used tip-enhanced Raman imaging and spectroscopy with a spatial resolution of sub 20 nm to study SWNT on glass. Nanoscale spectroscopy of the radial breathing mode revealed structural variations that are hidden in confocal measurements. Images taken at the end of a SWNT show a spectroscopic contrast indicating localized phonon modes.

ACKNOWLEDGMENTS

The authors wish to thank N. Anderson, H. Qian and G. Schulte for experimental support. This research was supported by AFOSR through grant F-49620-03-1-0379.

REFERENCES

1. Rao, A. M., Richter, E., Bandow, S., Chase, B., Eklund, P. C., Williams, K. A., Fang, S., Subbaswamy, K. R., Menon, M., Thess, A., Smalley, R. E., Dresselhaus, G., and Dresselhaus, M., *Science*, **275**, 187 (1997).
2. Jorio, A., Saito, R., Hafner, J. H., Lieber, C. M., Hunter, M., McClure, T., Dresselhaus, G., and Dresselhaus, M., *Phys. Rev. Lett.*, **86**, 1118 (2001).
3. Kuzmany, H., Plank, W., Hulman, M., Kramberger, C., Grüneis, A., Pichler, T., Peterlik, H., Kataura, H., and Achiba, Y., *Eur. Phys. J. B*, **22**, 307 (2001).
4. Reich, S., Thomsen, C., and Maultzsch, J., *Carbon nanotubes*, Wiley-VCH, Weinheim, 2004.
5. Quabis, S., Dorn, R., Glöckl, O., Eberler, M., and Leuchs, G., *Opt. Comm.*, **179**, 1 (2000).
6. Novotny, L., Sánchez, E. J., and Xie, X. S., *Ultramicroscopy*, **71**, 21 (1998).
7. Bouhelier, A., Beversluis, M., Hartschuh, A., and Novotny, L., *Phys. Rev. Lett.*, **90**, 013903 (2003).
8. Hartschuh, A., Beversluis, M. R., Bouhelier, A., and Novotny, L., *Philosophical Transactions: Mathematical, physical and engineering science*, pp. 807–819 (2004).
9. Hartschuh, A., Sánchez, E. J., Sunney, X. S., and Novotny, L., *Phys. Rev. Lett.*, **90**, 095503 (2003).
10. Hartschuh, A., and Novotny, L., *Int. J. Nanotechnol., submitted* (2004).
11. Jiang, C., Zhao, J., Therese, H. A., Friedrich, M., and Mews, A., *J. Phys. Chem. B*, **107**, 8742 (2003).
12. Henrard, L., Hernandez, E., Bernier, P., and Rubio, A., *Phys. Rev. B*, **60**, 8521–8524 (1999).
13. Rubio, A., *Appl. Phys. A*, **68**, 275–282 (1999).

Vibrational spectromicroscopy of graphite and carbon nanotubes

L. Vitali, M.Burghard, M. A. Schneider, and K. Kern

Max-Planck Institute for solid state research, Stuttgart, Germany

Abstract. The vibrational properties of highly oriented pyrolitic graphite (HOPG) and single wall carbon nanotubes have been probed locally with atomic scale resolution by Inelastic Electron Tunneling Spectroscopy (IETS). The observed spectral features are in very good agreement with the calculated vibrational density of states (vDOS). Furthermore, the high spatial resolution of Scanning Tunneling Microscopy (STM) has allowed the unraveling of changes in the local phonon spectrum related to topological defects.

INTRODUCTION

The local vibrational density of states of surface supported nanostructures become accessible by inelastic electron tunneling spectroscopy (IETS) with a scanning tunneling microscope (STM). A few percent of the tunneling electrons are capable to excite vibrational modes of the sample under investigation increasing the total conductance at the onset of the excitation. The IETS-STM spectrum is obtained by recording with a lock-in technique the second harmonic component of the tunneling current, which is proportional to d^2I/dV^2. This quantity shows a peak whenever the energy of the tunneling electrons is in resonance with the energy of an inelastic excitation. In this way, the local vibrational density of states of single molecules, ranging in size from acetylene to C_{60}, adsorbed on metallic surfaces has become accessible [1, 2, 3].

In this paper, IETS-STM spectra obtained on HOPG and SWCNTs are presented. The direct comparison of the experimental data and theoreticaly predicted phonon modes of graphite shows the close similarity of the observed IETS-STM spectrum and the total density of vibrational states (vDOS). On isolated carbon nanotubes adsorbed on Au(111), the radial breathing mode (RBM) is detected locally, and its linear dependence on the inverse of the tube diameter is proven. By mapping the spatial dependence of this phonon feature across topological defects, such as an intermolecular junction and a tube cap, we demonstrate the ultimate spatial resolution of this technique.

CP723, *Electronic Properties of Synthetic Nanostructures*, edited by H. Kuzmany et al.
© 2004 American Institute of Physics 0-7354-0204-3/04/$22.00

RESULTS AND DISCUSSION

The inelastic tunneling spectrum of HOPG, shown in figure 1a, was measured over a topographically clean and flat terrace with a lock-in technique. The observed features are exclusively due to inelastic processes. Electronic contributions in the spectrum can be excluded since the local density of states of HOPG is constant in this energy range. Figure 1b shows the calculated phonon density of states of graphite obtained from first principles DFT calculations [4]. The IETS-STM spectrum appears to be directly proportional to the total density of phonon states. The experimentally observed inelastic features, in the energy range from the rigid layer shear mode at ≈ 5meV to the optical modes at ≈ 200 meV, can be clearly assigned to specific vibrational modes. In the energy range between 120 and 150meV the agreement between theory and experiment is less apparent. This discrepancy can be traced to a possible underestimation of the shear horizontal (SH) phonon branch in the theoretical calculations [4]. A phonon assisted tunneling process can also contribute to the enhanced phonon modes at the K point of the surface Brillouin zone [4].

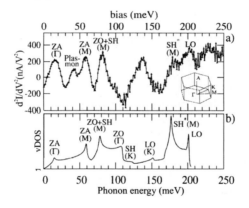

FIGURE 1. a) Inelastic electron tunneling spectrum of HOPG measured by STM at 6K. b) vibrational density of states of graphite calculated by DFT-LDA. The phonons at high symmetry points of the Brillouin zone (see insert) are specified as following: out of plane acoustic (ZA), acoustic shear (SH), longitudinal acoustic (LA), out of plane optical (ZO), optical shear (SH*), and longitudinal optical (LO). Furthermore, a plasmon excitation is identified at 40 meV [4].

In addition to the predicted phonon modes, the measured spectrum contains an energy loss feature at 40 meV, which can be associated to an out of plane oscillation of the electrons. The overlapping of the π and π^* band at at the Fermi level allows for such a low-energy plasmon excitation observed also with HREELS [5] at 6K.

In the following we show that the influence of localized defects on the vibrational modes of carbon nanotubes can be studied with IETS-STM. Individual nanotubes were studied, and the tube chiral index (n, m) was identified as previously shown [6]. Carbon nanotubes with no defects show a well-defined radial breathing mode (RBM) at low energies (Fig. 2a). The phonon energy is found, to scale inversely with the diameter (d_t) as documented in Fig.2b.

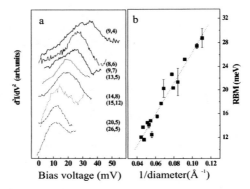

FIGURE 2. a) IETS-STM spectra of different isolated SWCNTs measured at 6K. b) Energy of the RBM mode of isolated SWCNTs plotted against the inverse diameter. It can be seen that the RBM energy scales inversely with the tube diameter, according to the relationship $E_{RBM} = A/d_t$, where $A = 262$ meV×Å.

The linear dependence of the RBM frequency on reciprocal tube diameter is well-documented from Raman spectroscopic studies on isolated SWCNTs and is here proven by the local characterization of individual tubes [7]. The RBM energies obey the relation $E_{RBM} = A/d_t$, where $A = 262$ meV×Å, in general agreement with values obtained by other theoretical calculations [8, 9, 10, 11].

Of particular interest is the detection of changes in the vibrational spectrum that arise from the presence of defects. Figure 3a shows the topographic image of an isolated metallic (17,14) tube which was selected for detailed spectroscopic investigations. The presence of an intramolecular junction connecting two different tubes can be identified by the abrupt reduction in tube diameter which decreases from 21.2 Å to 9.7 Å while the chiral angle is preserved. At a distance of ~5 nm from the intermolecular junction, the tube ends with a highly curved cap. Intermolecular junction and closed tube ending require the presence of at least one pentagon-heptagon pair and of six pentagons, respectively. The short tube segment between the intermolecular junction and the tube apex, denoted hereafter as "neck region", has a length of 2 nm.

The structural changes deduced from an inspection of the topographic image are further corroborated by the simultaneously acquired electronic LDOS. In particular, a peak in the differential conductance ascribed to a resonant state in tube cap region [12, 13], has been observed [6].

In Fig. 3b the RBM measured as a function of position along the (17,14) tube axis (indicated as white dashed line in Fig. 3a) is displayed as a gray scale map. On top of the (17,14) tube, the RBM is found at a constant energy of ~12 meV. One further observes the abrupt disappearance of this RBM at the intramolecular junction. Although one would expect a RBM to occur at 25 meV in the neck region according to its diameter, no such mode is evident. However, when the cap region is reached, a new vibrational feature appears at 29 meV.

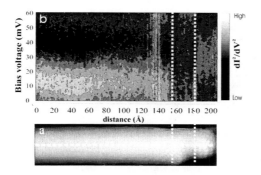

FIGURE 3. IETS-STM spectrum of a SWCNT across topological defects (a) Topographic image of a (17,14) tube containing an intramolecular tube junction and a closed ending. The white dashed line indicates the sampling positions along the tube axis where the IETS spectra were taken. (b) Gray scale map of the lock-in signal (d^2I/dV^2) as a function of energy and position along the tube axis. The white dotted lines delimit the different tube regions and highlight the changes in the vibrational spectrum. Data in (a) and (b) are taken at 6K.

A theoretical study aimed to verify the influence of the cap and of the finite length on the low frequency lattice dynamics of the tube was carried out on a model system [6]. This study revealed for a (5,5) tube capped at both ends by C_{60} hemispheres, a pronounced influence of the caps on the tube radial displacement. The RBM mode of the tube abruptly drops to zero in proximity of the cap region. A resonance mode between the "RBM" of the cap and a bulk (tube) mode appears at the tube apex. Furthermore, this study revealed that for (5,5) capped tubes of different lengths the tube is able to sustain a RBM only if its length exceeds 3.5 nm [6]. This scenario is completely reflected in the experimental observations that a cap region and a pentagon-heptagon pair suppress the RBM mode in their proximity. The neck region of the (17,14) tube shown in Fig. 2 behaves like a short capped tube.

For future studies, STM-IETS offers the unique possibility to correlate local vibrational changes arising from structural defects and modifications with the local electronic structure. For instance, the study of tube deformations like kinks or twists could deepen the understanding of the effect of stress fields on the electrical transport properties of SWCNTs. Moreover, the atomic scale structure of chemically functionalized nanotubes could be investigated with chemical specificity if characteristic group vibrations are selected.

REFERENCES

1. N.Lorente, and M.Persson, *Physical Review Letter*, **85**, 2997 (2000).
2. B.C.Stipe, M.A.Razei, and W.Ho, *Science*, **280**, 1732 (1998).
3. I.Pascual, J., Got'mez-Herrero, J., D.Sat'nchez-Portal, and H.-P.Rust, *J.Chem. Phys.*, **117**, 9531 (2002).

4. L.Vitali, M.A.Schneider, K.Kern, L.Wirtz, and A.Rubio, *Physical Review B*, **69**, R121414 (2004).
5. E.T.Jensen, R.E.Palmer, W.Allison, and J.F.Annett, *Physical Review Letter*, **66**, 492 (1991).
6. Vitali, L., Burghard, M., M.A.Schneider, Liu, L., Jayanthi, C., and Kern, K., *Phyiscal Review Letter*, **93**, accepted (2004).
7. Dresselhaus, M., Dresselhaus, G., Jorio, A., Filho, A. S., and Saito, R., *Carbon*, **40**, 2043 (2002).
8. Rao, A. M., Richter, E., Bandow, S., Chase, B., Eklund, P. C., Williams, K. A., Fang, S., Subbaswamy, K. R., Menon, M., Thess, A., Smalley, R. E., Dresselhaus, G., and Dresselhaus, M. S., *Science*, **275**, 187–191 (1997).
9. Jishi, R. A., Venkataraman, L., Dresselhaus, M. S., and Dresselhaus, G., *Chem. Phys. Lett.*, **209**, 77–82 (1993).
10. Kuerti, J., Kresse, G., and Kuzmany, H., *Phys.Rev.B*, **58**, R8869 (1998).
11. Sanchez-Portal, D., E.Artacho, Soler, J., A.Rubio, and P.Ordejon, *Physical Review B*, **59**, 12678 (1999).
12. Kasahara, Y., Tamura, R., and Tsukada, M., *Physical Review B*, **67**, 115419 (2003).
13. Yaguchi, T., and Ando, T., *J.Phys.Soc.Jpn*, **71**, 2224 (2002).

Atomically Clean Integration of Carbon Nanotubes with Silicon

P. M. Albrecht and J. W. Lyding

Department of Electrical and Computer Engineering and Beckman Institute, University of Illinois,
405 N. Mathews Ave., Urbana, IL 61801, USA

Abstract. We have optimized an *in situ* dry contact transfer (DCT) technique for the ultra-clean deposition of single-walled carbon nanotubes (SWNTs) onto H-passivated and clean Si(100) surfaces. Following DCT in ultrahigh vacuum (UHV), a scanning tunneling microscope (STM) is used to elucidate the topology and electronic structure of individual SWNTs and the Si substrate. Moreover, we demonstrate controlled SWNT manipulation with the STM probe. Reproducible DCT of mostly isolated SWNTs with minimal contamination is extended to UHV-cleaned Si(100)-(2×1), a reactive surface incompatible with ambient solution-phase processing.

INTRODUCTION

Previous atomic-scale studies of single-walled carbon nanotubes (SWNTs) with the ultrahigh vacuum scanning tunneling microscope (UHV-STM) incorporated a variety of supporting substrates, including Au(111) [1,2], highly oriented pyrolytic graphite (HOPG) [3], and SiC [4]. In the case of SWNTs on Au(111), atomically-resolved STM combined with scanning tunneling spectroscopy (STS) confirmed that the SWNT electronic structure is highly sensitive to subtle fluctuations in diameter and chiral angle. SWNT-substrate registration and preferential nanotube alignment were investigated on HOPG and SiC, respectively.

Our UHV-STM investigation is focused upon SWNTs directly interfaced with Si(100). We recently demonstrated the *in situ* dry contact transfer (DCT) of SWNTs to Si(100)-(2×1):H [5]. This deposition method avoids solution-phase processing of SWNTs and precludes ambient degradation of the H-terminated Si(100) surface [6]. Furthermore, the formation of a virtually pristine interface between predominantly isolated SWNTs and Si(100) was consistent over many experimental trials employing the UHV DCT technique. Both semiconducting and metallic SWNTs were subsequently identified by room temperature STS, with spectral features unique to the nanotube evident within the substrate band gap.

At present, we show STM topographic and current images of an isolated SWNT with atomic resolution. We report both current-voltage (*I-V*) and conductance-voltage (*dI/dV-V*) spectra for a semiconducting SWNT supported by Si(100)-(2×1):H. In addition, precise manipulation of a 15-nm-long SWNT is achieved using the STM tip. Finally, we extend the DCT technique to the deposition of SWNTs onto Si(100)-(2×1), a clean surface that cannot withstand *ex situ* wet chemical processing.

CP723, *Electronic Properties of Synthetic Nanostructures*, edited by H. Kuzmany et al.
© 2004 American Institute of Physics 0-7354-0204-3/04/$22.00

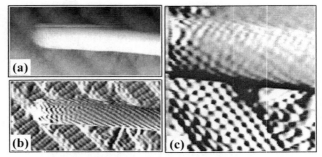

FIGURE 1. (a) 160 × 84 Å^2 STM topographic image showing an isolated SWNT spanning several terraces of a vicinal Si(100)-(2×1):H surface. Tunneling parameters were -1.50 V, 25 pA. (b) STM current image acquired simultaneously with the STM topograph shown in (a), revealing the chiral lattice of the nanotube. (c) 66 × 66 Å^2 STM topographic image showing simultaneous atomic resolution of the SWNT and Si substrate. Tunneling parameters were +1.20 V, 20 pA.

EXPERIMENTAL

Si(100)-(2×1) surfaces were prepared *in situ* from *p*-type (B doped, 5-20 mΩ-cm) samples using a standard recipe [7], and the subsequent H-passivation procedure was outlined previously [5]. Room temperature UHV-STM was performed with a homebuilt microscope [8] at a base pressure of 7×10^{-11} Torr. W and Pt/Ir STM probes were used interchangeably.

In the UHV DCT method, a fiberglass applicator [9] is first prepared with a visible coating of as-produced HiPco [10] SWNTs. The applicator is subjected to a mild degas upon entry into the UHV chamber. Nanotube deposition occurs upon gentle mechanical contact between the applicator and a freshly prepared Si(100) surface.

I-V curves are recorded by momentarily interrupting the feedback loop and sweeping the substrate bias with the tip held fixed. *dI/dV-V* data is obtained by lock-in technique, with a 10 kHz, 100 mV$_{rms}$ ac signal added to the dc tunneling bias.

STM AND STS OF ISOLATED SWNTS ON Si(100)-(2×1):H

Figure 1(a) shows a 160 × 84 Å^2 STM topograph of an isolated SWNT spanning several terraces of a vicinal Si(100)-(2×1):H surface. One end of the tube is visible near the left edge of the scan window. The current image [Fig. 1(b)] was acquired in parallel with the topographic data, and reveals the SWNT chiral lattice and the Si dimers. The absence of surface contamination is attributed to the cleanliness of the *in situ* DCT. Figure 1(c) shows a 66 × 66 Å^2 STM current image, with simultaneous atomic resolution achieved for the Si substrate and the adsorbed SWNT.

A 154 × 63 Å^2 STM topograph from a different DCT experiment is shown in Fig. 2(a), depicting an individual SWNT and the proximal Si(100)-(2×1):H substrate. Rich atomic-scale detail is provided by the complementary current image [Fig. 2(b)]. Tunnel current (I_{tunnel}) versus substrate bias is plotted in Fig. 2(c) for the SWNT and the *p*-type Si, respectively. The conduction (empty states) and valence (filled states) band edges of the SWNT are both evident within the Si band gap.

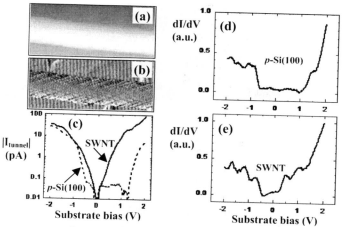

FIGURE 2. (a) 154 × 63 Å² STM topograph showing an individual SWNT adsorbed onto Si(100)-(2×1):H. Tunneling parameters were -1.70 V, 15 pA. (b) STM current image recorded in parallel with (a). (c) *I-V* spectra for the *p*-Si(100) substrate (dashed) and SWNT (solid) depicted in (a) and (b). (d) *dI/dV* for the substrate, showing *p*-type semiconducting character in agreement with the *I-V* curve shown in (c). (e) *dI/dV* recorded for the SWNT, consistent with a semiconducting tube.

dI/dV-V spectra for the H-passivated Si(100) substrate and the SWNT are shown in Figs. 2(d) and 2(e), respectively. The *p*-type character of the substrate *dI/dV* is in agreement with the independently measured dc *I-V* curve depicted in Fig. 2(c). The nanotube *dI/dV* reveals three peaks in the valence band (-1.7, -1.2, and -0.7 V) and two peaks in the conduction band (0.5 V and 1 V), in addition to a low bias conductance gap. Consequently, we identify this SWNT as a semiconductor. The peaks at -0.7 V and 0.5 V are ascribed to the first van Hove singularities in the valence and conduction band (comprising the E_{11} transition), respectively. Similarly, the next set of peaks (-1.2 V and 1 V) correspond to the E_{22} transition. An unambiguous interpretation of the SWNT *dI/dV* peak at -1.7 V is elusive in the context of a semiconducting tube [11], and is potentially an artifact of the -1.8 V peak apparent in the substrate *dI/dV*. Notably, the onset of the SWNT conductance inside the band gap of the Si substrate is consistent with the independent dc *I-V* measurement. This phenomenon is routinely observed in STS of semiconducting SWNTs on Si(100) [5].

SWNT MANIPULATION WITH THE UHV-STM

Avouris *et al.* demonstrated the precise manipulation of multi-walled carbon nanotubes (MWNTs) on H-passivated Si surfaces using an atomic force microscope (AFM) [12]. For a ~10-nm-diameter MWNT, a binding energy of ~0.8 eV/Å was determined. The magnitude of van der Waals interaction was sufficient to stabilize MWNTs in strained conformations induced by the AFM. We have extended this technique to the manipulation of SWNTs on Si(100)-(2×1):H with the UHV-STM.

A 600 × 600 Å² STM topograph containing four SWNTs is shown in Fig. 3(a). Three of the tubes are butted together near the bottom of the image, and the shortest of

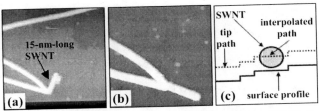

FIGURE 3. (a) 600×600 Å^2 STM topograph showing four SWNTs and the Si(100)-(2×1):H substrate. Tunneling parameters were -2 V, 50 pA. (b) 410×410 Å^2 scan after manipulation of the 15-nm-long SWNT, showing a 95° clockwise rotation of the tube compared to (a). (c) Schematic depiction of the STM manipulation routine. During manipulation, the tip is not retracted upon encountering the SWNT.

the three SWNTs served as the subject of the STM manipulation experiment. Figure 3(b) focuses upon the aforementioned region after manipulation, revealing a 95° clockwise rotation of the 15-nm-long SWNT relative to its orientation in Fig. 3(a).

First, the STM tip was scanned in constant-current mode along a 27 nm path that included the midpoint of the 15-nm-long SWNT. The tunneling parameters were -2 V and 0.1 nA, and the contour of surface topography was stored in memory. Next, the feedback loop was disabled, the tip was positioned 5 Å closer to the surface, and the stored contour was traced out with 0 V sample bias and 100 Å/s tip velocity. However, the tip trajectory in close proximity to the nanotube was modified to an interpolated path, as illustrated schematically in Fig. 3(c). By inhibiting tip retraction upon approaching the left sidewall of the SWNT, tip-nanotube interactions are enhanced and the SWNT is displaced laterally across the H-passivated Si substrate.

UHV DCT AND STM OF SWNTS ON CLEAN Si(100)-(2×1)

The fabrication of a pristine interface between SWNTs and clean semiconducting surfaces, including Si(100)-(2×1) and cleaved GaAs(110), requires an alternative to the conventional solution deposition of SWNTs in an ambient environment. UHV DCT is one potential approach for surfaces unable to withstand ambient exposure.

Figure 4 shows a 375×180 Å^2 STM topographic image of an isolated SWNT and a single terrace of the UHV-cleaned Si(100)-(2×1) surface after *in situ* DCT. A faint artifact adjacent to the primary nanotube feature is attributed to multiple tip imaging.

FIGURE 4. 375×180 Å^2 STM topograph showing an individual SWNT and a single terrace of the clean Si(100)-(2×1) surface following *in situ* DCT. Multiple tip imaging is the cause of the faint tube artifact parallel to the true SWNT feature. Tunneling conditions were -1.75 V, 0.1 nA.

The STM investigation of SWNTs on clean Si(100) is relevant to the interpretation of nanolithography experiments on H-passivated Si(100) [13], where a short (~4 nm) section of the nanotube was interfaced directly with the depassivated Si substrate. Interestingly, *ab initio* density functional calculations by Orellana *et al.* predict the formation of chemical bonds between preferentially aligned SWNTs and Si(100) [14].

CONCLUSION

Dry contact transfer (DCT) of SWNTs in UHV enables atomically-resolved STM and STS of pristine SWNT/Si(100)-(2×1):H surfaces. The extension of DCT to clean Si(100)-(2×1) reinforces the generality of the deposition method. Engineering the SWNT/Si(100) interface with atomic-scale precision may provide a new approach towards modifying the mechanical and electronic properties of carbon nanotubes.

ACKNOWLEDGMENTS

Funding was provided by the Office of Naval Research under grant N000140310266. PMA acknowledges support from a NDSEG graduate fellowship.

REFERENCES

1. J. W. G. Wildöer, L. C. Venema, A. G. Rinzler, R. E. Smalley, and C. Dekker, *Nature* (London) **391**, 59-62 (1998).
2. T. W. Odom, J.-L. Huang, P. Kim, and C. M. Lieber, *Nature* (London) **391**, 62-64 (1998).
3. C. Rettig, M. Bödecker and H. Hövel, *J. Phys. D: Appl. Phys.* **36**, 818-822 (2003).
4. V. Derycke, R. Martel, M. Radosavljević, F. M. Ross, and Ph. Avouris, *Nano Lett.* **2**, 1043-1046 (2002).
5. P. M. Albrecht and J. W. Lyding, *Appl. Phys. Lett.* **83**, 5029-5031 (2003).
6. P. M. Albrecht, R. M. Farrell, W. Ye, and J. W. Lyding, *Third IEEE Conference on Nanotechnology* **1**, 327-330 (2003).
7. B. S. Swartzentruber, Y.-W. Mo, M. B. Webb, and M. G. Lagally, *J. Vac. Sci. Technol. A* **7**, 2901-2905 (1989).
8. J. W. Lyding, S. Skala, J. S. Hubacek, R. Brockenbrough, and G. Gammie, *Rev. Sci. Instrum.* **59**, 1897-1902 (1988).
9. #16 Natural Fiberglass Sleeving (SPC4917), SPC Technology, Chicago, IL.
10. HiPco single-walled carbon nanotubes were purchased from Carbon Nanotechnologies, Inc.
11. For a semiconducting SWNT, the theoretical energy separation between the second and third van Hove singularities (vHS) in the valence band is twice the spacing between the first and second vHS. An expression for the latter is given by $a_{C-C}\gamma_0/d_{NT}$, where a_{C-C}=0.142 nm is the nearest-neighbor C-C distance, γ_0 is the C-C tight-binding overlap energy (typically taken to be ~2.9 eV), and d_{NT} is the diameter of the semiconducting SWNT.
12. Ph. Avouris, T. Hertel, R. Martel, T. Schmidt, H. R. Shea, and R. E. Walkup, *Appl. Surf. Sci.* **141**, 201-209 (1999).
13. P. M. Albrecht and J. W. Lyding, *Superlattice Microst.*, in press (2004).
14. W. Orellana, R. H. Miwa, and A. Fazzio, *Phys. Rev. Lett.* **91**, 166802 (2003).

FUNCTIONALIZATION OF CARBON NANOTUBES

NMR on cesium intercalated carbon nanotubes

M. Schmid[1,2,3], C. Goze-Bac[2], M. Mehring[3],
S. Roth[1], P. Bernier[2]

[1]Max-Planck-Institut für Festkörperforschung, Heisenbergstr. 1, D-70569 Stuttgart, Germany
[2]GDPC, Univ. Montpellier II, F-34095 Montpellier cedex5, France
[3]2. Physikal. Inst., Universität Stuttgart, Pfaffenwaldring 57, D-70550 Stuttgart, Germany

Abstract. Intercalation of single wall carbon nanotube (SWNT) bundles with alkali metals is expected to modify the electronic band structure and to raise the Fermi level. We report results from temperature dependent ^{13}C- and [133]Cs-NMR measurements on Cs intercalated SWNT. Cs was reversibly intercalated with different stoichiometries. NMR lineshapes as well as relaxation effects are studied and discussed in context of dynamics of alkali ions in SWNT bundles. The results are compared with structural simulations of Cs-ions intercalated in SWNT.

INTRODUCTION

Single wall carbon nanotubes (SWNT) are being investigated in many fields of scientific research. Intercalation of alkali has been shown to modify the intrinsic electronic and mechanical properties of SWNT [1, 2, 3, 4]. Potential applications are the use of Li intercalated SWNT in secondary-ion batteries or the use of cesium intercalated nanotubes in field emitter devices. However, the detailed spatial structure and dynamics of most intercalation species in the nanotube lattice host is yet unknown. Only from intercalation of iodine into SWNT bundles the formation of well defined linear iodine chains has been reported [5]. As nuclear magnetic resonance (NMR) is able to provide detailed information about interaction between nuclei and their environments, we performed [133]Cs-NMR measurements on cesium intercalated SWNT bundles. NMR should be able to reveal the cesium dynamic properties as well as the structural properties of such a system.

SAMPLE PREPARATION

Electric arc SWNT were Cs intercalated in the vapor phase by applying a temperature gradient between a cesium source and SWNT. We used as grown nanotubes, as any purification process is supposed to induce defects to SWNT in an uncontrollable way. In analogy to graphite intercalation compounds (GIC), we expect a saturated intercalation phase with a stoichiometry of about CsC_{7-8}. After intercalation, the sample was sealed in NMR glass tubes under vacuum. As the SWNT remain chemically or mechanically untreated, we expect most nanotubes to have closed end caps. Therefore the only possibility for the Cs atoms is intercalation in the interstitial sites of a nanotube bundle.

CP723, *Electronic Properties of Synthetic Nanostructures*, edited by H. Kuzmany et al.
© 2004 American Institute of Physics 0-7354-0204-3/04/$22.00

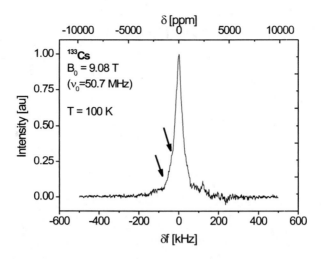

FIGURE 1. ^{133}Cs NMR at magnetic field of 9 T and a temperature of 100 K. The arrows indicate quadrupolar satellites.

EXPERIMENTAL

We performed our ^{133}Cs NMR experiments using a home built pulsed NMR spectrometer at a magnetic field of 9 T. The ^{133}Cs-frequency was 57 MHz. We used a solid-echo $(x - y)$ pulse sequence of $\frac{\pi}{2} - \beta$. β was chosen to be a 30°-pulse, as cesium is a quadrupolar nucleus with spin 7/2 and the maximum echo height is obtained for that rotation angle β. So all NMR transitions are properly excited [6]. Temperature dependent spectra between 100 K and 295 K were taken by a Fourier transform of the second half of the echoes.

In Fig. 1 a typical ^{133}Cs low temperature spectrum of cesium intercalated SWNT is shown. The ^{133}Cs central transition is located at 0 ppm (referenced to CsNO$_3$). As indicated by arrows, two satellite transitions are observed with a quadrupolar spacing of about 70 kHz. At higher temperatures the satellites get averaged (not shown here) and a narrowing of the central line occurs. The line position at 0 ppm gives evidence for a fully ionized state of the cesium atoms. As reported in [7], we were able to determine the charge transfer from cesium to the carbon nanotube host and calculate the density of states at the Fermi level. The appearance of quadrupolar satellite transitions is a direct indication for an ordered cesium structure in the sample as a disordered Cs phase would completely average out those satellite transitions and a featureless broad line would appear. For GIC (CsC$_{24}$) a similar satellite spacing of about 76 kHz at 115 K and 74 kHz for CsC$_8$ at 1.3 K was reported [8]. So, the electric field gradient (EFG) is of the same order of magnitude in our cesium intercalated SWNT as in cesium GICs. At higher temperatures, the EFG gets averaged due to the onset of dynamics of the cesium ions and the quadrupolar structure disappears.

Fig. 2 shows the homogeneous spin-spin relaxation rate (T_2) versus temperature. At

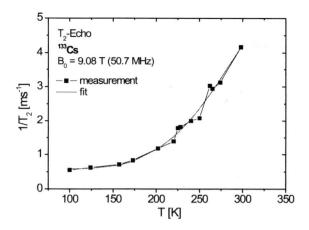

FIGURE 2. ^{133}Cs spin-spin relaxation rates versus temperature. Solid line: fit assuming an Arrhenius like activated dynamic process.

all temperatures the spin-spin relaxation consists of a distribution of relaxation rates with two main components that follow exactly the same temperature dependence. With increasing temperature, T_2 increases constantly up to 300 K. At 300 K the free induction decay (FID) of the second β pulse starts to become dominant and overlay our solid echo. So at this temperature, a clean analysis of the spin-spin relaxation rate cannot be performed any more. The solid line in fig. 2 is a fit using an Arrhenius like activated process. From the fit we can estimate an activation energy for a cesium dynamic process of about $E_A \sim 0.1$ eV. As we are sure that the nanotubes have closed caps, we assign this dynamic process to cesium diffusing in the interstitial sites or hopping via vacancies in the SWNT bundles. As at higher temperatures the dynamics become very strong, the solid-echo pulse sequence is not capable any more to create a proper echo. As observed, around 300 K the FID starts to dominate.

SUMMARY

^{133}Cs NMR was performed on cesium intercalated SWNT. The low temperature ^{133}Cs spectrum consists of a line centered around 0 ppm indicating full ionization of the cesium atoms. The line shows a quadrupolar structure with a satellite spacing of about 70 kHz at 100 K. As this value is comparable to the satellite spacing in cesium GICs we deduce a similar EFG in our sample. However, it is yet unclear if a similar structural cesium atom arrangement is possible in SWNT as in cesium GIC. From T_2-relaxation measurements we deduce a cesium dynamic process with an activation energy of about $E_A \sim 0.1$ eV. We assign this process to hopping of cesium ions via vacancies in the SWNT bundle.

ACKNOWLEDGMENTS

We thank J. Conard and P. Lauginie for helpful discussions and the FUNCARS (No. HPRN-CT-1999-00011) European Project for funding and support.

REFERENCES

1. Suzuki, S., Maeda, F., Watanabe, Y., and Ogino, T., *Physical Review B*, **67**, 115418–115424 (2003).
2. Sauvajol, J.-L., Bendiab, N., Anglaret, E., and Petit, P., *Comptes Rendus Physique*, **4**, 1035–1045 (2003).
3. Liu, X., Pichler, T., Knupfer, M., and Fink, J., *Physical Review B*, **67**, 125403–125411 (2003).
4. Jeong, G.-H., Farajian, A. A., Hirata, T., Hatakeyama, R., Tohji, K., Briere, T. M., Mizuseki, H., and Kawazoe, Y., *Thin Solid Films*, **435**, 307–311 (2003).
5. Fan, X., Dickey, E. C., Eklund, P., Williams, K. A., Grigorian, L., Buczko, R., Pantelides, S. T., and Pennycook, S. J., *Phys. Rev. Lett.*, **84**, 4621–4624 (2000).
6. Kanert, O., and Mehring, M., *Static Quadrupole Effects in Disordered Cubic Solids*, vol. 3 of *NMR-Basic Principles and Progress*, 1971.
7. Schmid, M., Mehring, M., Roth, S., Bernier, P., and Goze-Bac, C., *Mater. Res. Soc. Symp. Proc.*, **772**, M.3.4.1 (2003).
8. Estrade-Szwarckopf, H., Conard, J., Lauginie, P., Klink, J. V. d., Guerard, D., and Lagrange, P., "133Cs NMR studies of cesium GIC," in *Physics of Intercalation Compounds*, Springer series in Solid State Science, 1981, vol. 38, p. 274.

Purification and dispersion of carbon nanotubes by sidewall functionalization with single-stranded DNA

Sebastian Taeger*, Oliver Jost†, Wolfgang Pompe† and Michael Mertig*

*Max Bergmann Zentrum für Biomaterialien, Technische Universität Dresden, D-01069 Dresden, Germany
†Institut für Werkstoffwissenschaft, Technische Universität Dresden, D-01062 Dresden, Germany

Abstract. We report a simple and fast procedure to disperse single-wall carbon nanotubes (SWCNT) in an aqueous solution, using single-stranded DNA (ssDNA) as a dispersing agent. The procedure waives aggressive chemicals and provides the additional benefit of sidewall functionalization of the CNT with DNA. Atomic force microscopy (AFM) investigations of SWCNT from this suspension showed a high degree of isolation of tubes from the bundles present in the raw material and also a removal of almost all contaminating particles.

INTRODUCTION

Due to their geometry and extraordinary electronic properties [1], carbon nanotubes are the most promising candidates for bottom-up assembled molecular electronics [2–4]. Yet the main obstacles to an application of SWCNT in nanoelectronic devices are purification, dispersion, separation and positioning of the nanotubes. Recent work [5–8] raises hope that the problems of dispersion, separation and positioning can be solved by CNT functionalization with DNA molecules. A first step towards this goal may be sidewall functionalization of CNT with ssDNA, which is shown to be an easy and straightforward approach that yields the additional benefits of purification and isolation of single nanotubes.

EXPERIMENTAL

Production of ssDNA-CNT

The SWCNT were produced by laser ablation [9]. Details have been published elsewhere [10]. Following closely a protocol given in [5], we mixed 0.5 mg of raw CNT soot with 500 μl of 100 mM NaCl solution containing 0.5 mg of single-stranded salmon testes DNA fragments (purchased from Sigma, D9156). The mixture was sonicated for 90 min in an ultrasonic bath (Merck USR 30H, 80 W @ 35 kHz) and cooled with ice water simultaneously. The resulting suspension resembles black ink. It was centrifuged at 16,000 g for another 90 min to remove the insoluble material. The brownish supernatant

CP723, *Electronic Properties of Synthetic Nanostructures*, edited by H. Kuzmany et al.
© 2004 American Institute of Physics 0-7354-0204-3/04/$22.00

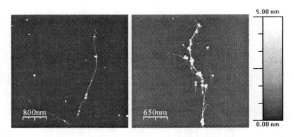

FIGURE 1. AFM images of CNT suspended in 1 wt-% SDS by the same procedure that was used to obtain the ssDNA-CNT suspension.

was used to prepare the AFM samples.

A control experiment was performed by suspending 0.5 mg of CNT raw material into 0.5 ml of 1 wt-% sodium dodecyl sulfate (SDS) solution. This suspension was produced by the same sonication and centrifugation procedure used for the ssDNA-CNT suspension.

AFM-measurements

Silicon and mica have been used as substrates for AFM samples. By cleaving mica, absolutely clean surfaces can be obtained easily. Therefore only samples prepared on mica will be discussed here.

For preparation of the AFM samples, the ssDNA-CNT suspension was diluted 1:20. $MgCl_2$ solution was added to a final concentration of 10 mM, in order to allow for good adsorption of the ssDNA-CNT at the surface [11]. A droplet of the suspension was set onto a freshly cleaved mica substrate and allowed to incubate for 4 min. After incubation the sample was thoroughly rinsed with twofold deionized water and dried with N_2. The samples were investigated by tapping mode AFM with a NanoScope IIIa (Digital Instruments, Santa Barbara). Image processing and analysis have been performed with WSxM [12].

RESULTS

On the control sample, obtained from the SDS dispersion, only bundles of SWCNT, contaminated with a large number of particles, were found (Fig. 1). The particles have diameters from 20 nm to 50 nm and are therefore most likely catalyst material.

Despite the rather simple treatment and moderate centrifugation, samples prepared from ssDNA-CNT suspension show a high degree of carbon nanotube isolation. Most of the observed CNTs are free from impurities within the experimental limits (Fig. 2). At least 75% of the observed nanotubes or bundles had no particles attached to them. None of the observed tubes was contaminated with more than two particles.

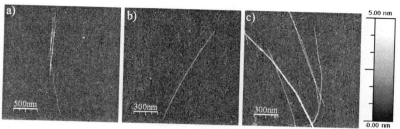

FIGURE 2. AFM images of ssDNA-CNT on mica. Typical are small groups of individual CNT (a). Often single CNT are found too(b). Larger bundles (c) are rare.

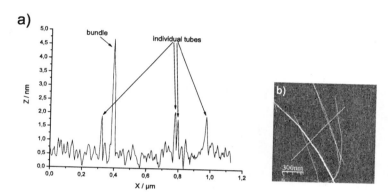

FIGURE 3. Height profile of ssDNA-CNT along the line in (b). The smaller objects have a hight of about 1.5 nm, which agrees well with the expected diameter of the used SWCNT. The bigger feature has to be a bundle of several nanotubes.

The SWCNTs in the suspension are rather long. Individual tubes longer than $3\,\mu m$ have been observed. The procedure has been successfully reproduced several times.

AFM allows exact measurement of height. All elongated objects with a height about 1.5 nm, which corresponds to the diameter of SWCNT can be unambiguously identified as single tubes. However if such an object is observed, the possibility that this object consists of two or more SWCNTs lying side by side next to each other can not be excluded due to tip convolution. Nevertheless we use the height as criterion and refer to objects with heights of 1.5 nm as individual SWCNTs and to objects with heights of more than 1.5 nm as bundles (Fig. 3).

Dispersion of CNT by ssDNA is accompanied by the additional benefits of unbundling, purification and sidewall-functionalization of CNT [5]. Compared to standard purification processes [13], this treatment is supposed to be non-destructive since it waives aggressive chemicals like H_2O_2 or HNO_3. By this, longer individual SWCNT with less purification induced defects will become available.

Compared to the synthetic oligomers used in [5] the random ssDNA fragments are more easily prepared and therefore available in larger quantities. However a drawback of using these fragments so far is, that their binding to the SWCNT seems to be less efficient, since we observe a lot of unbound DNA remaining in the suspension.

OUTLOOK

Several methods for alignment and area-selective binding of DNA-molecules are available [14–16]. The goal for the next future will be to apply and develop these methods in order to achieve positioning and alignment of ssDNA-CNT, which might be an important step towards DNA-templated molecular electronics. First experiments to remove the DNA from the CNT by enzymatic degradation and acidic hydrolysis showed promising but preliminary results.

ACKNOWLEDGMENTS

This work is supported by the BMBF (contract:13N8512) and the DFG (FOR335).

REFERENCES

1. Dresselhaus, M. S., Dresselhaus, G., and Avouris, P., editors, *Carbon Nanotubes-Synthesis, Structure, Properties, and Applications*, vol. 80 of *Topics in Applied Physics*, Springer, Berlin, 2001.
2. Tans, S. J., Verschueren, A. R. M., and Dekker, C., *Nature*, **393**, 49–52 (1998).
3. Postma, H. W. C., Teepen, T., Yao, Z., Grifoni, M., and Dekker, C., *Science*, **293**, 76–79 (2001).
4. Keren, K., Berman, R. S., Buchstab, E., Sivan, U., and Braun, E., *Science*, **302**, 1380–1382 (2003).
5. Zheng, M., Jagota, A., Semke, E. D., Diner, B. A., McLean, R. S., Lustig, S. R., Richardson, R. E., and Tassi, N. G., *Nature Mater.*, **2**, 338–342 (2003).
6. Zheng, M., Jagota, A., Strano, M. S., Santos, A. P., Barone, P., Chou, S. G., Diner, B. A., Dresselhaus, M. S., McLean, R. S., Onoa, G. B., Samsonidze, G. G., Semke, E. D., Usrey, M., and Walls, D. J., *Science*, **302**, 1545–1548 (2003).
7. Nakashima, N., Okuzono, S., Murakami, H., Nakai, T., and Yoshikawa, K., *Chemistry Letters*, **32**, 456–457 (2003).
8. Williams, K. A., Veenhuizen, P. T. M., de la Torre, B. G., Eritja, R., and Dekker, C., *Nature*, **420**, 761 (2002).
9. Guo, T., Nikolaev, P., Thess, A., Colbert, D. T., and Smalley, R. E., *Chem. Phys. Lett.*, **243** (1995).
10. Jost, O., Gorbunov, A., Möller, J., Pompe, W., Liu, X., Georgi, P., Dunsch, L., Golden, M., and Fink, J., *J. Phys. Chem. B*, **106**, 2875–2883 (2002).
11. Bustamante, C., Vesenka, J., Tang, C. L., Rees, W., Guthold, M., and Kellers, R., *Biochemistry*, **31**, 22–26 (1992).
12. Nanotec Electronica S. L., WSxM SPM Software, http://www.nanotec.es (2004).
13. Rinzler, A., Liu, J., H.Dai, Nikolaev, P., Huffman, C., Rodriguez-Macias, F., Boul, P., Lu, A., D.Heymann, Colbert, D., Lee, R., Fischer, J., A.M.Rao, Eklund, P., and Smalley, R., *Appl. Phys. A*, **67**, 29–37 (1998).
14. Mertig, M., Seidel, R., Ciacchi, L. C., and Pompe, W., "Nucleation and Growth of Metal Clusters on a DNA template," in *AIP Conference Proceedings Vol 633*, 2002, pp. 449–453.
15. Mertig, M., and Pompe, W., "Biomimetic fabrication of DNA-based metallic nanowires and networks," in *Nanobiotechnology - Concepts, Applications and Perspectives*, edited by Niemeyer and Mirkin, WILEY-VCH, 2004, pp. 256–277.
16. Opitz, J., Braun, F., Seidel, R., Pompe, W., Voit, B., and Mertig, M., *Nanotechnology*, **in press** (2004).

Transport Properties of Functionalized Single Wall Nanotubes Buckypaper

V. Skákalová, U. Dettlaff-Weglikowska and S. Roth

Max Planck Institute for Solid State Research, Heisenbergstr. 1, 70569 Stuttgart, Germany

Abstract. We have shown recently that ionic doping (with $SOCl_2$, iodine, HNO_3) of single wall carbon nanotubes (SWNT) buckypaper causes an increase of electrical conductivity (factor of 5 for $SOCl_2$-doping). Optical absorption spectra of ionic-doped SWNT show a vanishing of the first peak of van Hove singularities for semiconducting tubes, and at the same time a significant increase of the absorbance in the infrared region. In contrast, molecules forming covalent bonds with SWNT affect the properties of buckypaper so that electrical conductivity decreases (factor of 9 for aniline-functionalization) and the absorbance in the infrared region of the optical absorption spectra is reduced, unlike what is observed for ionic doping.

INTRODUCTION

Many exciting properties of SWNT, like ballistic electron transport combined with a large Young modulus [1-3] as well as a thermal conductivity higher than that of diamond [4,5], have been reported. But all these properties are not observed when the object of the study is a bucky-paper. Bucky-paper is a network of entangled single wall nanotubes (SWNT). Here the intermolecular Van der Waals interaction is the limiting factor for the macroscopic properties of a bucky-paper. Intermolecular interactions can be influenced by different treatments like chemical functionalization [6], physical doping [7], and gamma-irradiation combined with doping [8,9].

In this work, we report on changes of the electrical conductivity after ionic doping and chemical functionalization of bucky-papers made of HiPCO-SWNT.

EXPERIMENTAL

Samples preparation:

Purified SWNT produced by HiPCO method at CNI, Huston (USA) were used for the further treatment. The bucky-paper was prepared by ultrasonication of the nanotubes suspension in SDS solution and subsequent vacuum filtration on a membrane polycarbonate filter.

Ionic doping was performed as follows:

CP723, *Electronic Properties of Synthetic Nanostructures*, edited by H. Kuzmany et al.

a.   Thionyl chloride, SOCl₂ (99 % purity) was purchased from Fluka AG (Switzerland). The bucky-paper filtrated from the suspension of HiPCO-SWNT with SDS-water was treated in $SOCl_2$ at 45 °C for 10 hours.

b.   Iodine doping was performed treating the HiPCO bucky-paper in solution of iodine in toluene for 10 days. Afterwards, the sample was dried in a stream of argon.

The following procedure of chemical functionalization with aniline was used:

25 mg of Hipco SWNTs was stirred in 5 ml of aniline under reflux for 3 days. The dense slurry was then filtered on the nylon membrane filter (pore size 5 μ) and washed with iso-propanol.

Electrical conductivity measurements:

Electrical conductivity was measured using the four-probe method. The contacts were formed by silver-paint. The sample was placed in a long tube filled with helium gas. For temperature dependence measurements, the tube was fixed in a liquid helium bottle.

Optical absorption spectroscopy:

Optical absorption and IR spectra were measured evaporating a drop of CCl_4 suspension on a KBr pellet. For optical spectroscopy, UV-VIS/NIR Perkin-Elmer Lambda Spectrometer was used. The infrared spectra were measured by a Bruker IFS66 Spectrometer.

RESULTS AND DISCUSSION

Fig. 1 shows the temperature dependences of the electrical conductivity for pristine bucky-paper and for bucky-paper treated with $SOCl_2$, iodine and aniline, respectively.

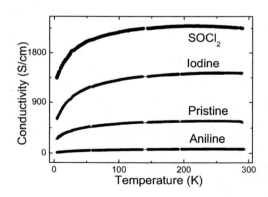

FIGURE 1. Temperature dependences of the electrical conductivity of SWNT bucky-paper doped with $SOCl_2$, iodine and aniline.

For SOCl$_2$ doping the electrical conductivity increases by a factor of 5, for iodine doping by a factor of 3. For aniline treatment the electrical conductivity decreases by factor of 8.

There is a correlation observed between the results obtained from the electrical conductivity measurements and the optical absorption spectra. Fig. 2 represents the optical absorption of those doped samples where electrical conductivity increased upon doping: SOCl$_2$-doped (fig. 2A) and iodine-doped (fig. 2B) compared to pristine SWNT. Significant changes are observed in two features at low energies: 1. vanishing of the first peak of van Hove singularity of the semiconducting tubes, 2. large increase of absorption in the infrared region. The effects are much more pronounced for SOCl$_2$-doping which shows also the highest electrical conductivity.

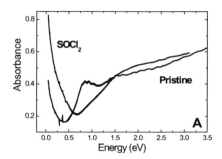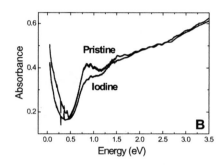

FIGURE 2. Optical absorption spectra of the SOCl$_2$-doped (A) and the iodine-doped (B) SWNT compared to the pristine SWNT.

A different effect can be observed for the optical absorption of the aniline-doped SWNT with low electrical conductivity (fig. 3). The typical van Hove singularities lose their features and there is a decrease of absorption in the infrared region.

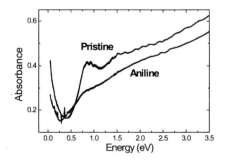

FIGURE 3. Optical absorption spectra of the aniline-doped SWNT compared to the pristine SWNT sample.

This correlation between enhancing of the electrical conductivity and the "filling" the electronic states close to the Fermi level upon ionic doping on one hand, and lowering of the electrical conductivity and the depleting the electronic states close to the Fermi level upon chemical functionalization on the other hand, was observed in all the cases we have studied so far.

CONCLUSION

The correlation between changes in the electrical conductivity and changes in optical absorption spectra of SWNT treated with different molecules was studied:
1. Upon ionic-doping ($SOCl_2$, iodine), the increase in electrical conductivity is accompanied by an enhancement of absorption in the infrared region.
2. Upon chemical-functionalization (aniline), the decrease in electrical conductivity is accompanied by a lowering absorption in the infrared region.

ACKNOWLEDGMENTS

This work was supported by the EU Project CARDECOM and a grant of the Slovak Ministry of Education VEGA 1/0055/03.

REFERENCES

1. Treacy, M. M. J., Ebbesen, T. W. and Gibson, T. M., *Nature* **381,** 680-687 (1996).
2. Wong, E. W., Sheehan, P. E. and Lieber, C. M., *Science* **277,** 1971-1975 (1997).
3. Yu, M. F., Files, B. P., Arepalli, S. and Ruoff, R. S., *Phys. Rev. Lett.* **84,** 5552-5555 (2000).
4. Che, J., Cagin, T., Goddard, W.A., *Theor. Chem. Acct.* **102,** 346 (1999).
5. Osman, M.A.,Srivastava, D., *Nanotechnology.* **12,** 21-24 (2001).
6. Holzinger, M., Vostrowsky, O., Hirsch, A., Hennrich, F., Kappes, M., Weiss, R., Jellen, F., *Angew. Chem. Int. Ed.* **40,** 4002 (2001).
7. Hennrich, F.,Wellmann, R., Malik, S., Lebedkin, S., Kappes, M., *Phys. Cem. Chem. Phys..* **5,** 178 (2003).
8. Skákalová, V., Hulman, M., Fedorko, P., Lukáč, P., Roth, S., *AIP Proceedings of the IWEPNM Conference, Kirchberg-Tirol, Austria, March 8-15, 2003,* **685,** 143-147 (2003).
9. Skákalová, V., Dettlaff-Weglikowska, U., Roth, S., *Diamond and Related Materials.* **13,** 296-298 (2004).

Controlled Functionalization of Carbon Nanotubes by *in Situ* Polymerization Strategy

Chao Gao*, Hao Kong, Deyue Yan

College of Chemistry and Chemical Engineering, Shanghai Jiao Tong University, 800 Dongchuan Road, P. R. China. Tel: +86-21-54742665. Fax: +86-21-54741297. E-mail: chaogao@sjtu.edu.cn

Abstract. The *in situ* ATRP (atom transfer radical polymerization) "grafting from" approach was successfully applied to graft covalently polymers such as poly(methyl methacrylate) (PMMA) and polystyrene (PS) onto the convex surface of multi-walled carbon nanotubes (MWNTs). The thickness of the coated polymer layers can be conveniently controlled by the feed ratio of vinyl monomer to MWNT-supported macroinitiators (MWNT-Br). The resulting MWNT-based polymer brushes were characterized and confirmed with FTIR, NMR, SEM, TEM, and TGA. Moreover, the approach has been extended to the copolymerization system, affording novel hybrid core-shell nanoobjects with MWNT as the hard core and amphiphilic copolymers as the soft shell. The approach presented here may open an avenue for exploring and preparing novel carbon nanotubes-based nanomaterials and molecular devices with tailor-made structure, architecture, and properties.

INTRODUCTION

Carbon nanotubes (CNTs) have been paid more and more attention to since their discovery in 1991 due to their unique electronic and mechanical properties and their interesting 1D tubular structure.[1] To improve their solubility and processibility, functionalization chemistry of CNTs was opened several years ago.[2] CNTs can be functionalized with organic moieties at tips (or ends), on exterior walls and in the interior cavity. Generally, both of the tips and the convex surface are covalently functionalized because of the statistical chemical reaction. In order to combine the merits of CNTs and polymeric materials together, the nanoobjects with CNT as the core and polymers as the shell were fabricated. To date, two main strategies have been developed to covalently functionalize CNTs with polymers. The first is to graft macromolecules onto the surface of CNTs, which is called "graft-to" strategy.[3] The other one is to grow polymers on the CNTs by *in situ* polymerization of monomers, which is called "graft-from" strategy.[4] The prerequisite for the "graft-to" strategy is functional groups-contained macromolecules. What's more, it is difficult to control the linked polymer amount and to realize high density grafting. Therefore, the focus shifts to "graft-from" strategy, by which the polymer content and structure can be controlled.

In this work, we functionalized multi-walled carbon nanotubes (MWNTs) by *in situ* atom transfer radical polymerization (ATRP)[5] approach via "graft-from" strategy.

CP723, *Electronic Properties of Synthetic Nanostructures*, edited by H. Kuzmany et al.
© 2004 American Institute of Physics 0-7354-0204-3/04/$22.00

Experimental Section

The ATRP-active initiating points (MWNT-Br) were firstly introduced onto the surface of MWNTs. Scheme 1 shows the specific steps for synthesis of the macroinitiators. By oxidization with concentrated HNO_3 aqueous solution, carboxyl groups were introduced onto the surface of MWNTs, giving MWNT-COOH. Reaction of MWNT-COOH with thionyl chloride ($SOCl_2$) afforded carbonyl chloride groups-contained MWNTs (MWNT-COCl). Then hydroxyl groups-functionalized MWNTs (MWNT-OH) was prepared by reaction of MWNT-COCl with superfluous glycol in reflux. MWNT-OH was used to react with 2-bromo-2-methylpropionyl bromides in the presence of tertiary amine catalyst, resulting in MWNT-Br. The details can be referred to the previous publications.[4,6,7]

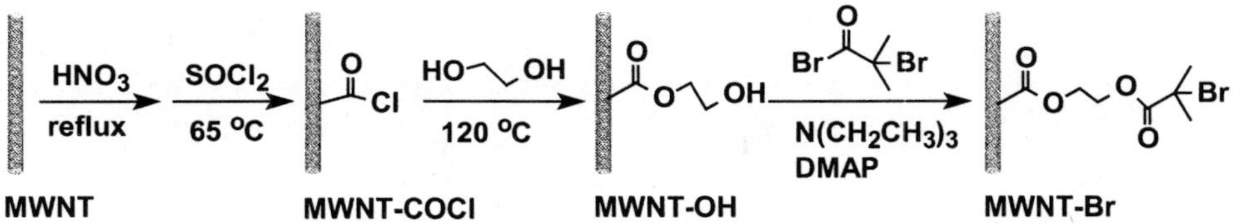

SCHEME 1. Synthesis of MWNT-supported initiators (MWNT-Br).

Scheme 2 displays the process of grafting polystyrene (PS) or poly(methyl methacrylate) (PMMA) onto the surface of MWNTs initiated from MWNT-Br by the *in situ* ATRP approach with Cu(I)Br/*N,N,N',N'',N'''*-pentmethyldiethylenetriamine (PMDETA) as catalyst system.

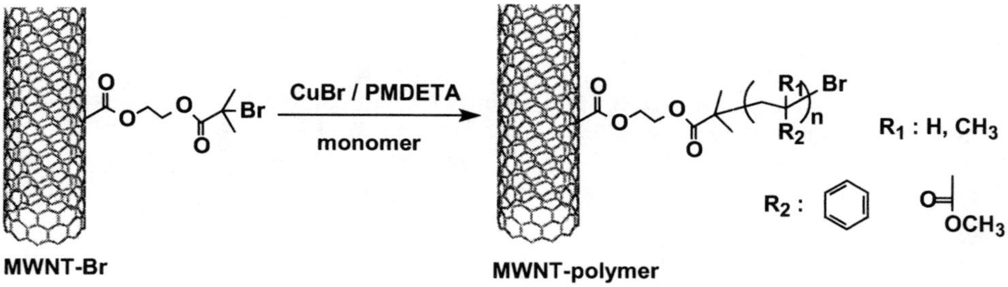

SCHEME 2. Growing polymer grafts on the surface of MWNTs by *in situ* ATRP approach.

Results And Discussion

ATRP, a living/controlled radical polymerization, is a powerful tool for design and constructing polymeric materials.[5] Compared with other living/controlled polymerization method (e.g., cationic and anionic polymerizations), ATRP shows at least four merits: (1) both styrene and acrylate/acrylamide monomers can be directly used as the raw materials, (2) the initiating sites in the reaction system remain constant after the graft polymerization and purification of the products, so it is convenient to further perform block copolymerization and chain extension, (3) it presents an access to prepare the polymers with a functional group in each repeating unit such as poly(hydroxyethyl methacrylate) (PHEMA), and (4) because of its radical polymerization feature, the influence of impurity and other masses on the reaction is much weaker.

The polymer-grafted MWNTs was characterized by FTIR, NMR, TGA, TEM and SEM. TGA measurements indicated that the grafted polymer content could be varied in a wide range (at least 15-90 wt%). For functionalized nanotubes with higher amount of polymer, core-shell structure can be clearly observed from the TEM images, and the thickness of the polymer shell can be controlled by the feed ratio of vinyl monomer and MWNT-Br. Figure 1 shows a TEM image of crude MWNT (a), two TEM images of MWNT-PMMA (b, c), and a image of MWNT-PS (d).

As a comparison, two individual phases, polymer's and MWNTs', were observed by TEM and SEM for the mixture of polymers and MWNTs. The parallel experiments also showed that the possibly adsorbed polymers could be completely removed from the filtered solid by washing with solvent.

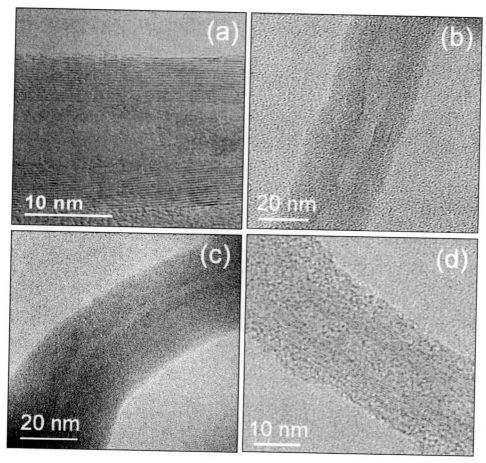

FIGURE 1. TEM images of MWNT (a), MWNT-PMMA (b, c), and MWNT-PS (d).

The polymer grafted MWNTs can be used as the initiators to initiate further copolymerization. The grafted copolymer content can be measured from the weight loss of corresponding TGA curves, and calculated from ^{1}H NMR spectrum with the original polymer block as standards. The results showed that the content of the second polymer blocks can be also simply controlled by the feed ratio of monomer and the polymer grafted MWNTs.[6,7]

If *tert*-butyl acrylate (*t*BA) was selected as the second building monomer, MWNT-PS-*block*-P*t*BA was prepared with MWNT-PS as macroinitiators. After hydrolysis of *tert*-butyl groups, the P*t*BA blocks were converted to poly(acrylic acid) (PAA) chains. Consequently, MWNT-PS-*block*-PAA, amphiphilic copolymer grafted hybrid nanowires, was obtained.[6]

Figure 2 displays a TEM image of MWNT-PMMA-*block*-PHEMA and an image of MWNT-PS-*block*-P*t*BA. Again, the core-shell structure of the copolymer grafted MWNT was obviously observed.

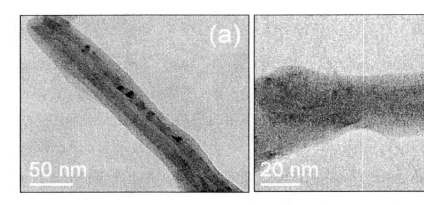

FIGURE 2. TEM images of MWNT-PMMA-*b*-PHEMA (a) and MWNT-PS-*b*-P*t*BA (b).

ACKNOWLEDGMENTS

Financial support from the National Natural Science Foundation of China (No. 20304007, 20274024) and Rising-Star Foundation of Shanghai (No. 03QB14028) was acknowledged.

REFERENCES

1. Ajayan, P. M., *Chem. Rev.* **99**, 1787-1799 (1999).
2. Chen, J., Hamon, M. A., Hu, H., et al., *Science* **282**, 95-98 (1998).
3. Sun, Y.-P., Fu, K., Lin, Y., and Huang, W., *Acc. Chem. Res.* **35**, 1096-1104 (2002).
4. Kong, H., Gao, C., and Yan, D., *J. Am. Chem. Soc.* **126**, 412-413 (2004).
5. Qin, J., Charleux, B., Matyjaszewski, K., *Prog. Polym. Sci.* **26**, 2083-2134 (2001).
6. Kong, H., Gao, C., and Yan, D., *J. Mater. Chem.,* 2004, in press.
7. Kong, H., Gao, C., and Yan, D., *Macromolecules,* 2004, accepted.

Effect of physical and chemical doping on optical spectra of SWNT's

K. Kamarás[*†], H.Hu[†], B. Zhao[†], S. Niyogi[†], M.E. Itkis[†] and R.C. Haddon[†]

*Research Institute for Solid State Physics and Optics, Hungarian Academy of Sciences,
P. O. Box 49, Budapest, H-1525, Hungary
†Center for Nanoscale Science and Engineering, Departments of Chemistry and Chemical and
Environmental Engineering, University of California, Riverside, CA 92521-0403, U.S.A.

Abstract. We discuss the use of far-infrared spectroscopy in the characterization of doped and functionalized nanotube derivatives.

INTRODUCTION

Optical spectroscopy is one of the most widely used methods in characterization of carbon nanotubes. Recently, it was demonstrated that absorption and especially fluorescence studies can detect individual nanotubes and identify them by chirality index [1]. Most studies concentrated on the NIR/VIS spectral range where transitions between van Hove singularities occur: based on these observations, selectivity by semiconducting/metallic character was reported both for ionic doping [2, 3] and covalent functionalization [4]. Relatively less attention was devoted to the low-frequency part of the spectrum. Here we want to emphasize the importance of far-infrared measurements as a sensitive indicator of intrinsic charge carriers in metallic tubes and extrinsic carriers in doped materials.

EXPERIMENTAL

We used arc-produced nanotube powders for the doping experiments and HiPCO nanotubes for functionalization. Details of functionalization are described in Ref. [5]. Nitric acid doping was performed under ambient conditions by exposing the films to saturated HNO_3 atmosphere for one hour. Thin films for transmission measurements were prepared on silicon and sapphire substrates from DMF suspensions using an airbrush [6]. Spectroscopic measurements were performed on a Bruker 120HR FTIR instrument working in the far and mid-infrared (silicon samples), and a Varian Cary 5000 NIR/VIS/UV dispersive spectrometer (sapphire samples). Transmission spectra were converted to absorbance without further correction and scaled together in the overlap region.

CP723, *Electronic Properties of Synthetic Nanostructures*, edited by H. Kuzmany et al.

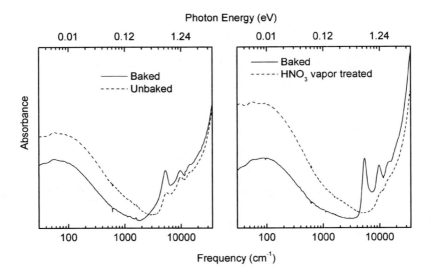

FIGURE 1. Change in absorbance of SWNT films on baking of HNO₃ purified sample (left panel), treating HCl purified sample with HNO₃ vapor (right panel). Note the logarithmic frequency scale to emphasize the far-infrared region.

RESULTS AND DISCUSSION

We present two examples for ionic doping and chemical functionalization, respectively: HNO₃ treatment (which occurs during standard steps of purification) and covalent functionalization by dichlorocarbene. Figure 1 shows optical spectra taken during two separate experiments: removal of HNO₃ from purified as-prepared SWNT's and intentional doping of another SWNT sample which was previously purified by HCl and then annealed at 450° C in vacuum. The effect of nitric acid is twofold: introduction of -COOH groups to the ends and defects of nanotubes, and increasing the concentration of mobile charge carriers by removing electrons from the filled states of semiconducting nanotubes. The latter process is similar to nitrate intercalation into graphite [7]. The addition of free carriers is reflected in the low-frequency absorption, but there are also more subtle changes in the interband transitions which have been observed before. In the purified sample (dashed line in the left panel of Fig. 1), the interband transitions are still discernible, but in the vapor-treated sample (dashed line in the right panel) they disappear. This difference, together with the higher far-infrared intensity in the vapor-treated material, indicates that the HNO₃-purified sample has a lower carrier concentration than the vapor-treated one. As in graphite, several equilibrium doped phases with varying dopant concentration seem to exist [2, 3]. The far-infrared oscillator strength change thus corroborates the model outlined by Kazaoui et al. [2] for the sequence of doping, also in the case of nitrate.

A very different result emerges when introducing covalent side groups to carbon nanotubes. This is shown in Figure 2, left panel. Dichlorocarbene addition transforms sp^2 carbon atoms to sp^3 carbons, and thus effectively interrupts the conjugation of double bonds on the tube surface. As a consequence, the one-dimensional electronic structure collapses, causing both interband transitions and free-carrier absorption to decrease and eventually disappear at high enough sidegroup concentration. The far-infrared region in this case reflects the effect of functionalization on metallic tubes. Selective functionalization of metallic tubes has been demonstrated by Strano et al. [4] who monitored the M_{11} interband transition and explained their findings along similar lines. Although we do not detect signs of selectivity in this reaction (because of the different mechanism), the present results support that explanation.

The right panel of Figure 2 illustrates the effect of annealing on dichlorocarbene functionalized material. Raman spectra of this substance [5] reveal that on annealing, both the C-Cl vibrational transition and the graphitic D-band decrease in intensity, indicating formation of intertube bonds with sp^2 carbons. However, no change occurs in the region of free carriers, proving that the one-dimensional collective electron system does not recover.

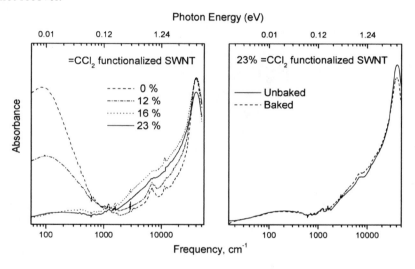

FIGURE 2. Left panel: Absorbance of HiPCO films functionalized with dichlorocarbene at different functionalization levels. Right panel: Absorbance of the highest functionalization dichlorocarbene product before and after thermal treatment at 450 C.

CONCLUSIONS

Far-infrared spectra give valuable information on both intrinsic free carriers in metallic nanotubes and extrinsic ones introduced by intentional or unintentional doping. While ionic doping increases the number of free carriers and thus the far-infrared absorption, covalent functionalization destroys the collective electronic system by localizing the carriers. The effect is much more sizable than the changes in the M_{11} transition, therefore extending the spectral range towards low frequencies can be very useful in chirality-dependent studies.

ACKNOWLEDGMENTS

This work was supported by the following grants: DARPA DMEA90-02-2-0216, OTKA T 034198.

REFERENCES

1. Bachilo, S. M., Strano, M. S., Kittrell, C., Hauge, R. H., Smalley, R. E., and Weissman, R. B., *Science*, **298**, 2361 (2002).
2. Kazaoui, S., Minami, N., Jacquermin, R., Kataura, H., and Achiba, Y., *Phys. Rev. B*, **60**, 13339 (1999).
3. Hennrich, F., Wellmann, R., Malik, S., Lebedkin, S., and Kappes, M. M., *Phys. Chem. Chem. Phys.*, **5**, 178 (2003).
4. Strano, M. S., Dyke, C. A., Usrey, M. L., Barone, P. W., Allen, M. J., Shan, H., Kittrell, C., Hauge, R. H., Tour, J. M., and Smalley, R. E., *Science*, **301**, 1519 (2003).
5. Hu, H., Zhao, B., Hamon, M. A., Kamarás, K., Itkis, M. E., and Haddon, R. C., *J. Am. Chem. Soc.*, **125**, 14893 (2003).
6. Itkis, M. E., Niyogi, S., Meng, M. E., Hamon, M. A., Hu, H., and Haddon, R. C., *Nano Lett.*, **2**, 155 (2002).
7. Ubbelohde, A. R., *Proc. Roy. Soc. London Ser. A*, **304**, 25 (1968).

Electrochemical Functionalization Of Single-Walled Carbon Nanotubes With Polyaniline Evidenced By Raman And FTIR Spectroscopy

S. Lefrant[a], M. Baibarac[a, b], I. Baltog[b], C. Godon[a], J. Y. Mevellec[a], O.Chauvet[a]

[a] *Institut des Matériaux Jean Rouxel, Lab. de Physique Cristalline, 2 rue de la Houssinière, B.P.32229, 44322 Nantes, Cédex 03, France*
[b] *National Institute of Materials Physics, Lab. Optics and Spectroscopy, Bucharest, P. O. Box MG-7, R-76900, Romania*

Abstract. Using Raman and Fourier transform infrared (FTIR) spectroscopy, we show that depending on the acid medium used, i.e. HCl and H_2SO_4, electrochemical polymerization of aniline on a single-walled carbon nanotubes (SWNTs) film leads to a covalent functionalization of SWNTs with polyaniline (PANI) and a PANI doped with carbon nanotube fragments, respectively. For polymer-functionalized SWNTs, the Raman scattering suggests an additional nanotubes roping with PANI as a binding agent. A post chemical treatment with the NH_4OH solution of polymer-functionalized SWNTs involves an internal redox reaction between PANI and carbon nanotubes. As a result, the polymer chain undergoes a transition from the semi-oxidized state into a reduced one. Besides, the FTIR spectrum of PANI-functionalized SWNTs display an intense absorption band with two components at 773 and 755 cm^{-1}, which are associated with the vibration of deformation of the benzene and the quinoid ring, respectively. This indicates a strong hindrance effect produced by the binding on the polymer chain of SWNTs.

INTRODUCTION

For the conducting polymers/ carbon nanotubes (CPs/CNTs) composites, it has been suggested that either the polymer functionalizes the CNTs [1, 2], or the CPs are doped with CNTs, i.e. a charge transfer occurs between the two constituents [1, 3, 4]. Three routes may be used to prepare PANI/CNTs composites: direct mixing of the CP with nanotubes (NTs), chemical polymerization of aniline in the presence of NTs and electrochemical polymerization of the monomer on a SWNT film. Recently, we have shown that the first two routes result in different materials [1, 5-7].

The purpose of this paper is to show, by Raman and FTIR spectroscopy, that the electrochemical polymerization of aniline on a SWNTs film results in composites of the type covalent functionalized SWNTs with PANI, when HCl is used as electrolyte. The significant role of the acide type used for the preparation of PANI/SWNTs composite is evidenced by the electrochemical polymerization of aniline in aqueous solution of H_2SO_4 on the SWNT film.

CP723, *Electronic Properties of Synthetic Nanostructures*, edited by H. Kuzmany et al.
© 2004 American Institute of Physics 0-7354-0204-3/04/$22.00

EXPERIMENTAL

PANI was electrochemically prepared via cyclic voltammetry using an aqueous solution of 0.5 M HCl or H_2SO_4 and 0.05 M aniline (AN) in the potential range (-200; +700) mV vs. SCE at a sweep rate of 100 mVs^{-1}. Electrochemical measurements were carried out using a potentiostat /galvanostat type Princeton Applied Research (PAR), model 173 and a PAR pulse generator, model 175. Raman and FTIR spectra have been recorded under the excitation wavelength of 1064 nm using an FT Bruker RFS 100 spectrophotometer and a Bruker IFS 28 spectrophotometer, respectively.

RESULTS AND DISCUSSION

Figs. *1a* and *1b* show the variation of the Raman spectra of the SWNT film during the electropolymerization of AN in HCl and H_2SO_4, respectively. Two spectral ranges are of particular importance in the Raman spectrum of SWNTs. The first one, 50 - 250 cm^{-1}, contains two Raman lines at 164 and 178 cm^{-1} associated with the radial breathing mode (RBM) of isolated and bundled nanotubes, respectively [8]. The second group, consisting of the G and D bands, is found in the interval from 1100 to 1700 cm^{-1}. The former, at 1595 cm^{-1}, is attributed to the in-plane E_{2g} vibration mode [9]. The D band (at 1277 cm^{-1}) is considered as an indication of disorder in graphitic lattices or defects in CNTs [9]. In accordance with spectra 2-4 and 6-8 from Fig. *1a*, the following changes are induced in the Raman spectrum of SWNTs: i) the Raman lines of the PANI-ES, at 1175, 1330-1380, 1506, 1589 and 1618 cm^{-1}, appear and increase with the growing of cycles number; ii) a sudden decrease of the RBM of isolated tubes accompanied of a gradual increase of the Raman line at 178 cm^{-1} is observed. It suggests that an additional roping is induced by the electro-polymerization. Comparing Figs. *1a* and *1b*, in the latter case the increase of the intensity of the Raman line at 178 cm^{-1} is no longer observed. The decrease in the intensity of the Raman line at 164 cm^{-1} is rather associated with a SWNTs breaking process. In Figs. *1a*, on the curves 5 and 9, one notice that the Raman lines of PANI no longer appear. In Fig. *1b* the modification of the ratio between the intensities of Raman lines at 1330 and 1380 cm^{-1} reveals that a partial dedoping of PANI-ES takes place. This fact may be explained taking into consideration that the breaking of SWNTs leads to the formation of charged CNTs fragments, which appear both as cation and anion radicals. A charge transfer reaction may thus occur between the polymer and charged CNTs fragments which is similar to a doping process. In this context, a dedoping reaction of PANI-ES involves the formation of both PANI-emeraldine base (EB) and PANI doped with CNTs fragments. In Fig. *2*, two new bands are seen at 770 and 740 cm^{-1}. They are assigned to the deformation vibration of the benzoid (B) and quinoid (Q) ring, respectively [10]. The transformation from PANI-salt functionalized SWNTs to PANI-base functionalized SWNTs (Fig. *2*) involves: i) the modification of the ratio between the FTIR bands intensity at 770 and 750 cm^{-1}; ii) the increase in the intensity of the 1219 cm^{-1} band, attributed to C-N stretch.+ ring def. (B) + C-H bending (B) vibration mode; iii) the appearance of new bands at 1458, 1540 and 1559 cm^{-1} attributed to C-C stretching, C-H and N-H bending vibrations [10] and iv) the disappearance of the complex band at 1560-1625 cm^{-1}.

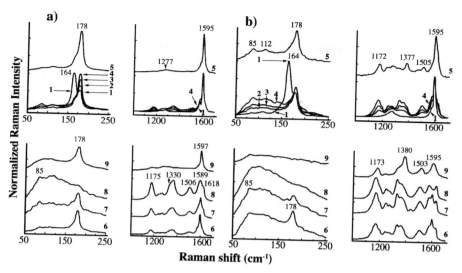

FIGURE 1. Raman spectra (λ_{exc} = 1064 nm) of PANI/SWNTs composites obtained by electropoly-merization of AN on a SWNT film in HCl 0.5 M (**a**) and H_2SO_4 0.5 M (**b**). Curve 1 corresponds to the SWNT Raman spectrum. Curves 2–4 and 6–8 show the evolution of the Raman spectrum after 25, 50, 75, 100, 150 and 300 cycles, respectively. The de-doping of the PANI-salt functionalized SWNT films (curves 4 and 8), as a result of the reaction with the NH₄OH 1M, is illustrated on curves 5 and 9.

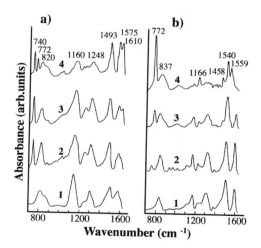

FIGURE 2. FTIR spectra of PANI-ES (curve 1, *a*) and PANI-salt functionalized SWNT composites (curves 2 - 4, *a*). These composites were electrochemically prepared by the achievement of 75, 150 and 300 cycles (curves 4, 3 and 2, respectively) on the SWNT film, immersed in the solution of AN and HCl 0.5M. FTIR spectra of PANI-EB (curve 1, *b*) and PANI-base functionalized SWNT composites (curves 2 – 4, *b*) obtained by subsequence reaction of PANI-salt functionalized SWNTs (curves 2 – 4, *a*) with NH₄OH 1M solution.

CONCLUSIONS

The main results can be summarized as follows: i) the covalent functionalization of SWNTs with PANI is obtained in two succesive ways. The first one corresponds to the electrochemical polymerization of aniline on SWNTs film using an HCl solution which results in composites of the type SWNTs functionalized with PANI-leucoemeraldine salt (LS) and PANI-ES, respectively. The second one is the result of a post treatment with NH_4OH on PANI-(LS and ES) functionalized SWNTs. It involves an internal redox reaction between PANI-EB and SWNTs which transforms the polymer chain from the semi-oxidized state into a reduced one; ii) the increase of the intensity of the Raman band at 178 cm^{-1} during the electrochemical polymerization of aniline on CNTs film, in an HCl solution, indicates an additional nanotube roping with PANI as a binding agent; iii) the binding of SWNTs as whole units on the polymer chain induces strong steric hindrance effects observed in FTIR spectra by the enhancement of bands at ca. 740 - 750 and 772 cm^{-1} associated to the deformation vibration of the benzoid and quinoid ring, respectively; iv) PANI doped with CNT fragments is obtained by the electrochemical polymerization of aniline on SWNT film using an aqueous H_2SO_4 solution. The NH_4OH post treatment of PANI doped with CNT fragments leads to a partial dedoping of the conducting polymer.

ACKNOWLEDGMENTS

Samples of SWNTs have been provided by the Groupe de Dynamique des Phases Condensées of the University of Montpellier II. A part of this work was performed in the frame of the Scientific Cooperation between the Laboratory of Crystalline Physics of the Institute of Materials "Jean Rouxel", Nantes, and the Laboratory of Optics and Spectroscopy of the National Institute of Materials Physics, Bucharest. Other support was provided in the frame of a European program COMELCAN (HRPN-CT-2000-00128).

REFERENCES

1. Baibarac,M., Baltog, I., Lefrant, S., Mevellec, J.Y., Chauvet, O., *Chem.Mater.* **15**, 4149-4156 (2003).
2. Sun, Y.P., Fu, K.,Li, Y., Huang, W., *Acc.Chem.Res.* **35**, 1096-1104 (2002).
3. Zengin, H., Zhou, W., Jin, J., Czerw, R., Smith., D.W., Echegoyen, L., Carroll, D.L., Foulger, S.H., Ballato, J., *Adv. Mater.* **14**, 1480-1483 (2002).
4. Huang, J.E., Li, X.H., Xu, Y.Q., Liang, J., Wang, X.G., Yoshino, K., *Carbon* **41**, 1551-1557 (2003).
5. Lefrant, S., Baltog, I., Baibarac, M., Mevellec, J.Y., Chauvet, O., *Carbon* **40**, 2201-2211 (2002).
6. Huang, J.E., Li, X.H., Xu, J.C., Li, H.L., *Carbon* **41**, 2731-2736 (2003).
7. Zhou, Y.K., He, B.L., Zhou, W.J., Huang, J., Li, X.H., Wu, B., Li, H.L., *Electrochim. Acta* **49**, 257-262 (2004).
8. Marcoux, P.R., Schreiber, J., Batail, P., Lefrant, S., Renouard, J., Jacob, G., Albertini, D, Mevellec, J.Y., *Phys.Chem.Chem.Phys.* **11**, 2278-2285 (2002).
9. Dresslhaus, M.S., Dresselhaus, G., Eklund, P.C., *Science of fullerenes and carbon nanotubes*, Publisher: Imperial College Press, 1998, pp.25-30.
10. Quillard, S., Louarn, G, Lefrant, S., MacDiarmid, A.G., *Phys. Rev. B.* **50**, 12496-12508 (1994).

Covalent interaction in Ba-doped single-wall carbon nanotubes

X. Liu*, T. Pichler*, M. Knupfer* and J. Fink*

*Leibniz-Institut für Festkörper- und Werkstofforschung Dresden, D-01069 Dresden, Germany

Abstract. Barium-doped single-wall carbon nanotubes (SWNTs) were investigated using high-resolution electron energy-loss spectroscopy in transmission. The core-level excitations show a splitting of the π^* states, which provides evidence of hybridization between SWNT π states and Ba valence states. Therefore barium is not fully ionized Ba^{2+}. The structural analysis indicates the lattice expansion in Ba-doped SWNTs to be much stronger than that in the case of doping with potassium. Since K^+ and Ba^{2+} have the same ionic radius this is further support that the charge transfer is not purely ionic in the alkaline-earth doped compounds. Regarding the electronic properties at maximum doping the energy position of the free charge carrier plasmon is about two times higher than that of the fully potassium doped nanotubes. This can be explained by a higher total charge transfer and a reduced screening of the plasmon.

INTRODUCTION

Due to their unique physical properties [1], single-wall carbon nanotubes (SWNTs) have attracted wide interests. One way to use these properties for technical applications is a modification of their intrinsic properties by manipulating their structure and/or their electronic properties. A promising route for modification of the electronic properties of SWNTs is to add electron acceptors or donors, so-called intercalation/doping [2, 3, 4]. In this case, many investigations were focused on alkali-metal doped SWNTs. A complete charge transfer from the alkali s electron to the nanotube conduction band was observed [4]. It is a natural idea that alkaline-earth metal doping such as Ba-doped SWNTs might be an even better metal if both $6s$ electrons were transferred to nanotubes. However, in the graphite intercalation compound (GIC) BaC_6 only partial charge transfer from Ba $6s$ to the graphite π bands, and a hybridization between Ba $5d$ and the graphite π bands were observed [5].

In this paper, we present recent results on the electronic and structural properties of Ba-doped SWNTs using electron energy-loss spectroscopy (EELS) in transmission.

EXPERIMENTAL

Thin films of SWNTs produced by laser ablation (a mean diameter of 1.37 nm) [6] with an effective thickness of about 100 nm were prepared by drop coating described elsewhere [4]. The doping was carried out in a UHV chamber by evaporation of Ba from commercial SAES barium metal getter sources. The SWNT film was kept at 150°C

CP723, *Electronic Properties of Synthetic Nanostructures*, edited by H. Kuzmany et al.
© 2004 American Institute of Physics 0-7354-0204-3/04/$22.00

during the evaporation, after that a 2 hour 230°C post-anneal treatment was performed. The measurements were performed using a purpose-built 170 keV EELS spectrometer [7] with the energy and momentum resolution being 180 meV (300 meV) and 0.03 Å^{-1} (0.1 Å^{-1}) for the low energy-loss function and electron diffraction (C 1s excitation). All EELS measurements are carried out at room temperature and under UHV conditions.

RESULTS AND DISCUSSION

Figure 1(a) depicts C 1s core-level excitations of pristine and fully Ba-doped SWNTs compared with GIC KC_8, and K-doped SWNTs with C/K~7. In pristine SWNTs (curve A), the two obvious features can be assigned to transitions from C 1s core electrons into unoccupied $\pi^\star$ and $\sigma^\star$ electronic states at energies of about 285.4 and 292 eV. Due to the strong influence of the core hole no peaks related to singularities of the density of states of SWNTs are observed. In the fully Ba-doped SWNTs (curve B), the spectral shape is obviously different from that of the pristine tubes. The $\sigma^\star$ spectral shape becomes flat upon doping. The $\pi^\star$ spectral weight decreases about 50% compared to that of the pristine tubes. The $\pi^\star$ excitations are characterized by two separate peaks at 285 eV and 283.8 eV. The former can be assigned to the downshift of $\pi^\star$ peak position from 285.4 eV (pristine) to 285 eV (full doping), the latter is the new doping-induced peak. With annealing up to 600°C, the $\pi^\star$ peak can be recovered which demonstrates that there is no Ba carbide formation upon doping. Furthermore, the corresponding transitions from Ba $3d_{5/2}$ and $3d_{3/2}$ to Ba $4f$ orbitals show two peaks at 785 and 800 eV respectively. Since in oxidized Ba compounds, the energy of the two transitions is about 3 eV higher, these data provide evidence that Ba doping in between SWNTs is not fully ionized Ba^{2+}. On the other hand, the spectrum of the fully K-doped SWNT (curve C) is very similar to that of the pristine one, only the intensity is lower due to the charge transfer to the conduction band [4]. It indicates that the unoccupied states are not greatly affected by the introduction of the dopants. However, in GIC KC_8 (curve D), transitions into $\pi^\star$ states just above the Fermi level are characterized by the extremely steep edge rise and two individual features clearly show up, indicated by two arrows. This has been rationalized by a hybridization of carbon and alkali-metal derived states [8]. The overall structure is considerably broader than that of pristine graphite. In Ba-doped SWNTs, the splitting of $\pi^\star$ peak is larger than for GIC KC_8 and the structure becomes very broad, but the excitation edge is not as steep as for GIC KC_8. It is reasonable to assume that there is hybridization between Ba $5d$ and SWNT π states in Ba-doped SWNTs. Such hybridization has also been demonstrated for Ba intercalated C_{60} compounds and graphite [5, 9].

Next we turn to the structural analysis. Since SWNTs produced by laser ablation always form bundles they provide interstitial channels which the dopants can occupy. Due to the insertion of dopants the intertube distance expands depending on the dopant size and nanotube diameter [4]. This expansion of the hexagonal lattice structure can be estimated from the position of the first diffraction peak. In the present sample, as shown in Fig. 1(b), with increasing doping a downshift of this peak from 0.426 to about 0.36 Å^{-1} is observed. This corresponds to the variation of the intertube distance from about

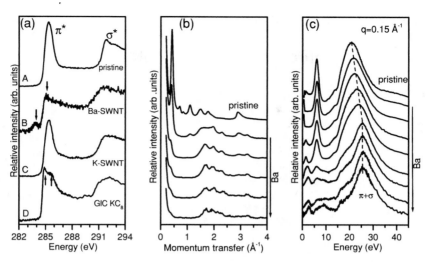

FIGURE 1. (a) The C 1s spectra of pristine (A), fully Ba-doped (B), fully K-doped (C) SWNTs and GIC KC$_8$ (D). (b) The change in electron diffraction with increasing Ba concentration from pristine (top) to fully doping (bottom). (c) The evolution of loss function with increasing Ba content. The dashed line guides the $\pi + \sigma$ position.

1.7 nm (pristine) to 1.98 nm (fully doped). Since the ion size of Ba^{2+} is the same as that of K$^+$, their expansion rates should be nearly the same if Ba^{2+} exists in the doped samples. However, the observed lattice expansion in fully Ba-doped SWNTs is much bigger than that in the fully K-doped SWNTs. Compared with other alkali-metal doped nanotubes, the observed expansion is similar to that for Cs-doped SWNTs in which Cs$^+$ has an ionic radius about 1.3 times bigger than K$^+$ [4]. Since a much higher doping level is very unlikely due to space limitation in between the tubes in the bundles, the larger expansion for Ba-doped tubes is a further support of the above mentioned formation of hybrid states between the Ba dopants and the nanotubes associated with only partial charge transfer.

Finally, we turn to an analysis of the optical properties of Ba-doped SWNTs which can be extracted from the measured loss function at low energies. In pristine SWNTs and similar to optical data, the low energy loss spectrum below 3 eV is dominated by the interband transitions between corresponding van Hove singularities in the nanotube density of states. In addition, two higher energy peaks at about 6 eV and 23 eV, the so-called π and $\pi + \sigma$ plasmons, can be assigned to $\pi - \pi^\star$ excitations and to $\sigma(\pi) - \pi^\star(\sigma^\star)$ and $\sigma - \sigma^\star$ excitations, respectively [10]. In the doped SWNTs, due to the charge transfer the conduction band of the SWNTs is filled which quenches the previously optically allowed transitions [4]. Figure 1(c) depicts the evolution of the loss function with increasing Ba concentrations from the pristine to the fully doped SWNTs within the energy range between 0.2 and 45 eV at $q = 0.15$ Å^{-1}. With increasing Ba content, the low energy interband transitions disappear and a new peak appears which is related to the collective excitations of the introduced conduction electrons, the so-called charge carrier plasmon [4]. Its position is related to the doping level and reaches its maximum at around 2.6 eV at saturation of the Ba doping, which is about two times higher than in potassium doped SWNTs at saturation doping. In addition to the appearance of the charge carrier plasmon, the π plasmon is strongly affected by the Ba doping. At lower doping levels, the π plasmon intensity only slightly decreases due to a change in the dielectric background whereas at high doping levels the π plasmon peak is strongly suppressed. Regarding the doping dependence of the $\pi + \sigma$ plasmon (dashed line in Fig. 1(c)) two different behaviors are observed; at lower doping levels, the $\pi + \sigma$ peak upshifts with doping. At the doping level when the π plasmon is suppressed, the position of $\pi + \sigma$ is nearly independent of doping. In comparison with the corresponding C $1s$ spectra we find that the suppression of the π plasmon is related to the appearance of the additional structure at the onset of the $\pi^\star$ peak. Therefore, the suppression of the π plasmon can be explained by a vanishing of the pronounced $\pi - \pi^\star$ transition due to the filling of the conduction band and by a hybridization between the Ba $5d$ and SWNTs states.

Regarding the optical conductivity in the fully Ba-doped compounds, the measured loss function was analyzed within the framework of a Drude-Lorentz model [4]. The evaluated unscreened plasmon is about 6.4 eV and the optical conductivity at zero frequency is about 3132 S/cm. This optical conductivity is comparable with that of the K-doped SWNTs [4], but the unscreened plasmon energy is much higher and has a three times bigger width. The former can be attributed to a charge transfer of more than one electron and the latter can be assigned to the hybridization between Ba and nanotubes giving rise to a stronger scattering of the electrons.

In conclusion, we have shown that Ba atom is not fully ionized in Ba-doped SWNTs and hybridization is formed in between Ba and SWNT which results in a splitting of the $\pi^\star$ states, a stronger lattice expansion, and wider width of Drude plasmon as compared with K-doped SWNTs.

ACKNOWLEDGMENTS

We acknowledge financial support from the DFG (PI440). Technical support by R. Hübel, K. Müller and S. Leger is appreciated.

REFERENCES

1. R. Saito et al., *Physical Properties of Carbon Nanotubes* (Imperial College Press, London 1998).
2. S. Kazaoui et al., Phys. Rev. B **60** 13339 (1999); T. Takenobu et al., Nature Materials **2**, 683 (2003).
3. R.S. Lee et al., Nature **338**, 255 (1997); A.M. Rao et al., Science **275**, 187 (1997).
4. T. Pichler et al., Solid State Commun. **109**, 721 (1999); X. Liu et al., Phys. Rev. B **67**, 125403 (2003).
5. M.E. Preil et al., Phys. Rev. B **30**, 3536 (1984); J.E. Fischer et al., Phys. Rev. B **36**, 4449 (1987).
6. O. Jost et al., Appl. Phys. Lett. **75**, 2217 (1999); H. Kataura et al., Synth. Met. **103**, 2555 (1999).
7. J. Fink, Adv. Electr. Electron Phys. **75**, 121 (1989).
8. J.J. Ritsko, Phys. Rev. B **25**, 6452 (1982); L.A. Grunes et al., Phys. Rev. B **28**, 3439 (1983).
9. M. Knupfer et al., Phys. Rev. B **49**, 7620 (1994).
10. T. Pichler et al., Phys. Rev. Lett. **80**, 4729 (1998).

Reaction Of Single-Wall Carbon Nanotubes With Radicals

A.S. Lobach[1*], V.V. Solomentsev[1], E.D. Obraztsova[2], A.N. Shchegolikhin[3] and V.I. Sokolov[4]

[1]*Institute of Problems of Chemical Physics RAS, 142432 Chernogolovka, Moscow Region, Russia*
[2]*Natural Sciences Center of General Physics Institute, RAS, 38 Vavilov Street, Moscow, 119991 Russia*
[3]*Institute of Biochemical Physics RAS, 4 Kosygina Street, 117997 Moscow, Russia*
[4]*Nesmeyanov Institute of Organoelement Compounds RAS, 28 Vavilov Street, 117813 Moscow, Russia*

Abstract. A method for functionalizing the sidewalls of HiPco SWNT via interaction with carbon- and metal-centered radicals is presented. A number of methods: UV-vis-NIR spectroscopy, thermogravimetric analysis, TEM and Raman spectroscopy provided a direct evidence of a chemical attachment of functional groups to the tubes. Functionalization was shown to be reversible: a thermal treatment led to the recovering of pristine structure of SWNT.

INTRODUCTION

Chemical functionalization of single-wall carbon nanotubes (SWNT) is very important for the development of new diagnostics methods and technological applications of nanotubes. The present work reports on the investigations of interaction process of SWNT with carbon- and metal-centered radicals and on analysis of products of radical-nanotube interaction. The radicals were extracted from m-chloroperoxybenzoic acid and a metal complex of (pentamethylcyclopentadienyl) chromiumtricarbonyl dimer, $[Cp*Cr(CO)_3]_2$. The SWNT samples prepared by the HiPco methods were used. A number of methods: UV-vis-NIR spectroscopy, TGA, TEM and Raman spectroscopy provided a direct evidence of a chemical attachment of functional groups to the tubes. Functionalization was shown to be reversible: a thermal treatment led to the recovering of pristine structure of SWNT.

RESULTS AND DISCUSSION

Chemical functionalization of HiPco SWNT is based on the reaction of radical addition to sidewalls of carbon nanotubes. The source of radicals were (pentamethyl cyclopentadienyl)chromiumtricarbonyl dimer, $[Cp*Cr(CO)_3]_2$, which at room temperature dissociates to two metal-centered radicals via reaction (1) in solution, and

$$[Cp*Cr(CO)_3]_2 \rightarrow 2\ Cp*Cr^{\bullet}(CO)_3 \qquad (1)$$

m-chloroperoxybenzoic acid, which is both an oxidant (2) and a source of radicals (3)

$$m\text{-}ClC_6H_4COOOH \rightarrow m\text{-}ClC_6H_4COOH + [O] \qquad (2)$$

$$m\text{-}ClC_6H_4COOOH \rightarrow m\text{-}ClC_6H_4COO\bullet + \bullet OH \rightarrow m\text{-}ClC_6H_4\bullet + CO_2 \ (3)$$

CP723, *Electronic Properties of Synthetic Nanostructures*, edited by H. Kuzmany et al.
© 2004 American Institute of Physics 0-7354-0204-3/04/$22.00

in catalytic decomposition in the presence of Fe remained in HiPco SWNT after purification.

Functionalization Procedures. 1. HiPco SWNT were suspended in oxygen-free toluene and a solid sample of [Cp*Cr(CO)$_3$]$_2$ dimer was added in an argon flow (1:62 mol C/mol dimer). The heterogeneous reaction mixture was stirred at 15°C for 7 hours. 2. HiPco SWNT were suspended in chloroform and a solid sample of m-chloroperoxybenzoic acid (1:8 mol C/mol acid) was added. The heterogeneous reaction mixture was stirred at 60°C for 12 hours. The reaction products were separated from the solution by vacuum filtration through a track membrane (0.2 µm), washed with solvents and dried in air. A black solid film of the resulting materials was easily peeled off from the filter, dried for 8 h in vacuum at T=100 °C and weighed. Table 1 summarize the data on X-ray microanalysis (EDAX) of the materials.

TABLE 1. The content of elements in pristine HiPco SWNT and functionalized carbon nanotubes determined from X-ray microanalysis (EDAX).

Sample	Content of Elements, % wt.				
	C	O	Cl	Fe	Cr
HiPco SWNT (**A**)	89.7	7.4	1.2	1.5	-
HiPco SWNT + m-chloro peroxybenzoic acid (**B**)	75.1	17.3	4.3	1.4	-
HiPco SWNT + [Cp*Cr(CO)$_3$]$_2$ (**C**)	82.3	14	0.5	0.8	1.4

It is seen from Table 1 that the interaction of **A** with m-chloroperoxybenzoic acid results in both the oxidation of nanotubes and the addition of chlorine. The oxidation of nanotubes seems to be realized similarly to the oxidation of fullerenes, which react to form epoxy compounds and m-chlorobenzoic acid that was found in the reaction products. Chlorine possibly adds as m-chloroaryl or m-chlorobenzoyloxyradicals by reaction (3). The data of X-ray microanalysis justified the addition of the Cr complex to nanotubes.

Fig. 1 show the UV-vis-NIR absorption spectra of **A, B,** and **C** suspended in dimethylformamide. The spectra of **B** and **C** show that the fine structure is indistinct as compared to that of **A** material and optical density of a solution increases for **B** and decreases for **C**. This indicates the sidewall functionalization of nanotubes via covalent addition.

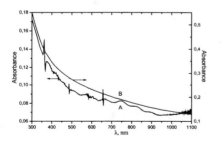 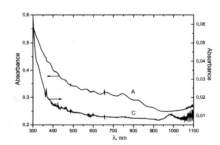

FIGURE 1. UV-vis-NIR absorption spectra of **A, B,** and **C** suspended in dimethylformamide

A comparative analysis of the Raman spectra of **A** and **B** (Fig. 2) shows that the breathing mode (266 cm^{-1}) almost completely disappeared in the spectrum of **B** and the intensity of the tangential mode (1586 cm^{-1}) strongly decreased to become equal in intensity with the disorder mode. Functionalization provides higher intensity of a phonon mode at 1339 cm^{-1} in the Raman spectrum. Its prominence corresponds to the conversion of sp^2 C to sp^3 C on the nanotube during the formation of new bands on the sidewall nanotubes. Higher power of laser irradiation upon spectrum measuring results in the appearance of a peak of breathing mode and the growth of intensity of the peak of tangential mode. Laser annealing of **B** at T = 450 – 500°C results in the full reduction of the Raman spectra of pristine HiPco SWNT. The changes in the Raman spectra indicate that thermal treatment led to the removal of functional groups from the surface of nanotubes and the recovery of pristine structure of HiPco SWNT. Thus it could be concluded that the functionalization of nanotubes through the reaction with m-chloroperoxybenzoic acid is reversible. In contrast to **B**, the Raman spectrum of **C** is almost similar to that of pristine HiPco SWNT: one observes a small widening of the peak of tangential mode and the peaks of breathing modes are well pronounced. The differences in the Raman spectra of functionalized **B** and **C** can be attributed to different mechanisms of nanotube reactions with the reagents studied and the degree of functionalization of nanotubes.

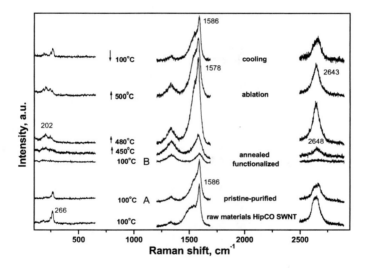

FIGURE 2. Raman spectra (λ_{ex}=514.5 nm) of **A** and **B**.

Thermal gravimetric analysis of **A, B** and **C** in air showed that the temperature of the beginning of weight loss falls in the row: **A** - 420 °C, **C** - 250 °C and **B** – 200 °C. Lower temperature of the beginning of weight loss of functionalized nanotubes can be associated with the violence in the nanotube structure as a result of covalent addition of functional groups, the increased number of groups results in a stronger temperature decrease for **B**. The plots of derivatives (weight loss with respect to temperature) for **A, B** and **C** (Fig. 3) show three maxima of temperatures of weight loss. A comparative analysis of the TGA curves and their temperature derivatives for **A** and **C** shows that they have the same character: weight loss is a three-step process and relative values of these steps (the area under the Gauss curve) are equal. However, the maxima of temperatures of weight loss for **C** are shifted by 160°C to lower temperatures. Lower temperatures of combustion of **C** can be as result of a catalytic action of metallic chromium remaining on the nanotube surface after complex decomposition. The behavior of the TGA curve and its temperature derivative for **B** is strongly different from those for **A** and **C** in both the positions of temperature maxima and the proportion of the steps of weight loss. This difference can be due to a higher degree of functionalization of nanotubes via the reaction with m-chloroperoxybenzoic acid.

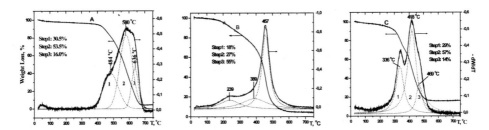

FIGURE 3. TGA, DTG and Gaussian fit of DTG of **A**, **B** and **C**. 25 °C/min, air.

All the samples were examined by transmission electron microscopy (TEM). Fig. 4 shows the TEM image of **A**, which consists of single bundles of the size varying from 20 to 34 nm. The profiles of the bundles are straight and the bundles are long. The TEM image of **B** appears as twisted bundles of approximately the similar size (10 – 15 nm) having a ball-like shape. The material of **C** appears as bundles thicker than those of pristine nanotubes. There are also metal particles and those of an undetermined material covering the nanotube surface in the image. The TEM images of the materials evidence that their structures are different and the differences are stipulated by different types of functionalization of pristine nanotubes.

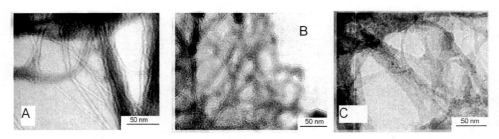

FIGURE 4. TEM images of the **A**, **B** and **C** samples.

ACKNOWLEDGMENTS

This work was supported by the RFBR – 03-03-32727 and the INTAS 01-254 grants.

A Raman Study of Potassium-Doped Double-wall Carbon Nanotubes

H. Rauf[1], T. Pichler[1], F. Simon[2], H. Kuzmany[2]

[1]Leibniz Institute for Solid State and Materials Research Dresden, P.O. Box 270016, D-01171 Dresden, Germany
[2]Institut für Materialphysik, Universität Wien, Strudlhofgasse 4, A-1090 Wien, Austria

Abstract. We report on studies of the n-type doping dependence of the Raman response of double-wall carbon nanotubes (DWCNT). The G-line is found to be shifting upon doping. The direction of the shift depends on if the exciting laser light is in resonance with mainly inner or outer tubes. The RBM response upon doping shows that a charge transfer from the dopant happens predominantly to the outer tubes at low doping. Charge transfer to the inner tubes occurs at higher doping levels. Charge transfer to the inner component is also shown to be starting earlier to the inner tube in DWCNT than to C_{60} in peapods.

INTRODUCTION

Double-wall carbon nanotubes (DWCNT) are a new member in the family of well-ordered carbon structures[1,2]. Due to their formation process from C_{60}-peapods as precursor in the clean interior of SWCNTs, the inner tubes are highly defect-free[3]. As it has been done extensively with other nanostructures [4,5], their electronic structure can be modified in a controlled way by doping with either electron donors or acceptors.

Resonant Raman spectroscopy is a powerful tool to monitor the electronic properties [3] and their changes due to the interaction between transitions in the electronic density of states and the vibrational modes. The intensity of the Raman response is increased dramatically if the excitation energy corresponds to a transition between van Hove singularities of the nanotube.

We present a study on the potassium intercalation (n-type doping) of these DWCNT investigated by Raman spectroscopy. We examine the effect of doping on the position of the G-line main component which is also influenced by the used excitation energy. Then, we investigate the differences in charge transfer to the outer and to the inner tube and compare the outcomes to known results on peapods.

EXPERIMENTAL

All Raman spectra were recorded at 25 K under high vacuum conditions with a Dilor triple monochromator, using different laser lines extending from 1.8 eV to 2.5 eV in 180° backscattering geometry. The response of the spectrometer was calibrated

CP723, *Electronic Properties of Synthetic Nanostructures*, edited by H. Kuzmany et al.
© 2004 American Institute of Physics 0-7354-0204-3/04/$22.00

using gas discharge lamps with well known line positions. The nanotubes used as a base material for the production of the peapods have a mean diameter of 1.5 nm and were prepared by arc-discharge, purified and filtrated into mats of bucky-paper before being filled with C_{60} as described previously [6]. Subsequent annealing in order to form DWCNTs was performed at 1250 °C in dynamic vacuum.

The intercalation was performed *in situ* in a purpose built cryostat by exposing the tube material to potassium vapor in front of the spectrometer at a sample temperature of 500 K. After the exposure, an additional equilibration of at least 30 min was performed in order to increase the homogeneity of the sample. Additionally, samples of pristine nanotubes and peapods were intercalated and measured simultaneously for comparison and estimation of the intercalation degree.

RESULTS AND DISCUSSION

Figure 1a. shows the G-line of pristine and doped DWCNT for exciting laser wavelengths of 488 nm and 568 nm respectively. An excitation with a laser wavelength of 488 nm leads to a Raman resonance with the E_s^{33} transition of the outer tubes, while 568 nm laser light is in resonance with the E_s^{22} transition of the inner tubes [7]. Accordingly, we associate the G-line main components to the outer tubes for the former and to the inner tubes for the latter case. Common for both excitation energies is the loss of Raman intensity in the doped samples (upscaled by a factor of five in the figure). This corresponds to a filling of the conduction band at least up to the corresponding van-Hove singularity in the electronic density of states due to a charge transfer from the potassium and consequently a loss of the Raman resonance.

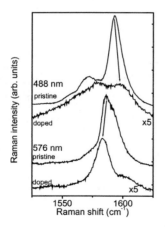
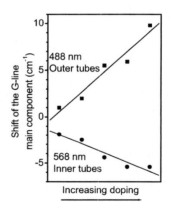

FIGURE 1. a) Raman spectra of pristine and doped DWCNT for laser wavelengths of 488 nm and 576 nm respectively. The shift is indicated by the arrows. b) Doping dependence of the G-line main component position for 488 nm and 576 nm laser wavelength and multiple intercalation steps.

Additionally, a shift of the G-line main component, indicated by the arrows in Figure 1a, can be observed. This shift is depicted in Figure 1b for multiple intermediate doping levels and both excitation energies. For excitation with 488 nm wavelength, the G-line, which we assign in this case to the outer tubes, shifts steadily to higher Raman frequencies with increasing doping. At the highest doping level, the shift adds up to 10 cm^{-1}. This upshift is in reasonable agreement with recent results [8] obtained with a similar excitation energy and p-type doping. Peapods also exhibit this upshift upon K-doping [5] while SWCNT are reported to show a downshift of the G-line main component with only a slight initial upshift at low doping [9]. In contrast to this, when excited with the 568 nm laser, the G-line main component, associated with mainly the inner tubes, exhibits a downshift, although with 5 cm^{-1} smaller than the upshift seen before.

In Figure 2, the radial breathing mode (RBM) is depicted for pristine as well as for low and medium doped DWCNT, this time with 676 nm (a) and 568 nm (b) excitation wavelength. In both cases, the RBM of the inner and outer tubes are clearly distinguishable, with the inner tubes exhibiting a more structured and outspread response and higher Raman shifts in the range from 250 to 350 cm^{-1}. Already at low doping, the outer tube signal decreases strongly for both wavelengths, meaning a loss of Raman resonance for the outer tubes. The resonance for the inner tube is only lost at further doping. This indicates that a charge transfer from the dopant, and as such a filling of the conduction band, happens first to the outer tubes while the inner tube is nearly not affected. This is similar to previous observations on peapods [5]. Only very thick inner tubes (d ≈ 1.0 nm) which are visible in the 568 nm spectrum at a shift of about 250 cm^{-1} (indicated by the arrows in Fig. 2b) are doped as soon as the outer tubes. Due to their larger diameter, the transitions between the van Hove singularities of these big inner tubes are narrower and thus filled up earlier.

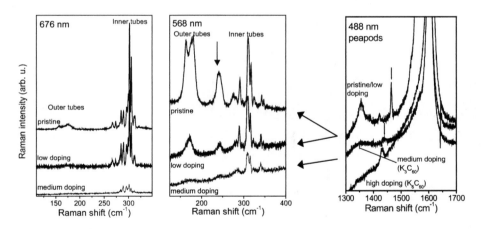

FIGURE 2. Raman response of the RBM of pristine, medium doped and low doped DWCNT excited with (a) 676 nm and (b) 568 nm wavelength laser light. c) Doping dependence of the simultaneously measured peapod sample in the pristine state and at medium and high doping. The arrows mark the position of the $A_g(2)$ mode. The arrows between the figures indicate similar doping levels.

In Figure 2c., the simultaneously intercalated peapod sample, measured with a 488 nm laser, is shown. Here, the charge transfer to the C_{60} can be derived from the shift of the $A_g(2)$ mode [10] indicated by the arrow in the figure. Due to the similar sample volume and exposure time to potassium vapor, this can also be used to give an estimate of the doping level achieved in the DWCNT sample. The upper peapod spectrum corresponds to the pristine state but also is not changed at low doping, where at least some of the inner tubes already start to dope. At a medium intercalation level, where the peapods are doped only up to K_3C_{60}, the inner tube signal of the DWCNT is already strongly supressed. This shows that in DWCNT charge transfer to the inner tube starts earlier than the charge transfer to C_{60} in peapods.

As a reason for this difference in charge transfer to the inner component in DWCNT and peapods, two scenarios can be imagined. The inner tube in DWCNT could have a higher electron affinity than the C_{60} in peapods. Alternatively, the charge transfer could start earlier in the DWCNT, because the distance of the inner tube to the outer tube is somewhere close to the van der Waals radius and therefore smaller than the mean distance between C_{60} and the nanotube in peapods. C_{60} peas only have a diameter of 7 Å, resulting in a pea to tube distance of about 4 Å which is slightly larger than the wall distance of 3.4 - 3.8 Å in DWCNT.

In summary, we found that upon n-type doping with potassium, the shift of the G-line main component depends on which tubes are mainly in resonance with the excitation. For the outer tubes, an upshift, similar to results in the literature, is found, while the G-line shifts down when the excitation is in resonance with mainly the inner tubes. The doping dependence of the RBM response proves that at the beginning of the doping process, charge is only transferred to the outer tubes. Charge transfer to the inner tube sets in later but still earlier than charge transfer to the C_{60}-filling in the peapod reference sample.

ACKNOWLEDGMENTS

This work was funded by the DFG PI440/1, the Austrian Science Funds (FWF) project Nr. 14893 and by the EU projects NANOTEMP BIN2-2001-00580 and PATONN Marie-Curie MEIF-CT-2003-501099 grants

REFERENCES

1. B. W. Smith et al., *Chem. Phys. Lett.* **315**, 31 (1999)
2. S. Bandow et al., *Chem. Phys. Lett.* **337**, 48 (2001)
3. R. Pfeiffer et al., *Phys. Rev. Lett.* **90**, 225501 (2003)
4. A.M. Rao et al., *Nature* **388,** 257 (1997)
5. T. Pichler et al., *Phys. Rev. Lett.* **87** , 267401 (2001)
6. H. Kataura et al., *Synth. Met.* **121** , 1195 (2001)
7. F. Simon et al., cond-mat/0404110
8. G. Chen et al., *Phys. Rev. Lett.* **90** , 257403 (2003)
9. N. Bendiab et al., *Phys. Rev. B.* **63** , 153407 (2001)
10. T. Pichler et al., *Phys. Rev. B.* **45** , 13841 (1992)

A Photoemission Study of Potassium-Doped Single Wall Carbon Nanotubes

T. Pichler[1], H. Rauf[1], M. Knupfer[1], J. Fink[1], H. Kataura[2]

[1]Leibniz Institute for Solid State and Materials Research Dresden, P.O. Box 270016,
D-01171 Dresden, Germany
[2] Nanotechnology Research Institute, National Instituteof Advanced Industrial Science and Technology
(AIST) 1-1-1 Higashi, Tsukuba, Ibaraki 305-8562, Japan

Abstract. We report on the doping dependence of the electronic structure of the valence band of mats of single wall carbon nanotubes (SWCNT) using high resolution photoemission spectroscopy as probe. For the pristine SWCNT about one third are metallic Tomonaga-Luttinger-Liquids (TLL) which is directly related to a power law scaling in the density of states. The changes in the valence band of these doped one dimensional nanostructures explicitly show the effect of the dimensionality on the electronic structure. At low doping we observe a crossover from a TLL behavior to a normal Fermi liquid whereas at high doping levels at low binding energy a Fermi level is found.

INTRODUCTION

One very efficient possibility to change the electronic properties by doping with electrons or holes is via intercalation. For single wall carbon nanotubes (SWCNT) the intercalation physics has been analyzed in some detail. In contrast to graphite intercalation compounds no distinct intercalation stages have been observed as yet. Alkali metal intercalation of mats of bundled SWCNT takes place inside the channels of the triangular bundle lattice and leads to a shift of the Fermi energy, a loss of the optical transitions [1] and an increase in conductivity by about a factor of thirty [2,3]. A complete charge transfer between the donors and the SWCNT was observed up to doping saturation, which was achieved at a carbon to alkali metal ratio of about seven [2]. However, much less has been reported on direct measurements of the low energy properties of the valence band as a function of doping. First results using photoemission revealed a Fermi edge at high doping [4].The charge transport properties of carbon nanotubes have been investigated intensively over the last years since they represent archetypes of a one dimensional system [5]. For such metallic systems, conventional Fermi liquid theory fails since it becomes unstable due to long range Coulomb interactions. Such a one dimensional paramagnetic metal is called a Tomonaga-Luttinger Liquid (TLL) and shows peculiar behavior such as spin charge separation and interaction dependent exponents in the density of states, correlation function and momentum distribution of the electrons [5,6]. Recently, the density of states (DOS) of the valence band electrons of mats of single wall carbon nanotubes

CP723, *Electronic Properties of Synthetic Nanostructures*, edited by H. Kuzmany et al.
© 2004 American Institute of Physics 0-7354-0204-3/04/$22.00

was directly monitored by angle integrated high resolution photoemission experiments [7].

In this contribution, we report on a detailed study of the change in the low energy electronic properties in mats of SWCNT as a function of potassium intercalation (n-doping) using high resolution photoemission as a probe. From an analysis of the valence band photoemission we observe that at low doping, first the metallic SWCNT are doped without change of the TLL parameter whereas the response of the van Hove singularity (vHs) shifts to a higher binding energy [8]. For further increases in the doping level the peaks of the vHs vanish and a pronounced Fermi edge is observed showing the transition from a TLL to a Fermi liquid.

EXPERIMENTAL

Mats of purified SWCNT which consist of a mixture of semiconducting and metallic SWCNT with a narrow diameter distribution which is peaked at 1.37 nm with a variance of about 0.05 nm [7] were produced by subsequent dropping of SWCNT suspended in acetone onto NaCl single crystals. The produced SWCNT film of about 500 nm thickness was floated off in distilled water and recaptured on sapphire plates. For the photoemission experiments the sample was mounted onto a copper sample holder and cleaned in a preparation chamber under ultra high vacuum (UHV) conditions (base pressure 9×10^{-11} mbar) by electron beam heating to 800 K. Electrical contact of the SWCNT film was established by contacting the surface to the sample holder via a Ta foil. Then the sample was cooled down to T=35 K and transferred under UHV conditions to the measuring chamber and its electronic properties were analyzed using a hemispherical high resolution Scienta SES 200 analyzer. For the angle integrated valence band photoemission spectra, using monochromatic HeIα (21.22 eV) excitation, the energy resolution was set to 10 meV. The core level photoemission measurements (XPS) were performed with an energy resolution of 400 meV using monochromatic AlKα excitation (1486.6 eV). The Fermi energy and overall resolution was measured using freshly cleaned Ta. The intercalation was performed *in situ* after heating the sample to 450 K using commercial SAES potassium getter sources. After subsequent exposure to the dopant vapor an additional equilibration for about 30 min at 450 K was performed to increase the homogeneity of the sample.

RESULTS AND DISCUSSION

We first discuss the analysis of the sample stoichiometry and purity using core level photoemission. The sample purity was checked by an overview XPS spectrum (see inset of Fig. 1). Only the signals of carbon could be observed with no contamination from oxygen or catalyst particles. The detailed analysis of the sample stoichiometry dependence was performed in the range of the C1s and K2p levels. The results are depicted in Fig. 1 for three different intercalation levels. The doping level was determined by the ratio of the C1s/K2p intensities taking into account the different photo-ionisation cross section. Consistent with previous experiments [4] the pristine mats of SWCNT show an asymmetric Donijac-Sunjic lineshape of the C1s core spectra with a width which increases by about a factor of three upon potassium intercalation. The binding energy of the C1s line is shifted by about 0.8 eV to higher values for the highest doping (here C/K=15). This can be explained by an up shift of the Fermi level into the conduction band in good agreement with results from electron energy-loss and Raman spectroscopy [2,9].

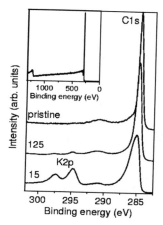

FIGURE 1. Core level photoemission of pristine and potassium intercalated mats of SWCNT at different doping levels as indicated by the C/K ratios. The inset shows a survey of the pristine sample.

We now turn to the detailed analysis of the high resolution valence band photoemission experiments at T=35K. The results are depicted in Fig. 2. In the left panel an overview of the valence band spectrum of the SWCNT mats is shown. Two broad peaks are observed at about 3 eV and at about 8 eV binding energy. These features are similar to angle integrated photoemission of graphite and to previous results on SWCNT mats [4,7,8] and correspond to the maxima in the DOS of the π and σ bands, respectively. In the right panel of Fig. 2 a detailed analysis of the photoemission response close to the Fermi level was performed. The photoemission peaks corresponding to the first and second van Hove singularities (vHs) of the semiconducting SWCNT (S_1, S_2) and that of the first vHs of the metallic SWCNT (M_1) are observed at binding energies of 0.44, 0.76 and 1.06 eV with a full width half maximum of about 0.11 eV, respectively. A key manifestation of the TLL state is the renormalisation of the DOS ($n(E)$) near the Fermi level which shows a power law dependence $n(E) \sim E^{\alpha}$ where α on the size of the Coulomb interaction and can be expressed in terms of the TLL parameter g as $\alpha=(g-g^{-1}-2)/8$ [5]. As can be seen in the inset of Fig. 2 we observed a power law scaling of $\alpha=0.43$, g=0.18 which within experimental error is identical to the previously reported value [7].

We now turn to an analysis of the doping dependence of the valence band of SWCNT as a function of potassium content. The results of the photoemission response in the vicinity of the Fermi level are depicted in Fig. 3. The left panel shows the evolution of the spectra at low doping (C/K > 100, charge transfer less than 0.01 e⁻/C), the right panel at high intercalation (up to C/K=15). With increasing doping, the peaks corresponding to the SWCNT vHs (S_1, S_2, and M_1) shift to higher binding energies due to a filling of the conduction band of the SWCNT with K 4s electrons. Interestingly, the conduction band of the semiconducting SWCNT is not filled at low doping (<0.002 e⁻/C, C/K= 500) since for this doping level the S_1 and S_2 peaks shift paralell up to 0.3 eV to higher binding energy which is not high enough to reach the first unoccupied vHs (S_1^*).

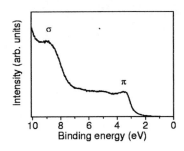 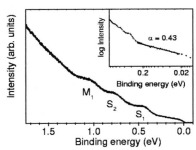

FIGURE 2. Left panel: Valence band photoemission of pristine SWCNT measured at T=35 K using HeIα excitation in the range of the π and σ electrons. Right panel: Photoemission response of SWCNT near the Fermi level. S_1, S_2 and M_1 assign peaks corresponding to the SWCNT vHs as mentioned in detail in the text. The inset shows an analysis of the power law scaling within the TLL theory close to the chemical potential.

For further a increase of the doping up to <0.008 e⁻/C, C/K= 125 the unoccupied S_1^* vHs is filled. Consequently, the S_1 and S_2 vHs only shift by an additional 0.07 eV to higher binding energies due to the presence of a high unoccupied DOS. Even higher doping leads to a shift of the S_1 peak beyond the position of the S_2 peak (see right panel of Fig. 3) concomitant with a smearing out and finally a vanishing of the peaks related to the SWCNT vHs. At very high doping levels the former unoccupied $S1^*$, S_2^*, and M_1^* bands are filled with electrons and corresponding peaks should show up in photoemissionbelow 1 eV binding energy. However, as can be seen in the figure there are no additional features in the spectra. This fact can be explained by an increasing number of scattering centers (K⁺ counter ions) and/or by an increasing intertube interaction within the SWCNT bundle in the intercalation compound [8]. The overall spectra of these highly doped samples are also very similar to equivalent GIC. The Fermi level shift can be extracted from the shift of the π band at about 3 eV. For the sample with C/K=15 (0.066 e⁻/C) we observe ΔE_F=1 eV.

Regarding the TTL scaling factor a we observe a distinct change to a Fermi liquid at doping levels of C/K=125 [8]. This is exactly the amount of doping which is high enough to drive all SWCNT within a bundle of SWCNT metallic. This also means that up to this doping level the long range Coulomb interaction is essentially unaffected by the potential of the counter ions. This is in good agreement with predictions showing that for a TLL the power law scaling parameter α is not affected until the first unoccupied vHs is reached and additional conduction channels are possible [10].

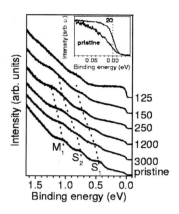 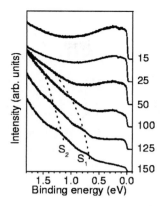

FIGURE 3. Doping dependence of the valence band photoemission spectra in the vicinity of the Fermi level (left panel: low doping, right panel: high doping levels). The numbers correspond to the C/K ratio derived from core level photoemission. The dotted lines are guidelines for the evolution of the S_1,S_2 and M_1 peaks with increasing doping. The inset shows the region around E_F (dotted line) in an extended range for pristine and highly doped (C/K=20) SWCNT mats.

In summary, we have studied the doping dependence of the SWCNT valence band in bundles of SWCNT using photoemission spectroscopy as probe. The character of the electron liquid changes as a function of dopant concentration, i.e. as a function of the position of the Fermi level.

ACKNOWLEDGMENTS

This work was funded by the DFG PI440/1, We thank S. Leger, R. Hübel, and K. Müller for technical assistance. H.K. acknowledges support by an Industrial Technology Research Grant Program in '03 from the New Energy and Industrial Technology Development Organization (NEDO) of Japan.

REFERENCES

1. S. Kazaoui et al. , *Phys. Rev. B* **60**, 13339 (1999).
2. T. Pichler et al., *Solid State Commun.* **109**, 721 (1999); X. Liu, et al., *Phys. Rev. B* **67**, 125403 (2003); T. Pichler, *New Journal of Physics* **5**, 23 (2003).
3. R.S. Lee et al., *Nature* **388**, 255 (1997).
4. S. Suzuki et al., *Phys. Rev. B* **67**, 115418 (2003)
5. R. Egger et al., *Phys. Rev. Lett.* **79**, 5082 (1997); C. Kane et al., *Phys. Rev. Lett.* **79**, 5086 (1997)
6. J. Voit, "One-dimensional Fermi liquids" in *Rep. Prog. Phys.* **58**, 977 (1995)
7. H. Ishii et al., *Nature* **426**, 540 (2003).
8. H. Rauf et al., cond-mat. 0404535
9. T. Pichler et al., *Phys. Rev. Lett.* **87**, 267401 (2001).
10. R. Egger, *Phys. Rev. Lett.* **83**, 5547 (1999)

Electronic Structure of Single-Wall Carbon Nanotubes and Peapods; Photoemission Study

H. Kataura[1], H. Shiozawa[1], S. Suzuki[1], Y. Achiba[1], M. Nakatake[2], H. Namatame[3], M. Taniguchi[3], H. Yoshioka[4], and H. Ishii[1]

[1]Graduate School of Science, Tokyo Metropolitan University, Tokyo 192-0397, JAPAN
[2]Photon Factory, High Energy Accelerator Research Organization, Tsukuba 305-0801, Japan
[3]Hiroshima Synchrotron Radiation Center, Hiroshima University, Higashi-Hiroshima, 739-8526, Japan
[4]Department of Physics, Nara Women's University, Nara 630-8506, Japan

Abstract. Electronic structure of single-wall carbon nanotubes was investigated by photoemission spectroscopy. Oscillations in electronic density of states due to one-dimensional van Hove singularities were clearly observed near the Fermi level, which can be well reproduced by simple tight-binding calculation with assumptions a broadening and an energy shift probably due to a charge transfer from the metal substrate. Furthermore, in the vicinity of the Fermi level, two kinds of typical power-law dependences were observed that can be regarded as evidence of Tomonaga-Luttinger liquid states. Both exponents show very similar values 0.46 and 0.48 good agreement with theoretical prediction. Results on fullerene peapod system are also discussed.

INTRODUCTION

Carbon nanotubes are regarded as quasi one-dimensional system because of the quantized wave vectors for the circumference direction of the cylindrical structure [1]. It is well known that in one-dimensional metals, because of the one-dimensional fluctuations, Fermi liquid is no more stable but Tomonaga-Luttinger liquid (TLL) state is coming up [2]. In TLL state, electronic density of states shows power-law dependences to the binding energy and the temperature. To date, these characteristics were tried to be observed mainly by transport measurements [3]. However, it is difficult to get good contact of between the electrode and the carbon naonotube. Furthermore, Coulomb blockade affects the transport properties at low temperatures. Photoemission spectroscopy has an advantage to measure the density of states without such problems, but it requires cleanliness of the surface of the sample. We have successfully prepared extremely high purity single-wall carbon nanotubes (SWCNTs) and then measure the photoemission spectra of them at low temperature [4].

EXPERIMENTAL

SWCNT sample was prepared by the laser ablation method by using NiCo alloy catalyst. Raw soot was refluxed in 15% hydrogen peroxide for two hours to burn out amorphous carbon particles. Then the soot was washed in sodium hydroxide water

CP723, *Electronic Properties of Synthetic Nanostructures*, edited by H. Kuzmany et al.

solution to dissolve byproducts. After the filtration, remaining metal catalyst particles in the sample was washed out by diluted hydrochloric acid. By the filtration, we got papers of high-purity SWCNTs. We used a small piece of the paper directly for the photoemission measurements. The SWCNT paper was heated in vacuum to remove all the other organic impurities before the measurements. Peapod samples were prepared by a vapor phase reaction. Some small pieces of SWCNT paper were installed in a quartz ampoule with fullerene powder. The ampoule was heated after evacuation to remove impurity molecules. Then the ampoule was sealed and heated in a furnace to 900 K to vaporize fullerenes. After keeping several hours, the ampoule was cooled down and cleaved. The obtained peapod papers were heated in dynamic vacuum to 900K to remove fullerenes outside SWCNTs. Filling factor of the sample was checked by a Raman spectrum that was already calibrated by X-ray diffraction measurements.

Angle integrated photoemission measurements were done at Photon Factory in High Energy Accelerator Research Organization, and at Hiroshima Synchrotron Radiation Center in Hiroshima University for high-resolution measurements. We have measured two kinds of SWCNTs with different mean diameter and peapods. A graphite crystal and a gold thin film were measured for references.

RESULTS AND DISCUSSION

1-D van Hove Singularities in SWCNTs

The photoemission spectra in the whole measured range up to 30 eV for SWCNTs and the graphite are shown in an inset of Fig. 1. They are very similar from each other except for in a range near the Fermi level. This is reasonable since both materials construct of common two dimensional grapheme sheets. Near the Fermi level, however, we found a significant difference between them. Figure 1 shows two kinds of typical photoemission spectra of SWCNT samples. In both cases, beautiful oscillations are observed around 0.4, 0.8 and 1.0 eV which were never observed in the graphite. The corresponding peak positions in two samples are slightly different from each other. Prof. Maruyama has calculated density of state of many kinds of individual SWNTs by simple tight-binding model and opened them in his web page. We used his results to the simulation indicated in Fig. 1 solid curves. The lower most spectrum shows simply summed density of states by assuming Gaussian diameter distributions for sample A1. The summed spectrum is shifted 0.1 eV to deeper to fit to the experimental results. This shift is probably caused by charge transfer from sample holder made of cupper or gold to SWCNT sample and is very important to discuss TLL states in SWCNTs later. By applying broadening due to spectroscopic resolution, life-time broadening, and bundle effects, finally we succeeded to reproduce photoemission spectra of both SWCNT samples. The difference in the oscillation in two SWCNT samples can be explained by the difference in the diameter distributions. These results are consistent with the optical absorption spectra but more important because photoemission spectra directly reflect the density of states instead of the joint density of states in the optical absorption. There are still disagreements between the

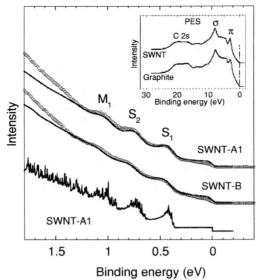

FIGURE 1. Typical photoemission spectra of two different diameter SWCNT samples. Circles indicate experimental results and solid curve the simulated spectra by simple tight-binding model. The lower most spectrum shows simulated density of states from that of individual SWCNTs by assuming a Gaussian diameter distribution. Inset shows whole range spectra of SWCNT sample and the graphite.

simulation and experimental result. In the higher binding energy region than 1.5 eV, experimental results show larger intensity. This additional intensity is probably a contribution from sigma bonding states that are not considered in the tight-binding model. The second disagreement is lower intensity in the experimental results than the simulation in lower binding energy region than 0.3 eV. This difference probably caused by TLL behavior in carbon nanotubes.

TLL States in Carbon Nanotubes

Since the metallic carbon nanotubes have two crossing linear band at the Fermi level, they can be regarded as an ideal TLL system. Left figure in Fig. 2 indicates calculated density of states in the carbon nanotube and simulated photoemission spectra by multiplying the Fermi distribution function. Right figure in Fig.2 indicates the high-resolution photoemission spectra of sample A1 for various temperatures. We found that the simulated photoemission spectrum at 150 K is quite similar to the experimentally obtained spectrum at 310 K. This is very interesting because theoretically determined temperature is a function of the cut off energy in the integration. Here we used a conventional cut off energy but the present results suggest that it should be modified to reproduce the experimental results.

In the whole temperature range, we found that the photoemission intensity of SWCNTs near the Fermi level is proportional to $\omega^{0.46}$. On the other hand, photo emission intensity at Fermi revel varied with the temperature as $T^{0.48}$. Interestingly, these two exponent show good agreement within experimental accuracy. We believe these two power law dependences are typical evidence of TLL states in SWCNTs.

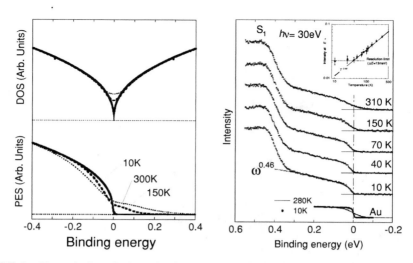

FIGURE 2. Theoretically calculated density of states and corresponding photoemission spectra of SWCNTs (left figure), and high-resolution photoemission spectra of SWCNT sample A1 at various temperatures (right figure). The inset in the right figure shows temperature dependence of the density of states at Fermi level.

Photo Emission Spectrum of Peapods

The most interesting issue of C_{60}-peapod system is about the position of LUMO band of C_{60}. Okada *et al.* calculated the electronic structure of C_{60}-peapods and showed the LUMO band of C_{60} cross the Fermi level that means metallic behavior of C_{60}-linear chain inside SWCNT [5]. Actually, Kavan *et al.* measured electrochemical doping effect on the resonance Raman scattering of C_{60}-peapods and found that LUMO band of C_{60} is stayed between the first gap states of semiconductor SWCNTs, which is consistent with the theoretical prediction [6]. It was expected that photoemission spectrum of peapod gives us more direct information about LUMO band position of C_{60} molecules inside SWCNTs.

Figure 3 indicates photoemission spectra of empty SWCNTs and C_{60}-peapods near the Fermi level. We could not find any difference between two spectra while we observed large differences in higher binding energy region that are attributed to the π- and σ- band of C_{60} (not shown here). As shown in Fig. 3, we reproduced both the spectra by the simple tight-binding calculation. This means that electronic structure of SWCNT was not largely affected by C_{60} molecules inside and that the position of LUMO band of C_{60} is higher than the Fermi level in this system.

Very recently, Otani *et al.* calculated the position of LUMO band of C_{60} as a function of diameter of SWCNTs [7]. According to their results, the mean diameter of our SWCNTs used in this work is too small to observe metallic C_{60} state. It is very interesting to measure the diameter dependent electronic structure of C_{60}-peapods including TLL state. If the LUMO band of C_{60} crosses the Fermi level, TLL state of SWCNTs should be modified since the LUMO band increases the metallic channel.

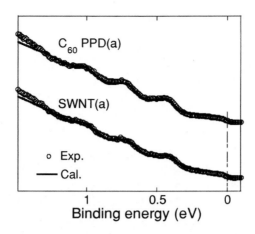

FIGURE 3. Photoemission spectra of SWCNTs and C_{60}-peapods. Solid curve indicates simulated spectrum calculated by tight-binding model.

ACKNOWLEDGMENTS

This work was in part supported by Grant-in-Aid for Scientific Research (A) (13304026) from the Ministry of Education, Culture, Sports, Science and Technology (MEXT) of Japan and Industrial Technology Research Grant Program in '03 from New Energy and Industrial Technology Development Organization (NEDO) of Japan.

REFERENCES

1. Saito, R. Dresselhaus, G. and Dresselhaus, M. S., Physical Properties of Carbon Nanotubes, (Imperial College Press, Imperial College, London, 1998.
2. Grioni, M. & Voit, J. in Electronic spectroscopies applied to low-dimensional materials. eds. Hughes, H. P.& Starnberg, H. I., Kluwer academic, Dordrecht, 2000, pp. 209-281.
3. Bockrath, M. et al. Luttinger-liquid behaviour in carbon nanotubes. Nature **397**, 598-601 (1999).
4. Ishii, H. *et al.*, Nature Nature, **426**, 540 - 544 (2003).
5. Okada, S., Saito, S. and Oshiyama, A., Phys. Rev. Lett. **86**, 3835 (2001).
6. Kavan, L., Dunsch, L. and Kataura, H., Chem. Phys. Lett. **361**, 79 (2002) .
7. Otani, M., Okada, S. and Oshiyama, S. Phys. Rev. B **68**, 125424 (2003).

NANOTUBE FILLING AND DOUBLE-WALL CARBON NANOTUBES

Metal-filled Nanotubes: Synthesis, Analysis, Properties and Applications

D. Golberg, Y. Bando, Y.B. Li, J.Q. Hu, Y.C. Zhu, Y.H. Gao, C.C. Tang

Advanced Materials Laboratory, National Institute for Materials Science,

Namiki 1-1, Tsukuba, Ibaraki 305-0044, Japan

Abstract. Carbon, Boron Nitride and novel inorganic, *e.g.* MgO, In_2O_3, Ga_2O_3 and SiO_2, nanotubes, were filled with various metals, ranging from low-melting point metals, *i.e.* Ga and In, to *3d*-ferromagnetic metals, like Co, Fe, Ni and alloys of those. The structures and properties of the tubes, and fillings were analyzed using high-resolution transmission electron microscopy. Intriguing "nanothermometer"-like behavior was displayed by Ga and In liquid columns inside inorganic tubes. Phase transformations in the Mg-O system within BN tubular cores, which result in O_2 outflow from the tubes, were documented. This creates first nanoscale oxygen burner/generator. Unusual crystal lattices were found in ferromagnetic metals crystallized within the tube channels. Potential applications of the discovered functional nanostructures and associated phenomena are discussed.

INTRODUCTION

Nanotube filling with metals represents an intriguing branch of nanomaterials research. An internal nanotube cavity, only several nanometers in diameter, promotes a confinement effect on metal crystallization, melting, flow, ·capillarity etc. Initial works on nanotube filling have been performed soon after the discovery of C nanotube in early 90th [1-4], but the lack of successful practical applications of the observed phenomenon has significantly hindered the research later on. Recently it has been realized that if a low melting point, wide-liquid range metal is encapsulated into a C nanotube, the resultant structure actually behaves as a nanoscale thermometer owing to a linear change in metal column height against temperature [5, 6]. A carbon nanotube-based thermometer, which has a diameter of only several tens of nanometers, has a serious drawback though: C nanotube is unstable in air or other oxidizing environment, if temperature reaches 400°C or more. Then, research has been carried out with respect to stable inorganic nanotube substitution for C nanotube in the thermometer [7-9]. This may significantly increase the range of possible temperature measurements using filled nanotubes. In addition, application of thermally and chemically tough BN nanotubes as nanocrucibles for various metallurgical and chemical high-temperature operations on a nanoscale has recently become possible [10, 11]; in fact, bulk crucibles made of BN have been used for material syntheses for many years. New phenomenon has recently become of an

CP723, *Electronic Properties of Synthetic Nanostructures*, edited by H. Kuzmany et al.
© 2004 American Institute of Physics 0-7354-0204-3/04/$22.00

increased attention: the change in thermodynamics and kinetics of metal nanowire/nanorod crystallization confined with a nanotube shape; this may result in the appearance of unexpected crystal structures within metals embedded [12, 13], which in turn, may create novel functional properties, i.e. electrical, magnetic, piezo- and ferroelectric etc.

The present paper consecutively documents all the above-mentioned breakthroughs with respect to nanotube filling with various metals. The resultant structure functionality is particularly highlighted.

EXPERIMENTAL

All nanostructures discussed herewith were synthesized in an induction furnace in a temperature range of 700-2000°C. Special attention has been paid to careful preparation of the precursors for a given nanomaterial synthesis. The details of each synthetic run were reported in our recent papers [5-16]. The nanostructures were examined in a high-resolution, transmission electron microscope (TEM) equipped with a field emission gun, an electron energy loss spectrometer and an energy dispersion X-ray detector. These analytical tools allowed us to precisely measure nanotube and filling chemical compositions. Heating experiments were performed during *in-situ* TEM and the series of TEM images were recorded using a CCD camera.

RESULTS AND DISCUSSION

Figure 1 displays various kinds of prepared nanoscale thermometers. Clearly, wide liquid range metals, *i.e.* Ga and In, were embedded into the nanotube channels.

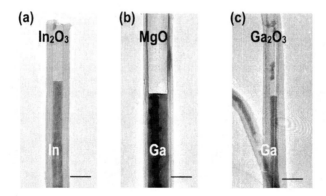

FIGURE 1. Some types of inorganic nanotube thermometers. Scale bars 50 nm. a) an In-filled In_2O_3 nanotube; b) a Ga-filled MgO nanotube; and c) a Ga-filled Ga_2O_3 nanotube. Both In and Ga columns expand linearly inside tubes, after melting, making possible assignment of a particular column height/length to a given temperature experienced by a filled nanotube inside a TEM.

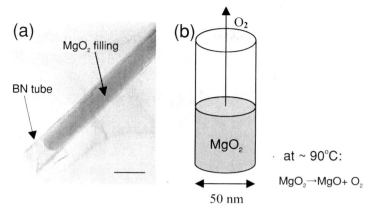

FIGURE 2. (a) TEM image of a MgO$_2$-filled BN nanotube exhibiting the open-tip-end. On moderate thermal or electron irradiation-induced heating, the unstable oxygen-rich MgO$_2$ perovskite phase transforms to stable MgO, thus pure molecular oxygen outflows from the open-tube-end, as schematically shown in (b). Scale bar in (a) 50 nm.

Figure 2 depicts another type of functional device made of a BN nanotube in which the oxygen-release compound, MgO$_2$, was encapsulated under a high-temperature synthesis followed by slow cooling to room temperature. The peroxide is unstable on heating and transforms to a stable MgO compound. It means that it is possible to locally release molecular oxygen from an open nanotube tip. Thus first ever nanoscale oxygen burner or generator comes into effect. The device may find smart technological applications in on-demand preparation of quantum dots, and may help to analyze living and deteriorating conditions of various viruses.

Another example of BN nanotube filling is presented in Figure 3. Iron fills the nanotube *via* capillarity, as evidenced by the specific morphology of a resultant nanostructure. The prepared nanocable is, in fact, ferromagnetic. A BN protecting tubular shield effectively prevents Fe nanowire oxidation and degradation of ferromagnetism when the material is under use in air and at high temperatures.

Surprisingly, *3d*-metal crystallization in the space confined with a nanotube shape may lead to unusual crystal structure appearance [12, 13]. For example, Co may take *f.c.c* rather than thermodynamically stable hexagonal crystal lattice at room temperature. Oppositely, Fe-based nanowires may exist with the *h.c.p.* crystal structure. Various crystallographic domains frequently alternate along and/or across the ferromagnetic fillings leading to numerous stacking faults within the nanostructures. These phenomena open up a new branch of the so-called "nanometallurgy." Detailed studies on the striking phase diagram changes in well-known metallic systems (when a material serves at the nanoscale and its growth is confined with the nanotube shape) is of our first priority with respect to the forthcoming research on metal-filled nanotubes.

Large-scale filling of nanotubes is, indeed, a challenging task, since it is extremely difficult to achieve a stable capillarity-driven action. Alternatively, the complex and not yet reliable filling technique may be substituted by a simultaneous growth of a nanowire, and a tubular shield [14], or a coating technique [15, 16].

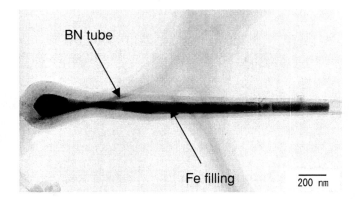

BN tube

Fe filling

200 nm

FIGURE 3. An iron-filled BN nanotube. The nanostructure was prepared *via* a two-stage process in which, firstly, Fe-particle filled C nanotubes were synthesized *via* chemical vapor deposition and then the material was heated in an atmosphere of boron oxide and nitrogen at the temperature equal to the melting point of Fe. Thus Fe fills the tube channel owing to capillarity.

The two methods may significantly improve the yield of the resultant filled nanostructures at the expense of unfilled ones, as we demonstrated by successful BN covering on SiC, Fe_3C, Si, ZnS and some other semiconducting nanowires.

Extensive analyses using electron diffraction have displayed the existence of structural orientation relationships between 3*d*-fillings and BN tubular shells. For cubic Co and Fe fillings, a prevailing structural configuration was observed. This reveals a certain {110} plane parallel to the tube axis. The same is true for hexagonally-packed fillings: we typically observed the (1121) planes of hexagonal lattices parallel to the BN tube axes. These facts imply the existence of an epitaxial crystallization mode of encapsulated ferromagnetic nanowires on the BN innermost wall templates.

It is finally noted that metal-filled inorganic nanotubes are of high interest from both fundamental (nanoscale crystal growth, stress-strain relationship within a nanotube channel) and practical viewpoints. A variety of functional materials through using metal-filled inorganic nanotubes may be awaited, including nanotube-based nanoscale thermometers, oxygen burners/generators and ferromagnetic nanowires thermally and chemically protected with the tough BN or other inorganic tubular shields.

SUMMARY

TEM studies on metal-filled inorganic nanotubes reveal their possible functionality as nanoscale thermometers (liquid metal fillings), oxygen releasing nanopipes (magnesium peroxide fillings) and ferromagnetic nanocables (3*d*-metal fillings). Applications of the observed phenomena for the growing demands of nanotechnology are in the agenda.

ACKNOWLEDGEMENTS

The authors thank Y. Uemura and K. Kurashima for a continuous technical support, M. Mitome and F.F. Xu for many fruitful discussions, and K. Fushimi, and M. Hasegawa for the assistance in experimental work.

REFERENCES

1. Ajayan P.M. and Iijima S. *Nature* **361**, 333-336 (1993).
2. Ajayan P.M., Ebbesen T.W., Ichihashi T., Iijima S., Tanigaki K., and Hiura H., *Nature* **362**, 522-524 (1993).
3. Dujardin E., Ebbesen T.W., Hiura H., and Tanigaki K., *Science* **265**, 1850-1853 (1994).
4. Tsang S.C., Chen Y.K., Harris P.J.F., and Green M.L.H., *Nature* **372**, 159-162 (1994).
5. Gao Y. and Bando Y., *Nature* **415**, 599 (2002).
6. Gao Y., Bando Y., Liu Z., and Golberg D., *Appl. Phys. Lett.* **83**, 2913-2915 (2003).
7. Li Y., Bando Y., and Golberg D., *Adv. Mater.* **15**, 581-585 (2003).
8. Li Y., Bando Y., Golberg D., and Liu Z., *Appl. Phys. Lett.* **83**, 999-1001 (2003).
9. Li. Y.B., Bando Y., and Golberg D., *Adv. Mater.* **16**, 37-40 (2004).
10. Golberg D., Bando Y., Fushimi K., Bourgeois L., Tang C.C., and Mitome M. *J. Phys. Chem.* **107**, 8726-8729 (2003).
11. Golberg D., Bando Y., Mitome M., Fushimi K., Tang C.C., *Acta Mater.* (2004), in press.
12. Golberg D., Xu F.F., and Bando Y., *Appl. Phys. A* **76**, 479-485 (2003).
13. Xu F.F., Bando Y., Golberg D., Hasegawa M., and Mitome M., *Acta Mater.* **52**, 601-606 (2004).
14. Tang C.C., Bando Y., Golberg D., Ding X., and Qi S., *J Phys. Chem.* B. **107**, 6539-6543 (2003).
15. Tang C.C., Bando Y., Golberg D., Mitome M., Ding X.X., Qi S., *Appl. Phys. Lett.* (2004), in press.
16. Zhu Y.C., Bando Y., Xue D.F., Xu F.F., and Golberg D., *J. Amer. Chem. Soc.* **125**, 14226-14227 (2003).

The Growth Process of Nanotubes in Nanotubes

M. Holzweber[1], C. Kramberger[1], F. Simon[1], R. Pfeiffer[1],
M. Mannsberger[1], F. Hasi[1], H. Kuzmany[1], H Kataura[2]

[1]Institut für Materialphysik,Universität Wien, Strudlhofgasse 4, A–1090 Wien, Austria
[2]Department of Physics, Metropolitan University of Tokyo, Tokyo, Japan

Abstract. We have investigated the growth process of carbon nanotubes in carbon nanotubes as a function of anihilation temperature and duration. Raman spectroscopy is applied to characterize the produced DWCNTs. The RBM scales as 1/d and renders the very narrow new inner CNTs easily detectable. We found a dramatic decrease of the fullerene concentration in the samples before the inner CNTs start to emerge. The growth of the new inner CNTs starts with the smallest diameter nanotubes, before it is uniformly extended to all possible diameters.

INTRODUCTION

Since their discovery in 1991, carbon nanotubes (CNTs) have raised a growing interest in the scientific community [1,2]. This interest is mainly based on their unique quasi one-dimensional structure and on their high application potential. Various ways of producing CNTs have been established since then. One of them is based on annealing C_{60} peapods [3] at high temperatures in dynamic vacuum. During this procedure the one dimensional chain of fullerenes is turned into a secondary inner SWCNT [4]. The outer SWCNT, that was filled with the fullerenes formerly, acts as a nano-reactor. It enables the coalescence of the fullerenes into a very narrow inner CNT. Moreover it acts as a physical barrier and guarantees cleanroom conditions for the growth process. As the reaction is effectively shielded the new inner CNT is made from chemically pure carbon.

Raman spectroscopy has been used extensively to study SWCNTs. This holds in particular for the tubes grown inside the primary tubes. Three main signatures are important in the Raman spectra. At low frequencies, between 150 and 450 cm^{-1}, one observes the response from the radial breathing mode (RBM), at high frequencies, between 1450 and 1600 cm^{-1} the graphitic lines (G-lines) appear and at medium high frequencies, between 1300 and 1400 cm^{-1}, a defect induced line (D-line) is seen.

For standard diameter tubes the RBM exhibits a broad structure of overlapping lines. For inner tubes sharp RBM lines are observed in spectra for all geometrically allowed chiralities [5,6]. The concentric system of the two grown nanotubes is now conveniently assigned as doublewall carbon nanotubes (DWCNTs).

To study the growth process of the inner tubes in detail, we analyzed various steps of the transformation process. The following factors are expected to have a significant influence: the annealing temperature, duration of annealing, and the details of the heating and cooling phase. In the following we concentrate on the annealing time. This

CP723, *Electronic Properties of Synthetic Nanostructures*, edited by H. Kuzmany et al.
© 2004 American Institute of Physics 0-7354-0204-3/04/$22.00

was done in order to see intermediate states of the transformation process. The Raman response from these states was compared with the response from fully transformed samples.

EXPERIMENTAL

For the heat treatment we used a horizontal tube furnace with a tube diameter of 38 mm. The sample was placed in a quartz tube and pumped to a dynamic vacuum better than 10^{-7} mbar. Then it was put into the hot furnace. After the treatment the sample was cooling down to room temperature by removing it from the furnace.

For our Raman measurement we use a Dilor xy triple monochromator spectrometer. After the sample was put into the cryostat it was tempered at 600 K to get rid of disturbing impurities. Spectra were excited at 90 K with 4 different laser lines as litsted in Table 1. The RBM was measured in normal dispersion mode with all 4 laser lines. Since the C_{60}-Molecules are best observed for blue laser excitation, we measured the high frequency region between D-line and G-line with 488 nm

TABLE 1 Summary of treatments and characterizations for four samples.

Steps	Treatment Time	Recorded spectra
A	Peapods (0 min)	RBM at 488 D-line, G-line at 488
B	30 min	RBM at 488,568,647,676, D-line, G-line at 488
C	60 min	RBM at 488,568,647,676, D-line, G-line at 488
D	90 min	RBM at 488,568,647,676, D-line, G-line at 488
E	120 min	RBM at 488,568,647,676, D-line, G-line at 488
F	180 min	RBM at 488,568,647,676, D-line, G-line at 488
G	360 min	RBM at 488,568,647,676, D-line, G-line at 488
H	540 min	RBM at 488,568,647,676, D-line, G-line at 488

RESULTS AND DISCUSSION

The duration of the annealing process has a significant influence on the resulting Raman spectra. In all experiments where a second tube has been grown inside the primary tube very narrow lines are observed in the spectral range of the RBM as demonstrated in Fig.1 (left).

After the first annealing step the signatures of the inner tubes are already present. They increase with the following annealing steps. Simultaneously with the growth of the inner tubes the signals from the peas inside the tubes located at 1425 and 1465 cm^{-1} decreased considerably after the first step as demonstrated in Fig. 1(right). This behavior continues with the following steps of the annealing process.

The decrease of the C_{60} signature with increasing annealing time is depicted in Fig. 2. At 1280 °C almost the whole signal is lost after the first 30 min of annealing. There are no traces of the signal left after 1 hour. At 1100 °C the signal drops to 20 % and is then gently reduced with succeeding steps. After 9 hours the lines match again at less than 1%.

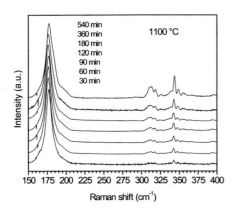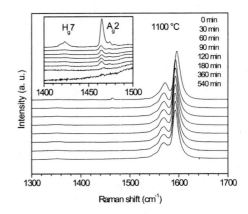

FIGURE 1. Left: RBM of DWCNTs excited with 568 nm. The duration of heat treatment increases from bottom to top. Right: Ag(2) and Hg(7) lines of C_{60} excited with 488 nm. The duration of the heat treatment increases from top to bottom. All spectra are normalized to the response of the RBM from the outer tubes.

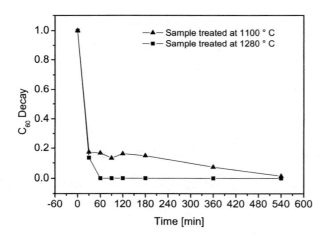

FIGURE 2.
C_{60} signal decay after the different treatment steps is plotted against time. Both samples were measured with 488 nm at 90 K.

At the same time the peapod signal decreases the nanotube lines of the inner tubes formed out during the annealing start to emerge which is shown in Fig 3. In contrary to the rapidly dropping peapod signal this signal increases slowly. After 30 min annealing at 1100 °C already 82% of the C_{60} signal is lost but anly 10% of the DWCNT signal is obtained. Furthermore the signal growth is not the same for all the different inner tubes. Figure 3 shows the signal evolution of two different peaks which correspond to two different tube diameters, 364 cm^{-1}, for the smaller tubes and 302 cm^{-1}, the larger tubes. The small tube growth shows a steep slope in the beginning and a smaller slope to the end. On the contrary the bigger one start to grow slower and shows a steeper slope at the final steps. This diamater dependence of the signal evolution is observed for all the laserlines.

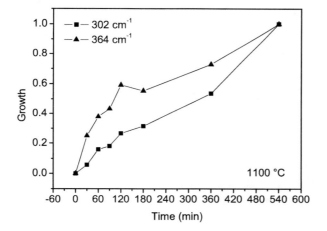

FIGURE 3.
The evolution of the signal of two different inner tubes is plotted against time. The The spectra are recorded with 676 nm at 90 K.

In summary, we have produced singlewall tubes from C_{60} peas with different annealing procedures. We have shown that the duration of the thermal treatment plays an important role in the transformation process. The growth process was found to start for tubes with the smallest diameters but is delayed with respect to the loss of the peapod signal. We conclude that there exists at least one intermediate state which cannot be observed by Raman with the excitation frequencies used. The line shape of the RBM of the outer tubes changes only marginally during the transformation process.

ACKNOWLEDGMENTS

This work was supported by the FWF in Austria, Project 14893 and by the EU RTN FUNCARS (HPRN-CT-1999-00011)

REFERENCES

1. Iijima, S., and Ichihashi, T., *Nature*, **363**, 603–605 (1993).
2. Thess, A. *et al.*, *Science*, **273**, 483–487 (1996).
3. Smith, B. W., Monthioux, M., and Luzzi, D. E., *Nature*, **396**, 323–324 (1998).
4. Bandow, S. *et al.*, *Chem. Phys. Lett.*, **337**, 48-54 (2001).
5. Pfeiffer, R. *et al.*, *Phys. Rev. Lett.* **90**, 225501 (2003).
6. Kramberger, Ch. *et al.*, *Phys. Rev. B* **68**, 235404 (2003).

^{13}C NMR on intercalated 2D-polymerised C_{60} and modified peapods

T. Wågberg [1]*, C. Goze-Bac [1], R. Röding [2], B. Sundqvist [2], D. Johnels [3], H. Kataura [4], and P. Bernier [1]

(1)GDPC, Université Montpellier II, Place E. Bataillon, 34095 Montpellier, Cedex 5, France
(2)Department of physics, Umeå University, 901 87 Umeå, Sweden
(3)Department of Chemistry, Umeå University, 901 87 Umeå, Sweden
(4)Dep. Of Physics, Grad. Sch. Of Science, Tokyo, Metropol. Univ. 1-1Minami-Osawa, Hachioji, Tokyo 192-0397 Japan
Corresponding author; E-mail: Thomas.Wagberg@physics.umu.se

Abstract. We present ^{13}C NMR results on the intercalated 2D C_{60}-polymers $Li_{4-x}Na_xC_{60}$ (x=0-4 with half integer steps) showing that while Li_4C_{60} forms a tetragonal polymer with double polymer bonds between the molecules, Na_4C_{60} form a monoclinic polymer with single polymer bonds. The chemical shift is much larger for Li-rich samples than for Na-rich samples, indicating perhaps a different type of interaction with the C_{60} molecules. The crossover between the two structures appears to be at the composition $Li_{2.5}Na_{1.5}C_{60}$. We also show NMR data on SWNT filled with C_{60} molecules, so called peapods, showing that the NMR line of C_{60} can not be observed when the C_{60} is inside the tubes. We interpret this as resulting from a very long relaxation time, indicating that the dynamics of the C_{60} molecule is hindered by the nanotube walls.

INTRODUCTION

One remarkable property of the C_{60} molecules is that they can be tailored into a large variety of polymeric structures, either by photopolymerisation, pressure polymerisation or intercalation of guest atoms into the structure. Here, we report on a study of materials polymerised by intercalation. This method has been shown to lead to several different types of structures but we are concerned with the compounds that form two-dimensional polymers. We have earlier studied Li_4C_{60} and Na_4C_{60}, where we have seen that while the former is built up by double polymer bonds forming a polymer bridge between the C_{60} molecules, the latter contains only single bonds [1]. To understand the reasons for this difference and to learn more about the mechanism that plays a role in the polymerisation we have extended our study to the compounds $Li_{4-x}Na_xC_{60}$ (x=0-4 with half integer steps) in which the number of alkali metal atoms per C_{60} molecule is always kept at four.

The samples were produced by mixing stoichiometric amounts of alkali metals and C_{60} and treating these at 523 K for several weeks, following a procedure described in ref [2]. We also report measurements on single-walled carbon nanotubes containing C_{60} molecules, so called peapods, that were produced by treating opened tubes together with C_{60} molecules in a sealed quartz tube [3]. To analyze the sample we have used ^{13}C NMR spectroscopy and x-ray diffraction. NMR spectroscopy is a very powerful method to probe the local environment of the carbon atoms in the C_{60} molecules and also to probe the relaxation time T_1 of the carbon nuclear spin. As T_1 is affected by for example charge

CP723, *Electronic Properties of Synthetic Nanostructures*, edited by H. Kuzmany et al.
© 2004 American Institute of Physics 0-7354-0204-3/04/$22.00

transfer to the region of the nuclear spin or the dynamics of the C_{60} molecules, valuable information from this parameter can be gained.

RESULTS AND DISCUSSION

Figure 1 shows the x-ray diffraction spectra of our samples. Li_4C_{60} can be interpreted as having a body centered tetragonal structure with lattice parameters a = 9.13 Å and c = 14.85 Å, while Na_4C_{60} perfectly matches the monoclinic I2/m structure determined by Ozlanyi et al [4]. We conclude that the former of these is formed by the creation of two polymeric between each pair of C_{60} molecules, while the molecules are connected by single bonds in the latter as discussed earlier [1]. For the samples containing mixtures of Na and Li, we see that the tetragonal structure is clearly dominating for a Li-content down to 2.5 atoms per C_{60} molecule, while for Na_4C_{60} and Na_3LiC_{60} the samples seem to be mainly monoclinic. For the samples having almost equal contents of Li and Na, the samples seem to be more inhomogeneous but they are not simple mixtures of the two "parent" structures, tetragonal and monoclinic. More effort is under work for a full interpretation of the x-ray diffraction patterns.

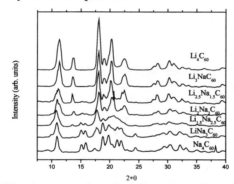

Figure 1. X-ray diffraction pattern of samples indicated in the figure, using the CuKα line.

Figure 2a and b show the NMR spectra of our intercalated C_{60} samples. As expected, the NMR spectra for the 2D-polymers now show a large number of lines, corresponding to different carbon sites. At least ten lines corresponding to sp^2 carbon sites and two lines corresponding to sp^3 carbon can be resolved for the Li-rich compounds, while there are significantly fewer for the Na-rich compounds. The measured NMR shift is the sum of the chemical shift and the Knight shift, where the chemical shift is due to perturbation by the external field at the nucleus. For sp^2 carbon this is expected to lie in the range 110-170 ppm [5] (all shifts are referring to TMS), where most peaks can be found in our spectra. The Knight shift arises from a hyperfine coupling of the nucleus to conduction electrons and the fact that some peaks fall outside the range of chemical shifts could indicate such a coupling. Further measurements of the temperature dependence of the relaxation time is under work to clarify the metallic properties of these samples. The large difference in the chemical shift between the Li-rich and the Na-rich compounds could indicate a different type of interaction between the C_{60} molecules and the Li and Na atoms, respectively. The number of different carbon sites can be compared with the six lines found for the undoped tetragonal polymer [6] and nine lines for the one-dimensional CsC_{60} polymer [7]. The peak positions and relative peak intensities of the Li-rich polymers agree well between the samples, although the line width increases when going to more equal amounts of Li and Na. Our interpretation is that the dominant structural phase in all these samples is the same, and referring to the x-ray diffraction results it should be the tetragonal one. However, it should be pointed out that, especially on the Na-rich side, we are still struggling with inhomogeneties in the samples. This can be seen, for example, from a sharp peak at 143 ppm for the Na_4C_{60} sample, corresponding to unreacted C_{60} and from a very sharp peak at 155 ppm for Li_3NaC_{60} which could be doped but unpolymerized C_{60}. Figure 2b shows that all the lines in the spectra are isotropic and none of them are sidebands. For all samples except Na_4C_{60} there seem to be two different lines corresponding to sp^3 carbon, one line at 58 ppm and one line at 73 ppm. The former is more pronounced in the Li-rich compounds and the latter gains in relative intensity for the Na-rich compounds. These positions correspond very well to the line positions found in the literature for Li_6C_{60} [8] (interpreted as a 2D-polymer) and Na_4C_{60} [9]. One can see that even in pure Li_4C_{60} the two different sp^3 carbon sites exist and as this is not the case for the undoped tetragonal polymer [6] it implies that the two sp^3 carbon atoms have a different chemical environment. The single sp^3-line at 73 ppm for Na_4C_{60} is in agreement with earlier reports [9]. The relaxation time corresponding to the two sites are different, less than a second for the 73 ppm-line and 3-4 seconds for the 58 ppm line as seen in fig. 2b. This is still orders of magnitude lower than the relaxation time for the sp^3-carbon in undoped tetragonal C_{60} [6]. The relative intensity of the sp^3 lines gets smaller when going from Li_4C_{60} towards the Na_4C_{60} composition, in good agreement with the proposed double bonded structure for the former and the single bonded polymer structure for the latter.

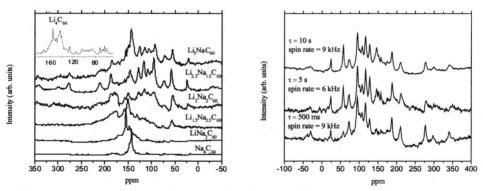

Figure 2. a) C-13, MAS NMR spectra of samples indicated in the figure, sampled at 4.7 T at spin rate ω=9 kHz and a repetition time, τ=10 s, inset shows Li_4C_{60} which was measured at 12 T with ω=7 kHz. b) C-13, MAS NMR spectra of $Li_{2.5}Na_{1.5}C_{60}$ with spin rate and repetition time indicated in figure.

In figure 3a we show the MAS-NMR spectrum for peapods, i.e SWNTs filled with C_{60} molecules (filling factor of 80-90% as determined from Raman spectroscopy and HR-TEM) together with that of normal SWNT which has been carefully mixed with C_{60} molecules but without making any attempts to get the C_{60} inside the tubes. No signal from C_{60} is observed when the C_{60} molecules are inside the tubes. This could be a screening effect from the SWNT walls, but it is more likely due to a dynamic effect; the C_{60} molecules are probably hindered in their free rotation. This effect would increase the relaxation time by several orders of magnitudes, as seen for example in undoped C_{60} polymers, and thus make the NMR signal from C_{60} very hard to observe. In fig. 3b we show NMR spectra for SWNTs and peapods, iodine doped until saturation. Both systems show a characteristic paramagnetic shift of about 10 ppm compared to the undoped systems but no difference can be found between SWNT and peapods.

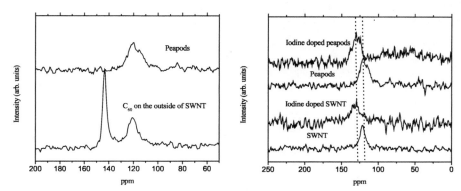

Figure 3a) C-13, MAS NMR spectra of samples indicated in the figure, sampled at 4.7 T at spin rate $\omega=9$ kHz and a repetition time, $\tau=30$ s, b) C-13, MAS NMR spectra of samples indicated in the figure, sampled at 4.7 T at spin rate $\omega=9$ kHz and a repetition time, $\tau=30$ s. for undoped samples and $\tau=5$ s for doped samples.

ACKNOWLEDGMENTS

This work was financially supported by the Swedish research Council. T. Wågberg also acknowledges support from the Wenner-Gren foundation and the Magnus Bergvalls foundation.

REFERENCES

1. T. Wågberg, P. Stenmark, and B. Sundqvist, J. Phys. Chem. Solids, **65**, 317 (2004).
2. R. Röding, T. Wågberg, and B. Sundqvist, J. Chem. Phys. Solids, **65**, 355 (2004).
3. H. Katuara, Y. Maniwa, T. Kodama, K. Kikuchi, K. Hirahara, K. Suenaga, S. Iijima, S. Suzuki, Y. Achiba, and W. Krätschmer, Synth. Metal, **121**, 1195 (2001).
4. G. Ozlányi, G. Baumgartner, G. Faigel, L. Gránásy, and L. Forró, Phys. Rev. Lett. **78**, 4438 (1997).
5. G. C. Levy, R. L. Lichter, and G. L. Nelson, Carbon 13 NMR spectroscopy (Wiley, New York, 1980).
6. A. Rezzouk, Y. Errammach, F. Rachdi, V. Agafonov, and V. A. Davydov, Physics E **8** 1 (2000).
7. T. M. De Swiet, J. L. Yarger. T. Wagberg, J. Hone, B. J. Gross, M. Tomaselli, J. J. Titman, A. Zettl, and M. Mehring, Phys. Rev. Lett, **84**, 717 (2000).
8. M. Tomaselli, B. H. Meier, M. Riccò, T. Shiroka, and A. Sartori, Phys. Rev. B, **63**, 113405 (2001).
9. A. Rezzouk, F. Rachdi, Y. Errammach, and J. L. Sauvajol, Physica E, **15**, 107 (2002).

The redox behavior of potassium doped C_{60} peapods

Martin Kalbáč[1,2], Ladislav Kavan[1,2], Hiromichi Kataura[3], Markéta Zukalová[1] and Lothar Dunsch[2]

[1]J. Heyrovský Institute of Physical Chemistry, Academy of Sciences of the Czech Republic, Dolejškova 3, CZ-182 23 Prague 8
[2]Leibnitz Institute of Solid State and Materials Research, Helmholtztstr. 20, D - 01069 Dresden
[3]Department of Physics, Tokyo Metropolitan University, 1-1 Minami-Ohsawa, Hachioji, Tokyo 192-0397, Japan

Abstract. The redox behavior of fullerene peapods C_{60}@SWCNT was studied by spectroelectro-chemistry at samples chemically n-doped by K vapor. Strong chemical doping was proven by vanishing of the RBM mode and the downshift of TG mode in Raman spectroelectrochemistry. The K-doped peapods were subsequently studied electrochemically and thus n- and p-doped, respectively. The $A_g(2)$ mode of intratubular fullerene in K-doped peapods contacting air was still red-shifted as referred to its position in a pristine peapod. An air-insensitive residual doping was found to be resistant also to cathodic charging. An explanation is given for this behavior.

INTRODUCTION

The Raman spectroscopy is widely used for studies of both SWCNT and fullerenes as a resonant process associated with allowed optical transitions dramatically enhances the signal intensities. There are only few studies on chemical doping of peapods [1,2]. Both n- and p-doping of SWCNT shift the Fermi energy level, which leads to a bleaching of optical transitions between Van Hove singularities. The intensity of resonance Raman spectra of SWCNT decreases accordingly. A similar effect could be achieved by electrochemical doping, which is favored due to precise and easy control of the doping level [2,3]. However there are few differences between chemical and electrochemical doping. The most interesting is that the characteristic softening of $A_g(2)$ mode, which would indicate the intratubular fulleride is not detectable during electrochemical doping.

Both the chemical and electrochemical doping are powerful tools for the variation of the electronic properties of carbon nanostructures as we have shown in our previous paper [4]. Here we present further data, which upgrade this first study.

CP723, *Electronic Properties of Synthetic Nanostructures*, edited by H. Kuzmany et al.
© 2004 American Institute of Physics 0-7354-0204-3/04/$22.00

EXPERIMENTAL SECTION

The sample of C_{60}@SWCNT peapods (filling ratio 85%) was available from our previous work [3]. The sample was outgased at $285°C/10^{-5}$ Pa (the residual gas was He) and subsequently exposed at $177°C$ to potassium vapor for 4-400 hours. The reaction took place in an all-glass ampoule interconnected to a Raman optical cell with Pyrex glass window. A part of this sample was treated in air saturated with water vapor at $90°C$.

A thin-film electrode was prepared by evaporation of the sonicated (approx. 15 min.) ethanolic slurry of the peapod sample on Pt electrode in air. The film electrode was outgased overnight at $90°C$ in vacuum (10^{-1} Pa), and then mounted in a spectroelectrochemical cell in a glove box. The cell was equipped with Pt-counter and Ag-wire pseudo-reference electrodes. 0.2 M $LiClO_4$ in dry acetonitrile was used as an electrolyte solution. Electrochemical experiments were carried out using PG 300 (HEKA) or EG&G PAR 273A potentiostats.

The Raman spectra were measured on a T-64000 spectrometer (Instruments, SA) interfaced to an Olympus BH2 microscope (the laser power impinging on the sample or cell window was between 1-5 mW). Spectra were excited by Ar^+ laser at 2.41 or 2.54 eV (Innova 305, Coherent). The spectrometer was calibrated before each set of measurements by using the F_{1g} line of Si at 520.2 cm^{-1}. The spot size was ca. 0.1x 0.1 mm^2.

RESULTS AND DISCUSSION

The overall intensity of the Raman spectra of potassium doped C_{60}@SWCNT is strongly lowered compared to the undoped sample as shown on Fig 1. The RBM band is fully vanished, which indicates complete filling of the c_S^3 states in SWCNT by our method of chemical doping.

The bands of inserted fullerene dominate the low frequency region of the spectra and, in agreement with previous results for K-doped samples [1], they are red shifted compared to their position in pristine peapod. The most intensive fullerene mode $A_g(2)$ of the heavily doped peapod is found at 1428 cm^{-1} which corresponds to ca. 6 extra electrons in C_{60} [5]. In contrast to other work [1] a down-shift of the TG mode frequency was observed and it is explained by "phase" transformation of crystalline SWCNT ropes [6].

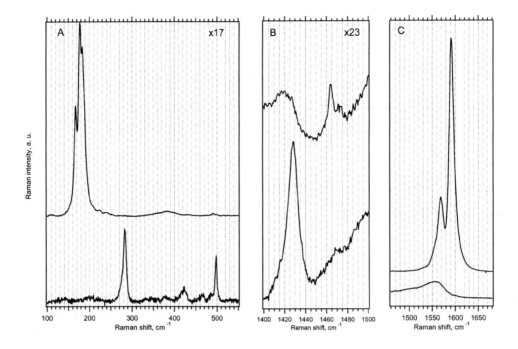

FIGURE 1. Raman response of pristine C_{60}@SWCNT sample (top) and K-doped C_{60}@SWCNT (bottom). The spectra of K-doped C_{60}@SWCNT are multiplied by a factor of 5 in the charts A and C. Intensities of spectra were normalised using the F1g line of Si at 520.2 cm^{-1}. Spectra are offset for clarity.

We have studied the development of the spectra at different doping time (Fig 2). The doping process is relatively fast. All the features of heavy doping occur already after 4h of potassium doping. The longer doping time leads only to small changes in the spectra. We have observed an increase of fullerene features intensity if the doping procedure is continued. This is in contrary to the work of other authors [1]. Probably the reason is that in the present work we achieved higher level of doping even after 4h as indicate the downshift of TG mode and almost complete bleaching of the RBM band.

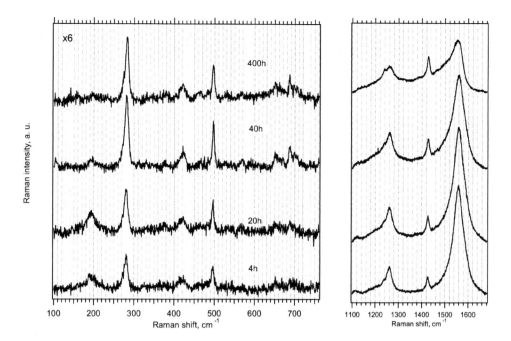

FIGURE 2. Raman response of K-doped C_{60}@SWCNT after different time of potassium doping. Curves are offset for clarity, but the intensity scale is identical for all spectra in the respective window.

The potassium doped samples were then washed with water. Surprisingly only partial de-doping was observed after this procedure. The intensities of nanotube modes are increased, however, they remain smaller than those of pristine peapods, which confirms a residual doping. The $A_g(2)$ band is shifted to 1440 cm^{-1} indicating that the fullerene remains still partly doped (by ca. 4 extra electrons per cage) upon exposure to water. We assume that potassium is located not only in between the peapods but it also penetrates inside of the peapods. The "internal" potassium is more resistant to water and therefore the sample exhibits residual doping even after exposition to water.

However, it is possible to remove this doping electrochemically. The anodic charging extracts electrons from the peapods and the fullerene $A_g(2)$ mode at 1440 cm^{-1} disappears. At 0.6 V the band at 1470 cm^{-1} starts to appear, which is close to the position of $A_g(2)$ in pristine peapods. As this band is known to increase significantly its intensity upon anodic charging, we assume that it belongs to the undoped fullerene. Going back to the zero potential the band around 1440 cm^{-1} reappears, however it is slightly decreased in intensity probably due to anodic burning. On the other hand, the $A_g(2)$ mode remains almost unchanged during the cathodic charging. This is in good agreement with our assumption. Electrons are injected to the peapods during cathodic charging but the potential is not sufficient to reach further reduction of fullerene and therefore no change in $A_g(2)$ is observed.

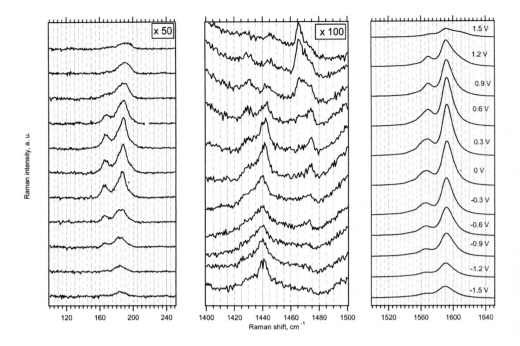

FIGURE 3. Raman response of K-doped C_{60}@SWCNT (after treatment with water) at different electrode potentials varied by 0.3 V from 1.5 to -1.5 V vs. Ag pseudoreference electrode. Curves are offset for clarity, but the intensity scale is identical for all spectra in the respective window. Spectra were excited by 2.54 eV laser energy. The new band around 1430 cm^{-1} which occurs at potentials > 0.9V belongs probably to H$_g$(7) mode of fullerene.

ACKNOWLEDGEMENTS

This work was supported by IFW Dresden, by the Academy of Sciences of the Czech Republic (contract No. A4040306) and by the Czech Ministry of Education (CZ-JP cooperation grant ME487).

REFERENCES

1. T. Pichler, A. Kukovecz, H. Kuzmany, H.Kataura and Y. Achiba, *Phys. Rev. B* **67**, 125416 (2003).
2. L. Kavan, L. Dunsch and H. Kataura, *Chem. Phys. Lett..,* **361**, 79-85 (2002).
3. L. Kavan, L. Dunsch, H. Kataura, A. Oshiyama, M.Otani, S. Okada, *J. Phys. Chem. B* **107**, 7666-7675 (2003).
4. M. Kalbac, L. Kavan, M. Zukalova and L. Dunsch, *J. Phys. Chem. B*, in press (2004).
5. T. Pichler, H. Kuzmany, H.Kataura and Y. Achiba, *Phys. Rev. Lett.* **87**, 267401 (2001).
6. N. Bendiab, L. Spina, A. Zahab, F. Poncharal, C. Marliere, J. L. Bantignies, E. Anglaret and J. Sauvajol, *Phys. Rev. B* **63**, 153407 (2001).

Distinct Redox Doping of Core/Shell Nanostructures: Double Wall Carbon Nanotubes

Ladislav Kavan[1,2], Martin Kalbáč[1,2], Markéta Zukalová[1], Matthias Krause[2], Hiromichi Kataura[3] and Lothar Dunsch[2]

[1]*J. Heyrovský Institute of Physical Chemistry, Academy of Sciences of the Czech Republic, Dolejškova 3, CZ-182 23 Prague 8*
[2]*Leibnitz Institute of Solid State and Materials Research, Helmholtztstr. 20, D - 01069 Dresden*
[3]*Department of Physics, Tokyo Metropolitan University, 1-1 Minami-Ohsawa, Hachioji, Tokyo 192-0397, Japan*

Abstract. Double wall carbon nanotubes (DWCNT) were prepared by pyrolysis of C_{60}, C_{70} and C_{78} peapods. Redox doping was followed by electrochemistry in acetonitrile electrolyte solutions and in ionic liquids. The doping-driven bleaching of optical transitions between Van Hove singularities causes reversible quenching of resonance Raman scattering of the tube related modes. The outer tube of DWCNT is more sensitive to doping-induced loss of Raman resonance. The inner tube is effectively screened from the double-layer charging and only available to doping after charging of the outer tube. Electrochemical p-doping of DWCNT allows deconvolution of the tangential mode features, which normally overlap each other in pristine DWCNT.

INTRODUCTION

The DWCNT are conveniently synthesized by pyrolysis of fullerene peapods. Due to their high structural perfection, the inner tubes exhibit narrow Raman lines of the radial breathing mode (RBM) [1]. The doping of nanotubes and peapods can be conveniently studied by electrochemical charging, interfaced to in-situ Vis-NIR and Raman spectroscopy [2-5]. The spectroelectrochemical approach is superior to chemical doping [6], as it allows easy and precise tuning of the electronic properties of nanocarbons. Our short communication [5] has confirmed the selective bleaching of RBM and tangential (TG) modes of inner/outer tubes in electrochemically doped DWCNT from C_{60}@SWCNT. Here we present further data, which upgrade this first study.

EXPERIMENTAL SECTION

The peapods C_{60}@SWCNT, C_{70}@SWCNT and C_{78}@SWCNT were prepared by the reaction of SWCNT with the corresponding fullerene at 550 °C. The conversion to DWCNT was carried out at 1200°C in vacuum. Depending on the particular precursor, the corresponding tubes will be further abbreviated DWCNT/60, DWCNT/70 and DWCNT/78. Ethanolic slurry of DWCNT was evaporated onto Pt or ITO (indium-tin oxide conducting glass) electrodes. Electrochemical experiments employed potentio-

CP723, *Electronic Properties of Synthetic Nanostructures*, edited by H. Kuzmany et al.
© 2004 American Institute of Physics 0-7354-0204-3/04/$22.00

static set-up with Pt auxiliary and Ag pseudo-reference electrodes. The potential of Ag electrode was calibrated against that of ferrocene (Fc/Fc$^+$), which was added to the electrolyte solution at the end of each set of measurements. The electrolyte solution was 0.2 M LiClO$_4$ + acetonitrile (<10 ppm H$_2$O). Butylmethylimidazolium tetra-fluoroborate and hexafluorophosphate were purified as reported elsewhere [3]. The ITO-supported film of peapods served for *in-situ* Vis-NIR spectroelectrochemistry (Shimadzu 3100 spectrometer). The Raman scattering was excited by Ar$^+$ laser at 2.54 eV and 2.41 eV and by Kr$^+$ laser at 1.83 and 1.91 eV (Innova 300 series, Coherent). Spectra were recorded on a T-64000 spectrometer (Instruments SA) interfaced to an Olympus BH2 microscope (objective 50x). The laser power impinging on the cell window or on the dry sample was between 1 and 5 mW.

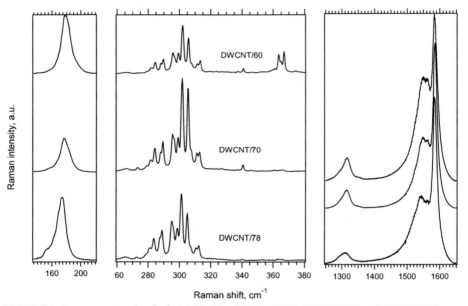

FIGURE 1. Raman spectra (excited at 1.83 eV) of DWCNT/60, DWCNT/70 and DWCNT/78, respectively. Spectra are offset for clarity, but the intensity scale is identical for all spectra. Intensities are normalized against the intensity of the TM band.

RESULTS AND DISCUSSION

Whereas the C$_{60}$/C$_{70}$ peapods show all the expected Raman features of the corresponding encapsulated fullerene [4], the intratubular C$_{78}$ is not detectable at the same conditions. In this case, the successful filling is evidenced only indirectly by the pyrolytic conversion of C$_{78}$@SWCNT into DWCNT (Fig. 1). The DWCNT/60 exhibits specific RBM features of very narrow inner tubes between 360 and 370 cm^{-1}.

The Vis-NIR spectra (data not shown) are characterized by an overlap of specific absorptions of inner and outer tubes. The outer tubes exhibit similar spectral features as those observed for SWCNT, the inner tubes demonstrate themselves by characteristic splitting of the NIR absorption band around 1.3 eV. Electrochemical charging causes reversible bleaching of optical transitions, while the behavior of outer tube resembles that of SWCNT. The inner tubes are less sensitive to electrochemical charging, since only small amount of doping charge is located at the inner tubes.

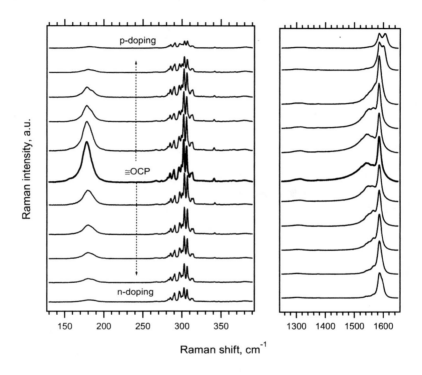

FIGURE 2. Raman spectra of DWCNT/70 (excited at 1.83 eV) in 0.2 M LiClO$_4$ + acetonitrile. The electrode potential varied by 0.3 V from 1.15 V vs. Fc/Fc$^+$ (highest p-doping) to -1.85 V vs. Fc/Fc$^+$ (highest n-doping) for curves from top to bottom. The spectrum at potential close to OCP is highlighted by bold. Spectra are offset for clarity, but the intensity scale is identical for all spectra.

The in-situ Raman spectroelectrochemistry further confirms different doping sensitivity of outer and inner tubes (Fig. 2). The charging-induced bleaching of RBM of inner tubes was sluggish compared to the bleaching of RBM of outer tubes. The attenuation of Raman spectra (excited at 1.83 eV) is due to the loss of resonance by quenching of optical transitions E$_{22}$ in semiconducting inner tubes and E$_{11}$ in metallic outer tubes. Consequently, the charging also erases the Fano broadening in metallic outer tubes at both cathodic and anodic potentials [5].

Comparison of inner tubes in DWCNT and HiPco tubes [2,7] confirms higher sensitivity of narrow tubes, asymmetrically stronger anodic bleaching and complex dependence in the middle region of diameters [2]. These features were ascribed to counterion insertion into the bundle of SWCNT [2,7]. Similar potential profile of the RBM intensity was confirmed even for isolated SWCNT, and was ascribed to solvation [8]. However, neither the solvation, nor the counterion effects are transferable to the discussion of inner tubes in DWCNT. The fast bleaching of wide SWCNT and outer tubes in DWCNT is due to their smaller vHs separations. Hence, their corresponding vHs are depleted/filled at less positive/negative potentials. This is further enhanced for outer tubes, which accommodate the great majority of doping charge [5,6].

Electrochemical charging further leads to the separation of the normally overlapping spectral features of inner/outer tubes in the region of D and TG modes ("electrochemical deconvolution"). It was possible both in acetonitrile medium (Fig. 2) and in ionic liquids. The latter medium allows the application of larger potentials [3], which is beneficial for deconvolution even at cathodic charging.

The blue shift of TG upon anodic charging reflects the stiffening of the graphene mode by insertion of holes into the π band. For SWCNT, the blue shift of TG was 20 cm^{-1} if the potential was increased from open circuit potentials (OCP) to 1.2V (data not shown). The corresponding shift for outer tubes in DWCNT was 22 cm^{-1} (Fig. 2). This again confirms that the doping charge is located mostly on the outer tube.

ACKNOWLEDGEMENTS

This work was supported by IFW Dresden, by the Academy of Sciences of the Czech Republic (contract No. A4040306) and by the Czech Ministry of Education (CZ-JP cooperation grant ME487).

REFERENCES

1. Pfeiffer, R., Kuzmany, H., Kramberger, C., et al. *Phys.Rev.Lett.* **90**, 225501-225501.4 (2003).
2. Kavan, L. and Dunsch, L. *Nano Lett.* **3**, 969-972 (2003).
3. Kavan, L. and Dunsch, L. *Chemphyschem* **4**, 100-106 (2003).
4. Kavan, L., Dunsch, L., Kataura, H., Oshiyama, A., Otani, M. and Okada, S. *J.Phys.Chem.B* **107**, 7666-7675 (2003).
5. Kavan, L., Kalbac, M., Zukalova, M., Krause, M. and Dunsch, L. *Chemphyschem* **5**, 274-277 (2004).
6. Chen, G., Bandow, S., Margine, E. R., et al. *Phys.Rev.Lett.* **90**, 257403-2574034 (2003).
7. Kukovecz, A., Pichler, T., Kramberger, C. and Kuzmany, H. *Phys.Chem.Chem.Phys.* **5**, 582-587 (2003).
8. Okazaki, K., Nakato, Y. and Murakoshi, K. *Phys.Rev.B* **68**, 035434(2003).

Highly Diameter Selective ^{13}C Enrichment in Carbon Nanotubes

C. Kramberger, F. Simon, R. Pfeiffer, M. Mannsberger, M. Holzweber, F. Hasi, and H. Kuzmany

Institute of Material Physics, University of Vienna
Strudlhofgasse 4, A-1090 Wien, Vienna, Austria

Abstract. We performed vapor phase filling of SWCNTs with various kinds of fullerenes. C_{60}, C_{70}, ^{13}C enriched C_{60} and mixtures of these were used. The resulting peapods as well as the DWCNTs obtained by annealing were characterized with multi frequency Raman spectroscopy. Simultaneous filling with C_{60} and C_{70} was found to take place with the same efficiency. When using the ^{13}C enriched C_{60} for DWCNT synthesis, mode softening for all inner shell lines was observed, while the outer shell response remained unchanged. The yield of inner shell CNTs is almost the same for C_{60} and C_{70} except that C_{70} is very inefficient in forming mid diameter inner CNTs with RBM frequencies close to 365 cm^{-1}. If inner CNTs are grown from mixed peapods of C_{70} and ^{13}C enriched C_{60} these particular CNTs are mainly built from C_{60}. This is confirmed by the observation of distinct line shifts in the inner shell RBM.

INTRODUCTION

Raman spectroscopy is a convenient and widely used technique in the investigation of different nano structured phases of carbon. Especially multi frequency Raman spectroscopy was demonstrated to be applicable for the observation of distinct inner shell CNTs in DWCNTs [1]. So far this was demonstrated for the strongly diameter dependent RBM, whereas the high frequency modes are on top of each other. By using ^{13}C enriched fullerenes and turning the obtained peapods into DWCNTs all inner shell modes are down shifted. This technique allows a clear separation of the outer and inner shell response throughout the whole spectrum. Furthermore, ^{13}C marked fullerenes in combination with Raman spectroscopy can be used to monitor the diameter dependency of the filling efficiency for different fullerenes in a mixture of these. This contribution is focused on an observed unusual behavior of the diameter dependence of the filling efficiency for C_{70}. This anomaly can be exploited to prepare very special DWCNT samples. The outer shell CNTs as well as any other carbonic impurities have the natural abundance (~1%) of ^{13}C. Only the inner shell CNTs are ^{13}C enriched. This enrichment is the same for almost all inner tubes, but in a very narrow diameter range corresponding to $v_{RBM} = 365 cm^{-1}$ it is twice as high compared to all other very thin CNTs. The general spectroscopic benefit of ^{13}C enriched inner shell CNTs in the high frequency domain (D,G and D* Band) is reported in detail later [2]

CP723, *Electronic Properties of Synthetic Nanostructures*, edited by H. Kuzmany et al.
© 2004 American Institute of Physics 0-7354-0204-3/04/$22.00

EXPERIMENTAL

Peapods were obtained through the vapor phase filling method. Fullerenes and SWCNTs material purified by oxidation and prepared as bucky paper were sealed in an evacuated quartz ampoule and heated at 650 °C for 2h. Excess fullerenes were removed by heating the peapods at 800 °C for 30 min in dynamic vacuum. Transformation to DWCNTs was performed by 1 h annealing at 1270 °C in dynamic vacuum. Raman spectra were recorded with a xy triple monochromator spectrometer from Dilor. The 180° backscattered light was analyzed. An ArKr laser was used for excitation at different laser lines.

RESULTS AND DISCUSSION

The above described vapor filling method for preparation of peapods works for either C_{60} or C_{70}. Figure 1 depicts Raman spectra of C_{60} and C_{70} peapods at 488 nm. In both cases the characteristic lines of the fullerenes are observed together with the D-band of the hosting SWCNTs. If a 50%:50% mass ratio mixture of C_{60} and C_{70} is subject to peapod preparation a superposition of the Raman spectra from Fig. 1 is observed, as shown in Fig. 2a. As expected this spectrum can be simulated by superimposing the two spectra of Fig. 1. Best agreement for this simulation is obtained for a 45% to 55% ratio of C_{70} and C_{60}. The resulting spectrum is shown in Fig. 2b.

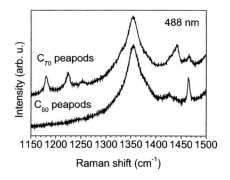 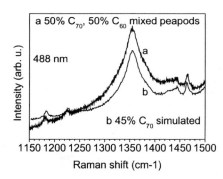

FIGURE 1. Raman spectra of vapor phase prepared C_{60} and C_{70} peapods at 488 nm.

FIGURE 2. Raman spectrum of C_{60}/C_{70} mixed peapods at 488 nm (a) and best fitting superposition of the spectra from Fig. 1 (b).

Thus, when the vapor filling method is applied to prepare mixed peapods from C_{60}/C_{70} mixtures, the initial mass ratio is almost retained.

When annealing C_{60} and C_{70} peapods and investigating the resulting DWCNTs by multi frequency Raman spectroscopy, it is found that both kinds of peapods have the same yield of inner shell CNTs, except for a very narrow spectral range around 365 cm^{-1}. In this range C_{70} peapod material exhibits a significantly lower yield of inner shell CNTs [3].

Figure 3 shows Raman spectra of the RBM of C_{60} and C_{70} based inner shell CNTs. While the spectra at 676 nm clearly show the suppressed signal around 365 cm^{-1} in C_{70} based DWCNTs, the spectra at 502 nm show that there is again a compareable yield at higher frequencies and even thinner inner shell CNTs. All spectra are normalized to the response of the outer shell RBM. Therefore the suppressed RBM signal at 365 cm^{-1} is not related to different minimum filling or transformation diameters of C_{60} and C_{70}. In both samples transformation and therefore filling is observed at higher and lower frequencies corresponding to smaller and bigger diameters. By now there are two possible scenarios explaining this anomaly, either the transformation of C_{70} into inner shell CNTs is hindered in the corresponding diameter range, or even more exciting, there are almost no C_{70} peapods in this diameter range. The crucial experiment to decide between these two scenarios is to prepare mixed peapods from ^{13}C enriched C_{60} and standard C_{70} and transform them to DWCNTs. Figure 4 depicts a compilation of the RBM of standard C_{60}, pure ^{13}C enriched C_{60} and of a mixture of ^{13}C enriched C_{60} and standard C_{70} based inner shell CNTs recorded at 647 nm.

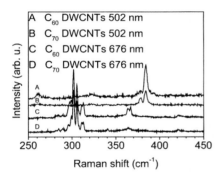 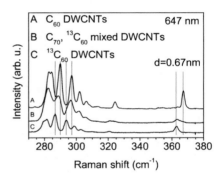

FIGURE 3. Raman spectra of the RBM of C_{60} and C_{70} based inner shell CNTs at 502 and 676 nm.

FIGURE 4. Raman spectra of the RBM of C_{60}, $^{13}C_{60}/C_{70}$ mixed and $^{13}C_{60}$ based inner shell CNTs at 647 nm.

Obviously the transformation of mixed peapods to DWCNTs works as well as the transformation of either pure C_{60} or C_{70} peapods. As expected the lower frequency RBM lines of the mixed material exhibit line positions only slightly down shifted as compared to the standard C_{60} based inner shell CNTs. In contrast the RBM line at 365 cm^{-1} shows almost the full downshift as compared to the pure $^{13}C_{60}$ material. Therefore, this particular CNTs are mainly built from the ^{13}C marked C_{60}. This proves that C_{70} is inefficient in filling the corresponding outer diameter CNTs. Taking into account the experimental wall to wall distance in DWCNTs obtained from X-ray diffraction [4] of 0.36 nm and different relationships of the RBM frequency and inverse diameter [5, 6, 7] the corresponding outer tube diameter is found in the range from 1.35 nm to 1.40 nm. This experimental result matches to the reported theoretical borderline between the preference of lying and standing, e.g. the longer axis either parallel or standing with respect to the CNT axis, arrangements in C_{70} peapods[8].

SUMMARY

^{13}C marked fullerenes can be used to prepare DWCNTs where the very thin inner shell CNTs are ^{13}C labeled. This selective labeling allows for a clear discrimination between these very thin CNTs from any other material in the samples. When using different kinds of fullerenes it is possible to determine which fullerenes contribute to which inner shell CNTs. In this way we have confirmed an anomalous filling behavior of the asymmetrically shaped C_{70} molecule. The latter is inefficient in filling CNTs with diameters very close to the borderline of standing and lying arrangements. Similar effects may be expected for other well defined asymmetric nano particles and thus provide a way to highly diameter selective reactions. This can open a new route to separation of SWCNTs.

ACKNOWLEDGEMENTS

This work was supported by the Austrian Science Foundation, project 14893 and the EU PATONN Marie-Curie MEIF-CTC2003-501099 grants.

REFERENCES

1. R. Pfeiffer et al. Phys. Rev. Lett. **90**, 225501 (2003)
2. F. Simon et al. Phys. Rev. Lett. (2004) cond-mat/0403179
3. F. Simon et al. Phys. Rev. Lett. (2004) submitted
4. M. Abe et al. Phys. Rev. B **68**, 041405(R) (2003)
5. A. Jorio et al, Phys. Rev. Lett. **86**, 1118 (2001)
6. S. Bachilo et al, Science **298** (2002)
7. C. Kramberger et al, Phys. Rev. B **68** 235404 (2003)
8. S.Okada et al. New J. Phys. **5**, 122 (2003)

Inserting Fullerene Dimers into Carbon Nanotubes: Pushing the Boundaries of Molecular Self-assembly

Kyriakos Porfyrakis [a]*, Andrei N. Khlobystov [a], David A. Britz [a], John J. L. Morton [a], Arzhang Ardavan [b], Mito Kanai [c], T. John S. Dennis [c] and G. Andrew D. Briggs [a]

[a] Department of Materials, University of Oxford, Parks Road, Oxford, OX1 3PH, UK. Fax: +44 1865 273789; Tel: +44 1865 273724; E-mail: kyriakos.porfyrakis@materials.ox.ac.uk
[b] Clarendon Laboratory, Parks Road, Oxford, OX1 3PU, UK
[c] Centre for Materials Research, Queen Mary, University of London, Mile End Road, London, E1 4NS, UK

Abstract. Carbon nanotubes can encapsulate several molecular species forming one-dimensional crystals. Using previously reported methods we produced directly-bonded, asymmetric C_{60}-C_{70} dimers and oxygen-bridged dimers of the type C_{60}-O-C_{60}. We present here microscopic evidence of filling single-walled carbon nanotubes (SWNTs) with the above fullerene dimers. The most important filling constraint is found to be the nanotube size. SWNTs with diameters around 1.6 nm incorporate dimers considerably more easily than SWNTs with smaller diameters. This kind of molecular self-assembly opens up the potential for using nanotubes and fullerenes for nanodevices.

INTRODUCTION

Carbon nanotubes have been extensively studied due to their promising electronic, transport and mechanical properties [1]. Due to their tubular nature, carbon nanotubes have been found to encapsulate several molecular species, thus forming quasi one-dimensional crystals [2]. The discovery of "peapods" (carbon nanotubes filled with fullerene cages) [3] has attracted more research and expanded the potential applications of these molecular structures in the field of nanotechnology.

The vast majority of peapods studied so far involve the filling of single-walled carbon nanotubes (SWNTs) with fullerenes of different cage size (such as C_{60} and C_{70}) or with endohedral metallofullerenes of the type $M_x@C_{82}$ (where x = 1-4). In this paper we present the first evidence of filling SWNTs with directly-bonded and oxygen-bridged fullerene dimers.

CP723, *Electronic Properties of Synthetic Nanostructures*, edited by H. Kuzmany et al.
© 2004 American Institute of Physics 0-7354-0204-3/04/$22.00

EXPERIMENTAL

Several types of fullerene dimers have been reported in the literature [4]. Using the solid-state mechanochemical reaction of C_{60} and C_{70} with K_2CO_3 as catalyst, according to the high-speed vibration milling (HSVM) technique pioneered by Komatsu and co-workers [5], we produced asymmetric C_{60}-C_{70} dimers. We also produced oxygen bridged dimers of the type C_{60}-O-C_{60} by thermally reacting C_{60} and $C_{60}O$ [6].

The reaction products were purified by high performance liquid chromatography (HPLC) using a Cosmosil 5PYE column, 20 mm × 250 mm, with pure toluene eluent and flow-rate of 15 ml min^{-1}. A combination of HPLC, mass spectrometry and UV-Vis spectroscopy have confirmed the formation of the different isomers of fullerene dimers. Our characterisation results are in good agreement with the literature [7].

The fullerene dimers were inserted in SWNTs by mixing them with purified, open-end SWNTs (Aldrich, diameters = 13-16 Å) in a quartz ampoule and heating the mixture at 350°C under 5·10^{-6} Torr for several hours. Non-encapsulated dimers were removed by extensive washing with CS_2 and the remaining material was examined by high-resolution transmission electron microscopy (HRTEM, JEOL JEM-4000EX, LaB$_6$, information limit <0.12 nm). The imaging conditions were set to minimize knock-on damage commonly occurring in peapod structures under exposure to the electron beam [8]. The accelerating voltage was set at 100 kV, the beam current on the specimen was reduced to minimum and exposure times were 1-2 sec to minimise the damage. Imaging of C_{60}@SWNT under these conditions showed no structural changes in peapods over 10 minutes.

RESULTS AND DISCUSSION

Approximately 30% of the total sample of SWNTs were found to be filled with fullerene structures. HRTEM micrographs (Fig. 1) show SWNTs filled with C_{60}-C_{70} dimers.

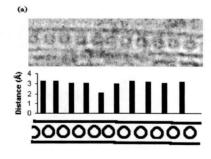
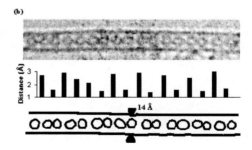

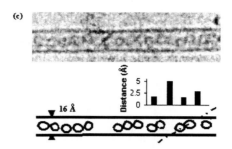

FIGURE 1. HRTEM micrographs of SWNTs filled with C_{60} and C_{60}-C_{70} dimers. The cartoons below each micrograph are used for clarity. The histograms show measured cage-to-cage distances between fullerene molecules. When the dimers have dissociated during filling, the nanotubes are filled with monomeric C_{60} or C_{70} molecules (a). In relatively narrow SWNTs the dimers are aligned with the nanotube axis (b). In wider SWNTs the dimers are tilted (as indicated by the dotted line) such that the van der Waals interactions between the molecules and the nanotube sidewalls are maximised (c).

Approximately 50% of the filled SWNTs appear to contain C_{60} or C_{70} molecules as in Fig. 1(a). Since there was no monomeric C_{60} or C_{70} in the purified dimer samples before filling, one must assume that some of the dimers have undergone dissociation during heating. Using differential scanning calorimetry, Wang *et al.* estimated the dissociation temperature of C_{120} to be between 150 and 175°C [9]. However, it seems that under our filling conditions some C_{60}-C_{70} dimers have remained intact. Indeed, extensive HRTEM imaging has shown that approximately 50% of the filled SWNTs contain dimers (Fig. 1(b) and 1(c)). Figure 1 shows clear microscopic evidence for the presence of fullerene dimers in SWNTs. In ordinary fullerene peapods, such as C_{60}@SWNT or C_{70}@SWNT, the typical van der Waals inter-fullerene spacing of ~ 3 Å is observed (Fig. 1(a)). When fullerene dimers are in the SWNTs, the constituent fullerene molecules of each dimer appear to be connected to each other by covalent bonds (Fig. 1(b) and 1(c)). We measured the cage-to-cage distance within a dimer to be 1.5±0.5 Å which is well below the van der Waals separation and corresponds to the length of a single C-C bond. Furthermore we observed that in relatively narrow SWNTs (d = 13-14 Å) the long axes of the dimers are aligned with the axis of the nanotube (Fig. 1(b)). In wider SWNTs (d = 16Å) the dimers are tilted such that the van der Waals SWNT···dimer interactions are maximised [10] (Fig. 1(c)). This tilting also indicates that the two fullerene cages comprising each dimer are linked by rigid chemical bonds. Both regular peapods and dimer-filled peapods were observed in the same sample using the same imaging conditions. Therefore, we cannot attribute the aggregated fullerenes to beam damage or transient charging effects seen in earlier peapod studies. Thus, these images show the first example of fullerene dimers inserted into SWNTs.

We employed a similar approach for filling SWNTs with oxygen-bridged C_{60}-O-C_{60} dimers. An excess of C_{60}-O-C_{60}, dissolved in CS_2, was dropped on freshly annealed SWNTs and the mixture was dried. It was then sealed in a quartz tube at 10^{-5} Torr and heated at 305°C for 3 days. Figure 2 shows a typical HRTEM micrograph of a SWNT containing C_{60}-O-C_{60} dimers.

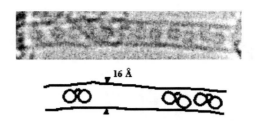

FIGURE 2. HRTEM micrograph of a SWNT containing C_{60}-O-C_{60} dimers. The dimers are tilted so that they retain the Van der Waals distance between them and the walls of the nanotube.

The oxygen-bridged dimers are more thermally robust than the directly bonded dimers and we observed that they do not dissociate as C_{60}-C_{70} does during filling. In a similar manner as the C_{60}-C_{70} dimers, the C_{60}-O-C_{60} dimers are tilted inside the nanotube so that the van der Waals distance between them and the walls of the nanotube is retained. Hence, it is shown that it is possible to fill SWNTs with molecules of relatively long aspect ratios. For dimers of both kinds, it appears that the most important constraint of the dimer peapod formation is the nanotube size. Wider SWNTs with diameters around 16 Å incorporate dimers more easily than SWNTs with diameters 13-14 Å. A likely explanation for this observation is steric hinderance due to the large size of the dimeric molecules.

ACKNOWLEDGMENTS

We are grateful to DTI, EPSRC and Hitachi Europe Ltd for the Foresight Link award *Nanoelectronics at the Quantum Edge* (www.nanotech.org), GR/R66029/01. TJSD thanks the Royal Society for a Joint Research Project (Japan), the EPSRC for grant GR/R55313/01. DAB thanks the Overseas Research Student Scheme for funding. AA is supported by the Royal Society.

REFERENCES

1. M. Dresselhaus, G. Dresselhaus and P. Eklund, *Science of fullerenes and carbon nanotubes: their properties and applications*, Academic Press, 1996.
2. K. Hirahara, K. Suenaga, S. Bandow, H. Kato, T. Okazaki, H. Shinohara and S. Iijima, *Phys. Rev. Lett.*, **85**, 5384, (2000).
3. B. W. Smith, M Monthioux and D. E. Luzzi, *Nature*, **396**, 323, (1998).
4. J. L. Segura and N. Martin, *Chem. Soc. Rev.*, **29**, 13, (2000).
5. K. Komatsu, G-W. Wang, Y. Murata, T. Tanaka and K. Fujiwara, *J. Org. Chem.*, **63**, 9358, (1998).
6. S. Lebedkin, S. Ballenweg, J. Gross, R. Taylor and W. Kratschmer, *Tetrahedron Lett.*, **36**, 4971, (1995).
7. N. Tagmatarchis, G. S. Forman, A. Taninaka and H. Shinohara, *Synlett*, **2**, 235, (2002); N. Dragoe, H. Shimotani, J. Wang, M. Iwaya, A. Dias, A. L. Balch and K. Kitazawa, *J. Am. Chem. Soc.*, **123**, 1294, (2001).
8. B. W. Smith and D. E. Luzzi, *J. Appl. Phys.*, **90**, 3509, (2001).
9. G-W Wang, K. Komatsu, Y. Murata and M. Shiro, *Nature*, **387**, 583, (1997).
10. M. Hodak, and L. A. Girifalco, *Phys. Rev. B*, **67**, 07541, (2003).

Properties Of N@C$_{60}$-Derived Peapods

A. Gembus*, F. Simon[+], A. Jánossy[#], H. Kuzmany[+], and K.-P. Dinse*

*) Chem. Dept., Darmstadt University of Technology
+) Institut für Materialphysik, Universität Wien
#) Budapest University of Technology and Economics

Abstract. Using pulsed EPR techniques, the basic spin relaxation properties of N@C$_{60}$-based peapods were determined. In contrast to narrow line spectra typical for N@C$_{60}$ in solid solution or in a C$_{60}$ matrix, substantial line broadening is observed for the SWCNT-encapsulated N@C$_{60}$ molecules, which might be indicative for uniaxial cage distortion by interaction with the nanotube.

INTRODUCTION

Recent progress in peapod preparation using a "solvent-based" procedure allows for the first time to consider the insertion of N@C$_{60}$ into single wall carbon nanotubes (SWCNT) [1,2]. Because it was recently shown that N@C$_{60}$ could be prepared as pure substance [3], it is now possible to use this exceptional spin probe to explore the electronic properties of nanotubes. In contrast to "seeding" with metallo-endofullerenes (MEF), N@C$_{60}$ can be expected to be an "inert" spin probe, not subject to charge transfer when inserted in the SWCNT. In the future it also seems possible for the first time to prepare a densely packed one-dimensional electronic spin system with negligible exchange interaction. Properties of the so formed one-dimensional super-paramagnet could be explored easily because the electronic Zeeman interaction can be readily adjusted to be larger than the thermal energy even at standard liquid helium temperatures. In the current contribution we report on the first results of a pulsed EPR study by which the basic spin relaxation properties of N@C$_{60}$ peapods were determined.

EXPERIMENTAL

Sample Preparation

N@C$_{60}$ was prepared using an ion implantation apparatus developed at the Hahn-Meitner-Institute in the group of Weidinger [4]. The already optimized relatively low yield of 50 ppm relative to "empty" C$_{60}$ obtained by this method necessitates the use of High-Performance Liquid Chromatography (HPLC) to obtain sufficiently high spin concentrations for a detailed EPR investigation. For this first study material enriched

CP723, *Electronic Properties of Synthetic Nanostructures*, edited by H. Kuzmany et al.
© 2004 American Institute of Physics 0-7354-0204-3/04/$22.00

to the 200 ppm level was utilized. At this concentration level spatially isolated spin probes are inserted into the CNT, thus preventing line broadening by dipole-dipole interaction from the spin labels.

N@C_{60}-doped SWCNT were prepared using the procedure recently described by Simon et al. [2]. In brief, a saturated solution of N@C_{60}/C_{60} in hexane is used to suspend opened SWCNT. After approximately 2 h, the solvent is removed and toluene is used to extract C_{60} and N@C_{60}, not encapsulated by the SWCNT.

Electron Paramagnetic Resonance (EPR) Spectroscopy

EPR spectra were measured either using a continuous wave (c. w.) (BRUKER ESP 300E) or pulsed (BRUKER ELEXSYS 680) spectrometer equipped with a variable temperature cryostat. C. w. spectra are used to check for any remaining non-inserted material, which can easily be discriminated from the peapod signal by their extremely narrow EPR lines. With a 2-pulse echo sequence echo-detected EPR spectra were recorded additionally, which instead of a field-modulated EPR signal allows to detect an undistorted EPR absorption spectrum if the spin dephasing time of the individual spin packets is larger than the time resolution of the spectrometer of approximately 100 ns. Under these conditions, narrow and broad signal components can be observed simultaneously. The echo signal decay as function of pulse separation was taken as measure of electron spin dephasing time.

RESULTS AND DISCUSSION

The c. w. EPR signal of the peapod preparation shows the characteristic 3-line spectrum of N@C_{60} superimposed on a broad unstructured signal of unknown origin in agreement with the previous observation [2]. Using electron spin echo detection, an additional broad resonance is detected at lower fields, as shown in Fig. 1. Such broad absorptions can be easily missed using c. w. techniques because of the reduced response with respect to field modulation. From the spectral characteristics we conclude that copper-based radicals of unknown origin contribute considerably to the total spin number in the sample. At higher spectral resolution the signal of N@C_{60} is clearly identified by its characteristic hyperfine coupling constant (hfcc) of a(N) = 0.57 mT (see Fig. 2). Final decomposition after numerically removing the broad signal results in a spectrum showing exclusively the EPR signal of N@C_{60}, is shown in Fig. 3. The lines can be fitted by using Gaussian line shape functions of width 3.1 ($m_I = 0$ hyperfine component (hfc)) and 4.8 MHz width ($m_I = \pm1$ hfc) (FWHM). Although it is tempting to interpret this difference in width by assuming a distribution of nitrogen hfcc as major source of inhomogeneous line broadening, numerical artifacts originating from spectral decomposition, which could contribute to this apparent difference, cannot be completely excluded at present. The width of the underlying individual spin packets was estimated from the echo signal decay. A bi-exponential echo decay with characteristic time constants of 600 and 8000 ns was observed ($B_0 = 346$ mT). These values can be translated to Lorentzians of width 500 and 40 kHz, thus indicating substantial inhomogeneous broadening of the observed N@C_{60} signals.

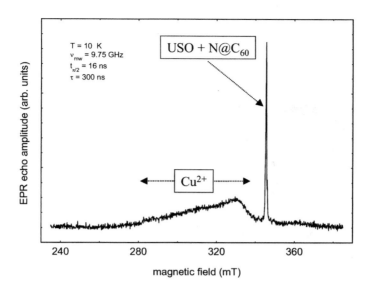

FIGURE 1. Echo-detected EPR of N@C_{60} peapods. EPR absorption originates from a Cu^{2+} radical and superimposed three lines from N@C_{60} (not resolved on this scale) with another strong signal from an "unidentified spin object".

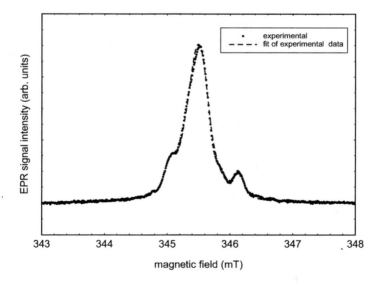

FIGURE 2. Part of the spectrum shown in Fig. 1 under higher spectral resolution. The experimental data were fitted by a superposition of an equidistant three-line spectrum of a(N) = 0.57 mT and a broader signal which is slightly down field shifted.

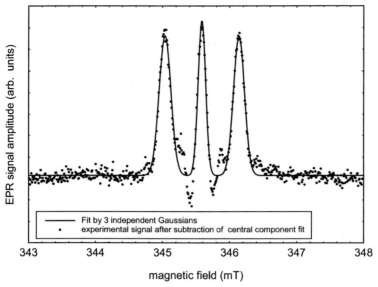

FIGURE 3. Echo-detected EPR absorption of $N@C_{60}$ peapods (T = 10 K).

Obviously the use of highly enriched $N@C_{60}$ material is mandatory for a detailed study aiming for the elucidation of the anticipated influence of itinerant spins in metallic CNT on the relaxation properties of the nitrogen spin probe. However, even the results presented in this preliminary study indicate that subtle effects of encapsulation might be detectable either via a variation of nitrogen hfc because of a change in effective volume seen by the atom or by Zero-Field-Splitting because of cage deformation.

ACKNOWLEDGMENTS

Financial support by the Deutsche Forschungsgemeinschaft, by a Hungarian state grant OTKA T 043255, and a PATONN Marie-Curie MEIF-CT-2003-501099 EU grant is gratefully acknowledged.

REFERENCES

1. Yudaska, M., Ajima, K., Suenaga, K., Ichihashi, T, Hashimoto, A., and Iijima, S., *Chem. Phys. Lett.* **380**, 42-46 (2003).
2. Simon, F. Kuzmany, H., Rauf, H., Pichler, T., Bernardi, J., Peterlik, H., Korecz, L., Fülöp, F., and Jánossy, A., *Chem. Phys. Lett.* **383**, 362-367 (2004).
3. Jakes, P., Dinse, K.-P., Meyer, C., Harneit, W., and Weidinger, A., *Phys. Chem. Chem. Phys.* **5**, 4080-4083 (2003).
4. Pietzak, B., Weidinger, A., Dinse, K.-P., and Hirsch, A.: "Group V Endohedral Fullerenes: N@C60, N@C70, and P@C60" in: "*Endofullerenes, A New Family of Carbon Clusters*", edited by Akasaka, T. and Nagase, S. , Kluwer Academic Publishers, Dordrecht, (2002), p. 13-66.

Interaction between Inner and Outer Tubes in DWCNTs

R. Pfeiffer*, Ch. Kramberger*, F. Simon*, H. Kuzmany* and V. N. Popov†

*Institut für Materialphysik, Universität Wien, Vienna, Austria
†Faculty of Physics, University of Sofia, Sofia, Bulgaria

Abstract. By annealing fullerene peapods at high temperatures in a dynamic vacuum it is possible to produce double-wall carbon nanotubes. A Raman investigation revealed that the inner single-wall carbon nanotubes are remarkably defect free showing very strong and very narrow radial breathing modes (RBMs). The number of observed RBMs is larger than the number of geometrically allowed inner tubes. This splitting is caused by the interaction of one type of inner tube with several types of outer tubes the inner tube may grow in.

INTRODUCTION

Single-wall carbon nanotubes (SWCNTs) [1, 2, 3] have attracted a lot of scientific interest over the last decade due to their unique structural and electronic properties. A few years ago, it was discovered that fullerenes can be encapsulated in SWCNTs, forming so-called peapods [4]. By annealing such peapods at high temperatures in a dynamic vacuum it became possible to transform the enclosed C_{60} peas into SWCNTs within the outer tubes, producing double-wall carbon nanotubes (DWCNTs) [5, 6]. The growth process of the inner tubes is a new route for the formation of SWCNTs in the absence of any additional catalyst.

A detailed Raman study of the radial breathing modes (RBMs) of the inner tubes revealed that these modes have intrinsic linewidths of about $0.4\,\mathrm{cm}^{-1}$ [7]. This is about a factor 10 smaller than reported for isolated tubes so far. This small linewidths indicate long phonon lifetimes and therefore highly defect free inner tubes, which is a proof for a nano clean room reactor on the inside of SWCNTs.

A closer inspection of the RBM response of the inner tubes revealed that there are more Raman lines than geometrically allowed inner tubes. In this contribution we will explain this splitting by the interaction of one type of inner tube with several types of outer tubes.

EXPERIMENTAL

As starting material for our DWCNTs we used C_{60} peapods (in the form of buckypaper) produced with a previously described method [8]. The outer tubes had a mean diameter of about 1.39 nm as determined from the RBM Raman response [9]. The filling of the tubes large enough for C_{60} to enter was close to 100% as evaluated from a Raman

CP723, *Electronic Properties of Synthetic Nanostructures*, edited by H. Kuzmany et al.
© 2004 American Institute of Physics 0-7354-0204-3/04/$22.00

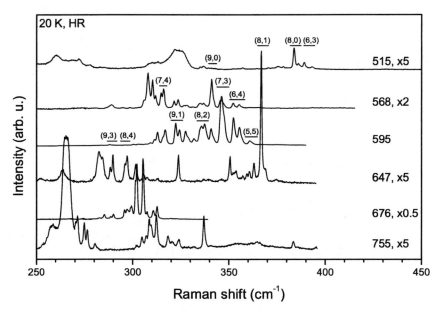

FIGURE 1. High resolution Raman response of the inner tube RBMs for several excitation wavelengths at 20 K. Selected chiralities after Kramberger et al. [13] and splitting widths are indicated.

[10, 11] and EELS analysis [12]. These peapods were slowly heated up to 1300 °C in a dynamic vacuum for 12 hours and were then slowly cooled down to room temperature.

The Raman spectra were measured with a Dilor xy triple spectrometer using various lines of an Ar/Kr laser, a He/Ne laser and a Ti:sapphire laser. The spectra were recorded at 20 K in high resolution mode ($\Delta\bar{\nu}_{HR} = 0.4\,\mathrm{cm}^{-1}$ in the red). In these measurements the samples were glued on a copper cold finger with silver paste.

RESULTS AND DISCUSSION

Fig. 1 depicts selected high resolution Raman spectra of the RBMs of the inner tubes. Using the frequency–diameter relation from [13], one should find the RBMs of 28 distinct inner tubes between 270 and 400 cm^{-1}. However, the observed number of lines in this region is about three times larger. This means, that the RBMs of the inner tubes are split into several components.

In order to determine the number of the split components and the width of the splitting we fitted six selected high resolution spectra with a number of Voigtian lines. Using our chirality assignment [13], each individual inner tube was assigned a number of RBM frequencies. For each tube we subtracted from these frequencies the mean value and plotted the split components vs. inner tube diameter in Fig. 2. It shows that the number of split components ranges between two and five and the width of the splitting is about

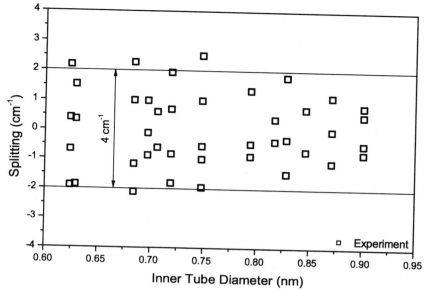

FIGURE 2. Number of split components and width of splitting for individual inner tubes from the spectra in Fig. 1 after subtracting the mean value of the RBM frequencies.

$4\,\mathrm{cm}^{-1}$.

In the following, we will show that the splitting is caused by two reasons: due to the interaction with the outer tube the RBM frequency of the inner tube depends on the diameter of the outer tube and one type of inner tube may form in several types of outer tubes. To calculate the RBM frequency of the inner tube of a DWCNT, we used a continuum model with two elastic concentric cylinders of the appropriate diameters. The interaction of these two cylinders was modeled with a Lennard-Jones potential [14].[1] The possible diameter differences between inner and outer tubes were assumed to be in the range 0.66–0.70 nm. This range was centered around 0.68 nm which is about twice the layer distance in graphite.

Fig. 3 demonstrates the dependence of the calculated RBM frequency of a $(6, 4)$ inner tube on the diameter of the outer tube. The isolated $(6, 4)$ tube has a diameter of $d_{\mathrm{inner}} = 0.691$ nm and a frequency of $v_{\mathrm{inner\,RBM}}^{\mathrm{isolated}} = 337.4\,\mathrm{cm}^{-1}$ (excluding C_2) [13]. This RBM frequency increases with decreasing outer tube diameter.

In a first step, we assumed that in all outer tubes in our sample (Gaussian distribution with $\bar{d} = 1.39$ nm and $\sigma = 0.1$ nm) only the best fitting inner tubes are formed.[2] The splitting calculated with this assumption is depicted in Fig. 4 (left). The number of split

---

[1] Dobardžić et al. [15] performed similar calculations of the inner tube RBM frequencies using a harmonic interaction between the inner and outer cylinder.
[2] "Best fitting" means $((d_{\mathrm{outer}} - d_{\mathrm{inner}}) - 0.68\,\mathrm{nm})^2 \rightarrow \min$.

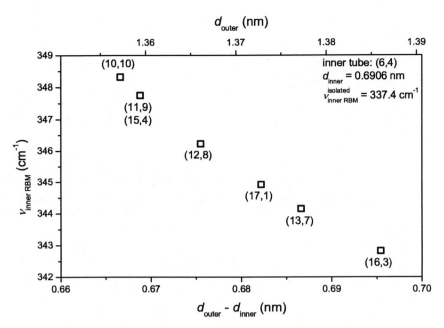

FIGURE 3. RBM frequency of a $(6,4)$ inner tube depending of the diameter d_{outer} of the encapsulating tube. The diameters were calculated from the interpolation formula from Kramberger et al. [13].

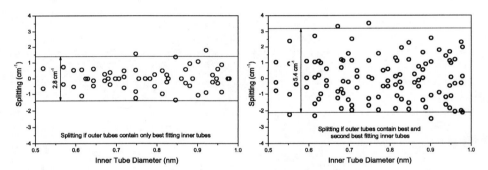

FIGURE 4. Number of split components and width of splitting for individual inner tubes from the theory after subtracting the mean value of the calculated RBM frequencies. Left: Only the best fitting inner tubes are formed. Right: The best and second best fitting inner tubes are formed.

components and the width of the splitting are smaller than the experimentally observed values. Therefore, in a second step, we assumed that also the second best fitting inner tubes may be formed. As Fig. 4 (right) shows, this assumption results in a larger splitting than observed. This suggests that also second best fitting inner tubes are formed but not in all cases.

In summary, we have shown that the RBMs of the inner tubes of DWCNTs are split into several components. This is attributed to the interaction between inner and outer tubes that causes a change of the inner tube RBM frequency. Since it is possible that one type of inner tube forms in several types of outer tubes (with slightly different diameters) every inner tube gives rise to more than one RBM in the Raman spectrum.

ACKNOWLEDGMENTS

The authors acknowledge financial support from the FWF in Austria, project P14893, and from the EU, project PATONN Marie-Curie MEIF-CT-2003-501099. VNP was partly supported by a fellowship from the Federal Science Policy Office for promoting the S&T cooperation with Central and Eastern Europe and by a NATO CLG. We thank Prof. H. Kataura for providing the peapod samples used for the production of the DWCNTs.

REFERENCES

1. Iijima, S., *Nature*, **354**, 56–58 (1991).
2. Iijima, S., and Ichihashi, T., *Nature*, **363**, 603–605 (1993).
3. Thess, A., Lee, R., Nikolaev, P., Dai, H., Petit, P., Robert, J., Xu, C., Lee, Y. H., Kim, S. G., Rinzler, A. G., Colbert, D. T., Scuseria, G. E., Tománek, D., Fischer, J. E., and Smalley, R. E., *Science*, **273**, 483–487 (1996).
4. Smith, B. W., Monthioux, M., and Luzzi, D. E., *Nature*, **396**, 323–324 (1998).
5. Bandow, S., Takizawa, M., Hirahara, K., Yudasaka, M., and Iijima, S., *Chem. Phys. Lett.*, **337**, 48–54 (2001).
6. Bandow, S., Chen, G., Sumanasekera, G. U., Gupta, R., Yudasaka, M., Iijima, S., and Eklund, P. C., *Phys. Rev. B*, **66**, 075416 (2002).
7. Pfeiffer, R., Kuzmany, H., Kramberger, C., Schaman, C., Pichler, T., Kataura, H., Achiba, Y., Kürti, J., and Zólyomi, V., *Phys. Rev. Lett.*, **90**, 225501 (2003).
8. Kataura, H., Maniwa, Y., Kodama, T., Kikuchi, K., Hirahara, K., Suenaga, K., Iijima, S., Suzuki, S., Achiba, Y., and Krätschmer, W., *Synthetic Met.*, **121**, 1195–1196 (2001).
9. Kuzmany, H., Plank, W., Hulman, M., Kramberger, C., Grüneis, A., Pichler, T., Peterlik, H., Kataura, H., and Achiba, Y., *Eur. Phys. J. B*, **22**, 307–320 (2001).
10. Pfeiffer, R., Pichler, T., Holzweber, M., Plank, W., Kuzmany, H., Kataura, H., and Luzzi, D. E., "Concentration of C_{60} Molecules in SWCNT," in *Structural and Electronic Properties of Molecular Nanostructures*, edited by H. Kuzmany, J. Fink, M. Mehring, and S. Roth, American Institute of Physics, Melville, New York, 2002, vol. 633 of *AIP Conference Proceedings*, pp. 108–112.
11. Kuzmany, H., Pfeiffer, R., Kramberger, C., Pichler, T., Liu, X., Knupfer, M., Fink, J., Kataura, H., Achiba, Y., Smith, B. W., and Luzzi, D. E., *Appl. Phys. A*, **76**, 449–455 (2003).
12. Liu, X., Pichler, T., Knupfer, M., Golden, M. S., Fink, J., Kataura, H., Achiba, Y., Hirahara, K., and Iijima, S., *Phys. Rev. B*, **65**, 045419 (2002).
13. Kramberger, C., Pfeiffer, R., Kuzmany, H., Zólyomi, V., and Kürti, J., *Phys. Rev. B*, **68**, 235404 (2003).
14. Lu, J. P., and Yang, W., *Phys. Rev. B*, **49**, 11421–11424 (1994).
15. Dobardžić, E., Maultzsch, J., Milošević, I., Thomsen, C., and Damnjanović, M., *phys. stat. sol. (b)*, **237**, R7–R10 (2003).

Single wall carbon nanotube specific ^{13}C isotope enrichment

Ferenc Simon, Christian Kramberger, Rudolf Pfeiffer and Hans Kuzmany

Universität Wien, Institute für Materialphysik, Strudlhofgasse 4, 1090, Wien, Austria

Abstract. We report on the single wall carbon nanotube specific ^{13}C isotope enrichment. The high temperature annealing of isotope enriched fullerenes encapsulated in SWCNTs yields double wall carbon nanotubes with a high isotope enrichment of the inner wall. The methods helps to discriminate the Raman signal of the outer and inner tubes. The vibrational spectra evidences that no carbon exchange occurs between the two walls.

INTRODUCTION

The characterization of the electronic and structural properties of SWCNTs, that is neccessary for both fundamental and applied studies, can be conveniently performed with Raman spectroscopy. It yields a variety of information ranging from the SWCNT diameter distribution[1] to the electronic structure of SWCNTs due to the photoselectivness of the tube Raman response. However, the full identification of the large number of Raman active modes[2] is to date still not a completed task[3]. A simplifying approach for this task could be the synthesis of fullerene peapod based double-wall carbon nanotubes (DWCNTs). In DWCNT samples, the outer tubes are left intact and inner, smaller diameter tubes are grown. The comparison of the starting SWCNT and the final DWCNT samples could in principle greatly simplify the identification of the Raman active components. In addition, the study of the inner tubes in C_{60} peapod[4] based DWCNTs[5] enables the synthesis and study of small diameter tubes. This approach has been successfully applied for the identification of the inner tube radial breathing modes (RBMs)[6], which enabled refinement of the empirical constants relating the RBM frequencies with the tube diameters[7]. However, for other modes such as e.g. the tangential G mode, the diameter dependence of the Raman response is weaker thus separation of inner and outer tube Raman signals is difficult.

Here, we report a ^{13}C isotope enrichment method of single-wall carbon nanotubes that is selective for inner shell nanotubes only. We synthesized SWCNTs encapsulating ^{13}C$^{.}$ enriched C_{60} and C_{70} fullerene molecules. A high temperature treatment transforms the fullerenes into a second smaller diameter inner tube that is isotope enriched reflecting the enrichement level of the fullerenes. The identification of inner tube vibrational modes is presented, based on different levels of ^{13}C enrichment using Raman spectroscopy.

CP723, *Electronic Properties of Synthetic Nanostructures*, edited by H. Kuzmany et al.

EXPERIMENTAL

We prepared fullerene peapod C_{60},C_{70}@SWCNT based DWCNTs. Arc-discharge prepared commercial SWCNT material (Nanocarblab, Moscow, Russia), ^{13}C isotope enriched fullerenes (MER Corp., Tucson, USA), and natural carbon containing fullerenes (Hoechst AG, Frankfurt, Germany) were used for the synthesis. The SWCNT was purified by the supplier to 50 %. The tube diameter distribution was determined from Raman spectroscopy[1] and we obtained $d = 1.40$ nm, $\sigma = 0.10$ nm for the mean diameter and the variance of the distribution, respectively. We used two degrees of ^{13}C enrichment: 25 and 89 %. These enrichment factors were determined using mass spectroscopy by the supplier and are slightly refined according to our studies. The fullerenes used were C_{60}/C_{70}/higher fullerene mixtures containing 75%:20%:5% ($^{25\ \%,\ 13}C_{60}$ in the following) and 12%:88%:<1 % ($^{89\ \%,\ 13}C_{70}$ in the following) for the 25 and 89 % enriched materials, respectively. The fullerenes containing natural carbon (^{Nat}C) had a purity of > 99 %. SWCNTs were sealed under vacuum in a quartz ampoule with the fullerene powder and annealed at 650 °C for 2 hours for the fullerene encapsulation[8]. Non-reacted fullerenes were removed by dynamic vacuum annealing at 650 °C. The peapod samples were transformed to DWCNT by a 2 h long dynamic vacuum treatment at 1250 °C following Ref. 5. Multi frequency Raman spectroscopy was used for the vibrational analysis on a Dilor xy triple axis spectrometer in the 1.64-2.54 eV (676-488 nm) energy range at 90 K. The spectral resolution as determined from the spectrometer response to the elastically scattered light was 0.5-2 cm^{-1} depending on the laser wavelength and the resolution mode used (high or normal resolution).

RESULTS AND DISCUSSION

In Fig. 1, we show the spectra of the DWCNTs based on the $^{Nat}C_{60}$, $^{25\ \%,13}C_{60}$ and $^{89\ \%,\ 13}C_{70}$ for the radial breathing mode (RBM) (Fig.1a.) and the D and G mode spectral ranges Fig. 1b. at 676 nm excitation and 90 K. The narrow lines in Fig. 1a.) were previously identified as the RBMs of the inner tubes[6]. An overall down shift of the inner tube RBMs is observed for the ^{13}C enriched materials accompanied with a broadening. The down shift is a clear evidence for the effective ^{13}C enrichment of inner tubes. The magnitude of the enrichment and the origin of the broadening is discussed below.

The RBMs are well separated for inner and outer tubes due to the ν_{RBM} $1/d$ reciprocal relation between the RBM frequency, ν_{RBM}, and the tube diameter, d. However, other vibrational modes such as the defect induced D and the tangential G modes are strongly overlapping for inner and outer tubes. Therefore, isotope substitution is a valuable too to separate the lines. This holds in particular for the D-line as shown in Fig. 1b. The broad line profile does not immediately allow to separate the response from the outer and inner tubes. The arrows in Fig. 1b. indicate a gradually downshifting component of the observed D and G modes. This mode is assigned to the D and G modes of the inner tubes. The sharper appearance of the inner tube G mode

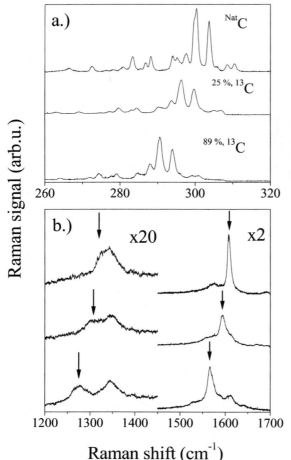

FIGURE 1. Raman spectra of DWCNT with ^{Nat}C and ^{13}C enriched inner walls at 676 nm laser excitation and 90 K. a.) RBM Raman shift range recorded with high resolution, (0.5 cm^{-1}), b.) D and G mode ranges recorded with normal resolution (1.5 cm^{-1}). Arrows indicate the D and G modes corresponding to the inner tubes.

as compared to the outer components is related to the excitation of semiconducting inner and the Fano lineshape broadened metallic outer tubes[6,9].

The relative magnitude of the inner and outer tube D modes are comparable as seen best for the $^{89\%, 13}C$ enriched sample. This is a surprising observation as the D mode is known to originate from a double resonance process and is related to the number of defects. However, the inner tubes were shown to contain significantly less defects than the outer ones as proven by the narrow RBM phonon linewidths[6]. The enhanced electron-phonon coupling in small diameter inner tubes can explain for the large D mode of inner tubes although they contain less defects. The electron-phonon coupling was experimentally found to be roughly 10 times larger for inner than for the outer tubes[6,9]. The presence of superconductivity in 4 Å diameter tubes[10] is also thought to be related to a large electron-phonon coupling in smaller diameter tubes.

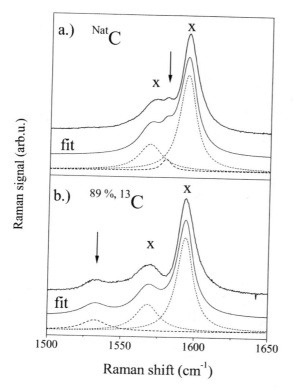

FIGURE 2. G modes of DWCNTs with a.) ^{Nat}C and $^{89\ \%,\ 13}C$ enriched inner walls at 488 nm laser excitation and 90 K. Crosses indicate the non-shifting components of the outer tube G modes. Arrows show the only shifting componenet, identified as the inner tube G modes. Dashed curves show the deconvolution of the observed spectra.

Identification of the different signals also allows to address whether any carbon exchange between the inner and outer tubes occurs during the DWCNT synthesis. In Fig. 2. we compare the G mode spectra of ^{Nat}C and $^{89\ \%,\ 13}C_{70}$ enriched DWCNTs at 488 nm excitation. For this excitation energy, the G mode of the outer tubes dominate the spectra[9]. Indeed, the only shifting component observed is very small (arrows in Fig. 2.) compared to the non-shifting components (crosses in Fig. 2.). A deconvolution using one Lorentzian line for the shifting and two for the non-shifting componenents allows to analyze whether any shift is observable for the outer tube G modes. We found that within the experimental uncertainty of our experiment the outer tube G modes are not shifting proving that no C exchange appears between the two walls at the 1250 °C of the inner tube synthesis. This, however, does not rule out the possibility of such exchange at even higher temperatures or at longer reaction times.

The shift of the RBM, D and G modes were analyzed for the two levels enrichment. We found that $(v_0 - v)/v_0 = 0.0109 \pm 0.0003$ and 0.0322 ± 0.0003 for the 25 and 89 % samples, respectively. Here, v_0-v is the shift of the mode at v_0 Raman shift in the material containing natural carbon only. Under the simplest assumption, the shift originates from the isotope effect such that $(v_0 - v)/v_0 = (c - 0.011)*(\sqrt{13/12} - 1)$, where c is the concentration of the ^{13}C enrichment on the inner tube. This formula also takes into account the 1.10 % natural abundance of ^{13}C in carbon. This gives 27.8±0.7 % and 80.0±0.7 % ^{13}C enrichement for the 25 and 89 % samples. The difference between the supplier provided values and those determined herein underlines the difficulties in estimating the nominal ^{13}C enrichment in the starting material.

CONCLUSIONS

We reported the synthesis of ^{13}C enriched SWCNTs. ^{13}C builds into the inner shell tube only without enriching the inevitable other carbon phases such as amorphous carbon or graphite. The method is based on the transformation of ^{13}C enriched fullerenes encapsulating SWCNTs into double wall carbon nanotubes. We showed that high levels of isotope enrichments can be achieved without significant carbon exchange between the two walls. The ^{13}C enriched inner tubes facilitate the identification of the vibrational modes of inner and outer tube components. The synthesis method opens the way for the controllabel isotope labelling of SWCNTs without labelling the non-wanted sideproducts. It is also hoped that the described isotope engineering may find application for the controllable doping of SWCNTs similar to the isotope engineering applied for Si in the semiconducting industry.

ACKNOWLEDGMENTS

This work was supported by the Austrian Science Funds (FWF) project Nr. 14893 and by the EU projects NANOTEMP BIN2-2001-00580 and PATONN Marie-Curie MEIF-CT-2003-501099 grants.

REFERENCES

1. H. Kuzmany, W. Plank, M. Hulman, C. Kramberger, A. Gruneis, T. Pichler, H. Peterlik, H. Kataura, Y. Achiba, Eur. Phys. J. B **22** (2001) 307.
2. M. Damnjanovic, I. Milosevic, T. Vukovic, and R. Sredanovic, Phys. Rev. B **60**, 2728 (1999).
3. M. S. Dresselhaus, G. Dresselhaus, A. Jorio, A. G. Souza, G. G. Samsonidze, and R. Saito, J. Nanosci. and Nanotechn. **3**, 19 (2003).
4. B. W. Smith, M. Monthioux, and D. E. Luzzi, Nature **396**, 323 (1998).
5. S. Bandow *et al.*, Chem. Phys. Lett. **337**, 48 (2001).
6. R. Pfeiffer *et al.*, Phys. Rev. Lett. **90** 225501 (2003).
7. Ch. Kramberger *et al.*, Phys. Rev. B **68**, 235404 (2003).
8. H. Kataura *et al.*, Synth. Met. **121**, 1195 (2001).
9. F. Simon *et al.*, cond-mat/0404110.
10. Z. K. Tang, L. Y. Zhang, N. Wang, X. X. Zhang, G. H. Wen, G. D. Li, J. N. Wang, C. T. Chan, and P. Sheng, Science **292**, 2462 (2001).

Thin films of C_{60} peapods and double wall carbon nanotubes

F. Hasi, F. Simon, H. Kuzmany

Institut für Materialphysik, Universität Wien, Strudlhofgasse 4, A-1090 Wien, Austria

Abstract. Thin films of SWCNTs were prepared from toluene suspension of single wall carbon nanotubes. The material was dropped on silicon and platinum surfaces with a controlled thickness. Filling with C_{60} and DWCNT – transformation of the films was performed. Multifrequency Raman spectroscopy was used to follow the peapod formation process and double wall transformation of the tubes. An efficient filling with C_{60} and DWCNT – transformation of the films was observed on platinum surface. Silicon reacts with carbon nanotubes forming SiC at the high temperature of the DWCNT formation, 1270 °C. The studies demonstrate the possibility for DWCNT formation process on individual tubes. In addition, it enables the production of homogeneous DWCNT films for several applications such as e.g. field emission devices.

INTRODUCTION

Since their discovery in 1991, carbon nanotubes (CNT) have attracted a lot of interest in the scientific community [1] due to their unique structural and electronic properties. Recently it was discovered that the C_{60} molecules can be combined with the single wall carbon nanotubes (SWCNTs) to forme the so-called peapods (C_{60} @ SWCNT) [2]. Annealing the peapods at high temperatures for several hours, in dynamic vacuum yields double wall carbon nanotubes (DWCNTs) [3] without any additional catalyst. Also, it has been shown that inner tubes formed by the fused C_{60} molecules inside the SWCNTs are higly defect free [4].

We show in the following that thin films of SWCNTs can be produced on different surfaces such as silicon and platinum. The filling with C_{60} fullerene and the transformation to DWCNTs is characterized using Raman spectroscopy. The study of the possible reaction between carbon nanotubes and different surfaces is also of practical interest. The aim of this work is to produce double wall carbon nanotube thin films for application devices, to study individual DW tubes, and to investigate possible reactions between carbon nanotubes and different substrates which is important for nanoreactor devices. A simple shema of a possible nanoreactor which uses the interior of a CNT as a "nanoclean room" to produce defect free, C_{60} based inner tubes is depicted below, in Fig 1.

CP723, *Electronic Properties of Synthetic Nanostructures,* edited by H. Kuzmany et al.
© 2004 American Institute of Physics 0-7354-0204-3/04/$22.00

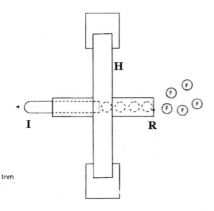

Figure 1. Simple schema of a nanoreactor. The heating zone is located at the crossover between the reactor tube (R) and the heating tube (H). F: fullerenes, I: inner tube, grown by fullerene fusion, ← : direction of growth or pulling.

EXPERIMENTAL

Comercial SWCNT (NCL-SWCNT from Nanocarblab, Moscow,Russia) suspended in toluene were used for the thin films preparation. Two different substrates, silicon and platinum, were used in this study. Likewise two different thickness of the films were studied.

One film was prepared by suspending 5 mg SWCNT in 10 mL toluene.The mixture was sonicated for 1 hour. 50 µL from the suspension was dropped on silicon and platinum platelets with 30 mm^2. The thickness of the film is approximately 1 µm.

The second film was prepared with 2 mg SWCNT in 10 mL toluene, sonicated for 1 hour and 10 µL dropped on platinum platelet. The calculated thickness was about 0.2 µm.

C$_{60}$ fullerene filling of the SWCNT-films was performed according to the method of Kataura et al. [5] which involves sealing the material with C$_{60}$ in a quartz ampoule and baking it at 650 °C for 2 hours.

Dynamic vacuum treatment at 800 °C was used to remove residual fullerene from the samples. DWCNT-transformation was performed by heating the peapods at 1270 °C under dynamic vacuum for 1 hour.

Multi frequency Raman spectroscopy was performed on a Dilor *xy* triple spectrometer using an Ar-Kr mixed-gas laser. All spectra were recorded for normal resolution and at ambient conditions.

RESULTS AND DISCUSSION

C$_{60}$ peapods and DWCNT on Si surface:.The effect of the filling with C60 molecules and the following heat treatements of the SWCNT- film on silicon is clearly demonstrated by the Raman spectra shown below in Fig. 2. and Fig. 3.

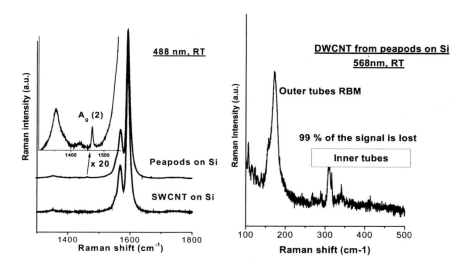

Figure 2. Raman spectra of a SWCNT film on Si (bottom) and of a C$_{60}$ peapod film (top).

Figure 3. Raman spectra of C$_{60}$ based DWCNT film grown on Si substrate.

In Figure 2., we compare Raman spectra of a SWCNT- film and C$_{60}$ based peapods for the films on the Si substrate. The well known Raman active line of C$_{60}$ peapods, the A$_{g}$ (2) mode at 1466 cm^{-1}, is shown enlarged. This Raman line is evidence that the carbon nanotubes on the film are filled with C$_{60}$. The Raman spectra after transformation of the peapods to a DWCNT- film on Si is shown on Figure 3. The line around 170 cm^{-1} corresponds to radial breathing mode (RBM) of outer tubes and the group of lines between 250 and 350 cm^{-1} to the RBM of the inner tubes. The relative low Raman intensity of the inner tubes evidences a low level of DWCNT formation from the peapod film. We supose the carbon nanotubes react with silicon at the high temperature treatement, 1270 °C, forming silicon carbide composites.

C$_{60}$ peapods and DWCNT on Pt surface: In Fig 4. and Fig 5. we show Raman spectra of SWCNT film on Pt and as-transformed DWCNTs respectively.

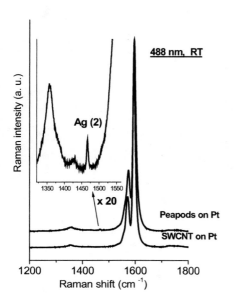 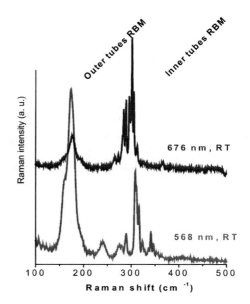

Figure 4. Raman spectra of SWCNT film on Pt (bottom), C$_{60}$ peapods (top).

Figure 5. Raman spectra of C$_{60}$ based DWCNT film on Pt substrate for two different laser exitation.

Raman characterization of the 1μm film was performed with a blue laser (488 nm). It clearly shows the A$_g$ (2) mode at 1466 cm^{-1} (Fig 4., top spectra). The average integrated intensity ratio A$_g$ (2) / G-mode = 2.6 * 10^{-3} which is consistent with the maximum attained level of filling as reported in the literature [6].

In Fig 5., we present the Raman shift of the DWCNT film obtained from annealing the peapod film at 1270 °C for 1 hour under dynamic vacuum. Yellow (568 nm) and red (676 nm) lasers were used to record the spectra. For the two excitations we see sharp and distinct lines between 230 and 350 cm^{-1} which correspond to the RBM of the inner tubes. For the red excitation (Fig 5., top spectra) the response of the inner tube RBM is even stronger than that of the outer tube RBM. This is a clear evidence that the transformation of the peapod film on platinum into DWCNTs was successful and the sharp RBM structure is a sign for defect free inner tubes.The defect free groth conditions inside the tubes makes them a perfect "nanospace". Simmilar patern were seen also for the 0.2 μm thick film on Pt, (not shown).

In summary, we have produced thin films of C_{60} peapods and DWCNT on silicon and platinum substrates. We have shown that transformation of peapod films on silicon in DWCNT films has a very low efficiency of DWCNT formation because of the reaction between carbon nanotubes and silicon substrate, forming SiC. Films of peapods on platinum substrate were transformed successfully to DWCNT films as shown by Raman investigation. This assigns a platinum substrate as a possible candidate for nanoreactor devices. Investigations on thinner films and other substrates is the ongoing work.

ACKNOWLEDGMENTS

The autors acknowledge financial support by the Austrian Science Foundation, project 14893 and the PATONN Marie-Curie MEIF-CT-2003-501099 grants.

REFERENCES

1. Iijima, S., *Nature*, **354**, 56 (1991).
2. Smith, B. W., Monthioux, M. and Luzzi, D. E., *Nature,* **396**, 323 (1998).
3. Bandow, S., Takizaw, M., Hirahara, K., Yudasaka, M., Ijima, S., *Chem. Phys. Lett.*, 48, **337** (2001).
4. Pfeiffer, R., Kuzmany, H., Kataura, H, Pichler, T., Kürti, J. et al, *Phys. Rev. Lett.* (2003).
5. Kataura, H., Maniwa, Y., Kodama, T., Kikuchi, K., Hirahara, K., Suenaga, K., Iijima, S., Suzuki, S., Achiba, Y., Kratschmer, W., *Synth. Met.* **121**, 1195, (2001).
6. Kuzmany, H. et al., *Appl. Phys. A*, **76**, 449, (2003).

Raman spectroscopy of PbO-filled single wall carbon nanotubes

Martin Hulman[a], Pedro Costa[b], Malcolm L.H. Green[b], Steffi Friedrichs[b] and Hans Kuzmany[a]

[a]*Institut für Materialphysik, Universität Wien, Austria*
[b]*Inorganic Chemistry Laboratory, University of Oxford, United Kingdom*

Abstract. We investigated single wall carbon nanotubes filled with lead oxide, PbO, by transmission electron microscopy and Raman spectroscopy. It is concluded from the positions of the Raman lines that PbO crystalizes in the orthorombic phase inside the nanotubes. The line position of the most prominent PbO line is downshifted in about 5 cm^{-1} as compared to the bulk material as a result of the reduced dimensionality. As a consequence of the filling, nanotubes become sensitive to the laser irradiation. At the higher laser power densities, they break up and the free PbO nanowires are left in the sample.

INTRODUCTION

The nano-sized "laboratory" within the cavities of single wall carbon nanotubes (SWNTs) has attracted a lot of attention in recent years. It provides an opportunity to synthesize and investigate materials under conditions not available so far. Carbon nanotube peapods made of fullerene molecules trapped inside SWNTs are the most explored [1] but the variety of the encapsulated solids are not limited to carbon-based materials. Nanotubes have been successfully filled with metals, metal oxides, alkali halides, etc. Enclosed materials form nanowires as revealed by transmission electron microscopy [2].

For fullerene peapods, Raman spectroscopy is a very useful tool enabling to probe vibrational and electronic properties of both the encapsulated fullerenes and nanotubes. In the case of non-fullerene fillers, the available Raman experiments probe exclusively nanotube shells whereas Raman lines of the encapsulated materials have not been observed [3]. On the other hand, lead oxide is a good Raman scatterer whose signal may be detected even it is inside the nanotubes. One can also expect that a close contact of carbon atoms and PbO solid gives rise to changes in physical properties of both the host and guest materials.

CP723, *Electronic Properties of Synthetic Nanostructures*, edited by H. Kuzmany et al.
© 2004 American Institute of Physics 0-7354-0204-3/04/$22.00

EXPERIMENTAL

Single wall carbon nanotubes were produced using an arc discharge apparatus. The as-prepared material was not subjected to further purification or chemical treatment. A 1:1 molar ratio of tubes:PbO was ground in a mortar to a fine powder. The powder sealed in an ampoule was heated in a furnace. Heating cycle regimes were used with target temperatures of around 30 °C above and below the melting point of PbO which is at 888 °C. Each cycle took in average 30-48 hours to complete.

For Raman measurements, the nanotubes were further dissolved and sonicated in chlorbenzene for 1 h to get rid of extraneous PbO. The solution was filtered and the black buckypaper was dried on air and peeled off from the filter. Raman experiments were carried out in the backscattering geometry using a triple monochromator equipped with a microscope. Raman spectra were excited by an Ar^+/Kr^+ ion laser. The diameter of the laser spot on the sample was between 2 and 4 µm. The power of the incident laser beam was kept between 1 and 6 mW leading to the laser power density of several tens of kW/cm^2 on the sample surface.

RESULTS AND DISCUSSION

The as- prepared material was characterized by high resolution transmission

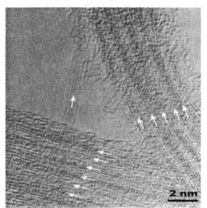

FIGURE 1. HRTEM picture of the as-prepared PbO@SWNT sample. White arrows point to nanotubes filled with lead oxide.

electron microscopy. The result is shown in Fig.1. SWNTs with a diameter of about 1.5 nm are filled with lead oxide forming well ordered nanowires inside the nanotubes. The filling yield was estimated to be about 50-60 % from the measurements. Since the sample was not purified after filling, there is extraneous PbO between the nanotubes.

In course of Raman measurements we observed a noticeable dependence of the shape of the spectra on the laser excitation. It turned out that the laser power and than the laser wavelength plays a decisive role. The laser power dependence of the Raman spectra was investigated systematically. Two examples of out of six different laser wavelengths used are displayed in Fig.2.

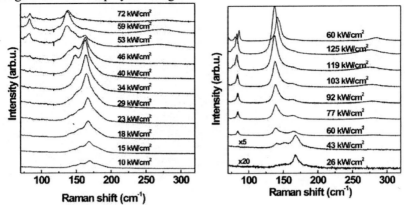

FIGURE 2. The Raman spectra excited with 514.5 nm (left) and 647.1 nm (right) for various laser power densities as indicated. The uppermost spectrum in the right panel was measured after two hours of irradiation of the sample at 125 kW/cm².

At low power densities (< 50 kW/cm²), the spectra show the radial breathing mode (RBM) of nanotubes in the range $150 - 200$ cm⁻¹. For the higher power densities, the RBM vanishes gradually from the spectra and the lines of the orthorombic β–phase of PbO appear at 84, 138 and ~280 cm⁻¹ [4]. Since the temperature during the nanowire growth was above the $\alpha \rightarrow \beta$ transition at 489 °C, the β-phase dominates over the α-phase. Unless the PbO samples are equilibrated for long time below the $\alpha \rightarrow \beta$ phase transition temperature the former persists even at the room temperature [5].

Quenching of RBM is irreversible as demonstrated by the uppermost spectrum in the right panel of Fig.2. No RBM of nanotubes was detected even though the laser power was reduced to 60 kW/cm². Moreover, lines belonging to the α-phase of PbO appear at 81 cm⁻¹ and as a new component at 145 cm⁻¹ of the strongest Raman line. Likely, laser heating induce a transition from the metastable β–phase to the stable α-phase.

An inspection under an optical microscope showed that the nanotubes were indeed destroyed. Light spots appeared on the sites where Raman spectra were taken. The PbO nanowires formed within the nanotube cavities become free. Nevertheless, the wires are stable and retain their 1D character. This is evident from the Raman measurements. The positions of the Raman lines are downshifted by about 5 cm⁻¹ as compared to the line positions for a bulk PbO powder. The reason for this might be twofold. The first is phonon confinement in nano-sized semiconductor crystals [6].

The frequency of transversal and longitudinal modes depends on the crystal dimensions when the size of a crystal is reduced down to a few nanometers. The second is lattice expansion observed for alkali halides inserted into SWNTs [2]. The expansion softens the interatomic bonds and decreases the phonon frequencies.

A similar conclusion can be drawn from the widths of the Raman lines. For the PbO nanowires, the latter are about three times broader as compared to bulk PbO. The diameters of the PbO nanowires should follow the diameter distribution of SWNTs.

TABLE 1. The positions (ν) and the line widths (γ) for the bulk PbO and the PbO nanowires

	ν (cm^{-1})	γ (cm^{-1})	ν (cm^{-1})	γ (cm^{-1})	ν (cm^{-1})	γ (cm^{-1})
bulk β-PbO, 64 kW/cm^2	86.8	3.2	142.6	7.7	289.9	25.5
β-PbO, 60 kW/cm^2	84.3	6.6	138.9	16.2	281.4	80.9
β-PbO, 125 kW/cm^2	83.1	7.0	137.6	20.1	277.4	106.1
β-PbO, 60 kW/cm^2, after 2 h.	86.3	4.5	141.2	12.5	284.5	46.0

Referring back to the phonon confinement, the PbO nanowires of different diameters contribute to the Raman spectrum broadening the spectral lines. Another contribution to the line width comes from the phonon lifetime. The latter can be decreased because of the presence of defects and grain boundaries in the PbO nanowires.

Following this line of reasoning, it is useful to look more carefully at the uppermost spectrum in the right panel of Fig.2. The spectrum was taken after the sample was irradiated at 125 kW/cm^2 for two hours. The Raman lines shift up and become narrower but still do not reach the values for the bulk PbO. The results are summarized in Table 1. The fact that the values for the two spectra taken at 60 kW/cm^2 are not the same rather excludes the changes of the line position and the line width to be a consequence of lattice anharmonicity. Likely, the PbO lattice relaxes back to its bulk configuration by laser beam heating. Moreover, the number of defects and grain boundaries decreases leading to the narrowing of the Raman lines.

As a summary, we investigated PbO-filled SWNTs by Raman spectroscopy. We observed an unexpected structural instability of the nanotubes when a sufficient laser power density was applied. Consequently, the nanowires of PbO formed inside the nanotubes become free. They retain their 1D character as confirmed from the position and width of the Raman lines.

Acknowledgments

M.H. acknowledges a support from the EU project NANOTEMP (HPRN-CT-2002-00192) and P.C. is grateful to FCT for a scholarship (SFRH/BD/3103/2000).

REFERENCES

1. Smith, B.W., Monthioux, M., Luzzi, D.E., *Nature* **396**, 323 (1998)
2. Sloan, J., Kirkland, A.I., Hutchinson, J.L., Green, M.L.H., *C.R. Physique* **4**, 1063 (2003)
3. Corio, P., Santos, A.P., Santos, P.S., Temperini, M.L.A., Brar, V.W., Pimenta, M.A., Dresselhaus, M.S., *Chem. Phys. Lett.* **383**, 475 (2004)
4. Baleva, M., and Tuncheva, V., *J. Solid State Chem.* **110**, 36 (1994)
5. Adams, D.M., Christy, A.G., Haines, J., Clark, S.M., *Phys. Rev.* **B 46**, 11358 (1992)
6. Krauss, T.D., Wise, F.W., Tanner, D.B., *Phys. Rev. Lett.* **76**, 1376 (1996)

NON-CARBONACEOUS NANOTUBES

Thermally Induced Templated Synthesis for the Formation of SiC Nanotubes and more

Mark. H. Rümmeli[1], Ewa Borowiak-Palen[1,2], Thomas Gemming[1], Martin Knupfer[1], Kati Biedermann[1], Ryszard J. Kalenczuk[2], Thomas Pichler[1]

1) IFW Dresden, Helmholtzstr.20, D-01069 Dresden, Germany
2) Institute of Chemical and Environment Engineering, Technical University of Szczecin, Poland

Abstract: A thermally induced templated synthesis for SiC nanotubes and nanofibers using ammonia or nitrogen as a carrier gas, single wall carbon nanotubes (SWCNT) as templates as well as gaseous Si is presented. The bundles of SWCNT act as both the carbon source and as a nanoframe from which SiC structures form. Depending on the duration of the thermally induced templated reaction, for a fixed temperature, carrier gas, and gas pressure, various SiC nanostructures are obtained. These structures include SiC nanorods coated in C, SiC nanorods, SiC nanotubes and SiC nanocrystals. From our analysis using transmission electron microscopy (TEM) and scanning electron microscopy (SEM), electron energy-loss spectroscopy (EELS), electron diffraction (EDX), optical absorption spectroscopy and Raman spectroscopy as probes we prove that H has a key role on the morphology and stochiometry of the different SiC nanostructures.

INTRODUCTION

The mechanical, chemical, thermal and electronic properties (wide band gap) of SiC make it an attractive material to include in the family of nanotubes. Potential applications for SiC nanostructures include nanosensors and nanodevices operable at high temperature, high frequency and high power [1, 2] and as a support material in the catalysis field[3]. SiC nanotubes have been obtained through various techniques [3-7].

Here, a thermally induced templated synthesis, is presented, in which a reaction of SWCNT at high temperature in the presence of gaseous Si in a reduced carrier gas atmosphere leads to a variety of SiC nanostructures depending on the conditions. The results show a variety of SiC based nanostructures, such as nanorods and nanotubes. In addition, the results help elucidate what formation processes take place during the reaction and highlight the distinctive role of H.

EXPERIMENTAL

The starting material was a black powder containing 70% of SWCNT with a mean diameter of 1.25 nm produced by a laser ablation process optimised for high SWCNT yield. The carrier gas was ammonia gas (or nitrogen), and the source of silicon was commercial silicon powder (Goodfellows, 99.99%). The SWCNT and Si were placed in an Al_2O_3 crucible to a ratio of 1:1. The crucible plus contents was then placed at the centre of a horizontal tube furnace (Carbolite STF 16/180). A schematic view of the experimental set-up is described elsewhere[8]. A highly sensitive needle valve then controls the carrier gas entry that in this instance was at a base pressure of $5*10^{-5}$ mbar. A reaction temperature of 1450°C was used throughout. Dwell times of 30 min., 4, 8 and 20 h were used with a heat up period of 30 min. and a cool down time of several hours. After the thermally induced templated synthesis the resultant material

CP723, *Electronic Properties of Synthetic Nanostructures*, edited by H. Kuzmany et al.
© 2004 American Institute of Physics 0-7354-0204-3/04/$22.00

was removed from the crucible. The product consisted of a portion that was light grey and the remainder was black unreacted SWCNT. The light grey material was then carefully separated (manually) for analysis.

The morphology of the reacted nanostructures were studied using scanning electron microscopy (SEM, Hitachi S4500) and transmission electron microscopy (TEM, FEI Tecnai F30) which also allowed EELS mapping of the nanostructures. In addition, electron diffraction measurements were performed in a purpose built high-resolution EELS spectrometer described elsewhere[9]. The energy loss was set to zero. These EELS measurements probe an area of about 1 mm^2 and thus present information for a bulk average of the produced nanostructures. Raman measurements were conducted on a Bruker FTRaman spectrometer with a resolution of 2 cm^{-1}.

For the measurements tiny quantities of the product were pressed onto standard platinised microscopy grids where upon they were annealed at 450 °C for 12 h in vacuum prior to measurement so as to remove contaminants.

RESULTS AND DISCUSSION

Reaction times ranging from 0.5 – 20 h were used. The produced samples consisted of a soot comprised of two species; a grey material in the centre and the remainder was black concomitant with the starting SWCNT.

These discussions are based solely on the grey product (unless otherwise stated). The amount of grey material increased with reaction time and many changes are observed in the produced nanostructures depending on the reaction time. Some of these changes are observable from their surface morphology whilst others are not. The shortest reaction time of 0.5 h yielded SiC nanorods with a C coating. With increasing reaction time one obtains pure SiC nanorods that then transform into SiC nanotubes, then SiC nanocrystals, which, with yet longer reaction times disappear altogether. This rather striking observation is indicative of some decomposition process. These various nanostructures were only observed when using NH$_3$ as the carrier gas. When using N$_2$ as the carrier gas only SiC nanorods were obtained which were narrower in diameter suggesting H plays a key role in the formation of the transformed nanostructures.

Bulk stochiometric studies of the samples began with electron diffraction studies that showed all samples to be a mix of SiC α and β phases with random orientation and Raman studies showed the α phase consists of many polytypes.

High-resolution EELS studies confirmed the samples altered bonding environment (from SWCNT to that of SiC). In addition the Si edges show no superposition of spectral features from SiC and Si [10].

Studies on a local scale of the produced samples were conducted using TEM, including cross-sectional EELS. These studies not only revealed the structural nature of the SiC nanostructures but also the role of H in the reaction process that begins with SWCNT bundles and Si powder (fig. 1. i). Analysis of the black material showed that it consisted primarily of SWCNT with small localized regions where SiC crystallization has begun and illustrates the earliest stages of the reaction process are localized (fig. 1 ii). These local crystallization sites can then grow laterally and also outward due to excess C that diffuses out from crystallized regions forming a C outer layer. This stage (0.5 h) is observed as SiC nanorods with a carbon cladding with diameters ranging from 19 – 60 nm (fig. 1 iii). Eventually the C source is used up from the C coating and the lateral SWCNT bundle, yielding a SiC nanorod (fig 1. iv). One might then expect that at this stage the reaction ceases and this is indeed the case for samples prepared in N_2. However, when NH_3 is used a remarkable transformation process begins. At the reaction temperature used NH_3 decomposes to N and H indicating that H is responsible for the transformation process.

The sample prepared with a 5 h reaction time contained a small fraction of nanocrystals (< 5%), and an even quantity of SiC nanorods and porous nanorods (fig. 1 iv & v respectively). In addition, the diameters of the porous nanorods (d = 40 – 100 nm) tended to be larger than those of the nanorods (d = 20 – 60 nm) and both nanorod species were on the whole larger than the C coated nanorods. Increasing the reaction time to 8 h yielded 3 types of species (aside from a small fraction of SiC nanocrystals); nanorods (d = 40 – 80 nm) and porous nanorods (d = 70 – 150 nm) as with the 5 h sample and now, additionally, hollow nanorods or SiC nanotubes (diameter range 100 – 250 nm) which where closed ended (fig. 1 iv, v & vi respectively). The transformation of the SiC nanorods to porous nanorods and then into nanotubes has not previously been reported and raises the interesting question as to what exactly is the role of H in this process? It is well known that amorphous SiC can be hydrogenated (a-SiC:h) and it has been shown that H substitutes $Si^{[11]}$. That H substitutes Si atoms in this reaction is highly probable. It may also substitute C atoms, although H preferentially bonds to C as opposed to Si [11]. Furthermore at 1450°C C-H bonds can easily break[12]. This substitution process then in essence leads to the decomposition of SiC, which will occur at a higher rate in the centre of the nanorods as this is where the concentration of H is greatest. It follows then that the decomposition process will provide a source of Si and C atoms that can diffuse outward. This we observe as the outer diameters of the SiC nanorods increasing with reaction time whilst at the same time becoming porous and then closed ended nanotubes. Not all the outwardly diffused Si and C will reform on the surface of the SiC nanorods/nanotubes such that eventually the nanotubes disintegrate leaving only SiC crystals which we observed with the sample from a 12h reaction and, with a sufficiently long reaction time, one obtains the total decomposition of the SiC species, which we observed with a 20 h reaction sample where virtually no material remained. Thus, the formation order when H is present in the reaction, is from SWCNT to SiC nanorods cladded in C, to SiC nanorods, to porous SiC nanorods to SiC nanotubes to eventual total decomposition. This process is illustrated in Fig. 1

To conclude, this is the first report of a reaction process that combines the substitutional ability of H on SiC that leads to the novel transformation of SiC nanorods to nanotubes. It is also the first report of SiC nanostructure formation via gaseous Si in the presence of SWCNT.

The reaction will not only help in our knowledge of hydrogenated SiC were currently very little is understood, it may also open the path, through a modification of this process, for the controlled etching of SiC films in micro/nano electronic applications.

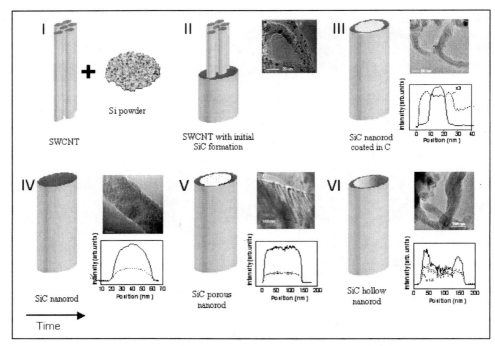

Figure 1. The various stages of the reaction. i. Starting materials. ii. Initial SiC formation with TEM image of the nanostructure. iii. SiC nanorod coated in C iv. SiC nanorod v. SiC porous nanorod. vi. SiC nanotube. iii includes an SEM micrograph of the structures, iv, v & vi include a TEM image and iii to vi include an EELS map across the nanorod/nanotube. Reaction time increases from i through vi. (see text).

REFERENCES

[1] Y. H. Mo, M. D. Shajahan, Y. S. Lee, Y. B. Hahn, K. S. Nahm, *Synth. Met.*, **2004**, *140*, 309

[2] Z. Pan, H.-L. Lai, F. C. K. Au, X. Duan, W. Zhou, W. Shi, N. Wang, C.-S. Lee, N.-B. Wong, S.-T. Lee, S. Xie, *Adv. Mater.*, **2000** ,*12*, 1186

[3] J. M. Nhut, R. Vieira, L. Pesant, J.-P. Tessonnier, N. Keller, G. Ehet, C. Pham-Huu, M. J. Ledoux, *Catalysis Today*, **2002**, *76*, 11.

[4] N. Keller, C. Pham-Huu, G. Ehet, V. Keller, M. J. Ledoux, *Carbon*, **2003**, *41*, 2131.

[5] C. Pham-Huu, N. Keller, G. Ehet, M. J. Ledoux, *Journal of Catalysis*, **2001**, *200*, 400.

[6] http://www.grc.nasa.gov/WWW/RT2002/5000/5510lienhard.html.

[7] X. H. Sun, C. P. Li, W.K. Wong, N.B. Wong, C.S. Lee, S.T. Lee, B.K. Teo, *J. Am. Chem. Soc.*, **2002**, *124*, 14464

[8] E. Borowiak-Palen, T. Pichler, G. G. Fuentes, A. Graff, R. J. Kalenczuk, M. Knupfer, and J. Fink, *Chem. Phys. Lett.*, **2003**, *378*, 516

[9] J. Fink, *Adv. Electron. Electron Phys.*, **1989**, *75*, 121

[10] N. Asaoka, S. Muto, T. Tanabe, *Diamond and Rel. Mat.*, **2001**, *10*, 1251

[11] A. E. Kaloyeros, R. B. Rizk, J. B. Woodhouse, *Phys. Rev. B*, **1988**, *38*, 13099

[12] H. Y. Wang, Y. Y. Wang, Q. Song, T. M. Wang, *Mat. Lett.*, **1998**, *35*, 261

Raman Spectra of $B_xN_yC_z$– Nanotubes: Correlation between B, N – Content and Frequency Shifts of the G-band

T. Skipa[a], P.Schweiss[a], K.-P. Bohnen[a], S. Lebedkin[b], B.Renker[a]

Forschungszentrum Karlsruhe, [a]Institut für Festkörperphysik and Physikalisches Institut, Universität Karlsruhe, [b]Institut für Nanotechnologie, D-76021 Karlsruhe, Germany

Abstract. Boron and nitrogen doped carbon nanotubes ($B_xN_yC_z$ - NTs) were synthesized with high doping levels up to x ~ 0.4 and y ~ 0.3, respectively, using the substitution reaction between single-wall CNTs and B_2O_3 in NH_3 atmosphere. Along with the intensity increase of the D-peak in Raman spectra, we observe for B and N doping considerable changes in the G-peak positions. In a first approximation these effects might be attributed to a simple mass-dependence of the phonon frequency and a lower symmetry of the system due to the incorporation of B and N atoms. We derive empirical correlations between the frequency shifts of the G-band and the B, N-content using a combination of micro-Raman technique and EDX element analysis.

INTRODUCTION

Nanotubes and nanomaterials attract substantial interest with respect to microelectronics due to the wide range of their electronic properties from metallic to insulating behavior in combination with unique mechanical, temperature and optical characteristics. $B_xN_yC_z$ nanotubes play a special role due to their "tunable" electronic properties. The band gap energies of these nanotubes do not strongly depend on the diameter and chirality. They show, however, essential changes for different atomic compositions [1, 2], which points to a way to control its electronic properties.

In order to obtain the B and N concentrations for $B_xN_yC_z$ nanotubes one can use direct methods such as electron energy loss spectroscopy (EELS) or energy dispersive X-ray analysis (EDX) in combination with high resolution transmission electron microscopy (HRTEM).

Since the atomic composition affects not only the electronic but also the vibrational properties of the nanotube, Raman spectroscopy provides indirect access to the nanotube's stoichiometry determination.

We present a systematic study on the influence of various B, N- doping levels on the vibrational and electronic properties of $B_xN_yC_z$ nanotubes, and explore the possibility of using micro-Raman spectroscopy for estimating locally variable B, N concentrations [3].

EXPERIMENTAL

$B_xN_yC_z$ - NTs were synthesized in a substitution reaction between single-wall carbon nanotubes (SWCNTs) and boric acid H_3BO_3 in NH_3 atmosphere [4].

SWCNTs were produced by pulsed laser vaporization (PLV) of carbon targets containing a Ni/Co catalyst in an inert atmosphere. This method provides defect-free SWCNTs with a narrow spread of the diameter distribution between ~1.0 and ~1.5 nm. For our experiments we used as-prepared SWCNTs as well as so-called "bucky paper" (resembling black paper of about 50-100 μm thickness) produced by a treatment of as-prepared SWCNTs with boiling sulphuric acid followed by

CP723, *Electronic Properties of Synthetic Nanostructures*, edited by H. Kuzmany et al.
© 2004 American Institute of Physics 0-7354-0204-3/04/$22.00

neutralization, filtration and drying. The acid treatment removes most of the Ni/Co catalyst and amorphous carbon particles, but causes numerous structural defects in the carbon tubes which are apparently responsible for the different B, N- doping behavior discussed below.

All our measurements were performed at room temperature. The micro-Raman spectra were obtained with a diode-pumped Nd:YAG laser (λ_{exc}= 532 nm). HRTEM was performed by a Philips Tecnai F20 S-Twin TEM with a field emission gun operating at 200 kV, equipped with an EDX Data acquisition module (EDAX) that is suitable for the detection of elements with order number $Z \geq 5$.

RESULTS AND DISCUSION

High Resolution Transmission Electron microscopy (HRTEM)

Varying the temperature and using as-prepared CNTs or bucky paper as starting materials, we synthesized $B_xN_yC_z$ nanotubes with various B and N concentrations (Figure 1).

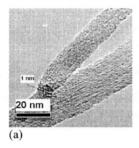 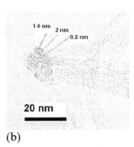

(a)   (b)

FIGURE 1. HRTEM images of the $B_xN_yC_z$ – NTs synthesized (a) at 1250 ^{0}C using bucky papers (x = 0.07, y = 0.06, z = 0.87) and (b) at 1200 ^{0}C using as-prepared CNTs (x = 0.12, y = 0.04, z = 0.84) as starting materials.

The major interest of this study was directed towards $B_xN_yC_z$ - NTs produced from bucky papers since their higher purity (without hexagonal BN). The doping level in general depends on the synthesis temperature. The mean concentration of B and N in the $B_xN_yC_z$ - NTs did not exceed x = 0.15 and y = 0.13 for samples synthesized at 1100 ^{0}C, or x = 0.37 and y = 0.28 at 1250 ^{0}C, respectively. At higher temperature, the bucky papers burned out. We found a carbon concentration z ~ 0.4 (x + y ~ 0.6) as the minimum possible carbon content at which we still obtained Raman spectra similar to those of pure CNTs.

Energy Dispersive X-ray Analysis (EDX)

For micro-Raman spectroscopic studies as well as EDX element analysis we placed a very small amount of our samples onto standard gold grids (mesh 300) with markers which allowed us to select certain areas for our investigations.

For our micro-Raman measurements we can focus the laser to a beam diameter of about 1μm. For the EDX-analysis the electron beam can be focused onto a 1nm-spot which allows then to determine the element composition of small tube ropes within this 1μm-area. In a study of our materials by HRTEM and EDX, we observed over the areas of about 1μm more or less homogeneous doping with a variation $\leq 5\%$ (x, y ± 0.05).

Raman Spectroscopy: G – band

Raman spectra obtained for the $B_xN_yC_z$ - NTs produced from bucky papers and as-prepared CNTs are shown in Figures 2 (a, b) respectively.

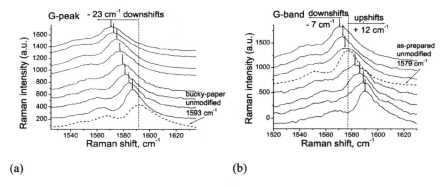

(a)   (b)

Figure 2. Raman spectra (G-bands) of $B_xN_yC_z$ – NTs with different doping levels produced from (a) bucky papers at 1100 ^{0}C and 1250 ^{0}C, and (b) as-prepared CNTs at temperatures from 1100 ^{0}C up to 1500 ^{0}C (λ_{exc} = 532 nm).

The lowest plot in Fig.2 (a) (dashed line) corresponds to unmodified bucky paper with the G-band centered at 1593 cm^{-1}. There is a clear trend for a G-peak shift to lower frequencies, by up to (-23) cm^{-1}. We attribute this downshift to different B and N concentrations incorporated into the carbon nanotube structure.

Figure 2 (b) shows the Raman spectra for as-prepared carbon nanotubes modified with boron and nitrogen at different temperatures. Here we see both downshifts and upshifts with respect to the G-band position (1579 cm^{-1}) of unmodified as-prepared CNTs (dashed line).

A simple explanation for the G-band upshifts regards the holes and broken bonds in original unmodified bucky paper tubes as "lighter" particles compared to the carbon atoms. Any atom incorporated into such defective structure would then lead to G-band downshifts via the related mass increase.

The observation of both up- and downshifts in the case of samples produced from as-prepared CNTs with originally well-graphitized structure could be explained by the mass argument, since nitrogen atoms are heavier and boron atoms are lighter then carbon. The actual G-band shifts for as-prepared CNTs would then be determined by the ratio of the competing boron-caused upshifts and nitrogen-caused downshifts.

Correlation between the G-peak Position and B, N – Concentration

Figure 3 (a, b, c) shows the experimental correlations between the G-band shifts in Raman spectra with respect to unmodified tubes, and the carbon, boron, and nitrogen contents for $B_xN_yC_z$ - NTs produced from bucky papers at 1100 ^{0}C and 1250 ^{0}C (spectra from Fig.2, a). For the $B_xN_yC_z$ – NTs produced from as-prepared CNTs the analogous correlations must look different.

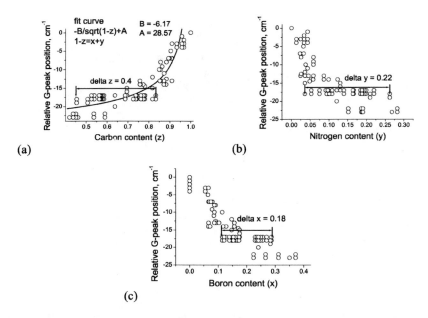

(a)   (b)

(c)

FIGURE 3. Correlations between G-band shifts in Raman spectra and B and N-concentrations inside the CNTs' cages for $B_xN_yC_z$-NTs produced from bucky papers at 1100 ^{0}C and 1250 ^{0}C. The arrows indicate our estimates with respect to stoichiometry variations within the Raman laser scan area (1μm).

In spite of the large stoichiometry variations for larger doping (delta x, y, z in Fig. 3) we clearly observe a downshift of the G-peak for B – as well as for N – doping where the latter seems to have a much stronger impact.

The largest experimental shifts of the G-bands correspond to the determined concentrations $(x + y) \sim 0.6$ (or $z \sim 0.4$). These concentrations are within the range obtained by the tight-binding calculations [2, 5].

ACKNOWLEDGEMENTS

The authors are gratefully acknowledge the group of Prof. M. Kappes (University of Karlsruhe) for the original SWCNT materials used in sample preparation.

REFERENCES

1. Blase X, Charlier J-Ch, De Vita A, and Car R. *Appl Phys A* **68**, pp. 293-300 (1999).
2. Yoshioka T, Suzuura H, Ando T. *Journal of Phys Society of Japan*, submitted (available online).
3. Skipa T., Schweiss P., Bohnen K.-P., Lebedkin S., Renker B. *Carbon*, 2004 (to be published).
4. Golberg D, Bando Y, Bourgeois L, Kurashima K, Sato T. *Carbon* **38**, pp. 2017-2027 (2000).
5. Bohnen K-P, Heid R. *Ab initio* density functional calculations for small radii boron carbonitride nanotubes (unpublished).

EELS measurements in single wall Boron Nitride nanotubes

R. Arenal[1], O. Stephan[2], M. Kociak[2], D. Taverna[2], C. Colliex[2], A. Rubio[3], A. Loiseau[1]

[1] LEM, Onera-Cnrs, 29 Avenue de la Division Leclerc, BP 72, 92322 Châtillon, France
[2] Laboratoire de Physique Solides, Université Paris-Sud, 91405 Orsay, France
[3] Dpto. Fisica de Materiales, UPV-DIPC, 20018 San Sebastian, Spain

Abstract. We present here the results of an electron energy loss spectroscopy (EELS) study in scanning transmission electron microscopy (STEM) on boron nitride nanotubes (BN-NTs). The low and core-loss regions have been analyzed to provide by the same technique a combined information about chemical bonding in the different materials in the sample and the electronic properties of individual BN-NTs. In particular, we deduce an optical gap value of about 5.8 eV for single walled nanotubes, which is independent on diameter.

INTRODUCTION

Carbon nanotubes have been the subject of an enormous number of studies concerning their synthesis, electronic, mechanics, optical properties... Although carbon nanotubes were the first to be discovered and remain the most well-known, they are not the unique form. Boron nitride nanotubes are an attractive alternative to carbon. They possess the same hexagonal structure, with boron and nitrogen atoms on alternate lattice sites. The result is a highly polar dielectric material, with a predicted wide band gap close to 5.5 eV, which is supposed to be independent on the tube diameter and helicity [1].

In the work reported here, we have used Transmission Electron Microscopy (TEM) imaging in combination with dedicated Scanning Transmission Electron Microscopy (STEM) experiments to investigate the chemical composition and electronic properties of the BN-NTs.

EXPERIMENT

BN-Nts were synthesized using a laser vaporization technique as described elsewhere [2].

The collected raw powder is dissolved in ethanol and ultrasonically dispersed. A drop of the solution is placed on a copper grid covered by a holey carbon film for electron microscopy study.

TEM images were performed using a Philips CM20 (at 200KV) microscope. The EELS measurements were carried out in a STEM VG-HB501 operated a 100 kV. It is

CP723, *Electronic Properties of Synthetic Nanostructures*, edited by H. Kuzmany et al.
© 2004 American Institute of Physics 0-7354-0204-3/04/$22.00

equipped with a cold field emission source (CFE), and a Gatan 666 parallel- EELS spectrometer. EEL spectra were recorded with a charge-coupled device camera optically coupled to a scintillator in the image plane of a Gatan magnetic sector. The high brightness CFE source allows the formation of small probes (0,5 nm) containing a primary current of 0.1 nA and with an energy spread of 0.3-0.4 eV. That provides appropriate conditions to record a spectroscopic information at the nanometer scale on individual nanostructures in the Spectrum-imaging mode. This mode consists of the acquisition of one EEL spectrum for each position of the probe. As the probe scans over a 2D region of the sample, then, it is possible to obtain not only the data to be analyzed but also the spatial statistics of a collection of spectra to be exploited.

RESULTS AND DISCUSSION

Description of the samples

The samples display a relatively heterogeneous morphology [3]. In figure 1 which shows a general view of the samples, it is possible to distinguish NTs (isolated or in bundles) and particles of various sizes and shapes. The NT diameter distribution is centered at 1.4 nm (FWHM=0.6 nm) and 1.6 nm (FWHM=0.3 nm) for isolated tubes or bundles, respectively. The bonding state and chemical composition of the individual nanostructures were determined by EELS. Tubes were found to be made of sp^2 bonded BN with a boron to nitrogen ratio equal to unity. Within the detection and resolution limit of the detector, no carbon was detected to be in substitution to either boron or nitrogen. The particles are in general of pure boron. The surface of a few of them is enriched in oxygen forming a thin boron oxide layer and covered by several sheets of h-BN [3,4], as emphasized below.

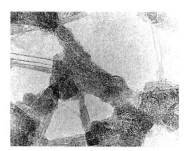

FIGURE 1. High Resolution TEM image, showing bundles and individual NTs and some particles.

Core losses – NNLS fitting

The NNLS (Non Negative Linear Square) fitting method was used to understand the complex situations in our samples. This technique consists of the reconstruction of the experimental EEL spectrum from a linear combination of reference spectra. The NNLS fit was applied only to the B-K edge. The reference spectra (B, B_2O_3, h-BN) were recorded at 0.5 eV dispersion and the fitting window considered was about 30 eV to distinguish the differences between electronic states in the different B compounds. Chemical maps can be obtained from the weights associated to the different references in the total reconstructed spectrum. The example of figure 2 illustrates the structure of a set of boron particles covered by a boron oxide layer and encapsulated in h-BN cages.

h-BN is an anisotropic layered material. The fine structures in B-K edge depend strongly on the orientation of the scattering vector with respect to the lattice. These effects are shown in figure 2 (d) and e)). The NNLS maps were obtained for two references of h-BN, which correspond to the two extreme orientation situations [4].

FIGURE 2. a) High Annular Dark Field image, showing a cluster of h-BN cages. b), c), d), e) NNLS maps for pure Boron, boron oxide and h-BN for the 2 orientations (edge and center), respectively.

Low losses

We next turn to the low-loss region study. In this energy region, the physical phenomena involve excitations of valence electrons (collective plasma oscillations and/or interband transitions), which define the structure of the band gap in the case of semiconducting or insulating materials. Low-loss EELS spectra were obtained by summing 500 unsaturated spectra acquired on isolated individual NTs with a dwell time of 40 ms. In order to disentangle the inelastic signal at low energy from the tail of the zero-loss peak (ZLP), a deconvolution procedure based on a Richardson-Lucy algorithm [5] was applied in combination with a subtraction operation of the ZLP. This spectroscopic information was combined with a structural one (diameter and number of layers) deduced from a bright field image systematically acquired on the investigated individual nano-object. Some processed spectra acquired on individual SWNTs of various diameters are displayed on the left of figure3. Such spectra contain the signature of surface modes excited in a near-field geometry where the electron beam is focussed at a grazing incidence of the nano-object and does not intersect with it. In this geometry and for a SWNT, it has been shown that the energy loss suffered by the incident electron is proportional to the imaginary part of the polarisability (by unit length) and writes as a function of $Im(\varepsilon_\perp) + Im(-1/\varepsilon_{//})$ where $\varepsilon_\perp$ and $\varepsilon_{//}$ are respectively the in-plane and out-of-plane components of the dielectric tensor of a planar h-BN sheet [6]. All the three spectra display two groups of modes, a first one in the 6-9 eV range and a second one centered at 16 eV. In agreement with the experimental dielectric constants published in [7], these modes can mainly be attributed to the $Im(\varepsilon_\perp)$ contribution in the dielectric response of the tubes. The intermediate feature in the 12 eV range is a contribution from $\varepsilon_{//}$. As the onset of the spectrum mainly reflects the contribution of $Im(\varepsilon_\perp)$, one can in principle deduce a value for the optical gap of the different investigated NTs. We find an homogeneous value around 5.8 eV [8]. Calculations are currently in process to interpret the slight variations in the lowest excitations modes and in particular the sharp feature at the onset of the spectra as a function of the diameter of the tubes. For comparison, spectra acquired in a penetrating geometry on other types of nano-objects (double and triple wall nanotubes and ropes) are displayed on the right of figure 3. The main differences with the SWNTs spectra (a higher intensity in the second mode of the doublet at the onset of the spectrum and a small extra intensity above 20 eV) come from extra contributions of volume modes excited in these objects for a penetrating geometry [8].

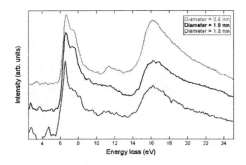 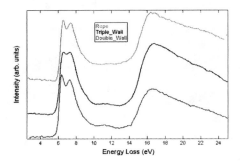

FIGURE 3 Inelastic signal in the low-loss energy region after removal of the zero-loss peak acquired on individual single walled nanotubes (left) and multiwalled and ropes of nanotubes (right).

CONCLUSION

We have presented a combined study of the EELS core and low-loss regions on BN-NTs and Boron nanoparticles. The mapping of the electronic states of boron was performed in order to push the limits of the usual information delivered by conventional chemical mapping. Preliminary results on the optical properties of individual BN-SWNTs were obtained, indicating an homogeneous optical gap value as a function of the tube diameter of about 5.8 eV.

ACKNOWLEDGMENTS

This work has supported by a TMR contract COMELCAN (HPRN-CT-2000-00128).

REFERENCES

1 X. Blase, A. Rubio, S.G. louie, M.L. Cohen *Europhys. Lett.* **28**, 335 (1994).
2 R. S. Lee, J. Gavillet, M. Lamy de la Chapelle, A. Loiseau, J.-L. Cochon, D. Pigache, J. Thibault, F. Willaime, *Phys. Rev. B* **64**, 121405 (2001).
3 R. Arenal, O. Stephan, J.Y. Mevellec, S. Lefrant, A. Rubio, A. Loiseau, submitted 2004.
4 O. Stephan, A. Vlandas, R. Arenal de la Concha, A. Loiseau, S. Trasobares, C. Colliex, EMAG conference Proceedings (2003) and R. Arenal, O. Stephan, A. Loiseau, (to be published).

5   A. Gloter, A; Douiri, M. Tencé, C. Colliex, *Ultramicroscopy* **96**, 385 (2003).
6   O. Stephan, D. Taverna, M. Kociak, K. Suenaga, L. Henrard, C. Colliex, *Phys. Rev. B* **66**, 155422 (2002).
7   G. G. Fuentes, E. Borowiak-Palen, T. Pichler, X. Liu, A. Graff, G. Behr, R. J. Kalenczuk, M. Knupfer, *Phys. Rev. B* **67**, 35429 (2003).
8   R. Arenal, O. Stephan, D. Taverna, M. Kociak, A. Rubio, C. Colliex, A. Loiseau, (to be published).

Magnetic Properties of TiO$_2$ Based Nanotubes

R. Blinc, P. Umek, P. Cevc, D. Arčon, B. Zalar, Z. Jagličić*, T. Apih and J. Dolinšek

''Jožef Stefan'' Institute, Jamova 39, 1000 Ljubljana, Slovenia
*Institute for Mathematics and Physics, Jamova 19, 1000 Ljubljana, Slovenia

Abstract. The Ti NMR and the field swept pulsed EPR spectra of bulk powder anatase TiO$_2$ and TiO$_2$ based nanotubes have been found to be radically different. The same is true for the magnetic properties measured by SQUID. Whereas no EPR line could be detected for the TiO$_2$ powder, a rather strong temperature dependent line has been seen for TiO$_2$ based nanotubes. The magnetization versus magnetic field curve of uncompressed TiO$_2$ based nanotubes at 5 K shows the presence of a huge jump-wise increase in the magnetization at about 4 T, which disappears at 1 T as well as in the compressed sample.

1. INTRODUCTION

TiO$_2$ is a wide gap semiconductor. It has been reported to become ferromagnetic on Co-doping [1]. Its properties are of great interest for spintronic and related applications [2]. In view of that, we decided to investigate the magnetic properties of the recently discovered TiO$_2$ based nanotubes.

Here we report on Ti NMR, EPR and SQUID magnetization measurements of hydrothermally synthesized TiO$_2$ based nanotubes and the differences with the starting material, i.e. TiO$_2$ powder.

2. EXPERIMENTAL

TiO$_2$ based nanotubes with a diameter of 10-20 nm and a length of 10-500 nm have been prepared hydrothermally at 130^0 C via a reaction of anatase TiO$_2$ powder with a concentrated NaOH solution [3]. Both open and closed nanotubes were obtained. The samples were characterized by TEM and elemental analysis. The Fe content is less than 80 ppm.

3. NMR

The ^{47,49} Ti NMR spectrum of tetragonal TiO$_2$ anatase powder at 9 T and room temperature is presented in Figure 1.

The separation between the Larmor frequencies for the ^{47}Ti and ^{49}Ti isotopes is 5 kHz at this value of the magnetic field. The observed spectrum (Fig. 1) has the characteristic line-shape expected for the second order quadrupole perturbed central

CP723, *Electronic Properties of Synthetic Nanostructures*, edited by H. Kuzmany et al.
© 2004 American Institute of Physics 0-7354-0204-3/04/$22.00

½ ↔ - ½ NMR transitions in a powder sample. From the shape of the powder spectrum we found that the asymmetry parameter η of the electric field gradient (EFG) tensor at the Ti-site is zero and the quadrupole coupling constant e^2qQ/h is of the order of 5MHz.

The surprising result is that no such spectrum could be observed for the TiO_2 based nanotubes. This demonstrates either a huge broadening of the Ti NMR spectra due to a distribution of internal magnetic fields in the nanotubes or a tremendous shift of the spectrum outside of the observation window of the spectrometer.

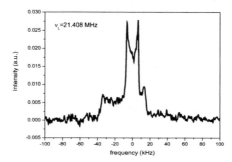

FIGURE 1. [47,49] Ti NMR powder spectrum of anatase TiO_2 at room temperature in a magnetic field of 9 T.

4. EPR

In contrast to the case of Ti NMR, no EPR spectrum could be obtained from the starting TiO_2 powder material. For TiO_2 based tubes, however, an EPR spectrum corresponding roughly to 10^{19} spins per sample (45 mg) could be measured in the field swept X-band pulsed EPR (Figure 2). At 20 K, the g-factor for the singularity is 2.09 and for the edge of the powder spectrum it is 2.49. The observed EPR spectrum is strongly T-dependent. At 280 K the peak position approaches g ≈ 2. A strong low field shift – corresponding to an increase in the effective g-factor – takes place on decreasing temperature. A similar but significantly broader spectrum has been observed in the Q band.

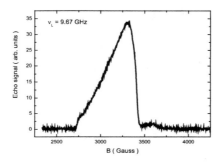

FIGURE 2. Field-swept X-band pulsed EPR spectrum of a random assembly TiO_2 based nanotubes measured at 10 K.

5. SQUID MEASUREMENTS

The magnetization of anatase TiO_2 powder is shown in Figure 3 a) as a function of the applied magnetic field at 300 K and at 5 K. At room temperature the magnetization is positive and linear in the magnetic field up to 5 T. At 5 K, on the other hand, the paramagnetic contribution dominates. The magnetization can be described by the sum of a paramagnetic contribution describable by a Brillouin function, which dominates at lower fields and a term linear in the magnetic field, which dominates at higher fields.

A completely different behavior is found for the case of TiO_2 nanotubes (Figure 3 b)). At 300 K the magnetization first increases with the magnetic field and then starts to decrease, going from positive to negative values. This demonstrates the presence of a dominant diamagnetic term. If the diamagnetic part of the magnetization is subtracted from the initial part (Figure 4 a)), a typical ferromagnetic magnetization versus field type response is obtained at 300 K. The average magnetization is however rather weak and corresponds to about 10^{-3} Bahr magneton per TiO_2 group. This seems to show a two-component behavior of the TiO_2 based nanotubes samples

The situation for the nanotube case is even more puzzling at 5 K. The magnetization versus magnetic field curve of uncompressed TiO_2 based nanotubes at 5 K shows the presence of a huge jump-wise increase in the magnetization at about 4 T (Figure 4 b)). On decreasing the magnetic field this additional magnetization disappears at about 1 T. This effect, which is seen also in the Q-band EPR is absent in compressed samples. It thus seems that these sudden jumps in the magnetization are due to reorientations of ferromagnetic type TiO_2 nanotubes in the applied magnetic fields. This would mean, that we indeed deal with a two-component type heterogeneous sample.

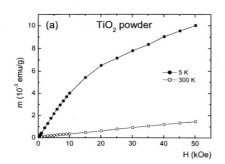 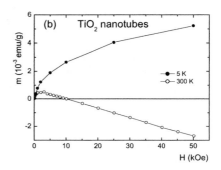

FIGURE 3. (a) Magnetization versus magnetic field for TiO_2 powder at 300 K and 5 K.;
(b) Magnetization versus magnetic field for TiO_2 nanotubes at 300 K and 5 K.

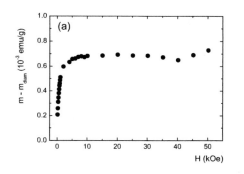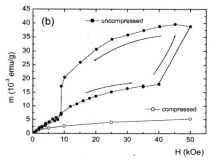

FIGURE 4. (a) Ferromagnetic contribution of the magnetization, M-M$_{dium}$, versus magnetic for TiO$_2$ nanotubes at 300K,

   (b) Magnetization versus magnetic field for uncompressed and compressed TiO$_2$ nanotubes at 5K. Note the jumps in the magnetization for uncompressed TiO$_2$ nanotubes, which are absent in the compressed sample.

ACKNOWLEDGMENTS

The authors gratefully acknowledge the financial support from the Slovenian Ministry for Education, Science and Sport.

REFERENCES

1. Fujishima A. and Honda, K., *Nature*, **37**, 238-239 (1972).
2. Higgins, J. S., Shinde S. R., Ogale, S. B., Venkatesen, T., and Green, R. L., *Phys. Rev. B,* **69**, 7-10 (2004).
3. Chen, Q., Zhou, W., Du, G. and Peng, L. M., *Adv. Mater.* **14**, 1208-1211 (2002).

EPR Study of TiO$_2$ Based Nanotubes and NO$_2$ Adsorption

P. Cevc, P. Umek, R. Blinc, A. Jesih, B. Jančar and D. Arčon

J. Stefan Institute, Jamova 39, 1000 Ljubljana, Slovenia

Abstract: TiO$_2$ nanotubes were synthesized via reaction of TiO$_2$ particles and NaOH solution. The average diameter of nanotubes was 15-20 nm and the active surface area increased by a factor of 13 in comparison to TiO$_2$ powder. When a TiO$_2$ nanotube material was exposed to NO$_2$ gas a strong adsorption of gas on the nanotube surface was observed. A strong echo signal was detected in pulsed EPR experiments below 200 K. A field sweep echo experiments revealed two distinct signals: three-lines at g=2, which are due to NO$_2$ radical, and a broad line with singularities at g= 2.09 and g= 2.49. The origin of the broad signal is at the moment not yet clear. Spin lattice relaxation time measurements demonstrated that at temperatures above 50 K the reorientation dynamics of adsorbed NO$_2$ molecules is thermally activated (E_a= 10 meV) while at low temperatures the molecules become nearly static on the EPR time scale. Upon extended heating at 60 $^{\circ}$C in air the NO$_2$ EPR signal almost completely disappeared.

1. INTRODUCTION

Titania (TiO$_2$) is a wide gap semiconductor that is found in many different applications such as gas sensors, new type of solar cells, electrochromic devices or and antifogging and self-cleaning devices [1-3]. It has been also researched throughout the chemical industry as a possible catalyst due to its powerful oxidizing properties [4]. The catalytic performance of titania critically depends on the surface-to-volume ratio and its band gap [5]. Preparation of TiO$_2$ in the form of nanotubes should dramatically increase the active surface area. In addition in small-diameter nanotubes one could expect that the band gap becomes strongly diameter dependent.

NO$_2$ is a primary component of NO$_x$ gases, which are beside CO and SO$_2$ considered as greenhouse gases. As a very reactive gas NO$_2$ in the air reacts readily with common organic chemicals and even ozone, to form a wide variety of toxic products [6]. Treatment of NO$_x$ gases is thus of extreme importance. In view of that, we decided to investigate the adsorption properties of recently discovered TiO$_2$ based nanotubes toward NO$_2$.

2. EXPERIMENTAL

TiO$_2$ based nanotubes were prepared hydrothermally at 130^0 C via a reaction of anatase TiO$_2$ powder with a NaOH solution [7]. Figure 1 shows a TEM image of as

CP723, *Electronic Properties of Synthetic Nanostructures*, edited by H. Kuzmany et al.
© 2004 American Institute of Physics 0-7354-0204-3/04/$22.00

prepared TiO_2 based nanotubes. From the TEM image it is evident that both, open and closed end nanotubes were obtained. Titania based nanotubes prepared in our process have a diameter between 10-20 nm and in length can reach up to 500 nm.

The specific surface area was studied by the BET technique. The specific surface area of TiO_2 based nanotubes increased by a factor of thirteen in comparison to anatase TiO_2 powder. In particular in the sample that has been used in our EPR studies the specific surface area was around 130 m^2/g.

Prior the adsorption of NO_2 on the TiO_2 based nanotubes, the material was dried for 48 hours at 200 °C. Adsorption process took place at 300 torrs at room temperature. During the adsorption the pressure in the reaction vessel dropped for 9 % (an equilibrium constant (K_p) for the equilibrium $2NO_2(g) \leftrightarrow N_2O_4(g)$ at adsorption conditions is 0.648). From the calculated amount of adsorbed $NO_2(g)$ and $N_2O_4(g)$ and the value of specific surface area we estimated that adsorbed NO_2 and N_2O_4 molecules form a monolayer.

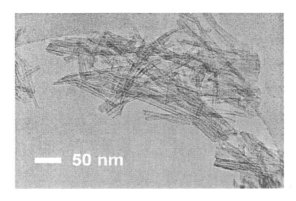

FIGURE 1. TEM image of TiO_2 based nanotubes. The estimated diameter of tubular nanostructures is found to be between 10 to 20 nm.

CW and pulsed EPR measurements were done on samples sealed in a standard 4 mm quartz tubes with X-band Bruker Elexsys E-580 spectrometer. Temperature was varied between RT and 5K using an Oxford Cryogenics liquid helium flow cryostat. Inversion recovery technique has been used for the spin-lattice relaxation time (T_1).

3. CW AND PULSED EPR RESULTS

A very strong echo signal was found in pulsed X band EPR measurements below 200 K. Field sweep echo experiments (Fig.2) revealed two distinct signals: three lines at $g=2$ due to the NO_2 center (Fig. 2b) and a very broad line. The broad signal has a characteristic powder lineshape with singularities at $g= 2.09$ and $g= 2.49$. Although the g-factor anisotropy and the estimated hyperfine constant $A \approx 90$ G of the broad line could arise from Cu impurities, electric thermal atomic absorption spectroscopy failed to prove the existence of Cu in our material. The origin of this broad line is thus still not yet clear.

	g_{xx}	g_{yy}	g_{zz}	A_{xx}	A_{yy}	A_{zz}	Ref.
Ads. on TiO$_2$	2.0061	1.9928	2.0031	52.9	48.7	67.6	this work
Ads. on ZnO	2.007	1.994	2.003	52.1	47.1	64.6	8

TABLE 1. Magnitudes of the principal values for g and A(Gauss) of NO$_2$ molecule.

The spectra corresponding to the NO$_2$ center have a typical powder-like lineshape. To simulate the spectrum measured at 18 K both g and hyperfine anisotropic contributions had to be taken into account (table 1). The obtained parameters are nearly identical to those measured on ZnO [8]. We can thus conclude that at low temperatures NO$_2$ molecules are rigid and physically adsorbed on the TiO$_2$ surface. Upon extended heating at 60 °C in air the NO$_2$ EPR signal almost disappeared.

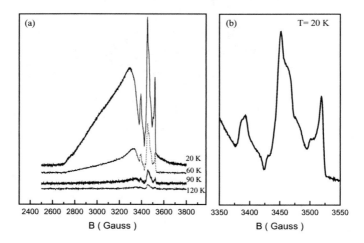

FIGURE 2. (a) Field sweep echo detected EPR spectra of NO$_2$ adsorbed on TiO$_2$ nanotubes at different temperatures. (b) Expanded region around g=2 corresponding to NO$_2$ centres.

To get additional information on the dynamics of NO$_2$ molecules adsorbed on TiO$_2$-based nanotubes we decided to measure the spin lattice relaxation time T_1 as a function of temperature. The magnetization curves are typically bi-exponential. Both relaxation times – assigned as T_{1a} and T_{1b} – show qualitatively similar temperature dependence (Fig. 3). For temperatures above 50 K T_1 has a temperature activated behavior with an activation energy of $E_a = 10$ meV. Such E_a is characteristic for NO$_2$ molecule reorientations. Below 50 K we found a change in the slope and at very low temperatures T_1 becomes nearly temperature independent indicating a possible tunneling dynamics of NO$_2$ groups.

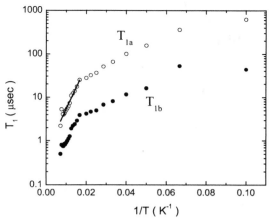

FIGURE 3: Temperature dependence of the spin-lattice relaxation times T_{1a} and T_{1b} of the NO_2 center adsorbed in TiO_2 based nanotubes. The solid line is a fit to activated behavior.

CONCLUSIONS

In conclusion, we have observed a strong adsorption of NO_2 gas on the surface of TiO_2-based nanotubes. A comparison with BET analysis shows that a NO_2 monolayer is formed on the walls of TiO_2-based nanotubes. The EPR powder lineshape demonstrates that the adsorbed molecules are nearly static on the EPR lineshape timescale at very low temperatures. At higher temperatures thermally activated molecular reorientations determine the spin-lattice relaxation time T_1. An additional very broad powder EPR line has been observed with shoulders at $g=2.09$ and $g=2.49$ whose origin is at the moment not yet clear.

ACKNOWLEDGMENTS

The authors gratefully acknowledge the financial support from the Slovenian Ministry for Education, Science and Sport.

REFERENCES

1. Graetzel M., *Nature* **353**, 736-739 (1991).
2. Hodes G., Howell I. D. J., Peter L. M., J. Electrochem. Soc. 139, 3136-3139 (1993)
3. Wang R., Hashimoto K., Fujishima A, Nature 388, 431-432 (1997).
4. Fujshima A., Honda K., *Nature* **37**, 238-240 (1972).
5. Narayanasamy, Maroni W.A., SiegelR. W., J. Mater. Res. 4, 1246-1250 (1989).
6. Cotton F. A., Willkinson, G. *Advanced Inorganic Chemistry*, **3rd ed**.; John Wiley & Sons: New York, 1972.
7. Chen Q., Zhou, W., Du, G. and Peng, L. M., *Adv. Mater.* **14**, 1208-1211 (2002).
8. Iyenger R., Subba Rao V., J. Am. Chem. Soc., **90**. 3267, (1969).

Mechanical Properties of Individual WS$_2$ Nanotubes

I. Kaplan-Ashiri[1], S.R. Cohen[2], K. Gartsman[3], R. Rosentsveig[1],
V. Ivanovskaya[4], T. Heine[4], G. Seifert[4], H.D. Wagner[1], R. Tenne[1]

[1]*Department of Materials and Interfaces, Weizmann Institute of Science, Rehovot 76100, Israel*
[2] *Surface Analysis Laboratory, Weizmann Institute of Science, Rehovot 76100, Israel*
[3]*Electron Microscopy Unit, Weizmann Institute of Science, Rehovot 76100, Israel*
[4] *Institut für Physikalische Chemie, Technische Universität Dresden, D-01062, Germany*

Abstract. The Young's modulus of WS$_2$ nanotubes is an important property for various applications. Measurements of the mechanical properties of individual nanotubes are challenging because of the small size of the tubes. Lately, measurements of the Young's modulus by buckling of an individual nanotube using an atomic force microscope[1] resulted in an average value of 171GPa. Further study of the buckling behavior is performed here using the scanning electron microscope and some preliminary results are shown. Furthermore, tensile tests of individual WS$_2$ nanotubes were performed experimentally (again using a scanning electron microscope) and simulated tensile tests of MoS$_2$ nanotubes were performed by means of a density-functional tight-binding (DFTB) based molecular dynamics (MD) scheme. Preliminary results for WS$_2$ nanotubes show Young's modulus value of ca.137GPa, tensile strength value of ca. 11 GPa and average elongation of ca. 12%. MD simulations resulted in elongation of 19% for zigzag and 17% for armchair MoS$_2$ single wall nanotubes. Since MoS$_2$ and WS$_2$ nanotubes have similar structures the same behavior is expected for both, hence there is a good agreement regarding the elongation of WS$_2$ nanotubes between experiment and simulation.

INTRODUCTION

WS$_2$ Nanotubes

The synthesis of inorganic fullerene-like nanoparticles and nanotubes (*IF*) was first reported in 1992[2]. At present, by using the fluidized bed reactor (FBR) WS$_2$ nanoparticles with fullerene-like structure enriched by 5% WS$_2$ chiral multiwall nanotubes are obtained in substantial amounts[3,4]. The length of the nanotubes varies from several hundred nanometers to several hundred microns. The nanotubes are also quite uniform in shape and are open-ended. Their outer diameter varies from 10 to 30 nm, with the main size distribution centered between 15-20 nm. The typical number of layers in a nanotube varies, between 5 and 8. The nanotubes were found to be perfectly crystalline and almost defect free.

CP723, *Electronic Properties of Synthetic Nanostructures*, edited by H. Kuzmany et al.
© 2004 American Institute of Physics 0-7354-0204-3/04/$22.00

A great deal of effort was devoted during the last few years to the study of the mechanical properties of nanotubes such as carbon and boron nitride, both experimentally[5-15] and theoretically[16,17]. Several techniques including direct tensile testing[11] were employed to measure the Young's modulus of carbon nanotubes. All the measurement techniques are microscopy based, making use of either electron microscopes or scanning force microscopes. The first measurement of the Young's modulus of carbon nanotubes, estimated from the amplitude of thermal vibrations of individual carbon nanotubes within a transmission electron microscope (TEM) was published in 1996[5]. Further studies of the Young's modulus of carbon nanotubes by other measurement techniques resulted in typical values in the range of 1.0-1.3GPa for multiwall nanotubes and 1.36-1.76TPa for single wall nanotubes[16].

The physical properties of individual WS_2 nanotubes or fullerene-like nanoparticles were not studied in detail so far. Recently, measurements of the shock wave resistance[18] and Young's modulus[1] of WS_2 nanotubes were carried out. It was found that WS_2 nanotubes are capable of withstanding stresses caused by shock waves of up to 21GPa. Preliminary measurement of the Young's modulus of individual WS_2 nanotubes resulted in a value of 171GPa. The Young's modulus was calculated from force measurements using Euler's buckling theory.

In the present paper a new method is introduced for studying the buckling behavior of individual nanotubes. Furthermore, experimental and numerical studies of WS_2 and MoS_2 nanotubes under tensile stress are presented.

EXPERIMENTAL

Two experimental routes are offered here for the measurements of Young's modulus of WS_2 nanotubes, presented in the next section, using a compression/buckling test and a tensile test.

Compression/Buckling Tests

Performance of the buckling experiment[19] within a high resolution SEM (HR-SEM) can provide information regarding the Young's modulus of the nanotubes as well as information regarding the lateral deflection of the nanotube and the actual buckling behavior. The nanotube was attached to an AFM cantilever within the Environmental-SEM (E-SEM model XL-30 FEG-FEI) by deposition of amorphous carbon[19]. The buckling experiment was done within a HR-SEM (LEO, model Supra55), which is equipped with a nano-manipulator (Nanotechnik). A silicon wafer with flat surface was attached to the nanomanipulator and the nanotube tip was attached to the microscope's stage. The silicon surface was pushed against the nanotube in fine steps of 10-20nm, and the lateral movement was recorded.

Tensile Tests

Direct tensile test of individual WS_2 nanotubes were performed according to a published procedure[10]. An individual nanotube was attached to two silicon cantilevers within the E-SEM by deposition of amorphous carbon as described before. The configuration of the experiment is shown in Fig. 1.

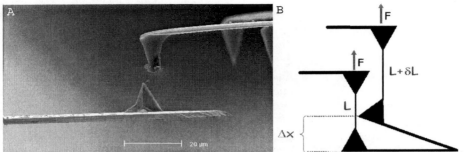

FIGURE 1. (A) SEM view of the tensile test configuration (B) schematic representation of the tensile test (adapted from M. Yu et al.[11])

In this configuration the upper cantilever (high k, ca. 1-5 N/m,) is actually moved by the SEM stage and the lower cantilever (low k, ca. 0.1-0.2 N/m) is deflected in response to the force, which is applied to the nanotube. The experiment is being recorded by a videocassette recorder (VCR). The Young's modulus and the strength of the nanotube are derived from this analysis.

Simulation Tests

Using the density-functional based tight-binding (DFTB) method, the Young's moduli of single-walled MoS_2 nanotubes have been determined earlier using a static approach[1]: For the (22,0) zigzag tube a Young's modulus of 230 GPa has been computed, which is similar to that of the (14,14) armchair tube (205 GPa). Here, a DFTB-based molecular dynamics (MD) techniques was employed to simulate the stretching process of these two tubes. The model structures contain 7 unit cells for the zigzag and 10 unit cells for the armchair tube, respectively. Some fixed rings of atoms were pulled and MD simulation of the remaining system was performed. This process was successively repeated until the point of rupture was encountered.

RESULTS

Compression/Buckling Tests

Preliminary results are presented as a series of frames (1-4) in Fig.2. The movement of the AFM tip can be seen clearly as well as the lateral deflection of the nanotube. By plotting the force applied on the nanotube versus the distance between the tip and the silicon surface, the buckling point can be determined. Following buckling, the compressed nanotube remains intact demonstrating its flexibility and strength.

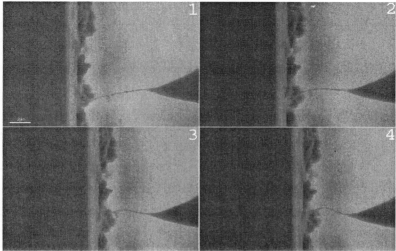

FIGURE 2. Buckling of WS$_2$ nanotube within the SEM

By using Euler's equation (for fixed-free beam)[20] the Young's modulus can be calculated. These results will be presented in a future publication.

Tensile Test

The diameter of the nanotube was measured at the beginning of each test; the nanotubes diameters range is 20-30nm. The nanotube's length and the cantilever's deflection were measured for each frame (Fig.3). These data were then used to calculate the elongation, and the tensile strength of the nanotube.

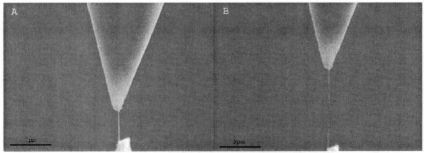

FIGURE 3. WS$_2$ nanotube before (A) and after (B) tensile test.

The stress vs. strain was plotted (Fig.4), and the Young's modulus was determined from the slope of the elastic region according to the equation[21]:

$$E = \frac{\sigma}{\varepsilon} \text{ (Pa)} \quad (1)$$

where σ is the stress and ε is the strain. The data collected so far is shown in Table (1).

TABLE 1. The results of tensile tests of individual WS_2 nanotubes.

Tensile strength (GPa)	Elongation (%)	Young's modulus (GPa)
5.2	9	
14.5	12	119
12.3	8.5	127
10.5	13.8	
10.4	19	156
9.4	14.8	145
	12.3	
Average value 10.4	Average value 12.8	Average value 136.8
Standard deviation 3.1	Standard deviation 3.6	Standard deviation 16.8

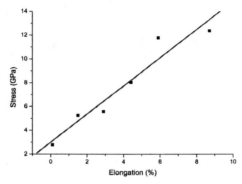

FIGURE 4. Stress-strain curve of WS_2 nanotube

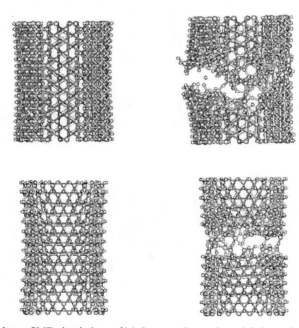

FIGURE 5. Snapshots of MD simulations of MoS_2 nanotubes under axial elongation just before (left-hand side) and after (right-hand side) rupture for *armchair* (14,14) (top) and *zigzag* (22,0) MoS_2 NT (bottom).

310

According to the MD simulations the point of rupture for MoS_2 is encountered after stretching of 19% for the zigzag and 17% for the armchair nanotube.

The simulations give a detailed insight into the structure deformation during the stretching process. Fig. 5 shows the model tubes just before rupture (left-hand side), and heavily deformed nanotubes (right-hand side). Further MD simulations for the quantitative determination of the strain-stress relation of these tubes are in progress and will be published elsewhere.

DISCUSSION

The compressive/buckling experiment within the SEM chamber is a promising technique to learn about the buckling behavior of the nanotubes. Nonetheless, some further improvements in the set-up are necessary in order to derive quantitative data from these experiments.

The Young's modulus for the WS_2 nanotubes, which was derived from both the tensile and the buckling tests (see Figs.3 and 4 and Table I and Ref. 1) is in good agreement with the value (150GPa) of the bulk material[1]. Previous calculations of the Young's modulus of MoS_2 nanotubes[1] resulted in a value similar to the bulk material (238GPa). These observations suggest a good agreement between theory and experiment, and between the nanotubes and the bulk material. Furthermore, the tensile strength of the nanotubes as determined from the tensile tests (Table 1) is about 10% of the Young's modulus, which comes close to theoretical estimations using a single rule of thumb. The shock resistance of the WS_2 nanotubes as determined in Ref. 16 (21 GPa) falls also in the same range of values. The elongation values determined in the tensile tests are slightly smaller compared to the results from MD simulations of MoS_2 nanotubes, but the scatter of the experimental data would require more experiments to improve the statistics. One should also note that while experiments were carried out on multiwall nanotubes, the simulation was performed on single wall nanotubes. Based on this set of data and on the similarity between the structure of MoS_2 and WS_2, a similar behavior for both kinds of nanotubes in tensile tests, is expected. Therefore, MoS_2 nanotubes are expected to be 50% stronger then their WS_2 analogous. The results also demonstrate the high degree of crystallinity of the present nanotubes.

CONCLUSIONS

The experimental values of the Young's modulus and the tensile strength combined with a relatively high elongation make WS_2 nanotubes unique and promising material for various mechanical applications. MD simulations show larger elongation values for single-wall MoS_2 nanotubes, similar values are expected for WS_2 nanotubes as well. In order to complete this study more data should be collected both experimentally and theoretically.

ACKNOWLEDGMENT

This work was supported by the Minerva foundation (Munich), the G.M.J. Schmidt Minerva Center for Supramolecular Chemistry, the German-Israeli Foundation (GIF), the Alfried Krupp von Bohlen and Halbach Stiftung, and Philip M. Klutznick research foundation.

REFERENCES

1. I. Kaplan-Ashiri, S.R. Cohen, G. Gartsman, R. Rosentsveig, G. Seifert, R. Tenne: J. Materials Research, **19**, 454, (2004).
2. R.Tenne, L. Margulis, M.Genut, G. Hodes: Nature, 360, 444, (1992).
3. R. Rosentsveig, A. Margolin, Y. Feldman, R. Popovich-Biro, R. tenne: Chemistry of materials, **14**, 471, (2002).
4. A. Margolin, R. Rosentsveig, A. Albu-Yaron, R. Popovitz-Biro, R. Tenne: J. Mater. Chem, **14**, 617, (2004).
5. M.M.J. Treacy, T.W. Ebbesen, J.M. Gibson: Nature, **381**, 678, (1996).
6. O.Lourie, D.M. Cox, H.D. Wagner: Phys. Rev. Lett., **81**, 1639, (1998).
7. O. lourie, H.D. Wagner: J. Materials Research, **13**, 2418, (1998).
8. J. Salvetat, G. Andrew, D. broggs, J. Bonard, R.R. Bacsa, A.J. Kulik, T. Stockli, N.A. Burnham, L. Forro: Phys. Rev. Lett., **82**, 944, (1999).
9. W. Shen, B. Jiang, B.S. Han, S. Xie, : Phys. Rev. Lett., **84**, 3634, (2000).
10. M. Yu, T. Kowalewski, R.S. Ruoff: Phys. Rev. Lett., **85**, 1456, (2000).
11. M. Yu, O.Lourie, M.J. Dyer, K. Moloni, T.F. Kelly, R.S. Ruoff: Science, **287**, 637, (2000).
12. C.A. Cooper, S.R.Cohen, A.H. Barber, H.D. Wagner: Appl. Phys. Lett., **81**, 3875, (2002).
13. S. Washburn, R. Superfine: Phys. Rev. Lett., **89**, 255502-1, (2002).
14. H.E. Troiani, M. Miki-Yoshida, G.A. Camacho-Bragado, M.A.L. Marques, A. Rubio, J.A. Ascenio, M. Jose-Yacaman: Nano Lett., **3**, 751, (2003).
15. A.H. Barber, S.R. Cohen, H.D. Wagner: Appl. Phys. Lett., **82**, 4141, (2003).
16. B.I. Yakobson, P. Avouris: Topics Appl. Phys., **80**, 287, (2001).
17. K.N. Kudin, G.E. Scuseria, B.I.Yakobson: Phys. Rev. B, **64**, 235406, (2001).
18. Y.Q. Zhu, T. Sekine, K.S. Brigatti, S. Firth, R, Tenne, R. Rosentsveig, H.W. Kroto, and D.R.M. Walton: J. Am. Chem. Soc, **125**, 1329 (2003).
19. H. Dai, J.h. Hafner, A.G. Rinzler, D.T. Colbert, R.E. Smalley: Nature, **384**, 147, (1996).
20. J.M. Gere: Mechanics of Materials, Australia:Brooks/Cole, 5th edition, (2001).
21. W.D. Callister: Materials Science and Engineering, New-York: Wiley, 5th edition, (2000).

THEORY OF NANOSTRUCTURES

Entanglement of Spin States in ^{15}N@C$_{60}$

W. Scherer[*], A. Weidinger[†] and M. Mehring [‡]

[*]2. Physikalisches Institut, Universität Stuttgart, Pfaffenwaldring 57, D-70550 Stuttgart, Germany
[†]Hahn-Meitner-Institut Berlin, D-14109 Berlin, Germany

Abstract. The endohedral fullerene ^{15}N@C$_{60}$ comprises an electron spin $S = 3/2$ coupled to a nuclear spin $I = 1/2$ and is therefore ideally suited for experimental testing of basic properties of quantum mechanics. We will show that the ^{15}N@C$_{60}$ molecule represents a multi qubit system where different kinds of entangled states can be generated.

INTRODUCTION

Within recent years quantum information theory has been developed extensively and led to exciting new ideas in fields like quantum cryptography [1, 2], quantum teleportation [3, 4] and quantum computation [5, 6]. A main reason for these new developments is that quantum systems can exhibit nonlocal correlations that violate our conception of the classical world. These so called entangled states are not separable i.e. they cannot be described in a way that distinguishes between the different parts of the quantum system. In spite of the large number of different quantum systems proposed as a potential hardware for a quantum computer , not many experimental realizations have been achieved so far. Entanglement has been demonstrated over large distances for polarization states of photons [7], among Ions confined in an electromagnetic trap [8], between an Ion and a photon [9] or in liquid state nuclear spin magnetic resonance (NMR), where even simple quantum algorithms have been demonstrated [10, 11, 12, 13, 14].

The endohedral fullerene ^{15}N@C$_{60}$ has been proposed as a qubit – the basic unit of a quantum computer – by Harneit [15]. In this proposal group V endohedral molecules (^{14}N@C$_{60}$, ^{15}N@C$_{60}$, ^{31}P@C$_{60}$) form an array of qubits which are coupled by dipolar forces. In this contribution we show that ^{15}N@C$_{60}$ should be regarded as a multi qubit system. After defining a two qubit subsystem we were able to prepare and detect the complete set of entangled states (Bell basis).

Endohedral fullerene ^{15}N@C$_{60}$

The endohedral fullerene ^{15}N@C$_{60}$ comprises one single ^{15}N atom, which resides in the center without charge transfer to the C$_{60}$ cage. The nitrogen atom is paramagnetic

---

[1] To whom correspondence should be addressed: m.mehring@physik.uni-stuttgart.de

CP723, *Electronic Properties of Synthetic Nanostructures*, edited by H. Kuzmany et al.
© 2004 American Institute of Physics 0-7354-0204-3/04/$22.00

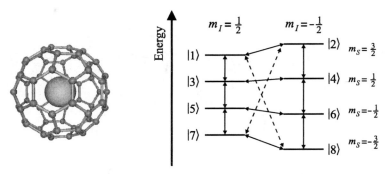

FIGURE 1. Left: Structure of the ^{15}N@C$_{60}$ molecule. Right: Energy level diagram of the eight spin states of the corresponding two spin system with electron spin $S = 3/2$ and nuclear spin $I = 1/2$.

due to three unpaired electrons resulting in an electron spin $S = 3/2$. In case of isotope ^{15}N the nuclear spin of the nitrogen atom is $I = 1/2$, which is coupled to the electron spin via hyperfine interaction. The spin system has already been discussed in detail by standard magnetic resonance methods [16, 17, 18]. The nitrogen atoms were implanted into the carbon cage by simultaneous evaporation and ion bombardement onto a cooled target. The two spin system can be expressed by the Hamiltonian

$$H = \hbar \left(\omega_S S_z + \omega_I I_z + 2\pi a S_z I_z \right) , \tag{1}$$

where second order hyperfine coupling terms have been neglected. The eight eigenvalues of this system are

$$E_{m_S m_I} = \hbar \left(\omega_S m_S + \omega_I m_I + 2\pi a m_S m_I \right) , \tag{2}$$

depending on the spin quantum numbers $m_S = \pm 1/2, \pm 3/2$ of the electron spin and $m_I = \pm 1/2$ of the nuclear spin. The eight eigenstates of the system are given by

$$|m_S m_I\rangle = |1\rangle, |2\rangle, \ldots, |8\rangle = \left| +\frac{3}{2} +\frac{1}{2} \right\rangle, \left| +\frac{3}{2} -\frac{1}{2} \right\rangle, \ldots, \left| -\frac{3}{2} -\frac{1}{2} \right\rangle . \tag{3}$$

The corresponding energy level scheme is shown in Fig. 1. Allowed transitions are indicated by arrows. There are six electron spin transitions ($\Delta m_S = \pm 1$) which are degenerate to first order for $m_I = +1/2$ and $m_I = -1/2$. Therefore within spectral resolution, all three transitions for $m_I = +1/2$ or $m_I = -1/2$ can only be excited simultaneously. Besides there are four nuclear spin transitions ($\Delta m_I = \pm 1$). The dotted arrows indicate forbidden transitions which are related to the entangled states discussed here.

PSEUDO ENTANGLED SPIN STATES

In ^{15}N@C$_{60}$ twenty four entangled states of the type $\Psi^{\pm}_{jk} = \frac{1}{\sqrt{2}}(|j\rangle \pm |k\rangle)$ with $jk \in \{14, 23, 36, 45, 58, 67, 16, 25, 38, 47, 18, 27\}$ are possible. In this sense one single

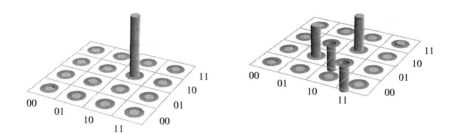

FIGURE 2. Experimental results for the density matrices of pseudopure state ρ_{10} (left) and pseudo entangled state Ψ^- (right). The states are given in qubit notation.

^{15}N@C$_{60}$ represents a multi qubit system. In this contribution we discuss the preparation of two special cases

$$|\Psi_{27}^-\rangle = \frac{1}{\sqrt{2}}(|2\rangle - |7\rangle) = \frac{1}{\sqrt{2}}\left(\left|+\frac{3}{2} -\frac{1}{2}\right\rangle - \left|-\frac{3}{2} +\frac{1}{2}\right\rangle\right) \qquad (4)$$

$$|\Phi_{18}^+\rangle = \frac{1}{\sqrt{2}}(|1\rangle + |8\rangle) = \frac{1}{\sqrt{2}}\left(\left|+\frac{3}{2} +\frac{1}{2}\right\rangle + \left|-\frac{3}{2} -\frac{1}{2}\right\rangle\right). \qquad (5)$$

Here we restrict ourselves to the fictitious two state subsystem of the electron spin with $|\pm 3/2\rangle$ as one qubit . For the second qubit we consider the $|\pm 1/2\rangle$ states of the nuclear spin. In this way the quantum states $|\pm 3/2 \pm 1/2\rangle$ represent the four states of our two qubit system.

The experiments were performed with a powder sample of ^{15}N@C$_{60}$ at $T = 50$ K. Therefore we are dealing with an ensemble of ^{15}N@C$_{60}$ molecules working in parallel. This situation has been discussed extensively in liquid state nuclear magnetic resonance (NMR) [10, 11, 12, 13]. These systems present mixed quantum states and can be described approximately by density matrices which are related to those of the corresponding pure quantum state. Those states are therefore called *pseudo pure* states.

The necessary unitary transformations for preparation and detection of the quantum states were achieved by transition selective microwave pulses (X-band) on electron spin transitions and by transition selective radio frequency pulses on nuclear spin transitions. The experimental approach is similar to [19] where entanglement was generated between an electron spin $1/2$ and a nuclear spin $1/2$. Initially the system is in thermal equilibrium described by the Boltzmann state. In order to start from an initial density matrix like in a pure quantum system we have invoked the pseudopure density matrix concept. Figure 2 (left) shows for example the experimental result of a density matrix tomography for the pseudo pure density matrix ρ_{10}. All elements are approximately zero except for the third diagonal element, which is one.

This state served as an initial state for the preparation of the pseudo entangled state Ψ_{27}^- in (4). Within the fictitious two qubit subsystem this was achieved by a two pulse sequence with a $\pi/2$ pulse on nuclear spin transition with $m_S = -3/2$ followed by a π

pulse on electron spin transitions with $m_I = -1/2$. The experimental result is shown in fig. 2 (right). The diagonal elements agree with the expected result $\{0, +0.5, +0.5, 0\}$. The off-diagonal elements are zero except for the characteristic elements $r_{27} = r_{72}$ which were determined to $r_{27} = r_{72} = -0.43$ which is somewhat smaller than the expected value of -0.50. Further details will be published in a forthcoming publication. The data represented in fig. 2 were obtained with a density matrix tomography discussed in the following section.

TOMOGRAPHY

In order to verify that the created pseudo pure and entangled quantum states have been properly prepared we have applied a special type of spin density matrix tomography introduced in [19]. There we have combined Rabi-oscillation measurements for the determination of the diagonal elements (occupation numbers) with unitary transformations which are based on phase rotated pulses. These were obtained by applying phase shifted detection pulses, which corresponds to a rotation around the quantization axis.

The idea behind this is the following phase dependence of the entangled states. Entangled states are by definition not separable into product states of its constituents. Local measurements on the individual spins do not provide any information on the entangled states as an entity. Only a combined approach by performing unitary transformations on on both spins allows to verify the degree of entanglement. Let us therefore consider the behavior the entangled states Ψ^- and Φ^- under a combined rotation around the z-direction by angles φ_1 for the electron spin S and φ_2 for the nuclear spin I.

$$U_{S_z} = e^{-i\varphi_1 S_z} \quad \text{and} \quad U_{I_z} = e^{-i\varphi_2 I_z}. \tag{6}$$

Under these rotations the quantum state $|m_S m_I\rangle$ exhibits the following phase variation:

$$
\begin{aligned}
U_{S_z} U_{I_z} |m_S m_I\rangle &= e^{-i\varphi_1 S_z} e^{-i\varphi_2 I_z} |m_S m_I\rangle \\
&= e^{-i(m_S \varphi_1 + m_I \varphi_2)} |m_S m_I\rangle
\end{aligned}
$$

Let us now apply these rotations to superposition and entangled states of the electron spin S (spin 1) and nuclear spin I (spin 2). The spin states are labelled $|\uparrow\rangle$ for $m_S = 3/2$ and $m_I = 1/2$ and $|\downarrow\rangle$ for $m_S = -3/2$ and $m_I = -1/2$. The rotation of the superposition state of spin S results in

$$
\begin{aligned}
(|\uparrow\uparrow\rangle - |\downarrow\uparrow\rangle)(\langle\uparrow\uparrow| - \langle\downarrow\uparrow|) &\xrightarrow{z-\text{rot.}} (|\uparrow\uparrow\rangle\langle\uparrow\uparrow| + |\downarrow\uparrow\rangle\langle\downarrow\uparrow| \\
&- e^{-i3\varphi_1}|\uparrow\uparrow\rangle\langle\downarrow\uparrow| - e^{i3\varphi_1}|\downarrow\uparrow\rangle\langle\uparrow\uparrow|).
\end{aligned} \tag{7}
$$

In case of an electron spin $S = 1/2$ the factor 3 in front of φ_1 would correspond to 1 [19]. Correspondingly the superposition state of nuclear spin states with $m_I = \pm 1/2$ (spin 2)transforms like

$$
\begin{aligned}
(|\uparrow\uparrow\rangle - |\uparrow\downarrow\rangle)(\langle\uparrow\uparrow| - \langle\uparrow\downarrow|) &\xrightarrow{z-\text{rot.}} (|\uparrow\uparrow\rangle\langle\uparrow\uparrow| + |\uparrow\downarrow\rangle\langle\uparrow\downarrow| \\
&- e^{-i\varphi_2}|\uparrow\uparrow\rangle\langle\uparrow\downarrow| - e^{i\varphi_2}|\uparrow\downarrow\rangle\langle\uparrow\uparrow|).
\end{aligned} \tag{8}
$$

Applying similar phase rotations to entangled states leads to a different behavior as is demonstrated here for the Ψ_{27}^- and Φ_{18}^- states:

$$(|\uparrow\downarrow\rangle - |\downarrow\uparrow\rangle)(\langle\uparrow\downarrow| - \langle\downarrow\uparrow|) \xrightarrow{z-\text{rot.}} (|\uparrow\downarrow\rangle\langle\uparrow\downarrow| + |\downarrow\uparrow\rangle\langle\downarrow\uparrow|$$
$$-e^{-i(3\varphi_1-\varphi_2)}|\uparrow\downarrow\rangle\langle\downarrow\uparrow| - e^{i(3\varphi_1-\varphi_2)}|\downarrow\uparrow\rangle\langle\uparrow\downarrow|) \tag{9}$$

$$(|\uparrow\uparrow\rangle - |\downarrow\downarrow\rangle)(\langle\uparrow\uparrow| - \langle\downarrow\downarrow|) \xrightarrow{z-\text{rot.}} (|\uparrow\uparrow\rangle\langle\uparrow\uparrow| + |\downarrow\downarrow\rangle\langle\downarrow\downarrow|$$
$$-e^{-i(3\varphi_1+\varphi_2)}|\uparrow\uparrow\rangle\langle\downarrow\downarrow| - e^{i(3\varphi_1+\varphi_2)}|\downarrow\downarrow\rangle\langle\uparrow\uparrow|) . \tag{10}$$

Performing the phase rotations in increments according to $\varphi_j = 2\pi\nu_j t$ with variable t results in an oscillatory behavior as has been observed before (see [19]). After Fourier transformation of the corresponding phase interferograms spectra are obtained which allow to distinguish the different entangled states. Spectral lines are observed at $3\nu_1 - \nu_2$ for the Ψ^- state and $3\nu_1 + \nu_2$ for the Φ^- state. More details will be presented in a forthcoming publication.

Since the quantum states presented here are mixed states due to the rather high temperature and low resonance frequency we call them *pseudo entangled states*. We will show, however, in a forthcoming publication that by applying similar pulse sequences as presented here that a true quantum state can be reached below the quantum critical temperature $T_Q = 7.7$ K for an ESR frequency of 95 GHz (W-band). The discussion of this phenomenon is rather involved and lengthy. Space does not permit to go into much detail. A few brief remarks may, however, guide the experienced reader. Starting from a Boltzmann state at high temperature and low resonance frequency where the density matrix of the pseudo pure states are *separable* the question arises is there a critical temperature at a given resonance frequency where the pseudo entangled state become *inseparable* or in other word turns into a really quantum entangled state? The answer is yes. By applying the so-called *partial positive transpose (PPT)* criterion one can show that the density matrix is no more separable below a critical temperature.

CONCLUSIONS

We have shown that pseudo entangled states corresponding to a two qubit subsystem of the multi qubit spinsystem in ^{15}N@C$_{60}$ can be prepared by proper unitary transformations.

ACKNOWLEDGMENTS

This work has been supported by the Bundesministerium für Bildung und Forschung and by the Landesstiftung Baden-Württemberg.

REFERENCES

1. Bennett, C. H., Besette, F., Brassard, G., Salvail, L., and Smolin, J., *J. Cryptology*, **5**, 3 (1992).
2. Bennett, C. H., Brassard, G., Crépeau, C., and Maurer, U. M., "Generalized Privacy Amplification," in *Proceedings of the IEEE Internatinal Conference on Computers, System and Signal Processing*, IEEE, New York, 1994, p. 350.
3. Bennett, C. H., Besette, F., Brassard, G., Salvail, L., and Smolin, J., *Phys. Rev. Lett.*, **70**, 1895 (1993).
4. Bouwmeester, D., Pan, J.-W., Mattle, K., Eibl, M., and Weinfurter, H., *Nature (London)*, **390**, 575 (1997).
5. Deutsch, D., *Proc. R. Soc. Lond. A*, **400**, 97 (1985).
6. Deutsch, D., *Proc. R. Soc. Lond. A*, **439**, 553 (1992).
7. Zeilinger, A., Horne, M., Weinfurter, H., and Zukovsky, M., *Phys. Rev. Lett.*, **78**, 3031 (1997).
8. Sackett, C. A., Kielpinski, D., King, B. E., Langer, C., Meyer, V., Myatt, C. J., Rowe, M., Turchette, Q. A., Itano, W. M., Wineland, D. J., and Monroe, C., *Nature*, **404**, 256 (2000).
9. Blinov, B. B., Moehring, D. L., Duan, L. M., and Monroe, C., *Nature*, **428**, 153 (2004).
10. Cory, D. G., Fahmy, A. F., and Havel, T. F., *Proc. Natl. Acad. Sci. U.S.A.*, **94**, 1634 (1997).
11. Cory, D. G., Price, M. D., and Havel, T. F., *Physica D*, **120**, 82 (1998).
12. Gershenfeld, N. A., and Chuang, I. L., *Science*, **275**, 350 (1997).
13. Knill, E., Chuang, I., and Laflamme, R., *Phys. Rev. A*, **57**, 3348 (1998).
14. Vandersypen, L. M. K., Steffen, M., Breyta, G., Yannoni, C. S., Sherwood, M. H., and Chuang, I. L., *Nature*, **414**, 883 (2001).
15. Harneit, W., *Phys. Rev. A*, **65**, 032322 (2002).
16. Almeida Murphy, T., Pawlik, T., Weidinger, A., Hoehne, M., Alcala, R., and Spaeth, J. M., *Phys. Rev. Lett.*, **77**, 1076 (1996).
17. Pietzak, B., Waiblinger, M., Almeida Murphy, T., Weidinger, A., H"ohne, M., Dietel, R., and Hirsch, A., *Chem. Phys. Lett.*, **279**, 259 (1997).
18. Weiden, N., Kaess, H., and Dinse, K. P., *J. Phys. Chem. B*, **103**, 9826 (1999).
19. Mehring, M., Mende, J., and Scherer, W., *Phys. Rev. Lett.*, **90**, 153001 (2003).

Omniconjugation

Marleen H. van der Veen, Harry T. Jonkman, and Jan C. Hummelen*

*Molecular Electronics, Materials Science Centre^Plus, University of Groningen,
Nijenborgh 4, NL-9747 AG, Groningen, The Netherlands; e-mail: j.c.hummelen@chem.rug.nl

Abstract. Omniconjugation is introduced as a new architectural concept for the design of complex molecular structures that allows for the interconnection of many functional entities in a fully conjugated manner. So far, such π-conjugated topologies have never been explicitly recognized or investigated from this point of view. A topological design method has been developed by which a large number of realistic omniconjugated structures can be constructed. This new class of π-conjugated systems can be divided in two sub-classes that differ in their 'switching' behavior upon passage of solitons. Furthermore, we found that the principle of omniconjugation may give rise to a pronounced and, sometimes, unique π-electron delocalization of the frontier orbitals. These preliminary results indicate that omniconjugation could be applicable in several ways (as passive or active elements) in future nanoelectronic devices.

INTRODUCTION

In the continuing race for faster computers and new electronic devices, enormous research effort is put into the miniaturization of classical semiconductor components and other crucial electronic devices. An intriguing alternative for the ongoing size reduction of silicon-based technology is the bottom-up approach based on molecules. The idea to implement molecules as elementary parts in electronic circuits stems already from the early seventies [1,2]. During the last decade, scientists have reported on successful fabrication of single molecular devices that indeed can function as, for example, wires, diodes or transistors [3–5]. However, a device or an electronic circuit is made up of many (complex) elements, wired in a specific way to make it operate. Up to now, the trivial aspect of interconnecting several elements has not been addressed on the molecular level. This of course will be crucial for the realization of fully integrated molecular circuits [6]. One of the simplest elements missing is the single-molecule version of a T-piece or an intersection of two wires. A first requirement for the transmission of charges through such a molecular junction is a complete π-electron delocalization between the three or four terminals. For organic systems it is the topology of the conductive path that determines its degree of delocalization. Charges can flow efficiently between two ends when the chain is linear conjugated, i.e., a strict alternation of single and double bonds. This condition is not met in cross-conjugated pathways where the strict alternation of single and double bonds is interrupted by an extra single bond (see model **1** and **2** in Fig. 1).

CP723, *Electronic Properties of Synthetic Nanostructures*, edited by H. Kuzmany et al.
© 2004 American Institute of Physics 0-7354-0204-3/04/$22.00

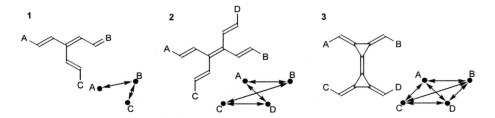

FIGURE 1. (Top) Schematic representation of three- and four-terminal cross-conjugation in **1** and **2**, and four-terminal omniconjugation in **3**. (Bottom) Corresponding topological connectivity schemes of conjugated pathways.

When using simple molecular architectures like **1** or **2**, which would mimic a T-piece or an intersection, respectively, it is not possible to create a situation in which transport can happen in all, preferable more than two or three, directions. Hence, we have sought for π-conjugated systems that do have direct linear conjugated pathways between *all* terminals. We call such systems, like model **3**, omniconjugated. In being truly conjugated, these systems could fulfill the simple function of an intersection between two molecular wires.

TOPOLOGICAL DESIGN OF OMNICONJUGATED MODELS

The fact that real molecules of the structure **3** are unstable makes it necessary to search for more realistic alternatives. A topological design scheme consisting of three steps has been developed for the systematic construction of molecules from small entities, the so-called, key-unit A or B (see Fig. 2) [7]. A large number of realistic models can be obtained by applying sequences of operations to one selected key-unit. The collection of eight operations all originate from one central argument: preservation of the existing conjugated pathways and avoiding the emergence of cross-conjugation.

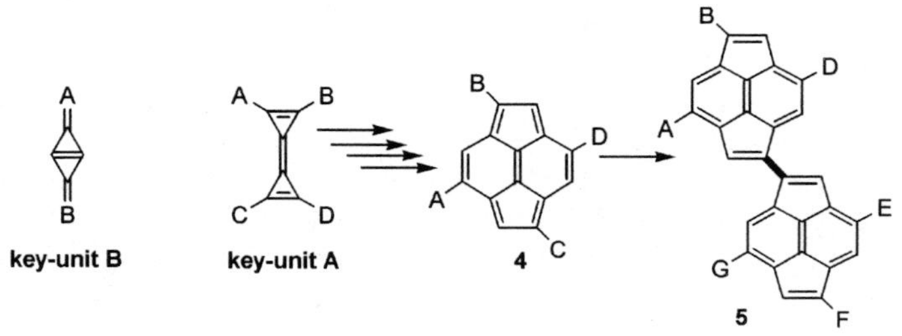

FIGURE 2. The key-units A and B, and the topological design of omniconjugated models from, for example, key-unit A. Model **5** is obtained after applying five operations (see Ref. [7]) and has fifteen conjugated pathways between the six terminals.

The design process allows for the number of interconnected molecules to be chosen anywhere between two and infinity, in principle. For example, model **5** is obtained by connecting two models of **4** through 'former' terminal C yielding a junction with six terminals (i.e., for six other moieties). The versatility of the design method is confirmed upon the emergence of complex though realistic models like **5**. In general, this class of molecules belongs to the group of nonalternant hydrocarbons since they all contain at least one odd-membered ring. As with any other organic system, one can draw the resonance contributors for an omniconjugated system. Interestingly, however, all resonance structures are also omniconjugated. This is a crucial property, because if this were not the case, omniconjugation would have been simply a topological curiosum based on the valence bond theory.

PROPERTIES OF OMNICONJUGATED SYSTEMS

As a result of some topological properties, we discriminate two classes of omniconjugated systems:

Type A omniconjugated systems, they stem from key-unit A, have the topological property that they remain omniconjugated upon a redox operation. A redox operation implies that all single bonds within a conjugated pathway become double and vice versa. The oxidation of an alternating pathway will induce a net change in the number of double bonds as shown in Fig. 3 for model **4**. Such a redox operation on any of the pathways (or combination thereof) is, regarding the topology of the path, analogous to the passage of a soliton [8]. In other words, a redox event does not influence the charge transport properties because the bond alternation pattern is not changing from linear to cross-conjugated.

Type B omniconjugated systems, defined as only obtainable from key-unit B (see Fig. 2), have the intriguing property that omniconjugation is *not* preserved upon a redox event. That is: in Type B systems a redox event between two terminals (say A and B in model **8** in Fig. 4) results in changing the topology between the two complementary terminals (C and D). Thus in *any* case the complementary path *always* changes from linear to cross-conjugated (hence, switched from 'on' to 'off': C × D in

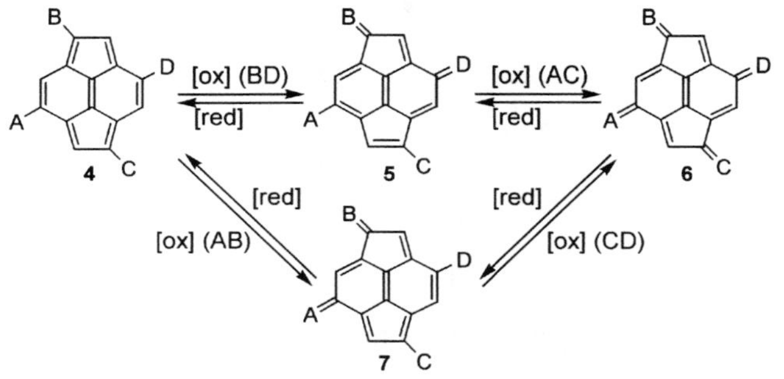

FIGURE 3. Topological property of Type A systems: omniconjugation is preserved upon redox operations as illustrated for repetitive oxidations.

323

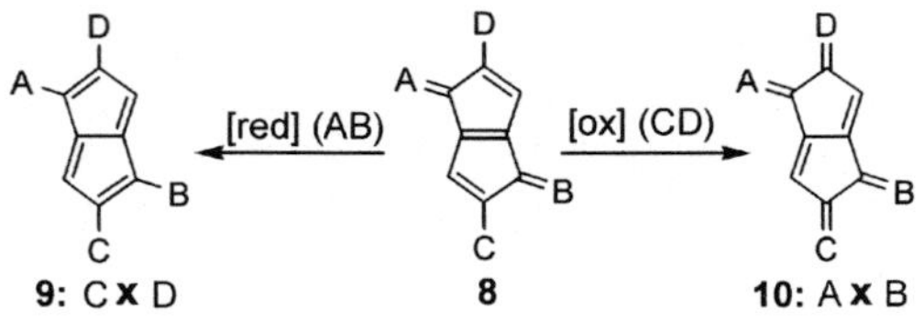

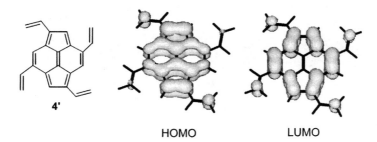

FIGURE 4. Topological property of Type B systems: in any case the complementary pathway to the redox event is not linear conjugated anymore: it is switched from 'on' to 'off'.

model **9**) while all other pathways (e.g., five in the case of **9**) remain linear conjugated. Note that, although Type B systems have a topology that is omniconjugated, they are closely related to ordinary conjugated systems in terms of their topological switching behavior upon redox events.

To determine whether or not these systems indeed can function as envisioned or not, we investigated the systems using quantum chemical calculations. One of the parameters to study is the charge delocalization between the four substituents, which should be strong due to the presence of fully conjugated pathways. For the qualitative evaluation we used simple π-conjugated moieties that represent the single and double bonded substituents at the terminals A, B, C and D (either, vinyl or methylene groups, respectively). Preliminary results indicate that the topological phenomenon of omniconjugation indeed may give rise to a pronounced delocalization of the (frontier) molecular orbitals into all terminals. As seen from Fig. 5, the π-electron density distribution suggests that this building block facilitates hole and electron transport between *all* moieties.

In some cases, a new and intriguing phenomenon was observed from the electronic structure that was not obvious from its topological framework. As a result of the local orbital symmetry, certain omniconjugated systems seem to show orthogonal preferential directionality for hole and electron transport. Hence, in the plane of the molecule, there could be different (preferential) transmission pathways for transport of either charge as illustrated in Fig. 6.

HOMO   LUMO

FIGURE 5. Spatial distribution of the π-electron density $|\psi(x,y)|^2$ in the frontier orbitals calculated by AM1 for the tetravinylene substituted model **4'**.

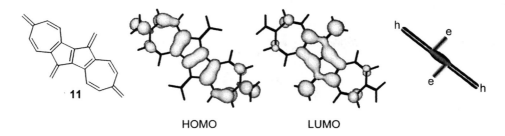

HOMO   LUMO

FIGURE 6. Shape of the HOMO and LUMO (AM1 calculated; $|\psi(x,y)|^2$) of the tetramethylene substituted molecule **11**. At the right a schematic representation is given of the preferred intramolecular transport direction of holes (h) and electrons (e).

From the above presented results we envision that the concept of omniconjugation could become important in the design of molecular electronic circuits. Omniconjugated entities are systems having conjugated pathways between all possible terminals. In going beyond three-terminal systems, this architectural concept is of interest to incorporate many logic functions inside a single molecule. Omniconjugation within a Type A system may become the single-molecule version of what is a quite trivial aspect of electronics, namely the wiring of various basic elements. The Type B systems have a π-conjugation topology that is closely related to that of the (passive) Type A systems, but are, on the other had, of use as active switching elements. These systems have the intriguing property that the complementary pathway to the redox event changes from linear to cross-conjugated.

ACKNOWLEDGMENTS

Financial support by the Dutch Ministries of EZ, O&W, and VROM through the EET program (EETK9115) is gratefully acknowledged.

REFERENCES

1. A. Aviram and M. A. Ratner, *Chem. Phys. Lett.* **29**, 277-283 (1974).
2. F. L. Carter, *Molecular Electronic Devices*, Marcel Dekker, New York, 1982.
3. C. Joachim, J. K. Gimzewski, and A. Aviram, *Nature* **408**, 541-548 (2000).
4. J. M. Tour, M. Kozaki, and J. M. Seminario, *J. Am. Chem. Soc.* **120**, 8486-8493 (1998).
5. A. Nitzan and M. A. Ratner, *Science* **300**, 1384-1389 (2003).
6. J. C. Ellenbogen and J. C. Love, *Proc. IEEE* **88**, 386-426 (2000).
7. M. H. van der Veen, M. T. Rispens, H. T. Jonkman, and J. C. Hummelen, *Adv. Funct. Mater.* **14**, 215-223 (2004).
8. S. Roth and H. Bleier, *Adv. Phys.* **36**, 385-462 (1987).

Quantum Chemical Study on La$_2$@C$_{80}$: Configuration of Endohedral Metals

Hidekazu Shimotani*,†, Takayoshi Ito‡, Atsushi Taninaka§, Hisanori Shinohara§, Yoshihiro Kubozono$^{\#,\dagger}$, Masaki Takata¶, Yoshihiro Iwasa*,†

*Institute for Materials Research, Tohoku University, Sendai 980-8577, Japan
†CREST, Japan Science and Technology Agency, Kawaguchi 330-0012, Japan
‡Japan Advanced Institute of Science and Technology, Tatsunokuchi, Ishikawa 923-1292, Japan
§Department of Chemistry & Institute for Advanced Research, Nagoya University, Nagoya 464-8602, Japan
$^{\#}$Department of Chemistry, Okayama University, Okayama 700-8530, Japan
¶JASRI-Spring-8, Koto 1-1, Hyogo 679-5198, Japan

Abstract. The molecular structure of La$_2$@C$_{80}$ was investigated theoretically and experimentally. The most stable configuration was found to be D_{3d} symmetry. La ions can move between ten equivalent D_{3d} configurations *via* D_{2h} configuration. The resulting dodecahedral trajectory of La ions agrees well with an X-ray powder diffraction experiment.

INTRODUCTION

La$_2$@C$_{80}$ has attracted considerable interest from the view point of its unique molecular structure. A nuclear magnetic resonance experiment [1] showed I_h symmetry of the molecule, which means two La ions rotate in a I_h symmetric [80]fullerene cage. I_h symmetry of the [80]fullerene cage and the rotation of the La ions in the cage was first predicted theoretically by Kobayashi *et al.*[2] They concluded D_{2h} structure shown in Fig. 1 is the most stable configuration.

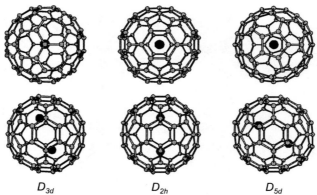

D_{3d} D_{2h} D_{5d}

FIGURE 1. Three configurations of La$_2$@C$_{80}$. Top and bottom diagrams show the views from the main axis connecting two La^{3+} ions and C_2 axis perpendicular to the main axis, respectively.

CP723, *Electronic Properties of Synthetic Nanostructures*, edited by H. Kuzmany et al.
© 2004 American Institute of Physics 0-7354-0204-3/04/$22.00

However, recent experimental results of X-ray powder diffraction [3] and XAFS [4] disagree with the D_{2h} structure. It is therefore necessary to construct an improved theoretical model of the motion of La ions in the [80]fullerene. This work was carried out to investigate the position of La ions in [80]fullerene cage.

MOTION OF LA IONS IN C_{80} CAGE

The total energies of three La ions configurations are summarized in Table 1, which were computed by density functional theory using Becke's three parameter functional with the nonlocal correlation functional of Lee-Yang-Parr (B3LYP). The employed basis sets were cc-pVDZ and LanL2DZ for C and La, respectively. All calculations in this work were carried out with Gaussian98 program package. This result shows the most stable configuration is not D_{2h} but D_{3d}. A difference of stability of the two configurations is very small, while the D_{5d} configuration is much less stable than them.

TABLE 1. The relative total energies of three configurations of $La_2@C_{80}$ computed by the B3LYP method.

Configuration	relative total energy / kcal mol^{-1}
D_{3d}	0
D_{2h}	0.1
D_{5d}	!1.4

In order to confirm the result experimentally, an experimental result of La K-edge XAFS of $La_2@C_{80}$ at 40 K was analyzed by use of the D_{3d} and the D_{2h} configuration as shown in Fig. 2. Details of the experiment were described in Ref. 4. The D_{3d} model fits well with the experimental data throughout the shown range, while the D_{2h} configuration does not fit especially in low wavenumber region. The computed geometrical parameters of the D_{3d} configuration agree better with those derived from XAFS experiment than the D_{2h} configuration as summarized in Table 2. These results support the validity of the calculation.

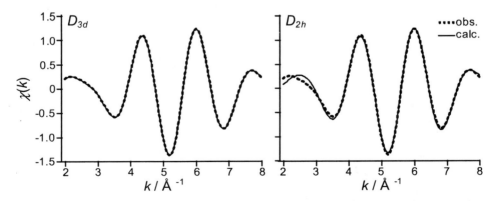

FIGURE 2. Experimental $\chi(k)$ (broken line) of $La_2@C_{80}$ at 40 K and calculated $\chi(k)$ (solid line) of the D_{3d} and the D_{2h} configuration against wavenumber (k).

TABLE 2. Experimental (XAFS at 40 K) and theoretical interatomic distances between a La ion to the closest (C(1)) and second closest (C(2)) carbon atoms and R-factor of XAFS

configuration	method	La-C(1) / Å	La-C(2) / Å	R-factor / %
D_{3d}	XAFS	2.39(1)	2.56(1)	1.2
	B3LYP[a]	2.43	2.55	
D_{2h}	XAFS	2.38(1)	2.94(2)	5.9
	B3LYP[a]	2.55, 2.58[b]	2.94, 2.98[b]	

[a]Employed basis sets were cc-pVDZ and LanL2DZ for C and La, respectively. [b]Carbon atoms having close interatomic distances were not resolved in the analysis of XAFS.

In order to investigate the motion of La ions between ten equivalent D_{3d} configurations, the potential energy surface of La ions rotation in [80]fullerene cage (Fig. 3) was computed by Hartree-Fock method employing 3-21G and LanL2DZ basis sets for C and La, respectively. This result demonstrates the existence of the path *via* the D_{2h} configuration with low energy barrier, through which La ions can move from a D_{3d} position to another equivalent D_{3d} position. By connecting the twenty equivalent D_{3d} positions of La ions with the path, vertexes and edges of an I_h symmetric dodecahedron are derived. This trajectory of La ions corresponds to that obtained by X-ray powder diffraction [5].

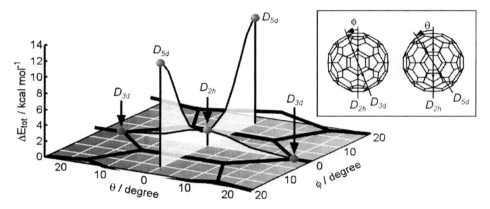

FIGURE 3. Potential energy surface of La^{3+} ions rotation from a D_{2h} configuration to D_{5d} and D_{3d} configurations. Relative total energies (ΔE_{tot}) were computed by the Hartree-Fock method. Definitions of two angles (θ and ϕ) are given in the inset. Geometrical parameters other than θ and ϕ were fully optimized to minimize a total energy. Thick lines in θ-ϕ plane show [80]fullerene cage structure below where La ions locate (see top diagrams of Fig. 1).

The Frontier orbitals of D_{3d}-symmetric $La_2@C_{80}$ computed by the B3LYP method are illustrated in Fig. 3. This shows a remarkable contrast between LUMO and HOMO. LUMO is mainly located on the La ions, while HOMO is almost located on the [80]fullerene cage. It is therefore proposed that the low mobility of *n*-type field-effect-transistor of $La_2@C_{80}$ [6] is due to a small overlap between LUMOs of neighboring $La_2@C_{80}s$. Furthermore, because LUMO has an antibonding character between La ions and [80]fullerene cage, destabilization of the D_{3d} configuration in anionic state is indicated. The computed HOMO-LUMO gap (1.43 eV) shows good agreement with the energy gap of $La_2@C_{80}$ multilayer islands (1.3–1.5 eV) measured by STS [8].

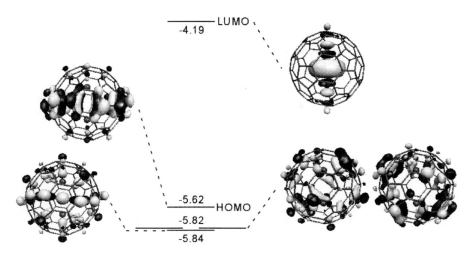

FIGURE 4. Frontier orbitals of $La_2@C_{80}$. Numbers show energy levels of the molecular orbitals in eV. Two molecular orbitals next to HOMO are degenerated.

CONCLUSIONS

The motion of La ions in $La_2@C_{80}$ was investigated theoretically and experimentally. The most stable configuration of La ions was concluded to be D_{3d} and the dodecahedral trajectory of La ions was derived. The result agrees well with experimental results of XAFS and x-ray powder diffraction.

ACKNOWLEDGMENTS

We thank the staff of the Center for Computational Materials Science at the Institute for Materials Research for computational assistance.

REFERENCES

1. Akasaka, T., Nagase, S., Kobayashi, K., Wälchli, M., Yamamoto, K., Funasaka, H., Kako, M., Hoshino, T., and Erata, T., *Angew. Chem., Int. Ed. Engl.* **36**, 1643-1645 (1997).
2. Kobayashi, K., Nagase, S., and Akasaka, T., *Chem. Phys. Lett.* **245**, 230-236 (1995).
3. Nishibori, E., Takata, M., Sakata, M., Taninaka, A., Shinohara, H., *Angew. Chem., Int. Ed. Engl.* **40**, 2998-2999(2001).
4. Kubozono, Y., Takabayashi, Y., Kashino, S., Kondo, M., Wakahara, T., Akasaka, T., Kobayashi, K., and Nagase, S., *Chem. Phys. Lett.* **335**, 163-169(2001).
5. Shimotani, H., Ito, T., Iwasa, Y., Taninaka, A., Shinohara, H., Nishibori, E., Takata, M. and Sakata, M., *J. Am. Chem. Soc.* **126**, 364-369 (2004).
6. Kobayashi, S., Mori, S., Iida, S., Ando, H., Takenobu, T., Taguchi, Y., Fujiwara, A., Taninaka, A., Shinohara, H., and Iwasa, Y., *J.Am. Chem. Soc.* **125**, 8116-8117 (2003).
7. Scott, A. P., and Radom, L., *J. Phys. Chem.* **100**, 16502-16513 (1996).
8. Taninaka, A., Shino, K., Sugai, T., Heike, S., Terada, Y., Hashizume, T., and Shinohara, H., *Nano Let.* **3**, 337-341 (2003).

Raman excitation profiles for the (n_1, n_2) assignment in carbon nanotubes

H. Telg[*], J. Maultzsch[*], S. Reich[†], F. Hennrich[**] and C. Thomsen[*]

[*]Institut für Festkörperphysik, TU Berlin, Hardenbergstr. 36, 10623 Berlin, Germany
[†]Department of Engineering, University of Cambridge, United Kingdom
[**]Institut für Nanotechnologie, Forschungszentrum Karlruhe, 76021 Karlsruhe, Germany

Abstract. The assignment of the chiral indices n_1 and n_2 in semiconducting and metallic nanotubes was performed comparing resonance Raman profiles and transition energies in a 3rd-order Kataura plot. We find several complete branches in the Kataura plot and are able to assign the resonance peaks to n_1 and n_2 without prior assumptions about the constant of proportionality between the radial-breathing mode frequency ω_{RBM} and the inverse diameter of the tube. We point out systematic differences in the transition energies and intensities in $+1$ and -1 nanotube families.

Ever since the discovery of how to keep isolated nanotubes from rebundeling in solution, it has become a realistic goal to identify the chirality of a nanotube by spectroscopic methods. Wrapping the nanotubes with SDS, Bachilo et al. [1] found that the luminescence from semiconducting nanotubes was large and not quenched by the presence of metallic tubes. They were able to map out luminescence energies of a sample as a function of laser-excitation energy and found a series of peaks presumably orginating from different chirality nanotubes. There were, however, several possibilities for matching the experimental transition energies E_{11} and E_{22} to the calculated energies.[2] To remove these ambiguities the authors had to define an anchoring element from which the chirality assignment of all other detected nanotubes followed. This anchoring element was found by minimizing the least-square error of detected RBM frequencies for several different constants in the well-known inverse-diameter dependence of nanotubes.

In this paper we study the excitation-energy dependence of the Raman intensity of the radial breathing mode. The resonance maxima yield the transition energies $E_{ii}^{S,M}$, the frequency of the RBM peak we take to be proportional to the inverse diameter of the tube (plus a constant), but we do not make *a priori* assumptions about the proportionality constant as is usually done. Instead, plotting the resonance maxima of the radial breathing mode *versus* $1/\omega_{RBM}$, we are able to resolve the individual branches of the radial breathing mode in a 3rd-order Kataura plot. The ambiguities in the assignment of the Raman peaks arising from the closely spaced ω_{RBM} are removed and a straightforward assignment of the RBM-peaks to chiral indices n_1 and n_2 follows. Our assignment agrees with the one given by Bachilo et al. [1], and partly also with the one of Strano et al. [3] but is different from many others. Different from luminescence measurements we also find metallic tubes and several zig-zag tubes.

From our results we can derive an independent value for the proportionality constant in the frequency–inverse-diameter relationship, generally taken to be between 200 and

CP723, *Electronic Properties of Synthetic Nanostructures*, edited by H. Kuzmany et al.
© 2004 American Institute of Physics 0-7354-0204-3/04/$22.00

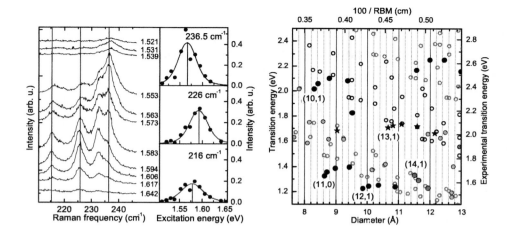

FIGURE 1. *left:* Resonance Raman spectra of the radial breathing mode excited with various laser energies of a dye laser as indicated. *right:* Kataura plot with the transition energies calculated in the 3rd-order tight-binding approximation,[2] *vs* theoretical diameter d (left and bottom axes, open symbols). Right and top axes refer to experimental values (black circles–strong, stars–medium, light gray–weak). The chirality of the tube with the smallest d within a branch is indicated just below or above that tube.

$250 \, cm^{-1}$ nm [4] and heavily debated in the literature [5]. The importance of this relationship lies in that it gives practical information about the diameter and even the chirality of an observed nanotube, both essential for most applications of carbon nanotubes. In contrast to its general importance, however, it was never established independently from experiment. There are many calculations putting the constant between $230\text{-}235 \, cm^{-1}$, at least for the larger diameters; see, *e.g.*, Ref. [6] for an overview.

In Fig. 1, *left*, we show one of the resonance profiles of several nanotubes with nearby resonance maxima. The largest intensity is seen to move through the peaks forming what looks similar to a laola wave.[7] There are several such laola-type resonances when exciting the nanotubes with different laser lines, each occurring over a fairly narrow energy and ω_{RBM} range. As we will see, each laola-series of resonances corresponds to a particular branch in the Kataura plot of transition energies *versus* diameter. Within each such a branch, there is a fixed relationship between the indices: n_1 decreases by one and n_2 increases by 2, going from smaller to larger diameters as long as $n_1 > n_2$. The tube with a smallest diameter in a branch is thus always a zig-zag or near–zig-zag tube.

We are now able to plot the transition energies obtained from the resonance profiles *versus* the inverse radial breathing mode frequency, both purely experimental values. In order to make the experiment (right and top axes in Fig. 1, *right*) match theory (left and bottom axes) in this plot, we stretch and/or shift the axes in the plot. For the x-axis this means varying the proportionality constant in the frequency–inverse-diameter relationship and introducing a constant offset. For the y-axis this adjusts the tight-binding energies to the experimental transition energies. We perform this transformation until we get a match between experiment and theory. Our stringent criterion is that in the vertical

TABLE 1. Chiral indices and RBM-frequencies ($\pm 0.5\,\mathrm{cm}^{-1}$) as determined from a comparison to several branches of the 3rd-order Kataura plot, see Fig. 1. The modes are grouped into branches and sorted in order of increasing smallest diameter d in a branch. (d is calculated from n_1 and n_2 with a graphite lattice constant of $a_0 = 2.461$Å.) Also given is the family $v = (n_1 - n_2)$ mod 3;[8] $v = 0$ corresponds to metallic nanotubes. The resonance maxima E_{22}^S and E_{11}^M are accurate to ± 3 (3 digits) or ± 30 meV (2 digits).

family v	chirality	$\omega_{\mathrm{RBM}}[\mathrm{cm}^{-1}]$	$d\,[\text{Å}]$	strength	E_{22}^S [eV]	E_{11}^M [eV]
0	10, 1	276.3	8.25	strong		2.38
	9, 3	272.7	8.47	strong		2.43
	8, 5	262.7	8.90	strong		2.47
	7, 7	247.8	9.50	strong		2.45
-1	11, 0	266.7	8.62	strong	1.657	
	10, 2	264.6	8.72	strong	1.690	
	9, 4	257.5	9.03	strong	1.72	
	8, 6	246.4	9.53	strong	1.73	
-1	12, 1	236.4	9.82	strong	1.551	
	11, 3	232.6	10.00	strong	1.570	
	10, 5	226.1	10.36	strong	1.578	
	9, l	216.0	10.88	strong	1.564	
0	13, 1	220.3	10.60	medium		2.057
	12, 3	217.4	10.77	medium		2.075
	11, 5	212.4	11.11	medium		2.084
	10, 7	204.0	11.59	medium		2.067
	9, 9	195.3	12.21	medium		2.01
+1	14, 1	205.4	11.38	weak	1.667	
	13, 3	203.3	11.54	weak	1.617	
	12, 5	198.5	11.85	weak	1.554	
	11, 7	–	12.31	–	–	
	10, 9	–	12.09	–	–	

direction there is a full correspondence between experiment and theory implying that all chiralities theoretically possible have actually been detected. Note that some chiralities give rather close diameters and hence closely lying peaks, *e.g.*, $d[(11,0)] = 8.6$ and $d[(10,2)] = 8.7$ Å whereas the next tube in the same branch is at a larger separation, $d[(9,4)] = 9.0$ Å. For the four branches with the smallest tubes (10,1), (11,0), (12,1), and (13,1) we observed all possible nanotubes, the irregularities in the diameter distances were oberved as well and put our assignment on firm grounds, see Table 1.

The requirement to find *all* experimental peaks in a given branch leads to a unique relationship between inverse diameter and ω_{RBM}. This makes clear the advantage of resonant Raman over luminescence spectroscopy where the zig-zag tubes were not observed, and hence the horizontal alignment in the Kataura plot could not be performed uniquely, consequently leading to ambiguities in the chirality assignment and/or somewhat far-fetched assumption about the chirality distributions in a sample.[1]

In Table 1 we summarize the radial breathing mode frequency and the chirality assignment as obtained from our experiment and as plotted in Fig. 1. We stress that the assignment is actually independent of the actual transition energies and their systematic deviations from the 3rd-order transition energies, but relies on the completeness of all possible chiralities within several branches of the Kataura plot. Our assignment agrees with that given by Bachilo et al.[1], but disagrees with several others in the literature.

As regards to the vertical axis in Fig. 1, right, we observe the following for tubes of the -1 and 0 families: Shrinking the vertical axis aligns well the nanotubes with near-armchair direction chiralities but leaves a systematically increasing deviation between experiment and theory when moving towards the zig-zag direction within a given branch. In other words, the experimental points within a Kataura branch curve away from the $1/d$ center of gravity of the transition energies more strongly than the calculated values. This systematic difference, also observed in luminescence spectroscopy, [1, 4] is apparently due to a chirality-dependent effect and currently under investigation,[9] together with a systematic dependence of the Raman intensity on chiral angle within a branch.[10] A second observation regarding the intensities of semiconducting tubes is at place: The lower branches in the Kataura plot are systematically much stronger experimentally than the upper ones. In terms of the chiralities this distinction translates into families with $(n_1 - n_2) \bmod 3 = -1$ (lower branch) having generally larger intensities than those with $+1$ (upper branch). This asymmetric behavior has been shown to originate from band renormalization effects[11] as predicted by Kane and Mele.[12] and a dependence fo the RBM matrix element on the nanotube family.[10].

In conclusion, we reported an assignment of RBM frequencies to carbon nanotube chiralities based on resonance Raman spectroscopy. The assignment is independent of prior assumptions about the frequency–diameter relationship; instead, including also RBM frequencies not shown here, this relationship is experimentally determined as [13]

$$\omega_{RBM} = \frac{(216 \pm 2)\,\mathrm{cm}^{-1}\mathrm{nm}}{d} + (17 \pm 2)\,\mathrm{cm}^{-1} \tag{1}$$

We also pointed out systematic differences of the measured and calculated transition energies and of the Raman intensities within a branch or a family of the Kataura plot.

We acknowlegde support by the Deutsche Forschungsgemeinschaft under grant number Th 662/8-2. SR was supported by the Oppenheimer Fund and Newnham College.

REFERENCES

1. S.M. Bachilo, M.S. Strano, C. Kittrell, R.H. Hauge, R.E. Smalley, and R.B. Weisman, Science **298**, 2361 (2002)
2. S. Reich, J. Maultzsch, C. Thomsen, and P. Ordejón, Phys. Rev. B 66, 035 412 (2002)
3. M.S. Strano et al., Nano Lett. **3**, 1091 (2003)
4. S. Reich, C. Thomsen and J. Maultzsch, Carbon Nanotubes: Basic Concepts and Physical Properties (Wiley-VCH, Weinheim, 2004)
5. Ch. Kramberger, R. Pfeiffer, H. Kuzmany, V. Zólyomi, and J. Kürti, Phys. Rev. B **68**, 235404 (2003)
6. J. Kürti, V. Zólyomi, M. Kertesz, and G. Su, New Journal of Physics **5**, 125 (2003)
7. I. Farkas, D. Helbing, and T. Vicsek, Nature **419**, 131 (2002)
8. S. Reich and C. Thomsen, Phys. Rev. B **62**, 4273 (2002)
9. S. Reich et al., to be published
10. M. Machón, S. Reich, J. Maultzsch, P. Ordejón, and C. Thomsen, to be published
11. S. Reich, M. Dworzak, A. Hoffmann, M.S. Strano, and C. Thomsen, to be published
12. C.L. Kane and E.J. Mele, Phys. Rev. Lett. 90, 207 401 (2003), and cond-mat 0403153 (2004)
13. H. Telg, J. Maultzsch, S. Reich, F. Hennrich, and C. Thomsen, to be published

Resonant Raman spectroscopy of nanostructured carbon-based materials: the molecular approach

Matteo Tommasini*, Eugenio Di Donato*, Chiara Castiglioni*, Giuseppe Zerbi*, Nikolai Severin†, Thilo Böhme† and Jürgen P. Rabe†

*Dipartimento di Chimica, Materiali e Ingegneria Chimica "G. Natta", Politecnico di Milano, Piazza Leonardo da Vinci 32 – 20133 Milano and INSTM UdR Milano – Italy
†Department of Physics, Humboldt University, Berlin – Germany

Abstract. π-conjugated molecules are unique models for studying confinement effects on a system of interacting $2p_z$ carbon electrons. When introducing edges in an extended periodic system, changes of the electronic structure in the vicinity of the edges are found. Scanning tunneling microscopy and Raman spectroscopy can be effectively used to characterize these effects.

One and two dimensional polyconjugated π systems are peculiar because the optical and vibrational properties do strongly depend on the size of the system. Specific Raman active vibrations are found which are selective fingerprints of the nanostructures examined. Both one and two dimensional π systems possess such vibrations which are coupled to $\pi \rightarrow \pi^*$ excitations [1]. Because the energies of those excitations are size-dependent, it turns out that multi-wavelength Raman spectroscopy is a powerful tool for the investigation of disordered and/or nanostructured materials where a distribution of possible excitation energies is present [2, 3]. In the following we will present new experimental evidences supporting the molecular approach.

STM measurements have been performed across a graphite armchair step edge. The typical experimental results are reported in Fig. 1. It is possible to clearly identify a series of intensity maxima and minima running parallel to the step edge. These stripes fade away when moving away from the edge. The average distance between two minima (maxima) is measured by the Fourier transform of the STM image, giving a spacing of 0.378 nm. This experimental finding is fully consistent with Hückel molecular orbital computations carried out on Polycyclic Aromatic Hydrocarbon (PAH) molecules possessing the armchair edge topology. In Fig. 1 we report the contour maps of the frontier orbitals of C5514, an armchair-edged D_{6h} PAH. The presence of nodal planes in the electron density contour maps can be immediately recognized. The nodal planes are parallel to the armchair edge and they tend to gradually disappear when moving to the graphitic core of the molecule (which corresponds to the origin of the cartesian system). These nodal planes in the frontier orbitals electron densities are the responsible for the observed stripe modulation in the STM image of Fig 1. The nodal planes are separated by a distance of $1.5 \times a = 0.369$ nm which nicely agrees with the experimental value of 0.378 nm. The frontier orbitals of increasingly large PAHs tend to become degenerate at the Fermi energy. This is the rationale for summing the HOMO and LUMO densities as reported in Fig. 1. Just because both HOMO and LUMO densities

CP723, *Electronic Properties of Synthetic Nanostructures*, edited by H. Kuzmany et al.
© 2004 American Institute of Physics 0-7354-0204-3/04/$22.00

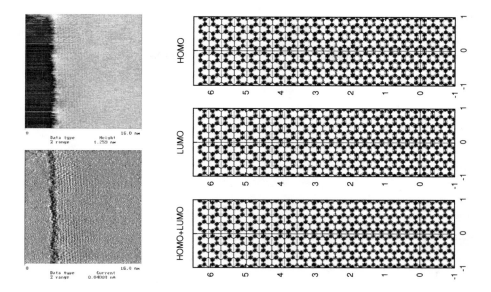

FIGURE 1. (a) STM image of an armchair step edge in graphite. Note the intensity modulation running parallel to the edge. (b) Contour maps, according to Hückel theory, of the electron densities of the frontier orbitals of a large PAH (C5514) with D_{6h} symmetry, overall hexagonal shape and armchair edge. A zoom of a rectangular region of the molecule comprising the center and the edge is shown. The cartesian axes are C_2 symmetry axes of the molecule. A few nodal lines are shown close to the edge. Units on the axes are nm.

do possess corresponding nodal planes, it turns out that such nodal planes are present also in the HOMO+LUMO density. The presence of these peculiar nodal planes is a characteristic of the armchair edge topology and not some accident related to the chosen molecular size. We performed a similar Hückel computation on a larger armchair-edged PAH of D_{6h} symmetry containing 15666 carbons. This can be considered as a representative model of a few nanometers long armchair step edge and it is closer to the experimental conditions of Fig. 1 than the smaller sized C5514. The map of the square of the coefficients of the HOMO of C15666 is reported in Fig. 2. Here again it is possible to observe the presence of nodal planes running parallel to the armchair edge. We therefore conclude that this pattern of the electronic density is a peculiar feature of the armchair topology. Moreover, we considered also the maps of several frontier orbitals of C15666 (both occupied and unoccupied[1]) in the energy range $-0.1\beta < \varepsilon < 0.1\beta$. Out of this analysis it turns out that all the orbitals in this range do have similar nodal planes

---

[1] The maps of the squares of the coefficients of occupied/unoccupied pairs of orbitals which have symmetric energies with respect to the Fermi level do coincide in Hückel theory and also in QCFF/PI theory as can be inferred from the HOMO/LUMO maps of the oligorylenes shown in Fig. 2.

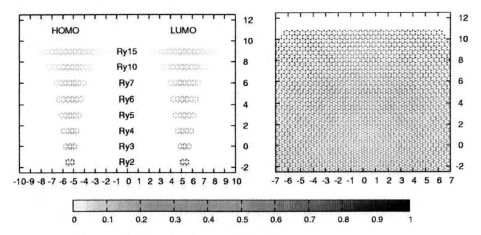

FIGURE 2. Frontier orbitals of oligo-rylenes. Comparison with the HOMO frontier orbital of a giant D_{6h} armchair-edged PAH molecule. Units on the cartesian axes are nm.

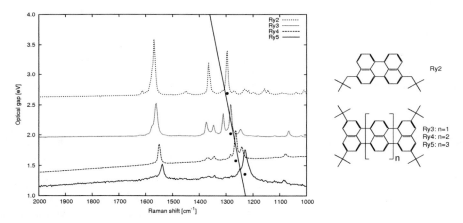

FIGURE 3. Experimental FT-Raman spectra of oligo-rylenes of increasing conjugation length ($\lambda_{exc} = 1064nm$). The linear dispersion of the most intense peak in the D region with respect to the optical gap is also shown (see text).

running parallel to the armchair edge. These nodal planes are indeed a peculiar feature of the armchair edge topology and are found also in the frontier orbitals of oligorylenes, a class of armchair-edged PAHs (see Fig. 2). The sketch of their chemical structure is shown in Fig. 3 together with their Raman spectra. Representatives of this class of molecules have been synthesized and characterized in the past by means of electronic and Raman spectrosopy [4, 5, 6, 7]. When considering the optical gap of oligorylenes as it is obtained from electronic spectroscopy, a softening of the excitation energy with respect to the size of the system is found, as expected in π-conjugated systems.

The optical gaps of the first members of the series (Ry2, Ry3, Ry4) are respectively 2.69, 2.02, and 1.58 eV [5]. According to ref. [5] the optical gap of Ry5 ($n = 3$) can be estimated by using an extrapolation formula to be 1.36 eV. The Raman spectra of oligorylenes have been analyzed in the past with respect to the vibrational contribution to their non-linear optical properties [6, 7]. More recently their Raman behavior has been reviewed in the framework of the Raman spectroscopy of PAHs [8]. From this work it results that oligorylenes exhibit Raman active collective breathing vibrations (in a region between 1350 and 1200 cm^{-1}) characterized by nuclear displacements very similar to those associated to the D line of disordered graphitic materials. The peculiar electronic structure induced by the confinement of π electrons in finite size domains is the responsible of the strong Raman activity of these collective breathing modes [8]. The D line is observed in graphite whenever small crystallites are present in the sample [2]. The analogy between the electronic structure of oligorylenes and that found in the vicinity of the armchair edges of very large PAHs (see Fig. 1 and 2) strongly suggests that, in the case of microcrystalline graphite, the edges are responsible of the activation of the D band in the Raman spectra. According to what is suggested by our present calculations and by STM measurements, the region affected by a rylene-like electronic structure extends in the direction of the bulk up to about 4 nm far from the edge. Collective breathing modes of six member rings in this region are related to Raman transitions in the D band region. In a real polycrystalline sample different kinds of edge are certainly present, and they are characterized by a different degree of perfection, according to the presence of conformational defects (*e.g.* deviations from the fully armchair conformation). By consequence, a distribution of excitation energies for these electronic edge states is expected. This gives rise to edge-selective resonance conditions in multi-wavelength Raman experiments on polycrystalline graphite. As a further support to this this interpretation, a dispersion of the D band is indeed obtained from the experimental Raman data on oligorylenes which is very close to the dispersion measured on microcrystalline graphite [2]. In Fig. 3 the FT-Raman spectra of the first members of the class of oligorylenes are overlaid according to their optical energy gaps. When considering the most intense peak in the D region and the optical energy gap of the molecule a clear dispersion is found with a slope of 44 cm^{-1}/eV^2. On the other hand, the observed dispersion of the D band of graphite [2] can be also theoretically explained by simulations of the multi-wavelength Raman response of disordered carbon materials [3].

CONCLUSIONS

The result from the STM measurements and the theoretical modeling reported above is that the armchair edge topology causes a change of the electronic density for the levels close to the Fermi energy with respect to bulk graphite. This fact is also confirmed

---

[2] There is a variability of the slope, according to the chosen representative value for the vibrational frequency. In fact, the coupling of the breathing vibration with CH wagging vibrations spreads the Raman activity in the D region over a few bands. This effect is more relevant for the shorter oligorylenes.

by previous work on the structure and Raman spectroscopy of PAHs [8]. Both the very presence and the dispersion of the D peak is the Raman signature of the structural change of PAHs with respect to graphite. The analysis of the frontier orbitals of oligorylenes and giant PAHs mimicking graphite's armchair edges show close similarities. Moreover oligorylenes possess Raman spectra with clear signals in the D region. These two facts support the explanation of the origin of the D peak as due to the change of the electronic structure induced by confinement effects upon the π system.

ACKNOWLEDGMENTS

This work was supported by a grant from the European Commission, Fifth Framework Programme, Growth Programme (Research Project "MAC-MES; Molecular Approach to Carbon based Materials for Energy Storage", G5RD-CT2001-00571) and by a grant from MURST (Italy) (FIRB project "Carbon based micro and nano structures", RBNE019NKS).

REFERENCES

1. C. Castiglioni, M. Tommasini, G. Zerbi, *Theme Issue of the Philosophical Transactions of the Royal Society A on "Raman Spectroscopy in Carbons: from Nanotubes to Diamond"*, to be published
2. Póksic, M. Hundhausen, M. Koós, L. Ley, *J. Non Cryst. Solids*, **227-230**, 1083 (1998).
3. C. Castiglioni, E. Di Donato, M. Tommasini, F. Negri, G. Zerbi, *Synth. Met.*, **139**, 885–888 (2003).
4. A. Bohnen, K. H. Koch, W. Lüttke, K. Müllen, *Angew. Chem. Int. Ed. Engl.*, **29**, 525–527 (1990).
5. A. Schmidt, N. R. Armstrong, C. Goeltner, K. Muellen, *J. Phys. Chem.*, **98**, 11780–11785 (1994).
6. M. Rumi, G. Zerbi, K. Müllen, G. Müller, M. Rehahn, *J. Chem. Phys.*, **106**, 24–34 (1996).
7. M. Rumi, G. Zerbi, K. Müllen, *J. Chem. Phys.*, **108**, 8662–8670 (1998).
8. F. Negri, C. Castiglioni, M. Tommasini, G. Zerbi, *J. Phys. Chem. A*, **106**, 3306 (2002).

Orientational Charge Density Waves and the Metal-Insulator Transition in Polymerized KC$_{60}$

B. Verberck, A. V. Nikolaev, and K. H. Michel

Department of Physics, University of Antwerp, Universiteitsplein 1, 2610 Antwerp, Belgium

Abstract. Polymerized KC$_{60}$ undergoes a structural phase transition accompanied by a metal-insulator transition around 50 K. To explain the structural aspect, a mechanism involving small orientational deviations of the valence electron density on every C$_{60}$ monomer — orientational charge density waves (OCDWs) — has already been proposed earlier. In the present work, we address the metal-insulator transition using the OCDW concept. We are inspired by the analogy between a polymer chain exhibiting an OCDW and a linear atomic chain undergoing a static lattice deformation doubling the unit cell: such a deformation implies a band gap at the zone boundary, yielding an insulating state (Peierls instability). Within our view, a similar mechanism occurs in polymerized KC$_{60}$; the OCDW plays the role of the lattice deformation. We present tight-binding band structure calculations and conclude that the metal-insulator transition can indeed be explained using OCDWs, but that the threedimensionality of the crystal plays an unexpected key role.

Among the A$_x$C$_{60}$ alkali metal doped fullerides (A = K, Rb, Cs) [1], the $x = 1$ compounds [2] are of particular interest, because they form stable crystalline phases with cubic rock-salt structure at high temperature ($T > 350$ K) and polymerized structures [3] at lower T. Polymerization occurs along the former cubic [110] direction and is realized via [2+2] cycloaddition reactions between neighboring C$_{60}$ molecules. The polymer phase of KC$_{60}$ is orthorhombic [4,5] with space group *Pmnn*: the orthorhombic $\vec{a}$ axis is parallel to the axis of polymerization, the $\vec{c}$ axis is parallel to the former cubic [001] direction, and the polymer chains have alternating orientations $\pm|\psi_0| = \pm50°$ in successive ($\vec{a}, \vec{b}$) planes [Fig. 1(a)].

In addition to the structural transition at $T = 350$ K, a second transition has been observed in KC$_{60}$ [6]: single crystal x-ray diffraction has revealed the appearance of a $(\vec{a} + \vec{c}, \vec{b}, \vec{a} - \vec{c})$ superstructure at 60 K $< T_c <$ 65 K, while ESR measurements have shown a metal-insulator transition at $T_c = 50$ K. As a mechanism for the structural change in polymerized KC$_{60}$, orientational charge density waves (OCDWs) were suggested [7]. The OCDWs model, introducing small orientational deviations $\pm\Delta\psi_0$ of the valence electronic density on the C$_{60}^-$ ions [Fig. 1(b)], successfully explains the observed structural phase transition in KC$_{60}$ and the absence of it in

CP723, *Electronic Properties of Synthetic Nanostructures*, edited by H. Kuzmany et al.
© 2004 American Institute of Physics 0-7354-0204-3/04/$22.00

polymerized Rb- and CsC$_{60}$, both in agreement with the present experimental knowledge of these compounds [8].

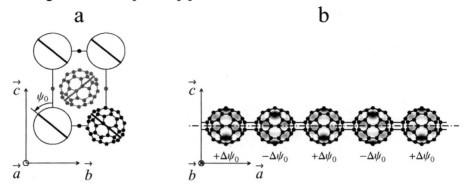

FIGURE 1. (a) Structure of polymerized KC$_{60}$, projected on the crystallographic $(\vec{b}, \vec{c})$ plane. The planes of cycloaddition are represented by thick bars. The cycloaddition planes have alternating orientations $\pm|\psi_0| = \pm 50°$ along the $\vec{c}$ axis. (b) An OCDW in a $(C_{60})_n$ polymer chain. The valence electronic charge densities (represented schematically by contour plot projections) have alternating rotational deviations $\pm\Delta\psi_0$ along the polymer.

In this contribution, we apply the OCDW concept to account for the metal-insulator transition at $T \approx 50$ K in polymerized KC$_{60}$. Our approach is inspired by the Peierls instability in a linear chain of atoms each contributing one electron to the associated 1D electron gas [9]: when the atoms are equally separated by a lattice constant a, the lowest electronic band is halfly filled, implying a metallic state, but when a lattice distortion with wave vector $q = 2k_F = 2\pi/2a$ (wave length $\lambda = 2\pi/q = 2a$ is applied, the unit cell volume is doubled, the first Brillouin zone reduced to half its original size, the lowest band completely filled, and the system becomes an insulator because of the band gap at the zone boundary — the presence of which one can demonstrate within the nearly-free electron approximation. Within our view, a similar mechanism is responsible for the experimentally observed metal-insulator transition: OCDWs imply periodicity doubling along both the $\vec{a}$ and the $\vec{c}$ direction and result in an insulating state.

To obtain the band structure $\{E(\vec{k})\}$ of the crystal, we follow the simplified Slater-Koster tight-binding approach of Ref. [10]. The electronic wave function $\psi_{\vec{k}}(\vec{r})$ is assumed to have the form

$$\psi_{\vec{k}}(\vec{r}) = \sum_{\vec{n}} e^{i\vec{k}\cdot\vec{X}(\vec{n})} \sum_{J=1}^{12} b_J \Psi_J(\vec{r} - \vec{X}(\vec{n})), \qquad (1)$$

where $\{\Psi_J(\vec{r}); J = 1,\ldots,12\}$ is the set of 12 molecular wave functions describing the unit cell: 3 t_{1u} functions for each of the 4 inequivalent C$_{60}^{-}$ monomers. (There are 4 inequivalent sites because of the two different polymer chain orientations $\pm\psi_0$ and the orientational deviations $\pm\Delta\psi_0$ of the valence electronic density within each polymer chain.) The lattice vectors read $\vec{X}(\bar{n}) = n_1\vec{t}_1 + n_2\vec{t}_2 + n_3\vec{t}_3$, with $\vec{t}_1$, $\vec{t}_2$ and $\vec{t}_3$ the primitive basis vectors and $\bar{n} = (n_1, n_2, n_3)$ a triple of integers. Requiring the electronic wave function to obey the Schrödinger equation $H(\vec{r})\psi_{\vec{k}}(\vec{r}) = E(\vec{k})\psi_{\vec{k}}(\vec{r})$ leads to a homogeneous system of equations for the coefficients b_J, having non-trivial solutions only if the secular equation is satisfied:

$$\det\left(\sum_{\bar{n}} e^{i\vec{k}\cdot\vec{X}(\bar{n})}\gamma_{JJ'}(\bar{n}) - (E(\vec{k}) - E_J)\delta_{JJ'}\right) = 0. \tag{2}$$

The overlap integrals $\kappa_{JJ'}(\bar{n}) = \int d\vec{r}\Psi_J^*(\vec{r})\Psi_{J'}(\vec{r} - \vec{X}(\bar{n}))$ are neglected for $\bar{n} \neq \vec{0}$; the interaction integrals $\gamma_{JJ'}(\bar{n})$ read $\gamma_{JJ'}(\bar{n}) = \int d\vec{r}\Psi_J^*(\vec{r})U(\vec{r})\Psi_{J'}(\vec{r} - \vec{X}(\bar{n}))$. Solving the secular equation yields the band structure $\{E(\vec{k})\}$. As for the potential, we extend the contact potential model of Ref. 10 to three dimensions:

$$U_{JJ'}(\vec{r}) = U_{JJ'}\sum_{\bar{m}}\sum_{t=1}^{16}\delta(\vec{r} - \vec{r}_t - \vec{X}(\bar{n})), \tag{3}$$

where $\vec{r}_t$ is the position of the t-th interaction center (IC). We work with 2 ICs on every cycloaddition bond, so that we have a total of 16 ICs in the unit cell. The interaction strengths $U_{JJ'}$ are assumed to be different for intrachain ($U_{JJ'} = U_0$), (000)-(½½½) interchain ($U_{JJ'} = U_1$) and (000)-(001) interchain ($U_{JJ'} = U_2$) interactions. For a C_{60} molecule in the standard orientation the three t_{1u} functions read,

$$\psi_j(r, \theta, \phi) = R(r)\sum_{\tau} a_j^{\tau}Y_{l=5}^{\tau}(\theta, \phi). \tag{4}$$

The coefficients a_j^{τ} can be found in Ref. 11. For the radial part, we have chosen model II of Ref. 12: $R(r) = Ce^{-K|r-R_0|}$, with $C = 1.928\times10^{-3}$, $K = 1.241\,\text{Å}^{-1}$, and R_0 the radius of the C_{60} molecule, $R_0 = 3.55\,\text{Å}$. The exponential decay allows to neglect $\gamma_{JJ'}(\bar{n})$ for large radial values.

Calculating the band structure for various interaction strength parameter combinations leads us to the following conclusions. First of all, we observe that a purely one-dimensional situation ($U_1 = U_2 = 0$) does not yield a band gap. This is due to canceling contributions coming from interactions of a molecular site with its "left" and its "right" neighboring sites, which are always equal, whether the OCDW is applied ($\Delta\psi_0 \neq 0$) or not ($\Delta\psi_0 = 0$). Secondly, we see that U_1 plays a key role: only if $U_1 \neq 0$, a band gap appears. Fig 2 shows the band structures for $U_1/U_0 = 100$ and $U_2/U_0 = 1$. Finally, we observe that for more realistic choices of parameters, e.g. $U_1/U_0 = 0.9$ and $U_2/U_0 = 0.8$, the band gap becomes extremely small. In summary: the metal-insulator transition in polymerized KC$_{60}$ can be explained by the OCDW concept, with the three-dimensionality as a key element, but including the alkali atoms (so far neglected), might be required to obtain realistic band gap magnitudes.

B. V. is a research assistant of the Fonds voor Wetenschappelijk Onderzoek – Vlaanderen. A. V Nikolaev acknowledges financial support from the Bijzonder Onderzoeksfonds, Universiteit Antwerpen (BOF – NOI).

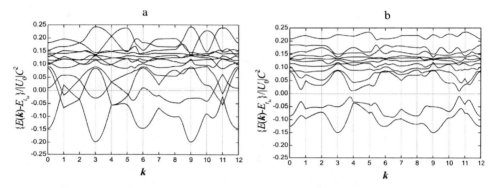

FIGURE 2. (a) Calculated band structure without OCDWs ($\Delta\psi_0 = 0°$). (b) Calculated band structure with OCDWs ($\Delta\psi_0 = 13°$). The interaction strengths are $U_1/U_0 = 100$ and $U_2/U_0 = 1$. For convenience, the representative $\vec{k}$ path is linearly mapped onto the interval [0,12[.

REFERENCES

1. For an overview, see Forró, L. and Mihály, L., *Rep. Prog. Phys.* **64**, 649-699 (2001).
2. Winter, J. and Kuzmany, H., *Solid State Commun.* **84**, 935-938 (1992); Zhu, Q. et al., *Phys. Rev.* B **47**, 13948-13951 (1993).
3. Pekker, S. et al., *Solid State Commun.* **90**, 349-352 (1994).
4. Stephens, P. W. et al., *Nature* **370**, 636-639 (1994).
5. Launois, P. et al., *Phys. Rev. Lett.* **81**, 4420-4424 (1998).
6. Coulon, C. et al., *Phys. Rev. Lett.* **86**, 4346-4349 (2001).
7. Verberck, B., Nikolaev, A. V., and Michel, K. H., *Phys. Rev.* B **66**, 165425 — 1-14 (2002); Verberck, B., Nikolaev, A. V., and Michel, K. H., *J. Electron Spectrosc.* **129**, 133-137 (2003).
8. Launois, P. and Moret, R., private communication.
9. Peierls, R. E., *Quantum Theory of Solids* (Clarendon, Oxford, 1974).
10. Nikolaev, A. V. and Michel, K. H., *Solid State Commun.* **117**, 739-743 (2001).
11. Nikolaev, A. V., Prassides, K., and Michel, K. H., *J. Chem. Phys.* **108**, 4912-4923 (1998).
12. Nikolaev, A. V. and Michel, K. H., *J. Chem. Phys.* **117**, 4761-4776 (2002).

First principles calculations for the electronic band structures of zone folding metallic single wall carbon nanotubes

Viktor Zólyomi* and Jenő Kürti*

*Department of Biological Physics, Eötvös University Budapest, Pázmány Péter sétány 1/A, H-1117 Budapest, Hungary

Abstract.
We present first principles calculations on the band structures of 20 different small diameter (d) single wall carbon nanotubes (SWCNTs), considering those tubes which are metallic in simple zone folding picture, including 6 chiral ones, employing density functional theory (DFT). The band gaps are calculated and discussed for all of the tubes. The Fermi wave-vector of armchair tubes shows a downshift from its ideal, zone folding expected value; this shift is proportional to $1/d^2$. Four zigzag and four chiral tubes show a small gap in the band structure. A higher order correction term appears in the diameter dependence of the gap for the zigzag tubes, the magnitude of which is apparently overestimated. This overestimation can likely be eliminated by considering many-electron effects.

INTRODUCTION

It is well known, that with simple zone folding (ZF) of the nearest neighbor (NN) tight binding (TB) dispersions of graphene, the large diameter single wall carbon nanotubes (SWCNTs) can be reasonably well described [1, 2]. In this approximation, the metallicity of the tubes is determined by their (n, m) chiral indices: if $(n - m)$ is a multiple of 3, the tube is metallic in the ZF picture (ZF-M), else it is semiconducting.

However, for small tubes, where the diameter (d) is less then 10 Å, the ZF/TB approximation fails to accurately describe the electronic properties of the tubes. Electronic application of these tubes requires the cognition of accurate theoretical band structures. In this work we present our results on the band structure of 20 different ZF-M SWCNTs (including 6 chiral tubes), based on first principles calculations of the actual nanotubes.

METHOD

We carried out first principles calculations on the density functional theory (DFT) level, using the 'Vienna *ab initio* simulation package' (VASP 4.6) [3], in the framework of the local density approximation and the projector augmented-wave method [4]. The geometries of all 20 SWCNTs were optimized as detailed in Ref. [5]. The electronic band structures were calculated with a high k-point sampling of 31 irreducible k-points ($1 \times 1 \times 61$ Monkhorst-Pack grid centered around, and including the Γ point) for most of the tubes. Even for chiral tubes, we used the highest k-point sampling possible.

CP723, *Electronic Properties of Synthetic Nanostructures*, edited by H. Kuzmany et al.
© 2004 American Institute of Physics 0-7354-0204-3/04/$22.00

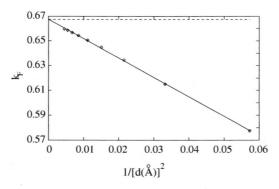

FIGURE 1. The k_F Fermi wave-vector of armchair tubes obtained from our DFT calculations (in units of π/a, where a is the lattice constant), as a function of the inverse diameter square. The straight solid line shows, that k_F keeps to $k_F(d \to \infty) = 2/3$ (dashed horizontal line) for large diameters, scaling as $1/d^2$.

To determine the gap for the $(7,4)$, $(8,2)$, $(6,3)$, and $(7,1)$ tubes, the use of a special method was necessary. For the zigzag tubes, the value of the gap is easy to determine very accurately, because the gap is located at $k = 0$. But in the case of chiral tubes, the exact location of the gap is not known. However, using an optimized charge density obtained with the k-point sampling listed in Table 1, a non-selfconsistent calculation can determine the band structure in any desired point of the Brillouin zone. With this method, we zoomed in on the band structure around the location of the gap, thus we were able to determine the gap with the same high accuracy as those of the zigzag tubes.

RESULTS AND DISCUSSION

The non-zero curvature of the rolled-up graphene sheet leads to the opening of a small band gap in the case of most ZF-M tubes, as it will be discussed below. However, in the case of armchair tubes, symmetry reasons forbid the opening of a gap; instead, curvature effects cause the Fermi wave-vector k_F to downshift from its ideal value. In ZF, the Fermi wave-vector for all armchair tubes is located exactly at $k_F(d \to \infty) = 2\pi/3$ (in units of inverse translational lattice constant of the given nanotube). For large diameters, this is indeed the case, but as the diameter decreases, this position slightly shifts down from $k_F(d \to \infty)$, proportional to $1/d^2$ (Figure 1). This diameter dependence is exactly what is expected even in TB [6].

Our results for the gaps are summarized in Table 1. Curvature leads to the opening of a small band gap for some of the non-armchair ZF-M tubes (see Table 1 and Figure 2), in agreement with previous results [6, 7, 8, 9]. This small gap appears already in the starting ZF/TB-predicted geometry, but its value is further modified in the optimized geometry. In particular, the $(9,0)$, $(12,0)$, $(15,0)$, $(18,0)$, $(7,4)$, $(8,2)$, $(6,3)$, and $(7,1)$ tubes prove to be small gap semiconductors, the latter $(7,1)$ being the one with the largest gap among all of these tubes. The band structures of these four chiral tubes,

TABLE 1. DFT band gaps for ZF-M tubes (excluding armchairs) compared to TB gaps. n, m: chiral indices, N_{at}: number of atoms in unit cell, θ: chirality angle d: tube diameter according to our DFT optimization, N_k: number of irreducible k-points in the band structure calculation, Δ_{TB}^{ZF}: band gaps expected from ZF/TB, Δ_0 and Δ: calculated values for the band gaps obtained from our DFT calculations in the starting ZF/TB-predicted geometry, and in the optimized geometry, respectively.

n	m	N_{at}	θ	$d(\text{Å})$	N_k	$\Delta_{TB}^{ZF}(eV)$	$\Delta_0(eV)$	$\Delta(eV)$
4	1	28	10.9	3.73	31	0.00	0.00	0.00
6	0	24	0	4.79	61	0.00	0.00	0.00
5	2	52	16.1	4.97	16	0.00		0.00
7	1	76	6.6	5.97	31	0.00	0.115	0.138
6	3	84	19.1	6.26	31	0.00		0.056
9	0	36	0	7.08	31	0.00	0.085	0.096
8	2	56	10.9	7.21	16	0.00	0.180	0.069
7	4	124	21.1	7.58	16	0.00	0.066	0.067
12	0	48	0	9.40	31	0.00	0.050	0.040
15	0	60	0	11.73	31	0.00	0.037	0.023
18	0	72	0	14.05	31	0.00		0.013

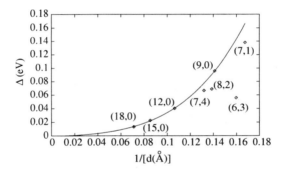

FIGURE 2. DFT calculated band gaps for the ZF-M SWCNTs, plotted against the inverse diameter. The solid line shows that a $\Delta = C_2/d^2 + C_4/d^4$ function fits well to the DFT calculation for the zigzag tubes with $C_2 = 1.99\ eV\text{Å}^2$ and $C_4 = 140.9\ eV\text{Å}^4$.

which have not been examined by DFT methods in the literature so far, are presented in Ref. [10].

The opening of this small gap follows already from simple π-electron calculations, which say, that the gap of the ZF-M zigzag tubes is proportional to $1/d^2$, and the gaps of chiral tubes fall below this curve. However, our calculations yield a different diameter dependence for the gap. A log-log scale plot of gap *vs.* diameter suggests, that the gap is approximately proportional to $1/d^3$; this is also what can be seen in the DFT work of Dubay and Kresse for the gap of the $(15,0)$ and $(18,0)$ tubes [6].

This diameter dependence is surprising however. The effects of the curvature on the matrix elements of the Hamiltonian can be generally written as a power series of the inverse diameter, containing only even power terms. This can be resolved by fitting the calculated gaps with a combination of different, but only even power terms; the simplest possible combination is a fit of $\Delta = C_2/d^2 + C_4/d^4$. This fit results in $C_2 = 1.99 \, eV\text{Å}^2$ and $C_4 = 140.9 \, eV\text{Å}^4$. Figure 2 shows, that this curve fits well to the DFT calculation for the zigzag tubes.

However, STM measurements show, that the band gaps of the $(9,0)$, $(12,0)$, and $(15,0)$ tubes are dominantly $\sim 1/d^2$[11], and higher order corrections are small[10]. While the presence of a $1/d^4$ correction term for small diameters is not surprising, this correction should obviously be smaller, than what was obtained in our DFT calculations. A well known shortcoming of DFT methods, is that DFT underestimates the absolute values of the band gaps. In many cases, simply scaling the DFT obtained gaps with an appropriate multiplicative scaling factor yields very accurate band gaps. On the other hand, such a scaling would leave the diameter dependence unchanged. It was recently shown, that the inclusion of many-electron (ME) corrections in DFT leads to very accurate absolute values for the optical gap of a few small diameter tubes [12, 13]. For example, comparing the one-electron band gap and the ME corrected band gap in Ref. [13], the gaps of the $(5,0)$ and $(8,0)$ tubes are increased by a factor of 1.15 and 3, respectively. This shows, that the ME correction cannot be described with a universal scaling constant, but instead, it depends strongly on the diameter. This suggests, that by including the necessary ME corrections in our presented DFT calculations, both the absolute values, and the diameter dependence of the band gap may be corrected.

ACKNOWLEDGMENTS

This work was supported by the OTKA T038014 and the OTKA T043685 grants in Hungary. The calculations were performed on the Schroedinger II cluster at the University of Vienna.

REFERENCES

1. R. Saito, G. Dresselhaus, and M. S. Dresselhaus, *Physical Properties of Carbon Nanotubes* (Imperial College Press, 1998).
2. S. Reich, C. Thomsen, and J. Maultzsch, *Carbon Nanotubes* (Wiley-VCH Verlag GmbH & Co. KGaA, 2004).
3. G. Kresse and J. Furthmüller, Comput. Mater. Sci. **6**, 15 (1996).
4. G. Kresse and D. Joubert, Phys. Rev. B **59**, 1758 (1999).
5. J. Kürti, V. Zólyomi, M. Kertesz, and G. Sun, New J. Phys. **5**, 125 (2003).
6. O. Dubay and G. Kresse, submitted to J. Chem. Phys.
7. X. Blase, L. X. Benedict, E. L. Shirley, and S. G. Louie, Phys. Rev. Lett **72**, 1878 (1994).
8. K. Kanamitsu and S. Saito, J. Phys. Soc. Jap **71**, 483 (2002).
9. S. Reich, C. Thomsen, and P. Ordejón, Phys. Rev. B **65**, 155411 (2002).
10. V. Zólyomi and J. Kürti, submitted to Phys. Rev. B.
11. M. Ouyang, J. L. Huang, C. L. Cheung, and C. M. Lieber, Science **292**, 702 (2001).
12. A. G. Marinopoulos, L. Reining, A. Rubio, and N. Vast, Phys. Rev. Lett. **91**, 046402 (2003).
13. C. D. Spataru, S. Ismail-Beigi, L. X. Benedict, and S. G. Louie, Phys. Rev. Lett. **92**, 077402 (2004).

Stabilizing Y-junctions and ring structures through nitrogen substitution

A. C. M. Carvalho and M. C. dos Santos

Instituto de Física Gleb Wataghin, DFMC, UNICAMP
CP 6165, 13081 - 970, Campinas-SP, Brazil

Abstract. In this work we theoretically investigate the conformational structure of nanotube junctions and bent tubules. Geometry optimizations were performed through the semi-empirical quantum chemical Parametric Method 3 technique. The defective regions of junctions and bends were built including five-, seven- and eight-membered rings in the otherwise hexagonal network of carbon bonds. In order to reduce the stress caused by the curvature, a chemical doping through nitrogen substitution is proposed. The energy associated to nitrogen incorporation was obtained. Results are consistent with the shortening of bonds within the junctions and bends and an increased chemical stability of the defects.

INTRODUCTION

Three-way or Y nanotube junctions have attracted the interest of researchers due to their unusual geometry and unique electronic properties. Experimental and theoretical studies have shown that electrical transport across these junctions can be nonlinear, with i x V characteristic curves exhibiting rectification properties. It was suggested that these systems are possible candidates for use in nanoscopic three-point transistors [1-3]. A synthetic method to systematically obtain these junctions has not yet been developed.

The creation of Y-junctions and bent structures involves the introduction of topological defects, for instance a pair of pentagon-heptagon, into the hexagonal network of carbon bonds [4]. In order to obtain more relaxed structures, defects formed by eight-membered rings or exclusively by heptagons have been proposed [5]. Square-shaped nanomaterials with large angle bends were experimentally observed in a hybrid multiwall nanostructures composed by carbon, boron, and nitrogen [6]. Also, bamboo-shaped nitrogen doped carbon nanotubes were synthesized [7]. Transmission electron microscopy and Electron energy loss spectroscopy characterizations are consistent with a high nitrogen concentration in the curved regions of the samples. These facts suggest that the presence of heteroatoms is important to the creation of nanostructures with accentuated curvature.

In the present work we report a semi-empirical quantum chemical study on nanotube-based Y-junctions and bent structures. We analyze the role played by nitrogen doping in the stability of these molecular systems. Our results indicate that substitutional nitrogen on pentagonal rings increase the chemical stability of defective and curved structures. This is in agreement with our previous calculations on small

CP723, *Electronic Properties of Synthetic Nanostructures*, edited by H. Kuzmany et al.
© 2004 American Institute of Physics 0-7354-0204-3/04/$22.00

azafullerenes [8]. In the following section we describe the model systems and the calculation methods. Next we present a discussion of the results. A final section contains the conclusions.

METHODOLOGY

The geometry of tubular, bent and branched structures composed by carbon were fully optimized through the semi-empirical quantum chemical method Parametric Method 3 (PM3) [9]. PM3 is a very reliable method to predict molecular geometries and heats of formation of carbon materials. It is also a very fast computational method when compared to *ab initio* techniques - clusters typically contained 100 to 500 atoms. Tube ends were saturated with hydrogen atoms or an appropriate fullerene. We focused in the most symmetrical zigzag and armchair nanotubes. The quantum chemical package GAMESS [10] was used.

These clusters were then nitrogen-doped and the geometries were re-optimized. Nitrogen atoms were randomly placed substituting carbons at given concentrations. For these substitutions, we adopted the following criteria: (i) adjacent atoms should not be substituted, since N_2 molecule is a much more stable configuration for the nitrogen atom than the inclusion in the carbon system; (ii) the substitution of even number of atoms is preferable because a closed shell system is formed. The substitutions in the branched and curved structures were not random, instead sites were chosen in order to give the most symmetrical molecules. The energy associated to nitrogen incorporation was calculated as the difference in formation enthalpy of N-doped and pure carbon systems divided by the number of nitrogens.

RESULTS AND DISCUSSION

We started by the computation of optimum geometries and respective formation enthalpies of tubular structures. The random replacement of carbon by nitrogen did not produce bends in nanotubes. The energy associated to nitrogen incorporation depends strongly on the tube helicity and diameter. N-doped zigzag tubes were found to be more stable (the formation enthalpy decreases) than armchair tubes of similar diameter, while the stressed small diameter tubes are more easily doped by nitrogen than the large diameter tubes, as shown in Fig. 1. This is consistent with the results reported on nitrogen-doping of graphene [11]: at small N concentration the formation enthalpy of N-doped graphene sharply increases due to the increased number of electrons compared to the pure carbon system.

The distorted bent and branched structures have defective rings and the energy associated to N incorporation depends on the substitution site. Substitution of nitrogen in pentagonal rings is more energetically favorable than the substitution in the hexagonal or higher order rings. Junctions and bends of zigzag tubes presented the largest decrease in energy when doped. Defective structures of armchair tubes presented a slight increase in formation enthalpy though the molecular structure improved, as discussed below.

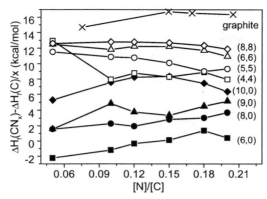

FIGURE 1. Relative heats of formation of CN_x nanotubes divided by the number of substituted N atoms as a function of the concentration [N]/[C]. Open (full) symbols are for armchair (zigzag) nanotubes. Crosses represent relative heats of formation of N-substituted graphite (Ref. 11).

The geometry of carbon nanotubes has not yet been experimentally measured. Currently accepted bond lengths are of the order of 1.43 Å. We adopted this value as a parameter to compare geometries. Bonds much larger or much shorter than 1.43 Å in defective regions indicate poor structural stability.

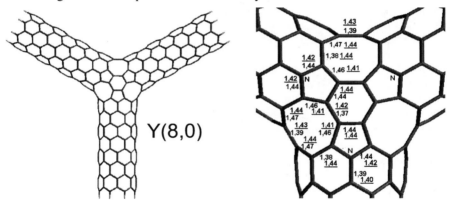

FIGURE 2. Fully relaxed nanojunction with (8,0) tubes forming all three arms. The structure to the right is a close-up of the junction, where the bond lengths of some bonds are shown, in Angstrom. Underlined values are the calculated bond lengths obtained on nitrogen substitution.

Figures 2 and 3 show an Y-branched (8,0) junction ($Y_{(8,0)}$) and a (4,4) bent nanotube ($D_{(4,4)}$), respectively. Structures are composed by nanotubes joined via six pentagons and six octagons (Fig. 2) and four pentagons and four heptagons (Fig. 3), without changing the tube chirality. The conformations of all-carbon $Y_{(8,0)}$ and $D_{(4,4)}$ molecules present a high degree of bond alternation, especially in the pentagonal and heptagonal rings. Some bonds have been quite stretched due to the curvature, reaching values of 1.48 Å. This is to be avoided in a junction because bond alternation produces localized states and stressed bonds are easier to break. Upon nitrogen doping, the degree of bond alternation decreases and the bonds as a whole present lengths closer to 1.43 Å, as indicated by the underlined bonds in Figs. 2 and 3. It is worth mentioning that the relaxation produced by nitrogen substitution is not restricted to the substituted ring, rather it extends over the junction.

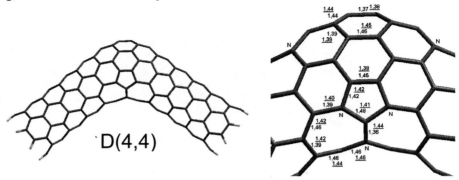

FIGURE 3. Optimized structure of a joint connecting two (4,4) armchair nanotubes. The scheme to the right shows a close-up of the junction. Bond lengths in the defect region are shown, in Angstrom. Underlined bond lengths are the optimized values calculated upon nitrogen substitution.

In chemical terms, this effect is explained by the polarity of the C-N bonds, which causes these bonds to be shorter than C-C bonds, and by the emergence of the nitrogen lone-pair, which provides a more pyramidal arrangement of bonds. It decreases the stress introduced by the curvature. Also, as nitrogen has one extra electron, the occupation of virtual states that are localized in the junction spreads the electronic density outside the substituted rings.

We reported results from semi-empirical calculations on nitrogen doping of strait and curved nanotube systems. The energy associated to nitrogen incorporation depends strongly upon the tube helicity and diameter. Substitution in hexagonal rings does not produce bends in strait tubes. However, experimental studies on carbon-nitride nanotubes suggested that the presence of the heteroatom induces the formation of bends. Nitrogen incorporation is shown to reduce the stress caused by the presence of pentagonal and heptagonal rings in the curved systems.

The authors thank the computational support from CENAPAD. This work was supported in part by the Agencies CNPq and FAPESP.

REFERENCES

1. Z. Yao, H. W. Ch. Postma, L. Balents, C. Dekker, *Nature* **402**, 273 (1999).
2. C. Papadopoulos, A. Rakitn, J. Li, A. S. Vedeneev, J. M. Xu, *Phys. Rev. Lett.* **85**, 3476 (2000).
3. A.N. Andriotis, M. Menon, D. Srivastava, L. Chernozatonskii, *Appl. Phys. Lett.* **79**, 266 (2001).
4 B. C. Satishkumar *et al., Appl. Phys. Lett.* **77**, 2530 (2000).
5. M. Menon, D. Srivastava, *J. Mat. Res.* **14**, 2357 (1998).
6. M.P. Johansson *et al., Appl. Phys. Lett.* **76**, 825 (2000).
7. R. Droppa Jr. *et al., Phys. Rev. B* **69**, 045405 (2004).
8. D. Schultz, R. Droppa, F. Alvarez, and M.C. dos Santos, *Phys. Rev. Lett.* **90**, 015501 (2003).
9. J. J. P. Stewart, *J. Comp. Chem.* **10**, 209 (1989).
10. M.W. Schmidt *et al., J. Comput. Chem.* **14**, 1347 (1993).
11. M.C. dos Santos and F. Alvarez, *Phys. Rev. B* **58**, 13918 (1998).

Sticking Effect of Carbon Nanotube Y-Junction Branches

L. A. Chernozatonskii and I. V. Ponomareva

Institute of Biochemical Physics, Russian Academy of Sciences, Moscow, 119991 Russia

Abstract. The sticking together of branches of single and double walled carbon nanotube Y-junctions is predicted to occur when the branches approach to each other under the action of a load which is applied to their ends.

INTRODUCTION

The interest in the multy-terminal junctions was stimulated primarily by the possibility of macroscopic synthesis of carbon nanotubes with Y junctions and by observations of the nonlinear current–voltage characteristics of such junctions [1-3]. However, mechanical characteristics of the Y junctions of carbon nanotubes were studied neither theoretically nor experimentally.

We have simulated by the molecular dynamics method the Y junctions of single- and double- wall carbon nanotubes (SJ and DJ) with a length of ~10nm and an acute angle between the branches and their behavior under the action of an external load. When a load is applied to the Y junction branch ends, the distance between the branches decreases to 3.4 Å. It was found that weak van der Waals forces between the branches are sufficient to keep them parallel upon unloading. This effect indirectly confirms the prediction, according to which nanomaterials will encounter the problem of attaching as a result of intermolecular interactions, which may lead to effects that are impossible on a macroscopic scale.

FORMULATION OF THE PROBLEM

Molecular dynamics (MD) simulation was carried out using the Brenner potential [4] for covalent bonds between carbon atoms and the Lennard-Jones potential for long-range interactions. It should be noted that the Brenner potential was used for the description of dynamics [5] and mechanical properties [6-8] of carbon nanotubes, because this potential allows the simulation of systems with large numbers of atoms ($\sim 10^4$).

Two types of Y junctions of the armchair SWCNs and DWCNs were considered: (I) comprising (20,20) SWCN stem branched into (13,13) branches Figs. 1(a)(b), (II) double wall nanotube junction containing from (15,15)@(20,20) nanotube stem and (8,8)@(13,13)} branches - Figs. 1(c)(d). Both structures possessed the similar on

CP723, *Electronic Properties of Synthetic Nanostructures*, edited by H. Kuzmany et al.

figuration of topological defects, representing six heptagons situated in the branching region, and with similar branch lengths about 10nm. Optimization of the geometry by MD modeling showed that this topology corresponds to an angle of ~23° between the arms, in good agreement with the observed geometry of Y junctions [1]. Prior to MD simulation, each Y junction was placed between two graphite planes parallel to the stem so that the initial distance from each plane nearest carbon atom of the nanotube branch was 8 A°. The branch moving procedure was expounded more detailed in [8], where the behavior of Y-junctions of short (10nm) and long (21nm) single-wall nanotube branches were considered.

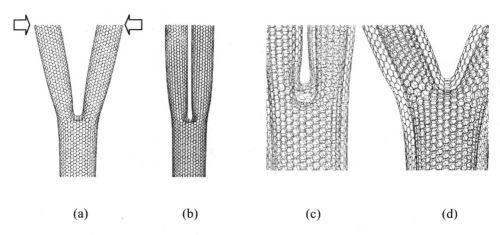

(a)   (b)   (c)   (d)

FIGURE 1. (a) SWCN Y-junction at the beginning of simulation, (b) this "fork" with stick together branches is in metastable state after unloading (up to 2000K). The DWNT Y-junction with (c) sticking branches, (d) opened after >1500K heating branches.

RESULTS AND DISCUSSION

In order to characterize the transition between these states, we (i) constructed the potential energy curve of the Y junction versus the distance between arms (Fig. 2) and (ii) studied the behavior of the Y junction in the states corresponding to the vicinity of the expected extremum on this curve. The total simulation time was ~30 ps.

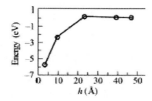

FIGURE 2. A plot of the potential energy of SJ versus distance between branches.

The results of simulations at 300K regime give existence of critical distance between arms (for SJ: 14.5– 23 Å), below which the arms spontaneously approach each other even in the absence of the external force. The possibility of such sticking is caused by a decrease in the potential energy of the deformed junction (see Fig. 2).

After 27.5 ps, the distance between SJ arms decreases to 3.4 Å, which corresponds to a Y junction with parallel arms. Then, the graphitic planes were removed and the MD simulation of this system was continued for 25 ps at 300 K. Figure 1(b) shows the SJ shape after this procedure. The energy of this parallel configuration is 5.5 eV lower than that of the initial configuration with an angle of 23° between arms.

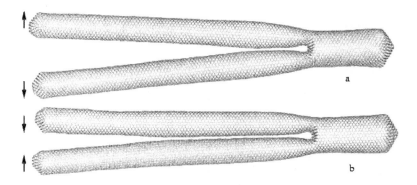

FIGURE 3. When the load of 10 nN is removed from the branches at an angle then (a) for $\theta > \theta_o$ the branches go away and vibrate, or (b) for $\theta < \theta_o$ they stick together by vdW forces.

We have observed critical branch connection angle of θ_o, depending on the branch length over which $(\theta > \theta_o)$ results to impossibility of the branch touching in normal room temperature condition (see (3a)). For the considered SJ structure these angle equals to $\sim 14°$. When we stop and unload branches with $\theta < \theta_o$ - starting from this moment van der Waals interaction acts like zipper-mechanism forcing branches to be stacked together –3b). So the Y-fork junction has at least two local energy minimums and transition between them provides pseudoplasticity that has been occurred in our simulation.

The similar picture of branch sticking are observed when we have investigated double-wall nanotube Y-junction (Fig.1c) using the same MD procedure. It was established that the DJ unloaded before the distance between their branches approach the threshold value $(\theta < \theta_o')$, angle θ_o' for DJ is smaller than the angle θ_o for SJ) execute oscillations also.

However the branches begin to open during the heat process. We have detected the effect of branch opening (Fig.1 (d)) between 1500-1800° C heating during the evolution time.

CONCLUSION

Thus, in contrast to the macroscopic systems where the Y-shaped configuration can behave as a spring with the ends always diverging upon unloading, whereby the system returns to the initial state. The molecular interaction between closely spaced nanotubes is an important factor in the analogous nanosystem. The examples considered above show that the intermolecular attraction between the arms of a Y junction can prevail over the restoring elastic force caused by the deformation of the system. This accounts for the effect of coalescence that is impossible on the macroscopic level. The new state with parallel arms bonded by a weak intermolecular interaction is stable up to a temperature of 2000 K and is energetically more favorable than the state of Y junction with diverging arms. These results show that the van der Waals interaction plays a key role in the nanotube Y junctions with closely spaced arms. It was established that the Y junctions unloaded before the distance between their branches (~10 nm long) approach the threshold value execute oscillations (like a tuning fork) at a frequency of ~100 GHz.

The effects of sticking at room temperature and temperature opening of "fork" multy-layered carbon nanotube Y-junctions may find use in different nanomechanisms and nanoelectric circuits.

ACKNOWLEDGMENTS

The work was supported by the Russion Federal Program "Currently Important Directions in Physics of Condensed Media" (Fullerenes and Atomic Clusters) and by the Federal Program "Low-Dimensional Quantum Structures".

REFERENCES

1. Papadopoulos C., Rakitin A., Li J., Vedeneev A.S., and Xu J.M., *Phys. Rev. Lett.* **85**, 3476-3479 (2000).
2. Satishkumar B. C., Thomas P. J., Govindraj A., and Rao C. N., *Appl. Phys. Letters* **77**, 2530-2533 (2000).
3. Andriotis A. N., Menon M., Srivastava D., and Chernozatonskii L., Phys. Rev. B **65**, 165416 (1-13) (2002).
4. Brenner D. W., Shenderova O. A., Harrison J. A., Stuart S. J., Ni B. and Sinnott S.B., *Phys.: Condens. Matter* **14** , 783-802 (2002).
5. Nardelli M. B., Yakobson B. I., and Berholic J., Phys. Rev. Lett. **81** , 4656-4659 (1998).
6. Roberston D. H., Brenner D. W., and Mintmire J. W., Phys. Rev. B **45** , 12592-12598 (1992).
7. Srivastava D., Wei C., and Cho K., Appl. Mech. Rev. (**2**), 215-221 (2003).
8. Chernozatonskii L. A., Ponomareva I. V., *JETP Lett.* 74, 467-470 (2003).

How (and why) twisting cycles make individual MWCNTs stiffer

A. DiCarlo*, M. Monteferrante†, P. Podio-Guidugli**, V. Sansalone‡ and L. Teresi*

*SMFM@DiS, Università "Roma Tre", Via Vito Volterra, 62 I-00146 Roma (Italy)
†Dip. Fisica E. Amaldi, Università "Roma Tre", Via della Vasca Navale, 84 I-00146 Roma (Italy)
**Dip. Ingegneria Civile, Università di Roma TorVergata, Via Politecnico, 1 I-00133 Roma (Italy)
‡ENEA Casaccia, Via Anguillarese, 301 I-00060 S. Maria di Galeria (Italy)

Abstract. Torsion of MWCNTs entails bumpy nanoscopic interwall phenomena, which gradually enhance the microscopic interwall coupling. This effect is likely to be confined at the ends of the suspended portions of the CNT, the structural changes required to link the outer wall to the inner ones being probably triggered by a complex interaction with the metal deposited onto the nanotube.

INTRODUCTION

Our modelling effort is motivated by results obtained by P.A. Williams et al. [1, 2], presented at IWEPNM 2003 by S. Washburn [3]. These authors fabricated nanometer-scale mechanical devices ("paddle" oscillators) incorporating multiwalled carbon nanotubes as torsional spring elements (Fig. 1a), and developed a method for measuring the torsional stiffness of the nanotubes (Fig. 1b). Arc grown MWCNTs were dispersed onto Si wafers having 500 nm of oxide. Large metal pads were patterned by electron-beam lithography over the ends of each MWCNT, and a stripe of metal over its center to form the paddle. The metal was thermally evaporated, 15 nm of Cr followed by 100 nm of Au. The oxide was etched such that the paddles were completely undercut, but the larger pads pinning down the MWCNT ends were not. Measurements revealed a remarkable stiffening behaviour of MWCNTs: after nearly 500 repetitions of a twisting cycle of

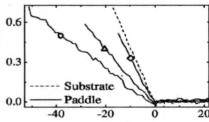

FIGURE 1. a) Paddle oscillator; b) Force-displacement traces on the substrate and on the paddle with three different eccentricities: photodiode signal (nA) vs. piezo-displacement (nm); the slope of the substrate trace yields the apparent overall stiffness of the AFM cantilever, equal to −41 pA/nm (from [3]).

CP723, *Electronic Properties of Synthetic Nanostructures*, edited by H. Kuzmany et al.
© 2004 American Institute of Physics 0-7354-0204-3/04/$22.00

small amplitude (with estimated in-wall strains less than 0.01), the stiffness of an individual MWCNT saturated to a value 12 times larger than its initial value (see Fig. 2a).

PRIMARY EXPERIMENTAL EVIDENCE

Measurements were performed [2] with an atomic force microscope (AFM) mounted inside the chamber of a scanning electron microscope. The AFM was used both to apply forces to paddles and to measure their displacement. All tested devices exhibited a sizeable increase in torsional stiffness, roughly correlated with the total number of previous twisting cycles. Torsional stiffness was measured with an estimated accuracy of $\sim 15\%$. Testing a two-paddle device, it was checked that MWCNTs twist uniformly along their length—at least in their initial conditions. The bending stiffness and the resonance frequency of the AFM cantilever were measured before and after each experimental run: no change was detected. In conclusion, the collected results are quantitatively consistent with the hypothesis that, while in the first twisting cycle only the outer wall is strained, successive cycles induce an increasing interwall coupling, until all the walls are strained and contribute to the overall torsional stiffness as if the nanotube were a solid rod. A noteworthy detail in Fig. 1b is the apparent correlation between torsion and *noise*: the steeper the force-displacement trace, the smoother; that is to say, less torsion, less noise. This fact suggests the hypothesis we advance in the abstract.

CONTINUUM MODELLING AND COMPUTATIONS

We assume that each wall behaves as a linearly elastic beam, interacting linearly with the neighbouring walls during each twisting cycle; and that, moreover, the interwall coupling *evolves with cycles*. We choose not to detail how the model parameters evolve within an individual cycle; consequently, we set up a *discrete-time problem* where, at any given axial location, each relevant field is interpreted as its total variation over a cycle (the **total variation over a cycle** of a quantity q is defined to be

$$\lfloor q \rceil := \int_{cycle} |\dot{q}| \, dt, \tag{1}$$

a superposed dot denoting time differentiation). As to space dependence, we take all quantities to be real-valued *even* distributions (regular or singular) over the closed interval $[-L, +L]$, the coordinates 0 and $\mp L$ labelling, respectively, the paddle and the anchored ends. For each cycle and each *wall*, the following equations prevail, all to be intended in the sense of distributions:

$$\lfloor T' + \mu_- + \mu_+ \rceil = 0, \quad \text{with } \lfloor T \rceil(-L_-) = \lfloor T \rceil(L_+) = 0 \quad \text{(torque balance)}, \tag{2}$$

$$\lfloor T \rceil = k \lfloor \vartheta \rceil' \qquad \qquad \text{(constitutive assignment)}, \tag{3}$$

where ϑ is the **rotation**, T the **twisting torque**, and a prime denote differentiation with respect to the axis coordinate. The *measures* μ_-, μ_+ represent the **interwall**

torques exerted on the wall under consideration by the preceding and the following one, respectively. For the outer wall, μ_- is the torque applied by the paddle and the anchors: three Dirac measures concentrated in 0 and $\mp L$, the first a control, the other two reactions to the constraint $\vartheta(\mp L) = 0$; for the inmost wall, $\mu_+ = 0$. The **wall stiffness** $k = 2\pi G r^3 h$ is *independent* of cycles: r is the wall radius, $h = 0.34$ nm the wall "thickness", taken equal to the interwall distance, and $G = 430$ GPa the in-wall shear modulus [4]. For each cycle and each *interwall*, we assume:

$$\lfloor \mu \rceil = C \lfloor \vartheta_- - \vartheta_+ \rceil \qquad \text{(interwall drag),} \qquad (4)$$
$$\lfloor C \rceil = \rho C \qquad \text{(coupling growth-law),} \qquad (5)$$

where C is the **interwall coupling**, ρ the **growth rate**, and $-$, $+$ signs refer to the immediately adjacent walls, respectively. The torque μ is defined as acted by the outward wall $(-)$ on the inward wall $(+)$; hence, the value prescribed by (4) enters the balance (2) for the outward wall as $-\mu_+$ and the one for the inward wall as μ_-. For simplicity, the growth rate ρ is assumed to be cycle-independent, and a space constant. Two different assignments for the *initial interwall coupling* C_0 will be introduced below.

The model above has been computed for a MWNT composed of 30 walls, with outer radius 16 nm, inner radius 6.1 nm, and suspended length $L = 260$ nm (these data presumably correspond to Paddle D in [3]). For each cycle, the **overall torsional stiffness** is evaluated as $\kappa = \lfloor \text{paddle torque} \rceil / \lfloor \text{paddle rotation} \rceil$; it ranges between $1.4 \cdot 10^{-14}$ Nm and $17 \cdot 10^{-14}$ Nm. The $\lfloor \text{mean torsion} \rceil$ of a wall is defined as $\lfloor \vartheta(0) - \vartheta(L) \rceil / L$. Fig. 2a compares our main results with the experimental data in [3]; the paddle stiffness corresponding to computed values of κ is evaluated using eq. (1) in [3], where the overall bending stiffness of the AFM cantilever $K_c = 1.1$ N/m and the eccentricity equals 650 (430) nm before (after) the 330th cycle. The crossed line corresponds to $\rho = 0.04$ and C_0 uniform over $[-L, +L]$, with line density 1 nN on all interwalls. The circled

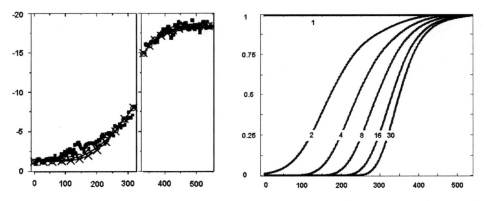

FIGURE 2. a) Progressive stiffening: paddle stiffness (pA/nm: photodiode signal intensity per unit piezo-dispacement) vs. number of cycles; the break in the experimental data [3] (squares) after about 330 cycles is due to a shift of the AFM tip, which is accounted for also by our calculations (circles and crosses), with obvious uncertainties; b) Relative $\lfloor$ mean torsion $\rceil$ of individual walls (w.r.t. the outer wall) vs. number of cycles; only the evolution corresponding to the concentrated interwall coupling is shown.

line corresponds to $\rho = 0.03$ and C_0 a triple of Dirac measures $(\gamma, 2\gamma, \gamma)$ concentrated respectively in $-L$, 0, $+L$ on each interwall; the mass $|\gamma|$ equals $3.3 \cdot 10^{-18}$ Nm on the outmost interwall and decays by a factor of 0.9 at each inward step. Both hypotheses fit the experimental data quite well, given the gaps and uncertainties in experimental data and the small number of adjustable parameters in our coarse model. Interestingly, we settled down for the simplest growth-law (5) after more complicate alternatives proved less effective to get a good fit with the data. In conclusion, our continuum model provides a mathematically precise description of the experimental evidence, thus enabling to test different constitutive assumptions. Our results, while not explaining *why* things go that way, do help extracting information out of available data and planning new experiments as well as more detailed and better focussed (atomistic, *ab initio*) simulations.

MORE (PUZZLING) EXPERIMENTAL EVIDENCE

Other experimentalists [5, 6] have manipulated MWCNTs very similar to the ones we referred to in previous sections, to fabricate nanoelectromechanical devices where an individual MWCNT, suitably engineered, provides a rotary bearing in which the sliding occurs between different walls. The interwall friction is extremely small and does *not* increase during operation: beyond doubt, the mechanism mimicked by (5) is not in action. Interestingly, the most effective technique for producing a nanorotor seems to be mounting a metal plate on a MWCNT as in Fig. 1a, and then breaking the outer wall off the anchors by a few twisting cycles of large amplitude [5]. This fact prompts the conjecture that the interaction between the CNT and the metal is crucial in determining both the fragile behaviour under a few large twists and the ratcheting effect under many small ones. To the best of our knowledge, this interaction is far from being understood.

ACKNOWLEDGMENTS

Our work was supported by PRIN MMSM 02, GNFM-INdAM and CASPUR. We thank S. Washburn and S.J. Papadakis for providing us with unpublished information. We also acknowledge helpful discussions with many people, both in Kirchberg and elsewhere.

REFERENCES

1. P.A. Williams, S.J. Papadakis, A.M. Patel, M.R. Falvo, S. Washburn and R. Superfine, Appl. Phys. Lett. **82**, 805 (2003).
2. P.A. Williams, S.J. Papadakis, A.M. Patel, M.R. Falvo, S. Washburn and R. Superfine, Phys. Rev. Lett. **89**, 255502 (2002).
3. S.J. Papadakis, P.A. Williams, M.R. Falvo, R. Superfine and S. Washburn, in *Molecular Nanostructures* (H. Kuzmany et al., eds.), AIP Conf. Proc. **685**, 577 (2003).
4. J.P. Lu, Phys. Rev. Lett. **79**, 1297 (1997).
5. A.M. Fennimore, T.D. Yuzvinsky, Wei-Qiang Han, M.S. Fuhrer, J. Cumings and A. Zettl, Nature **424**, 408 (2003).
6. B. Bourlon, D.C. Glattli, B. Plaçais, J.M. Berroir, L. Forró and A. Bachtold, in *these Proceedings*.

The electronic structure of achiral nanotubes: a symmetry based treatment

Eugenio Di Donato*, Matteo Tommasini*, Chiara Castiglioni* and Giuseppe Zerbi*

*Dipartimento di Chimica, Materiali e Ingegneria Chimica "G. Natta", Politecnico di Milano, Piazza Leonardo da Vinci 32 – 20133 Milano – Italy

Abstract. Analytic expressions for the electronic band structure and wavefunctions of zigzag and armchair nanotubes of any diameter are obtained within the Hückel theory. The explicit use of the C_n cyclic symmetry of the translational unit cell of the nanotube allows to obtain closed expressions for both the band structure and the Bloch wavefunctions. These analytic results can be successfully employed to predict the number and the energy of the Van Hove Singularities (VHSs) and to obtain a graphical representation of the associated wavefunctions.

DISCUSSION

The idea of this work is to use cyclic symmetry to describe any achiral nanotube by using a unit cell whose dimension is smaller than the traditional translational cell. This approch allows to obtain a powerful tool to visualize electronic wavefunctions, thus helping to identify electronic transition coupled with Raman active phonons [1]. Moreover, we will outline differences between the representation originating from the traditional zone folding procedure, and our present approach.

In Fig. 1 we can see the unrolled honeycomb lattice of an armchair nanotube. The whole nanotube can be thought as obtained by joining n polyacene stripes, where the single polyacene is obtained by a repeated translation of the structural unit consisting of 4 atoms. The Hückel matrix, which describes a single polyacene stripe is written on the basis of the carbon $2p_z$ orbitals $|2p_z^{k,m}\rangle$ and it is a cyclic matrix. We describe the electrons as Bloch functions:

$$|B_{\theta_j}^k\rangle = \frac{1}{\sqrt{N}} \sum_{m=1}^{N} |2p_z^{k,m}\rangle \, e^{i\theta_j m} \qquad (1)$$

where the index k labels the atom in the structural cell $(k = 1 \ldots 4)$ and m represents the index of the polyacene translational unit; θ_j is the phase factor associated to the Bloch wavefuntion and it is related to the quasi wavevector along the nanotube axis. By expressing the Hamiltonian in the Bloch functions basis set, we obtain N independent blocks in the form:

$$\mathcal{H}_B^{jj} = \mathbf{a} + \mathbf{b}^\dagger e^{-i\theta_j} + \mathbf{b} e^{i\theta_j} \qquad (2)$$

CP723, *Electronic Properties of Synthetic Nanostructures*, edited by H. Kuzmany et al.
© 2004 American Institute of Physics 0-7354-0204-3/04/$22.00

where:

$$\theta_j = \frac{2\pi}{N}j; \qquad j = 1\ldots N; \qquad N \to \infty \tag{3}$$

and:

$$\mathbf{a} = \begin{pmatrix} \alpha & \beta & 0 & 0 \\ \beta & \alpha & \beta & 0 \\ 0 & \beta & \alpha & \beta \\ 0 & 0 & \beta & \alpha \end{pmatrix}; \qquad \mathbf{b} = \begin{pmatrix} 0 & 0 & 0 & 0 \\ \beta & 0 & 0 & 0 \\ 0 & 0 & 0 & \beta \\ 0 & 0 & 0 & 0 \end{pmatrix}$$

The whole nanotube is described by introducing the suitable interactions between the n acene chains. This requires to introduce of a new phase factor which will be labelled ϕ. This implies a further change of the basis set, namely:

$$|P^k(\theta_j, \phi_s)\rangle = \frac{1}{\sqrt{n}} \sum_{r=1}^{n} |B_{\theta_j}^{k,r}\rangle \, e^{i\phi_s r} \tag{4}$$

ϕ_s represents the phase difference in the wavefunction between sites which are related by a $2\pi/n$ rotation around the tube axis. It results to be quantized: $\phi_s = (2\pi/n)s$, $s \equiv \{1, 2, \ldots, n\}$ where n is the order of the cyclic group of the tube. θ and ϕ are directly related to the quasi momentum and the quasi angular momentum of the electron, respectively. Both (n,n) armchair and $(n,0)$ zigzag nanotubes posses a C_n axis. The final expression of the diagonal blocks of the Hückel matrix for armchair nanotubes in the $|P^k(\theta_j, \phi_s)\rangle$ basis set is:

$$\mathcal{H}_P = \begin{pmatrix} \alpha & \beta(1 + e^{-i\theta_j}) & 0 & \beta e^{-i\phi_s} \\ \beta(1 + e^{i\theta_j}) & \alpha & \beta & 0 \\ 0 & \beta & \alpha & \beta(1 + e^{i\theta_j}) \\ \beta e^{i\phi_s} & 0 & \beta(1 + e^{-i\theta_j}) & \alpha \end{pmatrix} \tag{5}$$

An even simpler form of the Hückel matrix is obtained if we introduce the local symmetry of the structural unit (the mirror plane interchanging the atoms 2 and 3, 1

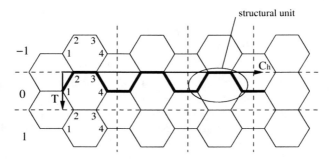

FIGURE 1. The unrolled honeycomb lattice of an armchair nanotube. In bold type the usual translational unit cell, among vertical dashed line the smaller cell which exploits the cyclic symmetry.

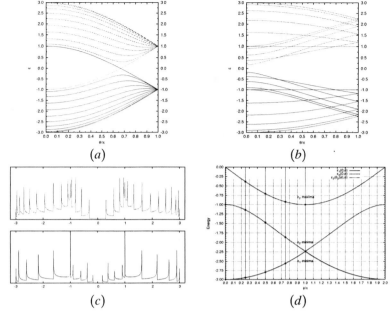

FIGURE 2. (a) Energy bands of the (10,10) nanotube; (b) Energy bands of the (10,0) nanotube; (c) Density of states of the (10,10) nanotube (top) and of the (10,0) nanotube (bottom); (d) graphical construction of the position of the VHSs for a generic armchair nanotube (a similar plot exists for zigzag tubes [1])

and 4, see Fig. 1). The transformed Hückel matrix results:

$$
\mathcal{H}_S = \begin{pmatrix}
-\cos\phi_s & -(1+e^{-i\theta_j}) & -i\sin\phi_s & 0 \\
-(1+e^{i\theta_j}) & -1 & 0 & 0 \\
i\sin\phi_s & 0 & \cos\phi_s & -(1+e^{-i\theta_j}) \\
0 & 0 & -(1+e^{i\theta_j}) & 1
\end{pmatrix}
\tag{6}
$$

where we have operated the usual substitutions $\alpha = 0$, $\beta = -1$ and the basis set change from $|P(\theta_j, \phi_s)^k\rangle$ to the symmetry adapted combination $|S'(\theta_j, \phi_s)\rangle$ [1]. The same treatment has been done for zigzag tubes, obtaining a $\mathcal{H}_S$ matrix showing a similar shape. The solution of the eigenvalue problem provides the band structure and the Bloch wavefunctions. We report in Fig 2a and Fig 2b the analytically determined energy bands for the (10,10) and the (10,0) nanonotubes. In Fig. 2c the DOS of the same tubes is reported where it is possible to distinguish the metallic and semi-conducting behaviors. These results are of course identical to those previously reported [2, 3]. Each nanotube has its own peculiar distribution of VHSs, which univocally identify the tube of a given diameter and chirality. Peaks appear in the DOS when the $\frac{\partial \varepsilon}{\partial \theta} = 0$, where $\varepsilon(\theta, \phi)$ is

---

[1] $|S^1\rangle = |P^1\rangle + |P^4\rangle$; $|S^2\rangle = |P^2\rangle + |P^3\rangle$; $|S^3\rangle = |P^1\rangle - |P^4\rangle$; $|S^4\rangle = |P^2\rangle - |P^3\rangle$

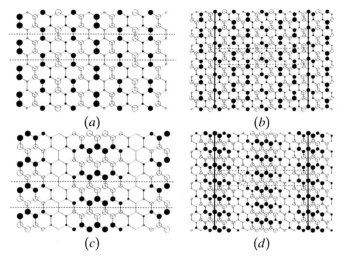

FIGURE 3. (a,c) E_2^y and E_4^y molecular orbitals symmetry adapted of (10,0) nanotube (b,d) E_2^y and E_4^y molecular orbitals of the (10,0) nanotube obtained with the zone folding procedure.

the energy band in the ε, θ plane [2]. The availability of the analytical expressions for ε, allows to obtain a function, $\varepsilon(\theta_0, \phi)$ [3] wich describes the position of maxima and minima of ε in the (ε, ϕ) plane, as reported Fig. 2d. For a given tube ϕ_s takes values which are determined by the n index of the tube. Fig. 2d can be used to determine (by the proper choice of the ϕ_s values) energy locations of the VHSs for any given armchair nanotube. A similar graph can be derived also for zigzag tubes [1] In Fig. 2d the procedure for the (8,8) tube is illustrated (the vertical solid lines identify the allowed ϕ_s values and thus the energy location of all the VHSs).

WAVE FUNCTIONS

The eigenvectors obtained by the diagonalization of the $\mathcal{H}_S$ matrix are the Bloch orbitals expressed in the symmetry adapted $|S^t(\theta_j, \phi_s)\rangle$ basis set. They can be immediately back-converted in the basis of the atomic $|2p_z^k\rangle$ orbitals of any structural cell. This is the necessary step for the graphical representation of the orbitals (see Fig. 3).

As an example we report in Fig. 3a, 3c a plot of the $|2p_z\rangle$ coefficients of the Bloch wavefunctions representing two different occupied states corresponding to the 2nd and

---

[2] $\varepsilon_{1,2} = -\left\{3 + 2\cos\theta \pm 2[(1+\cos\theta)(1+\cos\phi)]^{1/2}\right\}^{1/2}$

$\varepsilon_{3,4} = +\left\{3 + 2\cos\theta \pm 2[(1+\cos\theta)(1+\cos\phi)]^{1/2}\right\}^{1/2}$

[3] θ_0 is defined by: $\frac{\partial\varepsilon}{\partial\theta}(\theta_0,\phi) = 0$

4th VHS (*i.e.* E_2^y, E_4^y) of the (10,0) nanotube. These states are expressed by the choice $\theta_j = 0$, $\phi_s = (2\pi/10) \cdot 4$ (for E_2^y) and $\theta_j = 0$, $\phi_s = (2\pi/10) \cdot 1$ (for E_4^y). In Fig. 3b, 3d the alternative picture is reported, as it is obtained by following the zone folding procedure. This corresponds to the evaluation of the eigenfunctions obtained by diagonalization of the Hückel hamiltonian of graphene at peculiar **k** points in the first Brillouin zone. By expressing **k** as $\theta_1 \mathbf{b}_1 + \theta_2 \mathbf{b}_2$ the zone folding procedure for zigzag tubes gives $\theta_1 = \phi_s$ and $\theta_2 = (\phi_s - \theta_j)/2$. It can be immediately recognized that the orbitals sketched in Fig. 3a, 3c do fully reflect the symmetry of the specific nanotube, in particular the symmetry planes orthogonal to the tube axis. On the contrary, the sketches in Fig. 3b, 3d take onto account just the symmetry of the graphene sheet thus not reflecting the full symmetry of the tube; this can be seen in Fig. 3b, 3d when considering the highlighted coefficients.

CONCLUSIONS

Using the cyclic symmetry we have obtained analytical expressions which allow to describe the energy bands and DOS of any achiral nanotubes. We have also worked out a simple way to identify the number and the energy position of the VHSs. In this way a graphical representation of symmetry adapted wave functions is proposed, which allows to visualize the molecular orbitals. This is a useful tool for the identification of specific electronic transitions coupled with Raman active phonons [1].

REFERENCES

1. M. Tommasini, E. Di Donato, C. Castiglioni, G. Zerbi, in preparation.
2. R. Saito, M. Fujita, M. Dresselhaus, M. S. Dresselhaus, Phys. Rev. B, **46**, 1804 (1992)
3. R. Saito and H. Kataura: Carbon Nanotubes, Topics Appl. Phys. **80**, 213-246 (2001)

MD Simulations of Catalytic Carbon Nanotube Growth: Important Features of the Metal-Carbon Interactions

Feng Ding[*], Arne Rosén and Kim Bolton

Experimental Physics, School of Physics and Engineering Physics, Göteborg University and Chalmers University of Technology, SE-412 96, Göteborg, Sweden

Abstract: The nucleation of single-walled carbon nanotubes (SWCNTs) on catalyst cluster surfaces was studied by molecular dynamics (MD) simulation. It was found that SWCNTs grow from graphitic islands that precipitate on the cluster surface, and that the weak interaction between the catalyst metal particle and the bond-saturated island atoms (i.e., the carbon atoms at the center of the island that are bonded to three other carbon atoms) plays a critical role in the lifting of the carbon island off the particle surface.

INTRODUCTION

Much effort has been put into understanding the growth mechanism of carbon nanotubes (CNTs) since their discovery by Iijima in 1991 [1]. Understanding the growth of this one-dimensional (1D) structure is important for controlling and optimizing CNT growth, and may also help in improving the production of other 1D structures such as boron-nitride nanotubes [2]. These nanotubes have been produced using similar experimental methods to those used for CNT growth.

The layered structure that is found in graphite and CNTs is due to the sp^2 hybridization of C atoms. Due to this hybridization, each atom has three strong in-plane σ bonds and one weak out-of-plane π bond. For example, in graphite, the in-plane bonds between neighboring pairs of carbon atoms are about 6 eV and the out-of-plane bonds between atoms in different graphitic layers are about 0.01 eV – more than two orders of magnitude weaker. Similarly, all atoms that form part of a CNT wall are bonded, via strong σ bonds, to three other C atoms. However, C atoms at the open edge of CNTs and that are critically involved in CNT growth, are joined to at most two other C atoms. In order to illustrate the difference between the central C atoms (that form part of the nanotube wall) and those at the CNT end, 7 hexagons from a graphite sheet are shown in Figure 1. The central C atoms are σ bonded to three other carbon atoms and are therefore bond-saturated (C_S). These atoms are chemically inert and interact weakly with other atoms (that are not part of the carbon layer). In contrast, the edge atoms in Figure 1 are bonded to at most two other C atoms and are thus bond-unsaturated (C_{Uns}). These atoms are chemically

*Corresponding author: Email-fengding@fy.chalmers.se
Tel- +46-31-7723294
On leave from Department of Physics, Qufu Normal University, Qufu, 273165, Shandong, P. R. China

CP723, *Electronic Properties of Synthetic Nanostructures*, edited by H. Kuzmany et al.
© 2004 American Institute of Physics 0-7354-0204-3/04/$22.00

reactive and have strong interactions with other atoms.

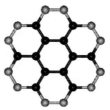

FIGURE 1. The bond saturated carbon atoms (the central black atoms) and the bond unsaturated carbon atoms (edge gray atoms) in a graphene island.

In this contribution we focus on the key role played by the weak interactions between C_S atoms and the metal catalyst particle in SWCNT nucleation.

MOLECULAR DYNAMICS METHODS

Previous computational studies of catalyzed SWCNT growth have included static calculations [3], MD simulations based on analytic force fields [4] or density functional theory (DFT) based Car-Parrinello simulations [5]. While the Car-Parrinello approach has the advantage of obtaining forces directly from electronic structure theory, it suffers from the disadvantage that it is computationally very expensive.

The MD simulation discussed here is based on an analytic potential energy surface (PES) that has been used to study the iron (Fe) catalyzed nucleation of SWCNTs [6]. Details of the PES and the SWCNT nucleation mechanism have been presented elsewhere [6]. For the present contribution it is necessary to note that the C atoms that are dissolved in, or that have precipitated from, the Fe_{100} cluster are treated differently in the PES description. In addition, two types of precipitated C atoms are distinguished. These are the C_S and C_{Uns} atoms and, as discussed above, the interaction between C_S and Fe atoms is weak, whereas the interaction between the C_{Uns} and Fe atoms is much stronger. These interaction energies were fit to DFT adsorption energies, which yield a C_S-Fe bond energy of 0.14 eV [7], which is about an order of magnitude weaker than the C_{Uns}-Fe bond energy of 1.5 eV [8]. Similarly to previous studies, the trajectories were initialized by thermalizing the pure Fe cluster to the desired temperature (1000 K in this study), and C atoms were subsequently added at a rate of one every 40 ps into the central part of the Fe cluster.

RESULTS AND DISCUSSION

The first (about 20-30) C atoms that are added to the Fe_{100} cluster dissolve in the metal, since the cluster is unsaturated in C. Continued addition of C atoms results in precipitation of these atoms on the cluster surface, but these atoms are unstable and can dissolve back into the cluster. Only when the cluster is supersaturated in C are there a sufficient number of precipitated C atoms to form C strings and small polygons on the surface. These polygons grow into small graphitic islands that, as shown in

Figure 2a-c for a temperature of 1000 K, nucleate graphitic caps that grow into SWCNTs.

The weak C_S-Fe interactions play a key role when the graphitic island (Figure 2a) lifts off the surface to form the cap (Figure 2b). To investigate the importance of these interactions we repeat the simulations (with the same temperature and rate of C addition) but using C_S-Fe bond strengths that vary between 0.14 and 1 eV. It must be noted that the C_S-Fe bond strength of 0.14 eV (used in the simulation shown in Figure 2a-c) is based on DFT energies and is the valid interaction potential. The other, higher bond strengths are unrealistic and are used merely to illustrate the importance of having the correct, weak C_S-Fe bond strength.

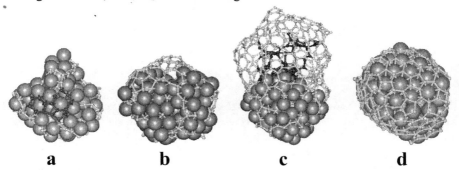

| a | b | c | d |

FIGURE 2. Snap shots showing a carbon island (small black balls in a) lifting off the Fe_{100}-C cluster to form a cap (b) that grows into the SWCNT (c) at 1000 K. When unrealistically strong C_S-Fe bond strengths are used in the simulation (e.g., 1 eV shown in d) the island does not lift off the surface but grows in size until it encapsulates the cluster (d).

Depending on the C_S-Fe bond strength, one of two graphitic structures nucleate on the cluster surface. SWCNTs, shown in Figure 2c, are formed when the bond energy is less than 0.5 eV, and graphene sheets that encapsulate the catalyst particle, shown in Figure 2d, are formed when the bond energy is higher than 0.7 eV. A bond energy between 0.5 and 0.7 eV can lead to either SWCNT growth or to particle encapsulation at 1000 K.

Two conclusions can be drawn from these results. First, the formation of SWCNTs on Fe_{100} clusters at 1000 K is not sensitive to moderate changes in the C_S-Fe bond strength. This is important since it indicates that valid growth mechanisms are obtained from simulations based on our analytic PES

Second, it is evident that the island cannot lift off the surface to form a cap when the C_S-Fe bond strengths are too large. This is because the kinetic energy that is available at 1000 K must be sufficient to overcome the interaction strength between the graphitic island and the cluster. This is possible when the cluster-island interaction is sufficiently weak (less than 0.5 eV) but not when the interactions are too strong.

It may also be noted that the strength of the bonding between the C_{Uns} and Fe atoms is important for SWCNT nucleation. As described above, it is known from DFT calculations that the C_{Uns}-Fe bonds are almost an order of magnitude stronger than the C_S-Fe bonds, so that even when the C_S-Fe bonds break and the island lifts off the particle to form the cap, the C_{Uns}-Fe bonds at the edge of the island are not broken. Thus, even once the (middle part of the) island lifts off the particle surface, precipitating C atoms can join at the edges of the cap. This leads to an increase in the cap diameter and length, which results in the growth of a SWCNT.

CONCLUSION

One of the critical steps in the nucleation and growth of CNTs is the lifting of graphitic islands off the metal particle surface to form graphitic caps. Results of the MD simulations presented here show that the weak interaction between the bond-saturated carbon atoms (in the graphitic islands) and the catalyst particle is key for this process. Carbon islands would not be able to lift off the cluster surface if this interaction were too strong, which would result in encapsulation of the cluster by a graphene layer.

REFERENCES

1. S.Iijima, *Nature* **354**, 56-58 (1991).
2. J.C. Charlier, X. Blase, A. De Vita, R. Car, *Applied Physics A: Materials Science & Processing* **68**, 267-273 (1999)
3. X. Fan, R. Buczko, A. A. Puretzky, D. B. Geohegan, J.Y. Howe, S.T. Pantelides, and S. J. Pennycook, *Phys. Rev. Lett.* **90**, 145501 (2003)
4. Y. Shibuta, S. Maruyama, *Physica B* **323,** 187-189 (2002)
5. J. Gavillet, A. Loiseau, C. Journet, F. Willaime, F. Ducastelle, and J. C. Charlier, *Phys. Rev. Lett.* **87,** 275504 (2001).
6. F. Ding, A. Rosén and K. Bolton, submitted.
7. E. Durgun, S. Dag, V. M. K. Bagci, O. Gulseren, T. Yildirim, and S. Ciraci, *Phys. Rev. B* **67**, 201401 (2003).
8. G. L. Gutsev, C. W. Bauschlicher Jr., *Chem. Phys.* **291,** 27-40 (2003).

DFT Investigation of Nanostructured Binary Compounds

S. Gemming*[†], G. Seifert* and M. Schreiber[†]

*Institut für Physikalische Chemie und Elektrochemie, Technische Universität, D-01062 Dresden, Germany
[†]Institut für Physik, Technische Universität, D-09107 Chemnitz, Germany

Abstract. Two-shell metallic nanowires from Au and from AgAu and PdAu alloys were investigated by density-functional band-structure calculations. The most stable structure has the composition MAu_8 (M = Au, Ag, Pd) and contains nine atoms in the repeat unit along the wire direction. The stability increases from $PdAu_9$ via Au_9 to $AgAu_8$. This trend follows the tensile stress acting on the central monatomic chain. According to an analysis of the electronic structure the binding between the two shells is not strongly directional, especially in the alloyed wires. However, the inter-atomic interaction along the central chain is weakened, which alleviates the tensile stress along this direction. In Au_9 and $AgAu_8$ eight s bands intersect the Fermi level and provide conductance channels for ballistic electron transport. For the Pd-centered wire only the seven conductance channels of the Au_8 shell are present, whereas the central, Pd-based one is depopulated. These findings rationalise the results of break-junction experiments, which yield a lower conductivity of PdAu contacts compared with AgAu contacts.

INTRODUCTION

PdAu alloys are prominent contact materials in microelectronic devices, thus the physical properties of this binary system [1, 2] have been studied in detail, and also some surfaces [3], and mixed, alloyed clusters [4] were investigated. For the AgAu system even elongated core-shell particles, $Ag_{core}Au_{shell}$ and $Au_{core}Ag_{shell}$, can be prepared [5], alloyed by irradiation with laser pulses [6] and employed as markers in biotechnology [7, 8]. Cluster-size effects on electronic and optical properties of the alloyed structures are more prominent than the dependence on the chemical composition [9]. Density-functional investigations on AgAu wires suggested a miscibility gap in the AgAu phase diagram for wires with high Ag content [10], whereas the bulk alloy exhibits complete miscibility. Quantised conductance was measured for AgAu and PdAu nanojunctions [11, 12]: The $1G_0$ peak in the conductance histogram is observed for any composition Ag_xAu_{1-x}, whereas the intensity of the $1G_0$ peak is substantially diminished for Pd-rich break junctions from a PdAu alloy. It was concluded that a Au-poor monatomic or diatomic nanocontact of Ag or Pd forms the break junction, and that the low s/p valence of a Pd atom dampens the s/p-type conductivity through an atomic Pd junction compared to Ag. The present study investigates these confinement effects by calculating the electronic band structure of thin nanowires within the density-functional framework provided by the program ABINIT [13]. The computational details were chosen in accordance with the settings described in more detail in refs. [10, 14].

CP723, *Electronic Properties of Synthetic Nanostructures*, edited by H. Kuzmany et al.
© 2004 American Institute of Physics 0-7354-0204-3/04/$22.00

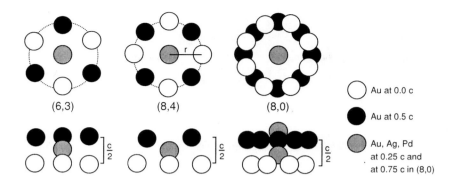

FIGURE 1. Schematic representation of the atom arrangement for the investigated (6,3), (8,4), and (8,0) wires with 7, 9, and 18 atoms in the repeat unit along the wire direction, respectively (from left to right).

GEOMETRY AND STABILITY

The investigated structures (cf. Fig. 1) contain 7, 9, and 18 atoms in the repeat unit along the wire direction, such that the central monatomic chain is surrounded by 6 or 8 direct neighbouring atoms. The nomenclature employed in the present study refers to the geometry of the outer shell and is adapted from the nomenclature of carbon nanotubes as described in ref. [14]. The structural units along the wire direction (see Fig. 1) are contained in flat supercells of size 12.7 Å x 12.7 Å x c, and repeated three-dimensionally by periodic boundary conditions.

The calculated formation energies with respect to the pure metallic bulk phases are between 0.40 and 0.61 eV/atom. Thus, all optimised structures are local minima of the formation energy, less stable than the unperturbed bulk phases, but more stable than the unreconstructed, planar Au(111) surface with a surface energy of 0.77 eV/atom. Especially for low Ag or Pd contents the alloyed wires are even more stable than the respective pure Au, Ag, or Pd wires, i.e., the energy of mixing is negative and favours the alloying process. The most stable wires have the composition MAu_8, where M = Au, Ag, or Pd. In their structural unit an outer (8,4)-type shell of eight Au atoms surrounds the central Au, Ag, or Pd chain at a distance of 3.02 Å to 3.05 Å, which is roughly 5% larger than the Au-Au distance in the bulk. The diameters of (6,3)-type and (8,0)-type wires differ from the bulk inter-atomic distance by more than 11% and 17%, respectively, thus those structures are less stable than the (8,4)-type ones. The least stable structures are Au-centered wires with high Ag or Pd content in the outer shell. They even exhibit a positive energy of mixing, thus, alloying is unfavourable for wires with low Au contents, although the bulk alloy is stable.

These deviations from the macroscopic behaviour are due to an internal stress balancing mechanism between the tensile stress on the central chain and the compressive stress on the outer shell. The ideal inter-atomic distance d_{M-M} of a free-standing monatomic chain increases from Pd (2.37 Å) via Au (2.49 Å) to Ag (2.57 Å), whereas the ideal

369

FIGURE 2. Difference between the electron density of the two-shell wires and the superposition of the densities of the outer Au_8 shell and the inner monatomic chain for Au_9, $AgAu_8$, and $PdAu_8$ from left to right. Electron accumulation occurs in white areas, electron depletion in black ones (11 contours from -0.033 (black) to +0.033 (white) electrons/$Å^3$).

length c of the free (8,4)-type shell amounts to 2.7 Å. Thus, an ideal bonding arrangement of chain and outer shell requires a stretching of the central chain by 12% (Pd), 8% (Au), or 5% (Ag). The resulting tensile stresses are reflected in the formation energies of 0.51 eV/atom (Pd) < 0.44 eV/atom (Au) < 0.40 eV/atom (Ag). As a consequence, the Ag-centered Au wire is more stable than the pure Au wire. This balancing is only possible for ultra-thin, extended nanostructures, but not in the isotropic bulk alloy.

ELECTRONIC STRUCTURE

The difference between the electron density of the two-shell wire and a superposition of the electron densities of the two shells from separate calculations was employed to analyse the inter-shell binding (cf. Fig. 2). The light grey spots between the atoms of the central chain and the atoms of the outer shell show that the inter-shell binding is not strongly directional, especially in the alloyed wires. Nevertheless, the binding electron density along the central chain is weakened compared to the binding in the free-standing monatomic wire. This electronic effect leads to an alleviation of the tensile stress along the wire direction.

The electronic band structure of the most stable structures, Au_9, $AgAu_8$, and $PdAu_8$ was analysed in more detail, because these structures are most likely realised in an experimental setup. In Au_9 and $AgAu_8$ eight s bands cross the Fermi level and constitute the conductance channels for ballistic electron transport; one s-band is completely filled. Qualitatively the same band structures are obtained also for higher Ag contents, only the degree of mixing with the occupied d-band manifold is lowered in going from Au_9 to Ag_9. For the Pd-centered wire only seven conductance channels and the completely filled s-band from the Au_8 shell are present, whereas the central s-type channel is depopulated and shifted above the Fermi level. These findings rationalise recent observations from break-junction experiments: good electronic conductivity occurs for very thin AgAu junctions, whereas ultra-thin PdAu junctions exhibit a significantly lower conductivity. According to our calculations a monatomic Ag contact would still provide a conductance channel along the junction, which is missing in the monatomic Pd contact.

CONCLUSIONS

Density-functional band-structure calculations were carried out for two-shell metallic nanowires from Au and from AgAu and PdAu alloys. The optimum structure for all three systems consists of an (8,4)-type outer shell, and a central monatomic chain of M = Au, Ag, Pd. The radii amount to about 3.05 Å and the structural unit along the wire is about 2.68 Å long. The stability of the wires increases from $PdAu_8$ via Au_9 to $AgAu_8$, in accordance with the tensile stress acting on the central monatomic chain. The latter is slightly alleviated by the inter-shell interaction. As a consequence, the alloying behaviour of thin wires differs from the bulk alloys: alloying is enhanced at high Au contents and suppressed at low Au contents. All three systems exhibit metallic conductivity along the wire. For Au_9 and $AgAu_8$ eight s-level derived bands constitute the conductance channels for ballistic electron transport, for the Pd-centered wire the central channel is depopulated. These findings rationalise the results from conductivity measurements on break junctions from AgAu and PdAu alloys: it is deduced, that a monatomic Ag contact would still provide a conductance channel along the junction, which is missing in the monatomic Pd contact. This result also supports the assumption that a a monatomic contact is formed in the break junction experiment.

ACKNOWLEDGMENTS

The authors acknowledge financial support by the Deutsche Forschungsgemeinschaft via the Graduiertenkollegs "Akkumulation von einzelnen Molekülen zu Nanostrukturen" and "Heterocyclen", and by the German-Israel Foundation (GIF).

REFERENCES

1. R. I. R. Blyth, A. B. Andrews, A. J. Arko, J. J. Joyce, P. C. Canfield, B. I. Bennett, Phys. Rev. B **49**, 16149 (1994).
2. T.-U. Nahm, R. Jung, J.-Y. Kim, W.-G. Park, S.-J. Oh, J.-H. Park, J. W. Allen, S.-M. Chung, Y. S. Lee, C. N. Wang, Phys. Rev. B **58**, 9817 (1998).
3. A. Roudgar, A. Groß, Phys. Rev. B **67**, 033409 (2003).
4. B. R. Sahu, G. Maofa, L. Kleinman, Phys. Rev. B **67**, 115420 (2003).
5. I. Srnovà-Zloufovà, F. Lednickỳ, A. Gemperle, J. Gemperlovà, Langmuir **16**, 9928 (2000).
6. J. H. Hodak, A. Henglein, M. Giersig, G. V. Hartland, J. Phys. Chem. B **104**, 11708 (2000).
7. Y.-W. Cao, R. Jin, C. A. Mirkin, J. Am. Chem. Soc., 10.1021/ja011342n (2002).
8. S. Link, Z. L. Wang, M. A. El-Sayed, J. Phys. Chem. B **103**, 3529 (1999).
9. E. Cottancin, J. Lerme, M. Gaudry, M. Pellarin, J.-L. Vialle, M. Broyer, B. Prevel, M. Treilleux, P. Melinon, Phys. Rev. B **62**, 5179 (2000).
10. S. Gemming, M. Schreiber, Z. Metallkd. **94**, 3153 (2003).
11. A. Enomoto, S. Kurokawa, A. Sakai, Phys. Rev. B **65**, 125410 (2002).
12. J. W. T. Hemskeerk, Y. Noat, D. H. J. Bakker, J. M. van Ruitenbeek, B. J. Thijsse, P. Klaver, Phys. Rev. B **67**, 115416 (2003).
13. The ABINIT code is a common project of the Universite Catholique de Louvain, Corning Inc., and other contributors.
14. E. Tosatti, S. Prestipino, A. dal Corso, S. Köstlmeier, F. di Tolla, Science **291**, 288 (2001).

Electron-Phonon Interaction and Raman Intensities in Graphite

A. Grüneis, R. Saito, J. Jiang*, L. G. Cançado, M. A. Pimenta, A. Jorio,
C. Fantini†, Ge. G. Samsonidze, G. Dresselhaus, M. S. Dresselhaus** and
A. G. Souza Filho‡

*Dept. of Phys. Tohoku Univ. and CREST JST, Sendai 980-8578 Japan
†Departamento de Física, Universidade Federal de Minas Gerais, Caixa Postal 702, Belo
Horizonte-MG 30123-970, Brazil
**Massachussetts Institute of Technology, Cambridge, MA, 02139-4307, USA
‡Departamento de Física, Universidade Federal do Ceará, Fortaleza-CE, 60455-760, Brazil

Abstract. We calculate the second order Raman spectra in graphite and compare it to recent measurements on highly oriented pyrolytic graphite (HOPG). Electron-phonon interaction is calculated as the inner product of the amplitude of vibration with the deformation potential vector. The deformation potential vector is calculated by atomic basis functions and the potential fitted by Gaussians. When we include the electron-phonon matrix element, the peak positions and the shapes agree well with the experimental results. The characteristic double peak structure of the G' band in HOPG is reproduced well.

INTRODUCTION

The Raman spectrum in graphite consists of first order and higher order processes. First order Raman processes include only one electron-phonon interaction process and higher order Raman modes include two or more electron-phonon or electron-defect interactions. In 2D graphite the only first order process is the G band. The two most prominent second order processes are the D band[1, 2] at about 1350 cm^{-1} and the G' band at about 2700 cm^{-1} for $E_{laser} = 2.2$ eV. The D band intensity depends on both electron-phonon and electron-defect interactions and thus also on the concentration and the type of defects. For investigating the electron-phonon interaction, the G' band is thus more suitable since it does not depend on defects. In this paper we will mainly focus on the position and lineshape of the G' band.

ELECTRON-PHONON INTERACTION

The electron-phonon matrix element $M_{vib}^{cc,v}(\mathbf{k}_i, \mathbf{k}_f)$ between an initial electron state $\mathbf{k}_i$ and a final electron state $\mathbf{k}_f$ in the conduction bands due to a phonon with the eigenfunction $\mathbf{S}_\sigma^v(\mathbf{q})$ of the mode v and at atom $\sigma = A, B$ in the unit cell is given by

CP723, *Electronic Properties of Synthetic Nanostructures*, edited by H. Kuzmany et al.
© 2004 American Institute of Physics 0-7354-0204-3/04/$22.00

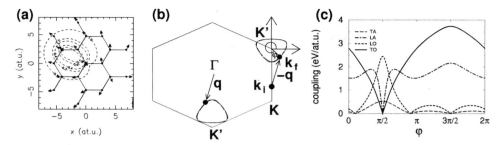

FIGURE 1. (a) The magnitude and direction of the deformation potential are indicated by dashed contour lines (value in eV per atomic unit) and unit vectors. The plot is made keeping two centers of the three center integral in Eq. (2) fixed. The potential and one electron wavefunction are indicated by a triangle and a box, respectively. One electron wavefunction is at variable positions In (b) the initial electron state is indicated by k_i and the angle φ for a final state k_f is shown. The phonon contour is shown around K' point. In (c) we show the electron-phonon coupling in accordance with Eq. (2) between a fixed electron state around the K point and states at an angle φ around the K' point is plotted for four in plane modes.

$$M_{\text{vib}}^{cc,v}(\mathbf{k}_i, \mathbf{k}_f) = \tag{1}$$
$$\sum_{l,l',\sigma,s,s'} c_{s'}^{c*}(\mathbf{k}_f)c_s^c(\mathbf{k}_i)\exp(-i\mathbf{r}_{l's'}^\sigma \cdot \mathbf{k}_f)\exp(i\mathbf{r}_{ls}^\sigma \cdot \mathbf{k}_i)S_\sigma^v(\mathbf{q}) \cdot \mathbf{m}_D(\mathbf{r}_{l's'}^\sigma, \mathbf{r}_{ls}^\sigma) \times$$
$$\delta[\hbar\omega \pm (E_i - E_f)]$$

For a Stokes process, we have $\mathbf{q} = \mathbf{k}_i - \mathbf{k}_f$ and for Anti-Stokes process $\mathbf{q} = \mathbf{k}_f - \mathbf{k}_i$ to ensure momentum conservation, which goes along with the "$\pm$" sign in the δ function, i.e. "$-$" corresponds to Stokes and "$+$" to Anti-stokes process. Here $\hbar\omega$ is the phonon energy and E_i and E_f are the energies of initial and final state, respectively. The $c_s^c(\mathbf{k})$ are coefficients of a Bloch function for the $s = A, B$ atoms in the conduction band. The summation is taken for $\mathbf{r}_{ls}^\sigma$ and $\mathbf{r}_{l's'}^\sigma$ in which l and s are the hexagon and atom index, respectively and both are within fourth shell of the origin and $\mathbf{m}_D$ is the deformation potential vector defined by

$$\mathbf{m}_D(\mathbf{r}_{l's'}^\sigma, \mathbf{r}_{ls}^\sigma) = \int \varphi^*(\mathbf{r} - \mathbf{r}_{l's'}^\sigma)\varphi(\mathbf{r} - \mathbf{r}_{ls}^\sigma)\nabla V_\sigma(\mathbf{r})d\mathbf{r}. \tag{2}$$

We fit both the ion core potential $V(\mathbf{r})$ and $2p_z$ atomic electron wavefunctions φ by a set of Gaussians to an LDA calculation for a carbon atom and we calculate $\mathbf{m}_D$ for all combinations of the two centers relative to the atomic potential $V(\mathbf{r})$. In Fig. 1(a) we show an example of $\mathbf{m}_D$ as a function of $\mathbf{r}_{ls}^\sigma$ where we kept the potential $V(\mathbf{r})$ at the center (triangle) and one electron wavefunction $\varphi(\mathbf{r})$ located on a site in the second shell (box).

An initial state and the contour line of final states and phonons for electron-phonon scattering are shown in Fig.1(b). In Fig. 1(c) we evaluate $M_{\text{vib}}^{cc,v}$ along the lines shown in Fig.1(b) for four in plane phonon branches. From Eq. (2) we can see that only in plane phonon modes can be Raman active when we consider a sheet of graphite because the phonon eigenfunction $S_\sigma^\nu(\mathbf{q})$ for the out of plane modes and the deformation potential $\mathbf{m}_D$ are perpendicular to each other. From Fig. 1(c) it can be seen that also in the electron phonon interaction a node occurs, similar to what has been reported for the electron-photon interaction [3, 4].

Double resonance theory [2] tells us that the G' band corresponds to a "$q = 2k$" mode, because the G' band frequency shifts with changing laser energy. We thus calculate $M_{\mathrm{vib}}^{cc,\nu}(\mathbf{k}_i,\mathbf{k}_f)$ for a fixed $\mathbf{k}_i$ and plot in Fig.1(c) as a function of the angle φ measured from the nearest K point in counterclockwise direction to $\mathbf{k}_f$ for the four in plane mode. We set the initial state at $\varphi = \pi/2$ and thus the final state that corresponds to a "$q = 2k$" mode appears at $\varphi = 3\pi/2$. It can be seen in Fig. 1(c) that a maximum at $\varphi = 3\pi/2$ occurs only for TO and LA branches and that LO and TA branches have zero coupling at this point. We calculated for different initial states and the situation, that for $\Delta\varphi = \pi$, the TO mode has a maximum does not change. Thus the G' band might in principle come from TO or LA branches. However a look at the corresponding phonon energies indicates that the TO branch is responsible, which was also the result of inelastic X-ray measurements on HOPG [5]. We are thus required to refit the phonon branches in order to be consistent withthe φ dependence of our matrix element. In the fitting, we used a previously employed fitting procedure[6] and only changed the assignment of the G' band data.

CALCULATION OF RAMAN INTENSITIES

The Stokes Raman intensity at phonon frequency ω due to double resonance processes with two phonons $\omega_1^\nu(-\mathbf{q})$ and $\omega_2^{\nu'}(\mathbf{q})$ from branches ν and ν' is calculated using

$$I(\omega) = \tag{3}$$

$$\sum_i \left| \sum_{a,b,c,\nu,\nu'} \frac{M_{\mathrm{op}}^{\nu c} M_{\mathrm{vib}}^{cc,\nu} M_{\mathrm{vib}}^{cc,\nu'} M_{\mathrm{op}}^{c\nu} \delta[\omega - \omega_1^\nu - \omega_2^{\nu'}]}{(E_{\mathrm{las}} - E_{ai} - i\gamma)(E_{\mathrm{las}} - E_{bi} - \hbar\omega_1^\nu - i\gamma)(E_{\mathrm{las}} - E_{ci} - \hbar\omega_1^\nu - \hbar\omega_2^{\nu'} - i\gamma)} \right|^2 .$$

Here E_{ai}, E_{bi} and E_{ci} are energies between the initial state and the three intermediate states and E_{las} is the laser energy. For simplicity we set the optical matrix elements for absorption and emission, $M_{\mathrm{op}}^{\nu c}$ and $M_{\mathrm{op}}^{c\nu}$, respectively, to be unity since our sample is polycristalline and a large number of different orientations contribute and smear out the node [3] and only a comparably small energy dependence of the average matrix element around the equi-energy contour has an effect on the intensity. For simplicity, we set the electron-photon matrix elements to unity in the present calculation. While the optical node position depends on the polarization vector, which is applied from outside, the phonon node position in Fig.1(c) is independent of an outside parameter and thus is not smeared out.For the numerical calculation of Raman spectra we use a 150×150 mesh around K points and include only terms that fulfill a double resonance condition, i.e. only transitions for which $E_{\mathrm{las}} - E_{ai} < \varepsilon$ or $E_{\mathrm{las}} - \hbar\omega_1 - \hbar\omega_2 - E_{ci} < \varepsilon$ are considered. ε is a tolerance and in our calculation, we used $\varepsilon = 5$ meV and for the broadening parameter, we chose $\gamma = 1$meV which corresponds to a lifetime of about

0.6ps. In both cases the second intermediate state E_{bi} must be resonant. Thus in total we have only two possibilities for a double resonance, unlike in the case of the D band, where four possibilities must be considered[7]. In Fig. 2(a) we compare a calculated Raman spectra for which the matrix elements are included to a calculation where we set the matrix elements are set equal to unity, i.e. only the joint density of states are included. The importance of including the matrix elements is shown by this plot, since the inclusion of matrix elements cancels many of the peaks in Fig.2(a) and only the experimentally observed G' band and a small peak around 2350 cm^{-1} survives, whose origin might be the LO branch, which has a maximum in $M_{vib}^{cc,v}$ for the "q=0" mode, as shown in Fig. 1(c). In this case, the phonon frequency would not shift with changing laser energy. However, since the LA and LO branches have the same energy at the K point, a contribution from the LA mode to the feature at 2350 cm^{-1} is also possible. For the LA mode, $M_{vib}^{cc,v} = 0$ for the "$q = 0$" mode and thus the LA contribution should cause a shift in the phonon frequency with laser energy. The calculated result shown in Fig.2(a) is plotted along with experimental data. It can be seen that the characteristic shoulder to lower frequency is well reproduced. The origin is both the phonon density of states and the matrix element that is decreasing quickly, if we move away from the "$q = 2k$" maximum. Figure 2(c) shows calculated results over a wider energy range for different laser energies, where also the overtone of the G band appears.

DISCUSSION

Due to the node in the electron-phonon matrix element, we can see that many peaks in the Raman spectra are suppressed. The G' band peaks gets pronounced because of the maximum in the matrix element as a function of angle around the K point. If we do not include the matrix element in the calculation of the Raman spectra, we get a large number of peaks that cannot be assigned experimentally. After including the matrix element we are left with only three peaks in the region 2300 cm^{-1} − 3300 cm^{-1}, including one large intensity peak at about 2750 cm^{-1} and one small peak at about 2350 cm^{-1}. Further the overtone of the G band appears at about 3200 cm^{-1}. All three peaks have corresponding experimental peaks.

ACKNOWLEDGMENTS

A.G. acknowledges financial support from the Ministry of Education, Japan. R.S. acknowledges a Grant-in-Aid (No. 13440091) from the Ministry of Education, Japan. A.J./A.G.S.F. acknowledge support from the Brazilian agencies CNPq/CAPES. The MIT authors acknowledge support under NSF Grants DMR 01-16042, and INT 00-00408.

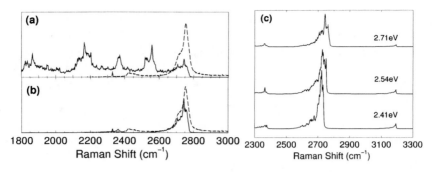

FIGURE 2. Calculated Raman spectra (solid line) compared to experimental spectra (dashed line) for $E_{las} = 2.71$ eV. In (a) we set all matrix elements equal to unity. In (b) the electron-phonon matrix elements are included according to Eq. (2). In (c) we show three features that appear in our calculation as a function of laser energy.

REFERENCES

1. Thomsen, C., and Reich, S., *Phys. Rev. Lett.*, **85**, 5214 (2000).
2. Saito, R., Jorio, A., Souza Filho, A. G., Dresselhaus, G., Dresselhaus, M. S., and Pimenta, M. A., *Phys. Rev. Lett.*, **88**, 027401 (2002).
3. Grüneis, A., Saito, R., Samsonidze, G. G., Kimura, T., Pimenta, M. A., Jorio, A., Filho, A. G. S., Dresselhaus, G., and Dresselhaus, M. S., *Phys. Rev. B*, **67**, 165402 (2003).
4. Zhao, Y., Wang, X., Ma, C., and Chen, G., *Chem. Phys. Lett.*, **387**, 149 (2004).
5. Maultzsch, J., Reich, S., Thomsen, C., Requardt, H., and Ordeon, P., *Phys. Rev. Lett.*, **92**, 75501 (2004).
6. Grüneis, A., Saito, R., Kimura, T., Cançado, L. G., Pimenta, M. A., Jorio, A., Souza Filho, A. G., Dresselhaus, G., and Dresselhaus, M. S., *Phys. Rev. B*, **65**, 155405 (2002).
7. Cançado, L. G., Pimenta, M. A., Saito, R., Jorio, A., Ladeira, L. O., Grüneis, A., Souza Filho, A. G., Dresselhaus, G., and Dresselhaus, M. S., *Phys. Rev. B*, **66**, 035415 (2002).

First principles calculations for the electronic band structures of zone folding non-metallic single wall carbon nanotubes

Jenő Kürti* and Viktor Zólyomi*

*Department of Biological Physics, Eötvös University Budapest, Pázmány Péter sétány 1/A, H-1117 Budapest, Hungary

Abstract.
  We present first principles calculations on the band structures of 20 different small diameter (d) single wall carbon nanotubes (SWCNTs), considering those tubes which are non-metallic in simple zone folding picture, including 8 chiral ones, employing density functional theory (DFT). The band gaps are calculated and discussed for all of the tubes. From small to large diameters, the gap of semiconducting zigzag tubes first increases, then reaches a maximum of about $1\,eV$ for $(11,0)$, after which it decreases, approximately as $1/d$, showing a buckling around this average behavior. The smallest diameter zigzag tubes are all metallic, due to $\sigma - \pi$ mixing caused by high curvature.

INTRODUCTION

It is well known, that with simple zone folding (ZF) of the nearest neighbor (NN) tight binding (TB) dispersions of graphene, the large diameter single wall carbon nanotubes (SWCNTs) can be reasonably well described [1, 2]. In this approximation, the metallicity of the tubes is determined by their (n, m) chiral indices: if $(n - m)$ is a multiple of 3, the tube is metallic in the ZF picture, else it is semiconducting (ZF-S).

However, for small tubes, where the diameter (d) is less then 10 Å, the ZF/TB approximation fails to accurately describe the electronic properties of the tubes. Electronic application of these tubes requires the cognition of accurate theoretical band structures. In this work we present our results on the band structure of 20 different ZF-S SWCNTs (including 8 chiral tubes), based on first principles calculations of the actual nanotubes.

METHOD

We carried out first principles calculations on the density functional theory (DFT) level, using the 'Vienna *ab initio* simulation package' (VASP 4.6) [3], in the framework of the local density approximation and the projector augmented-wave method [4]. The geometries of all 20 SWCNTs were optimized as detailed in Ref. [5]. The electronic band structures were calculated with a high k-point sampling of 31 irreducible k-points ($1 \times 1 \times 61$ Monkhorst-Pack grid centered around, and including the Γ point) for most of the tubes. Even for chiral tubes, we used the highest k-point sampling possible.

CP723, *Electronic Properties of Synthetic Nanostructures*, edited by H. Kuzmany et al.
© 2004 American Institute of Physics 0-7354-0204-3/04/$22.00

TABLE 1. DFT band gaps for ZF-S tubes compared to TB gaps. n, m: chiral indices, N_{at}: number of atoms in unit cell, θ: chirality angle, d: tube diameter according to our DFT optimization, N_k: number of irreducible k-points in the band structure calculation, Δ^{ZF}_{TB}: band gaps expected from ZF/TB, Δ^{ZF}_{DFT}: band gaps expected from ZF/DFT, Δ_0 and Δ: calculated values for the band gaps obtained from our DFT calculations in the starting ZF/TB-predicted geometry, and in the optimized geometry, respectively.

n	m	N_{at}	θ	$d(\text{Å})$	N_k	$\Delta^{ZF}_{TB}(eV)$	$\Delta^{ZF}_{DFT}(eV)$	$\Delta_0(eV)$	$\Delta(eV)$
4	0	16	0	3.34	61	2.52	1.96	0.00	0.00
3	2	76	23.4	3.53	16	2.41			0.39
5	0	20	0	4.04	61	2.32	2.03	0.00	0.00
4	2	56	19.1	4.25	61	2.11			0.25	0.25
5	1	124	8.9	4.46	15	1.87			0.13
4	3	148	25.3	4.83	13	1.76			1.31
6	1	172	7.6	5.22	9	1.74			0.41
7	0	28	0	5.55	31	1.50	1.21	0.47	0.21
5	3	196	21.8	5.55	6	1.59			1.18
6	2	104	13.9	5.70	15	1.47			0.67
8	0	32	0	6.32	31	1.42	1.23	0.56	0.59
6	4	152	23.4	6.86	12	1.27			1.09
10	·0	40	0	7.85	31	1.07	0.87	0.92	0.77
11	0	44	0	8.63	31	1.03	0.88	0.80	0.93
13	0	52	0	10.17	31	0.83	0.68	0.72	0.64
14	0	56	0	10.95	31	0.80	0.68	0.64	0.72
16	0	64	0	12.50	31	0.67	0.55		0.54
17	0	68	0	13.28	31	0.66	0.56		0.58
19	0	76	0	14.83	31	0.57	0.47		0.46
20	0	80	0	15.61	31	0.56	0.48		0.50

RESULTS AND DISCUSSION

Our results for the gaps are summarized in Table 1. The gaps of the zigzag tubes are shown in the upper panel of Figure 1. For the smallest diameter $(4,0)$ and $(5,0)$ ZF-S zigzag tubes, the band gap vanishes due to strong $\sigma - \pi$ rehybridization. The gap of the $(7,0)$ and $(8,0)$ tubes doesn't fall to zero, but it shows a clear decrease as compared to the TB values. Similarly, in the case of the 8 chiral tubes, the band gap shows a clear decreasing tendency with increasing curvature. For $d > 7$ Å , the gaps of the ZF-S zigzag tubes roughly follow an average $\Delta \sim 1/d$ dependence, and show a buckling, similar to the triad-structure observed in geometrical parameters [5, 6, 7]. This buckling is expected even in ZF/TB, but while the TB gaps decrease monotonically with increasing diameter, that is, the gaps for mod_2 $(n = 3p + 2)$ tubes are smaller than for mod_1 $(n = 3p + 1)$ tubes for the same value of p, this is reversed according to our DFT results: the mod_2 gaps are actually larger than the mod_1 gaps for two consecutive n values.

This can easily be explained in the ZF picture. If the electronic dispersion relation around the K point — the first Brillouin zone (BZ) corner — was isotropic, the mod_1 and mod_2 gaps would both show a $\Delta = C/d$ behavior, with a uniform value of C.

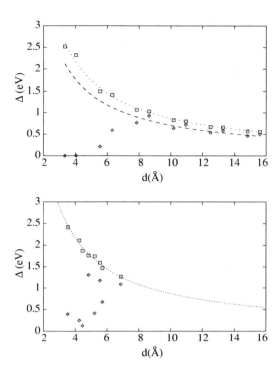

FIGURE 1. Band gaps of ZF-S zigzag (upper panel) and ZF-S chiral (lower panel) tubes versus diameter. The squares show TB values, while the diamonds show our DFT results, both from Table 1. Clearly, in TB, the ZF-S tubes follow an average $\Delta \sim 1/d$ behavior (the dotted line depicts $8.7/d$ on both panels). While in the DFT calculation, only those ZF-S zigzag tubes follow this behavior, where the diameter is larger than 7 Å (the dashed line shows $7.1/d$ in the upper panel).A clear buckling can be observed for the zigzag tubes both in TB and in DFT (see text).

However, the trigonal warping effect modifies this. The equi-excitation-energy contours of graphene around the K point are in fact not isotropic, but have a trigonal shape (with the pointed arches directed towards the three neighboring K' points), both in the case of TB, and in the case of DFT [8].

In the ZF approximation, the electronic dispersion relations of each nanotube are determined by equidistant parallel k-lines in the BZ. The value of the band gap is determined by the k-line which is closest to the K point. In the case of the zigzag tubes, the gap is always along the $M - K - \Gamma$ line. The excitation energy grows faster in the direction of $K\Gamma$ than in the direction of KM. Since the gap in the k-space is along the $K\Gamma$ line for mod_2, and along the KM line for mod_1 tubes, the value of C decreases for mod_1, while it increases for mod_2 tubes, as compared to the averaged isotropic values, leading to a buckling, which can be observed already in the case of TB gaps (squares on Figure 1). The magnitude, and even the "sign" of the buckling depends on the details of the actual electronic dispersion. If the trigonal warping is strong enough, the buckling will be so large that for the same p, the mod_1 gap can actually fall below the mod_2 gap. In TB, the trigonal warping is too weak to cause such a reverse in order, but the increased trigonal warping effect in ZF/DFT reduces the difference between the band gaps of e.g. $(10,0)$ and $(11,0)$, and even reverses the ordering, showing that the behavior of the gaps significantly differs from the ZF/TB expectations, already in the ZF/DFT approximation [8]. Thus, our DFT calculations show, that the trigonal warping effect in SWCNTs is actually larger than what is predicted by NN-TB. Furthermore, Table 1 shows that the actual values of the gaps (Δ) in the *optimized* geometries are somewhat different from the ZF/DFT obtained values, such that the reversed buckling is further increased in the optimized geometry, whereas using the starting ZF/TB-predicted geometry to obtain the gaps (Δ_0), the sign of the buckling is identical with what is obtained from ZF/TB. This buckling effect in the gaps of ZF-S tubes is discussed in more detail in Ref. [8].

ACKNOWLEDGMENTS

This work was supported by the OTKA T038014 and the OTKA T043685 grants in Hungary. The calculations were performed on the Schroedinger II cluster at the University of Vienna.

REFERENCES

1. R. Saito, G. Dresselhaus, and M. S. Dresselhaus, *Physical Properties of Carbon Nanotubes* (Imperial College Press, 1998).
2. S. Reich, C. Thomsen, and J. Maultzsch, *Carbon Nanotubes* (Wiley-VCH Verlag GmbH & Co. KGaA, 2004).
3. G. Kresse and J. Furthmüller, Comput. Mater. Sci. **6**, 15 (1996).
4. G. Kresse and D. Joubert, Phys. Rev. B **59**, 1758 (1999).
5. J. Kürti, V. Zólyomi, M. Kertesz, and G. Sun, New J. Phys. **5**, 125 (2003).
6. G. Sun, J. Kürti, M. Kertesz, and R. Baughman, J. Am. Chem. Soc. **124**, 15076 (2002).
7. G. Sun, J. Kürti, M. Kertesz, and R. Baughman, J. Phys. Chem. B **107**, 6924 (2003).
8. V. Zólyomi and J. Kürti, submitted to Phys. Rev. B.

The strength of the radial-breathing mode in single-walled carbon nanotubes

M. Machón*, S. Reich†, J. Maultzsch*, P. Ordejón** and C. Thomsen*

*Institut für Festkörperphysik, Technische Universität Berlin, Hardenbergstr. 36, 10623 Berlin, Germany
†University of Cambridge, Department of Engineering, Trumpington Street, Cambridge CB2 1PZ, United Kingdom
**Institut de Ciència de Materials de Barcelona (CSIC), Campus de la U.A.B. E-08193 Bellaterra, Barcelona, Spain

Abstract. We present calculations of the absolute Raman cross section of the radial breathing mode (RBM) of single-walled carbon nanotubes. We included all matrix elements explicitly as obtained from first principles calculations. Our results show a systematic dependence on diameter and chiral angle as well as on $v = (n_1 - n_2) \bmod 3$, which we explain with the help of a zone folding model. Thus, the comparison of relative Raman intensities can serve as an independent check for chirality assignments. The dependencies come mainly from the electron-phonon matrix elements $\mathcal{M}_{e-ph}$, which have to be taken into account when dealing with absolute Raman intensities. We compare our calculations to measurements of the absolute Raman cross section of individual nanotubes and find an agreement to within one order of magnitude. The obtained intensities are consistent with the fact that the Raman signal of a single nanotube can be detected experimentally.

One of the most interesting characteristics of single-walled nanotubes is the dependence of their physical properties on the particular geometry, defined through the diameter and chirality. Assigning chiral indices to carbon nanotubes is thus essential for research and application, but a reliable technique is not available yet. Interesting proposals were based on luminescence or Raman measurements [1, 2, 3, 4]. In particular, the frequency of the radial-breathing mode (RBM) with its roughly linear dependence on the nanotube diameter [5, 6] is widely used to determine the diameters present in samples. Here, we show how the Raman intensities of the RBM yield information about the chirality as well as the diameter, which can be used in an (n_1, n_2)-assignment.

The first-order Raman scattering cross section per unit length and solid angle can be expressed as

$$\frac{dS}{d\Omega} = \frac{\omega_l \omega_s^3 n_l n_s^3 V_c N}{(2\pi)^2 c^4 (\hbar \omega_l)^2} \left[n_{be}(\omega_{ph}) + 1 \right] \left| \int \frac{\rho(\hbar\omega)\mathcal{M}_{e-r}\mathcal{M}_{e-ph}\mathcal{M}_{e-r}\hbar}{(E_l - \hbar\omega - i\gamma)(E_l - \hbar\omega_{ph} - \hbar\omega - i\gamma)} d\omega \right|^2 .$$

A lot of work has been devoted to the resonance conditions expressed by the denominator of this equation [7]. In this work we rather concentrate on the numerator needed to determine the resonance intensity. In particular, we determined the electron-phonon coupling matrix elements using the *ab initio* package SIESTA [8, 9]. Using the calculated matrix elements we find absolute Raman intensities for single-walled carbon nanotubes.

CP723, *Electronic Properties of Synthetic Nanostructures*, edited by H. Kuzmany et al.
© 2004 American Institute of Physics 0-7354-0204-3/04/$22.00

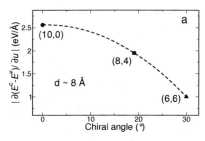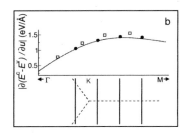

FIGURE 1. **a)** $|\partial(E^c - E^v)/\partial u|$ as function of the chiral angle for three nanotubes with diameter ≈ 8 Å. The dotted line is just a guide for the eye. **b)** $|\partial(E^c - E^v)/\partial u|$ for the (19,0) (dots) and (17,0) (squares) nanotubes mapped onto the Γ-K-M direction of the Brillouin zone of graphene. The gray line corresponds to a calculation for graphene in which we stretched the direction corresponding to the circumference of a zig-zag nanotube in order to simulate the RBM of a (19,0) nanotube. A part of the reciprocal space of graphene is shown below (dashed lines) together with some of the lines of allowed k-points which form the Brillouin zone of a (19,0) nanotube. The data points for the (19,0) nanotube are placed above the k-point lines they correspond to.

Calculations were performed within the local density approximation [10]. We replaced the core electrons by non-local norm-conserving pseudopotentials [11]. A grid cutoff of ≈ 270 Ry was used for real space integrations. A double-ζ, singly polarized basis set of localized atomic orbitals was used for the valence electrons, with cutoff radii of 5.12 a.u. for the s and 6.25 a.u. for the p and d orbitals [12]. 16 k points in the k_z direction were included for metallic nanotubes and 3 k points for semiconducting tubes. The matrix elements $\mathscr{M}_{e-ph}$ were calculated using the phonons obtained with a finite differences approach.

The absolute value of the matrix element tends to be smaller for bigger nanotubes [up to 0.017 eV for the (19,0) nanotube and 0.031 eV for the (10,0) nanotube]. This is expected since the RBM becomes a pure translation in the infinite-diameter limit, which cannot interact with the electronic system.

For totally symmetric Γ-point vibrations the electron-phonon matrix elements can be calculated by [13]:

$$\mathscr{M}_{e-ph} = C \sum_a \varepsilon_a^i \frac{\partial \left[E^c(\boldsymbol{k}) - E^v(\boldsymbol{k})\right]}{\partial \boldsymbol{u}_a}, \tag{1}$$

where $\boldsymbol{k}$ denotes the wave vector of the electronic state, c (v) denotes the conduction (valence) band which participates in a particular optical transition, i indexes the phonon with polarization vector ε_a^i normalized as $\sum_a \varepsilon_a^i \varepsilon_a^j = \delta_{ij}$, $\boldsymbol{u}_a$ is the atomic displacement, and C a normalization factor. In Fig. 1a we show an example of the chirality dependence of the electron-phonon coupling. The values of $|\partial(E^c - E^v)/\partial u|$ for the first transition of three nanotubes with diameter ≈ 8 Å are shown. As can be seen, the electron-phonon coupling for these three nanotubes decreases for increasing chiral angle. This invalidates the widespread assumption that the electron-coupling is constant for all nanotubes and means that it has to be taken into account when dealing with Raman intensities.

We find another systematic dependence of our data on the chiral indices: namely that on the value of $v = \pm 1$. If we denote by $\mathscr{M}_{1,2}$ the matrix elements corresponding to the

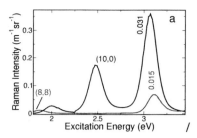 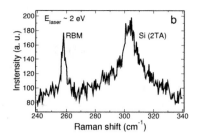

FIGURE 2. **a)** Raman resonance profile for the (10,0) and (8,8) nanotubes calculated including all matrix elements explicitly. Close to the ≈ 3 eV resonances, the corresponding electron-phonon matrix elements are written, in eV. **b)** measurement of the RBM of an isolated SWNT and the second order 2TA phonon of Si coming from the substrate. [15]

first and second optical transition, respectively, we obtain $|\mathscr{M}_1/\mathscr{M}_2| < 1$ for $\nu = -1$, but $|\mathscr{M}_1/\mathscr{M}_2| > 1$ for $\nu = +1$.

In Fig. 1b we show $|\partial(E^c - E^v)/\partial u|$ for a nanotube of each of these two families. The gray line shows a calculation in which we stretched a sheet of graphite in the direction corresponding to the circumference of zigzag nanotubes. This corresponds to the RBM of a (19,0) nanotube and yields an excellent agreement with the full calculation as can be seen in the figure. The x-axis corresponds to the Γ-K-M direction in the reciprocal space of graphene. This space is shown by the dashed lines in the lower half of the figure, together with some of the lines of allowed k-points which form the Brillouin zone of the (19,0) nanotube (solid lines). Each of these lines gives rise to an optical transition, to which an $\mathscr{M}_{e-ph}$ corresponds, [for the (19,0) nanotube each data point is placed above the k-point line it corresponds to]. Since the graphene bands are linear close to the K-point, the energies of the optical transitions are higher the closer from K they originate. Therefore, for the (19,0) nanotube the lowest transition E_{11} comes from the right from the K-point, E_{22} from the left, etc [14]. As can be seen in the figure, the electron-phonon coupling is lower on the left of the K point than on its right, thus, for the (19,0) nanotube $|\mathscr{M}_1/\mathscr{M}_2| > 1$. This applies for all nanotubes of the $+1$ family. For the (17,0) nanotube, on the other hand, the first transition is on the left, the second on the right, etc, yielding $|\mathscr{M}_1/\mathscr{M}_2| < 1$. Analogously for all nanotubes of the -1 family. We propose to measure and compare relative Raman intensities of consecutive optical transitions to discriminate this two families of nanotubes for which other physical properties like transition energies or RBM frequencies are too similar to allow the full characterisation.

We now calculate the absolute Raman cross section including explicitly all matrix elements as calculated *ab initio*. In Fig. 2 we show calculated Raman profiles for the (10,0) and (8,8) nanotubes. The resonance intensity at 3 eV differs by a factor of ≈ 4, which comes mainly from $|\mathscr{M}_{e-ph}|^2$ as can be seen from the $\mathscr{M}_{e-ph}$ values (in eV) indicated close to the peaks. We obtained an experimental value of the absolute Raman intensity of carbon nanotubes by comparing a measured RBM peak with the Si (2TA) peak at $300\,\mathrm{cm}^{-1}$ [15] (see Fig. 2b), yielding $dS_{RBM}/d\Omega \approx dS_{Si,2TA}/d\Omega \times 1.2\,10^6 = 1.1\,\mathrm{m}^{-1}\mathrm{sr}^{-1}$. In Fig. (2), the intensities at 2 eV are $\approx 0.01\text{-}0.03\,\mathrm{m}^{-1}\mathrm{sr}^{-1}$, two orders of magnitude lower than the experiment. In view of the difficulties of this type of measurements and calculations (the errors can amount up to one order of magnitude), the agreement is quite good. Both our calculated and measured values are very high if compared with measured Raman intensities for other materials (for instance the first order mode of Si: dS/dΩ=1.68 10^{-5} m^{-1}sr^{-1}) [16]. Our calculations thus explain why Raman scattering of a single, isolated nanotube can be observed experimentally. [15, 17]

We performed analogous calculations of the matrix elements $\mathscr{M}_{e-ph}$ for the high-energy mode. Comparing the matrix elements for the two phonons, we obtain $\mathscr{M}_{e-ph}^{HEM} / \mathscr{M}_{e-ph}^{RBM} \approx 4 - 6$ depending on the particular nanotube and electronic transition. The matrix elements for the HEM show a similar dependence on chirality and $(n_1 - n_2) \bmod 3$.

In conclusion, we studied the effect of the electron-phonon coupling of carbon nanotubes on the Raman intensity of the radial breathing mode. Our study shows systematic dependences on diameter and chirality which can be used in (n_1, n_2)-assignments of nanotube samples. We calculated absolute Raman intensities of the radial breathing mode and compared them to experimental results. The extremely large experimental and theoretical cross section explains why an individual nanotube can be observed by Raman scattering.

ACKNOWLEDGMENTS

We acknowledge the Ministerio de Ciencia y Tecnología (Spain) and the DAAD (Germany) for a Spanish-German Research action (HA 1999-0118). P. O. acknowledges support from Fundación Ramón Areces (Spain), EU project SATURN, and a Spain-DGI project. S. R. was supported by the Berlin-Brandenburgische Akademie der Wissenschaften, the Oppenheimer Fund, and Newnham College.

REFERENCES

1. Bachilo, S. M., Strano, M. S., Kittrel, C., Hauge, R. H., Smalley, R. E. and Weisman, R. B., *science*, **298**, 2361 (2002).
2. Strano, M. S. *et al.*, *Nano Letters*, **3**, 1091 (2003).
3. Kramberger, C. *et al.*., *Phys. Rev. B*, **68**, 235404 (2003).
4. Telg, H., Maultzsch, J., Reich, S., Henrich, F. and Thomsen, C. (submitted).
5. Rao, A. M. *et al.*, *Science*, **275**, 187 (1997).
6. Kürti, J., Kresse, G. and Kuzmany, H., *Phys. Rev. B*, **58**, 8869 (1998).
7. Thomsen, C. and Reich, S., *Phys. Rev. Lett.*, **85**, 5214 (2000).
8. Ordejón, P., Artacho, E. and Soler, J. M., *Phys. Rev. B*, **53**, R10 441 (1996).
9. Soler, J. M. *et al.*, *J. Phys. Condens. Mat.*, **14**, 2745 (2002).
10. Perdew, J. P. and Zunger, A., *Phys. Rev. B*, **23**, 5048 (1981).
11. Troullier, N. and Martins, J., *Phys. Rev. B*, **43**, 1993 (1991).
12. Junquera, J., Paz, O., Sánchez-Portal, D. and Artacho, E., *Phys. Rev. B*, **64**, 235111 (2001).
13. Khan, F. and Allen, P., *Phys. Rev. B*, **29**, 3341 (1984).
14. Reich, S., , Thomsen, C. and Ordejón, P., *Phys. Rev. B*, **65**, 155411 (2002).
15. Maultzsch, J., Reich, S., Schlecht, U. and Thomsen, C., *Phys. Rev. Lett.*, **91**, 087402 (2003).
16. Cardona, M., "Resonance Phenomena," in *Light Scattering in Solids II*, edited by M. Cardona and G. Güntherodt, Springer, Berlin, 1982, vol. 50 of *Topics in Applied Physics*, p. 19.
17. Jorio, A. *et al.*, *Phys. Rev. B*, **63**, 245416 (2001).

Quantum Mechanical Calculations of the Structure, Energetics, and Electronic Properties of the $(C_{60})_2$ and $(C_{60})_2{}^{2-}$ Fullerene Dimer

O.E. Kvyatkovskii [1], I.B. Zakharova [2], A.L. Shelankov [3], T. L. Makarova [3]

[1] *Ioffe Physico-Technical Institute of the RAS, St. Petersburg, 194021, Russia*
[2] *State Polytechnic University, 195251 St. Petersburg, Russia*
[3] *Umea University, 90187 Umea, Sweden*

Abstract. Using semiempirical AM1 (RHF and ROHF) quantum chemistry methods, optimized geometry, energetics and electronic structure of the $(C_{60})_2$, $(C_{60})_2{}^-$, and $(C_{60})_2{}^{2-}$ fullerene dimer for low (singlet) and high (triplet) spin states have been studied. The geometry optimization has been performed with the symmetry plane (C_s) or inversion symmetry (C_i) restrictions. We present numerical evidence in favor of high-spin ground state for $(C_{60})_2{}^{2-}$ dimers and argue that microscopic charge inhomogeneities may lead to magnetic order in pure carbon materials.

INTRODUCTION

It has been recently shown that pure carbon material may exhibit a ferromagnetic-like behavior: The magnetization of the phototransformed [1 - 3] and pressure-polymerized C60 fullerenes [4 - 7] shows hysteresis as a function of applied magnetic field, with a magnetic ordering temperatures ranging from 500 to 800 K.

In order to explain the origin of spontaneous magnetization in fullerene polymers, some unconventional mechanisms have been recently suggested such as interplay of carbon vacancies and sp3 hybridized interfullerene bonds [8], formation of negatively curved graphene surfaces [9], Stone-Wales rearrangements in fullerene spheres [10]; shortening of interfullerene bonds in fullerene polymers [11].

We analyze a set of the experimental data for ferromagnetic fullerenes and present a model based on this evidence. Here we show that the phenomenon can be understood within the model of charged fullerene dimers. The charge sources could be broken interfullerene bonds, partially destructed fullerene cages or various defects and impurities. It is important that the impurities do not have to be magnetic: even carbon impurity in the all-carbon network can act as a magnetic trigger.

EXPERIMENTAL DATA

Experiments on pressure polymerized fullerenes in several groups have shown that the synthesis conditions are of crucial importance for the formation of a ferromagnetic fullerene phase. The points at which the magnetic properties were found to be most

CP723, *Electronic Properties of Synthetic Nanostructures*, edited by H. Kuzmany et al.
© 2004 American Institute of Physics 0-7354-0204-3/04/$22.00

pronounced are situated at the special region of the pressure-temperature plane: they follow the line separating the polymerized fullerenes from graphitic carbon states, but on the fullerene side. Generation of the magnetic species starts at temperatures ~ 100 K less than the onset of cage collapse and increases until graphitization occurs. However, when noticeable amount of amorphous-like carbon is mixed with fullerene molecules, the ferromagnetic properties quickly decay.

Another method of preparing magnetic fullerene phase is the photopolymerization in the presence of oxygen, i.e. phototransformation. It should be emphasized that polymerization in vacuum does not lead to the appearance of non-linear magnetic behavior. In both cases a toluene-insoluble polymeric phase is formed, but the polymer structure is different. Under irradiation, molecular oxygen dissociates and interacts with the fullerene. Furthermore, the photoassisted reaction of oxygen with the fullerene may lead to the interfullerene bond breaking or partial opening and breaking of the C_{60} cages [12]. At very high fluencies, oxygen- exposed single C_{60} crystals can be laser-transformed to a glassy graphitic phase. Our experiments [13] show that this phase looses ferromagnetic features, similar to the graphitization phase for the pressure-polymerized fullerenes.

From these data we can formulate the necessary conditions for constructing the model of fullerene magnetism. The model should take into account the extreme conditions of sample preparation: excessively high temperature for pressure-polymerization or the presence of oxygen during photopolymerization.

As the first condition, we must consider the polymerized fullerene phase. The polymerization type is of less importance. For the pressure-polymerized fullerenes, the effect was observed on the rhombohedral phase [4 - 6] and also on the tetragonal phase [7]. Phototransformation created the orthorhombic phase: dimers and short linear chains.

As the second requisite, we must take into account that the ferromagnetic behavior appears quite close to the conditions at which the fullerene cages are destroyed. Two processes are possible: (i) high pressure at elevated temperatures or phototransformation in the presence of oxygen collapses some of the buckyballs, thereby generating unpaired electrons and localized spins; (ii) the buckyballs remain intact but the defects arise at the bonds between them.

According to the structural data, fullerene magnetism is a property associated with unbroken C_{60} structures connected in the polymeric network. However, we must take proper account of the defect nature of the effect. Both possibilities discussed above are considered. First, we suppose that part of the double interfullerene bonds are ruptured, and the cages are connected by the single bonds. Second, we assume that a minor part of fullerene molecules is decomposed into fragments producing microscopic charge inhomogeneities. Therefore, we examine magnetic properties of double bonded and single bonded polymers in the neutral and charged states.

RESULTS AND DISCUSSION

We study the effect of configuration and doping on the spin multiplicity in fullerene polymers using as an example neutral dimer $(C_{60})_2$ and negatively charged dimer $(C_{60})_2{}^{2-}$ with two additional electrons as model systems. The geometry was optimized for singlet and triplet dimer spin states, and two types of spatial symmetries, C_s (mirror) and C_i (inversion). To the best of our knowledge, the symmetry type has never been taken in consideration in previous treatments. Also, we have considered negatively charged dimer $(C_{60})^{2-}$ with one additional electron in doublet spin state. The results of the semiempirical AM1 calculations (RHF method for the singlet state and ROHF method for the triplet state) are collected in Tables 1 - 3.

Table 1. Total energies, lengths of the interfullerene bonds for a neutral fullerene dimer $(C_{60})_2$ (C_s and C_i isomers) in a low-spin (singlet) and high-spin (triplet) states for the optimized structures.

Dimer type	2+2 cycloadduct		Single bonded	
Symmetry	C_s (mirror)	C_i (inverse)	C_s (mirror)	C_i (inverse)
Spin state	S i n g l e t			
Total energy, a.u. E+562	- 0.411847	- 0.411847	-	- 0.359392
Bond length, Å	1.546	1.546	-	4.44
Spin state	T r i p l e t			
Total energy, a.u. E+562	- 0.348038	No convergence	- 0.347697	- 0.348467
Bond length, Å	1.528	-	1.547	1.540

Table 2. Total energies, lengths of the interfullerene bonds for a doubly charged fullerene dimer $(C_{60})_2{}^{-2}$ (C_s and C_i isomers) in a low-spin (singlet) and high-spin (triplet) states for the optimized structures.

Dimer type	2+2 cycloadduct		Single bonded	
Symmetry	C_s (mirror)	C_i (inverse)	C_s (mirror)	C_i (inverse)
Spin state	S i n g l e t			
Total energy, a.u. E+562	- 0.557219	- 0.557219	-0.573972	- 0.577810
Bond length, Å	1.547	1.547	1.543	1.538
Spin state	T r I p l e t			
Total energy, a.u. E+562	- 0.595211	-0.595580	No convergence	No convergence
Bond length, Å	1.547	1.547	-	-

Table 3. Total energies, lengths of the interfullerene bonds for a charged fullerene dimer $(C_{60})_2{}^{-1}$ (C_s and C_i isomers) in a doublet spin state for the optimized structure.

Dimer type	[2+2] cycloadduct			Single bonded	
Spin state	D o u b l e t				
Symmetry	C_s (x)	C_s (z)	C_i	C_s	C_i
Total energy, a. u. E+562	0.532683	No convergence	*	*	-0.491619
Bond length, Å	1.547	1.546	*	*	1.570

The results for a neutral dimer shown in Table 1 confirm a well known result that the singlet state is energetically preferable for a neutral dimer (shaded cells). It turns out that the ground state configuration is mirroring symmetric (C_s). Our results for the charged dimers are presented in Table 2. We observe that the charged dimer in the triplet state has lower energy compared to the singlet state (shaded cells), and, therefore, the ground state of a charged dimer is triplet. Furthermore, if we consider the singlet states, we find out that the most stable singlet state is the single bonded dimer (hatched cell). Finally, Table 3 shows that the stable charged fullerene dimer $(C60)_2^{-1}$ does exist and has a reasonable bond length.

CONCLUSIONS

On the basis of these numerical findings, we suggest the following physical mechanism of ferromagnetism in fullerene polymers: The high-spin ground state of fullerene units is stabilized by the presence of by donor impurities, or any other mechanisms of microscopic electric charge inhomogeneities. The fullerene clusters, dimer being the simplest possibility, turn out to be charged and their ground state is magnetic in accordance with our results. According to the experimental results, we deal with the elemental carbon impurity appeared due to (i) exceeded temperatures for the pressure polymerization; (ii) presence of oxygen under phototransformation.

ACKNOWLEDGEMENTS

Supported by the Swedish Research Council, Royal Swedish Academy of Sciences and Russian Federation for Basic Research (project 02-02-17617).

REFERENCES

1. Y. Murakami and H. Suematsu, Pure *and Appl. Chem.* **68**, 1463-1467 (1996).
2. T. L. Makarova, K.-H. Han, P. Esquinazi, R. R. da Silva, Y. Kopelevich, I. B. Zakharova, and B. Sundqvist. *Carbon* **41**, 1575 - 1584 (2003).
3. F. J. Owens, Z. Igbal, L. Belova, and K. V. Rao. *Phys. Rev. B* **69**, 033403 (2004).
4. T. L. Makarova, B. Sundqvist, R. Höhne, P. Esquinazi, Y. Kopelevich, P. Scharff, V. A. Davydov, L. S. Kashevarova, and A. V. Rakhmanina. *Nature* **413**, 718-721 (2001).
5. R. A. Wood, M. H. Lewis, M. R. Lees, S. M. Bennington, M. G. Cain, and N. Kitamura. *J Phys.: Condens. Matter* **14**, L385 - L390. (2002)
6. V. N. Narozhnyi, K.-H. Müller, D. Eckert, A. Teresiaka, L. Dunsch, V. A. Davydov, L. S. Kashevarova, A.V. Rakhmanina. *Physica B* **334**, 1217-1218 (2003).
7. T. L. Makarova, B. Sundqvist, and Y. Kopelevich. *Synth. Met.* **137**, 1335 - 1338 (2003).
8. A.N. Andriotis, M. Menon, R. M. Sheetz, L. Chernozatonskii. *Phys. Rev. Lett.*, **90**, 026801 (2003).
9. N. Park, M.Yoon, S. Berber, J.Ihm, E. Osawa, D. Tomanek. *Phys. Rev. Lett.* **91**, 237204 (2003)
10. Y.-H. Kim, J. Choi, and K. J. Chang. *Phys. Rev. B* **68**, 125420 (2003)
11. J. Ribas-Arino and J. J. Novoa. *Angew. Chem. Int. Ed.* **43**, 577 –580 (2004)
12. M. Wohlers, A. Bauer, and R. Schlögl. *Mikrochimica Acta, Suppl.* **14**, 267-270(1997).
13. Magnetic measurements were made by P. Esquinazi, Leipzig University, Germany.

Atomic pseudopotential model for wave packet tunneling through a carbon nanotube

Géza I. Márk*, László P. Biró*, Levente Tapasztó*, Alexandre Mayer†, and Philippe Lambin†

*Research Institute for Technical Physics and Materials Science, H-1525 Budapest
P.O.Box 49, Hungary, E-mail: mark@sunserv.kfki.hu
†Département de Physique, Facultés Universitaires Notre-Dame de la Paix
61, Rue de Bruxelles, B-5000 Namur, Belgium

Abstract. STM images of carbon nanotubes always contain the influence of both the geometrical and electronic structure. In our several former papers [1,2] we explored in detail the contributions of the geometrical factors including the effects caused by the STM tip curvature, point contacts between the tip and the nanotube, charge spreading along the nanotube during tunneling, and so on with a jellium model potential. Utilizing recent advances in computer power and a new carbon pseudopotential we can incorporate the atomic structure into our model. We investigate wave packet tunneling from the STM tip into the nanotube for n,m indices representing metallic and semiconducting nanotubes. First results of this calculation are presented and compared to *ab initio* calculations.

INTRODUCTION

Scanning Tunneling Microscopy (STM) is one of the main techniques to investigate carbon nanostructures [3] and devices fabricated from them. STM images, however, always contain both the effect of the geometry and the electronic structure. As proved by *ab initio* calculations [4], essential features of atomic resolution STM images of single wall carbon nanotubes (SWCNTs) can be successfully and effectively calculated [2,5,6] with the tight-binding method. Some of the features of the STM image, however, are of purely geometrical origin [1]. To investigate these geometrical effects without the effect of the specific atomic structure, formerly we performed wave packet (WP) scattering simulations [7–9] for jellium models of STM tip – CNT – support tunnel junctions. With this simple model, we were able to explain [9] that the STM tip causes apparent broadening [1] of CNTs, and that the displacement of the tunneling point on the surface of the tip during scanning of the CNT causes an apparent asymmetric distortion [5] of the atomic lattice.

CP723, *Electronic Properties of Synthetic Nanostructures*, edited by H. Kuzmany et al.
© 2004 American Institute of Physics 0-7354-0204-3/04/$22.00

Recent advances in computer power make it possible to include the details of the atomic structure into the 3D WP calculations. First results of these atomistic calculations are demonstrated in this paper.

CALCULATION METHOD AND RESULTS

The potential of the CNT was modeled by a local one electron pseudopotential [10] matching the band structure of graphite and graphene sheet. The STM tip and the conductive support surface was modeled by constant potential jellia. The STM tip is taken as a hyperboloid of $0.5\,nm$ apex radius. The CNT is floating above the support plane at a distance of $0.335\,nm$ and the tip – CNT gap is $0.409\,nm$. A zigzag $(10,0)$ and an armchair $(6,6)$ tube was analyzed. Their diameters are $0.783\,nm$ and $0.814\,nm$, respectively. The tip apex is above an atom in both cases.

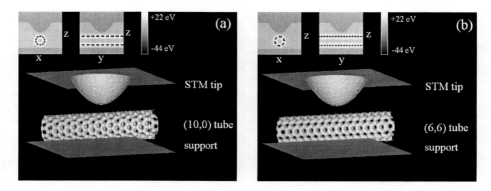

FIGURE 1. One electron local pseudopotential for the STM tip – nanotube – support system. The potential is visualized by the -2.7 eV isosurface. The upper insets show cross sections of the potential perpendicular to the tube and along the tube as grayscale plots. (a) (10,0) tube, (b) (6,6) tube. The apparent fine structure seen on the tip is a gridding artefact.

The time development of a Gaussian WP approaching the tunnel junction from inside the tip bulk was calculated by numerically solving the time dependent 3D Schrödinger equation with the split time FFT method [1] The $\rho(x,y,z,t) = |\psi(x,y,z,t)|^2$ time-dependent probability density function is visualized by snapshots of a constant density surface.

In the panel $t = 0.0\,fs$ of *Fig. 2* the initial WP is shown – still in the tip bulk region. The sphere surface is clipped at the upper boundary of the presentation box. At $t = 0.6\,fs$ the WP has already penetrated into the tip apex region. The part reflected back into the tip bulk forms interference patterns with the incoming wave. At $t = 1.8\,fs$ the WP has already tunneled from the tip into the tube and begins to flow around the tube circumference. The WP flows along the C-C bonds, there is negligible density at the centers of the hexagons because of the large positive value of the pseudopotential there. Note that for the $(6,6)$ tube the WP flows asymmetrically around the tube. This is because the $(6,6)$ tube is not symmetric to the yz plane (the plane containing the axis and the tip apex.) At $t = 3.02\,fs$ the WP already has flown around the tube circumference and begins to spread along the tube. The spreading speed is different for the two tubes.

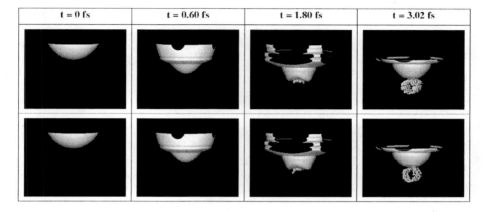

FIGURE 2. Time evolution of the probability density of the wave packet approaching the STM junction from the tip bulk and tunneling into the nanotube. Constant probability density surface is shown. *Upper row:* (10,0) tube, *lower row:* (6,6) tube. The isosurface is clipped at the presentation box boundaries.

DISCUSSION AND CONCLUSION

Fig. 3 shows the isodensity surface of the $(6,6)$ nanotube for $t = 6.34\,fs$. By this time the initial transient has already settled, and the WP is spread over the whole $3.84\,nm$ length of the presentation window. Note that the bonds are charged asymmetrically and the picture is qualitatively similar to that of the *ab initio* calculated isosurface seen on *Fig. 1.* of Ref. [4]. This is because, as is pointed out in Ref. [4], the wave functions at the Fermi level are a mixture of four different stationary states. The superposition parameters are changing in time.

In conclusion, the 3D WP tunneling simulation is a useful tool in interpretating experimental data and predicting the likely behavior of nanodevices built from carbon nanotubes. Next we plan to analyze the tunneling of the WP from the tube into the support but this needs somewhat longer simulation time.

FIGURE 3. Isodensity surface for $t = 6.34\,fs$ viewed from the side of the $(6,6)$ nanotube.

Acknowledgements: This work was partly supported by the EC, contract NANOCOMP, HPRN-CT-2000-00037, by the KFKI-CMRC project (no. ICA1-CT-2000-70029), by OTKA grants T 43685 and T 43704 in Hungary, and the Belgian PAI P5/01 project on "Quantum size effects in nanostructured materials". The calculations were done at the Sun E10000 supercomputer of the Hungarian IIF.

REFERENCES

1. Márk, G. I., Biró, L. P., and Gyulai, J., *Phys. Rev. B* **58**, 12645 (1998).
2. Lambin, Ph., Márk, G.I., Meunier, V., and Biró, L.P., *Int. J. Quant. Chem.* **95**, 493 (2003).
3. Biró L. P. and Lambin Ph., in *Encyclopedia of Nanoscience and Nanotechnology*, pp 1-12 (2003), edited by Nalwa, H. S.
4. Rubio, A., Sanchez-Portal, D., Artacho, E., Ordejón, P., and Soler, J.M., *Phys. Rev. Lett.* **82**, 3520 (1999).
5. Meunier V. and Lambin, Ph., *Phys. Rev. Lett.* **81**, 5888 (1998).
6. Kane C.L. and Mele, E.J., *Phys. Rev. B* **59**, R12759 (1999).
7. Márk, G.I., Koós, A., Osváth, Z., Biró, L.P., Benito, A.M., Maser, W.K., Thiry, P.A., and Lambin, Ph., *Diam. Rel. Mat.* **11**, 961 (2002).
8. `http://www.mfa.kfki.hu/int/nano/online/kirchberg2001/index.html`
9. Márk, G. I., Biró, L. P., Gyulai, J., Thiry, P. A., Lucas, A. A., and Lambin, Ph., *Phys. Rev. B* **62**, 2797 (2000).
10. Mayer, A., *Carbon*, in press.

Ab initio approach to superexchange interactions in alkali doped fullerides AC$_{60}$

Alexander V. Nikolaev[1,2] and Karl H. Michel[1]

[1] *Department of Physics, University of Antwerpen, UA, 2610 Antwerpen, Belgium*
[2] *Institute of Physical Chemistry of RAS, Leninskii pr. 31, 117915, Moscow, Russia*

Abstract. The superexchange interactions between the fullerenes arise as a result of the electron transfer from the C$_{60}$ molecule to the alkali atom and back. We present a scheme, which is a configuration interaction approach based on the valence bond (Heitler-London) method. The effect of superexchange is described together with chemical bonding by constructing and solving a secular equation, rather than by using a perturbation treatment. We have considered 180° and 90° superexchange for the C$_{60}$–Cs–C$_{60}$ pathways. The calculations account for unusual electronic properties of polymer orthorhombic and quenched cubic phases of CsC$_{60}$: two lines in nuclear magnetic resonance experiments, the development of a spin-singlet ground state and a decrease of magnetic susceptibility as T→0.

INTRODUCTION

The nature of the ground state of CsC$_{60}$ in the orthorhombic polymeric phase or in the metastable cubic ordered state attracted much experimental attention [1-4]. After a series of intense nuclear magnetic resonance (NMR) experiments [1,2] it was concluded that polymeric CsC$_{60}$ is not a conventional 3D metal. Spin-lattice relaxation time $^{13}(T_1)^{-1}$ for ^{13}C strongly suggests that electrons are localized. Another important conclusion was that the polymeric phase of CsC$_{60}$ is dynamically inhomogeneous. The local field at ^{133}Cs site fluctuates on the microscopic scale and statically one finds two different ^{133}Cs sites. At low temperature a transition to a spin-singlet state was established at T_S=13.8 K, which coexists with a magnetic order. A peculiar behavior was also observed for the quenched cubic phase of CsC$_{60}$ [3,4]. Similarly to the orthorhombic phase, the splitting of ^{133}Cs spectrum implies the existence of inequivalent ^{133}Cs sites and hence, inequivalent C$_{60}$ molecules, although from the crystallographic point of view all alkali and fullerene sites should be equivalent. Static singlets have been detected at T < 50 K, supporting the conclusion that CsC$_{60}$ is not a simple metal. It was pointed out that the spin gap is not associated with any long range magnetic ordering.

As we demonstrate in our approach many of these experimental findings can be rationalized with the help of the concept of superexchange, which so far has not been applied and considered for these materials. Originally the idea of superexchange was

CP723, *Electronic Properties of Synthetic Nanostructures*, edited by H. Kuzmany et al.
© 2004 American Institute of Physics 0-7354-0204-3/04/$22.00

applied to MnO [5]. There, is was assumed that the oxygen is in -2 state (O^{--}) so that its electronic shells are completely filled. Here we start with the alkali atom in the ionized state +1 (A^+), so that its electronic shells are also completely filled, although the charge of the mediator (i.e. an alkali atom) is positive in contrast with the negative charge of oxygen in MnO. However, in both cases it is the electron transfer to or from neighboring sites which is important for the model.

METHOD OF CALCULATIONS

The superexchange interaction operates through the excited states of the intermediate alkali atom. This implies a mixing of a few electronic configurations. This fact is very important because it shows that this kind of effect can appear only in a configuration interaction (CI) approach [6]. The three electronic configurations, which are relevant for our calculation are shown schematically in Fig. 1.

The first configuration (A) comprises two C_{60}^- monomers. Two electron basis ket-vectors are given by

$$\left| I_A \right\rangle = \left| i_1^t; i_2^t \right\rangle. \tag{1}$$

The indices $i_t=(k,s_z)$ stand for the t_{1u} orbitals ($k=1,2,3$) and the spin projection quantum number. In total, we find $6 \times 6=36$ independent vectors, or determinants for the first configuration. In order to describe the second configuration, Fig. 1B, we introduce the following basis states:

$$\left| I_B \right\rangle = \left| i_1^t; i^a \right\rangle. \tag{2}$$

Here $i^a=(k^a,s_z)$ stands for 12 states of one electron on the alkali site. The orbital index k^a comprises either one s-level or five orbital d-states. The total number of basis states is $6 \times 12=72$. Analogously, in the third electronic configuration, Fig.1C, the basis vectors are

$$\left| I_C \right\rangle = \left| i_2^t; i^a \right\rangle. \tag{3}$$

and the number of basis states is again 72. In the following we will operate in the quantum space which consists of these three electronic configurations, with the total number of basis states $36 + 2 \times 72 = 180$. Notice, that our quantum space is much bigger than that in the traditional molecular orbital approach. In the latter case we would have a single Hartree-Fock determinantal wave function with only $6+12+6=24$ independent states. The difference is accounted for by mutual correlations, which are omitted in the conventional molecular orbital approximation.

In order to solve the problem of superexchange we reduce it to the familiar secular equation:

$$\left(\hat{H} - E_i \cdot \hat{O} \right) \cdot \hat{V}_i = 0, \tag{4}$$

where H and O are 180 by 180 matrices of Hamiltonian and overlap, correspondingly, E_i and V_i stand for the 180 eigenvalues (electron energy spectrum) and eigenvectors. As follows from the basis, which comprises three configurations (Fig. 1), the matrices H and O have a block form:

$$
H = \begin{bmatrix} H_{AA} & H_{AB} & H_{AC} \\ H_{BA} & H_{BB} & H_{BC} \\ H_{CA} & H_{CB} & H_{CC} \end{bmatrix}, \qquad O = \begin{bmatrix} O_{AA} & O_{AB} & O_{AC} \\ O_{BA} & O_{BB} & O_{BC} \\ O_{CA} & O_{CB} & O_{CC} \end{bmatrix} \qquad (5)
$$

In constructing the Hamiltonian we took into account the attraction to the nuclei, core state repulsion and two valence electron repulsion. The details of the calculations will be given elsewhere [7].

RESULTS AND DISCUSSION

We start with considering the one electron energy of t_{1u} level, E_{t1u}, as a variable. From the experimental data we can estimate that E_{t1u} to which we add also ionic contributions (attraction and repulsion from the core densities of C_{60} and Cs) is of the order of -5 eV. This energy is in the region where one expects the hybridization with the s- and d- states of Cs. In order to analyze the results we have introduced three probabilities or statistical weights: $P(C_{60}\text{-}C_{60})$ for finding two electrons on the fullerene molecules, $P(C_{60}\text{-}Cs(s))$ for one electron on C_{60} and the other in the s- state

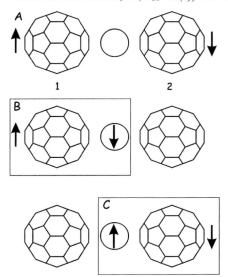

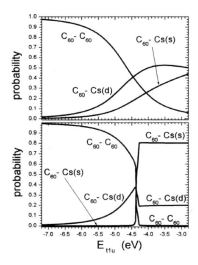

FIGURE 1. Three electronic configurations (A,B,C) responsible for superexchange in AC_{60}. **A.** $C_{60}^-\text{-}A^+\text{-}C_{60}^-$ (36 basis states); **B.** $C_{60}^-\text{-}A\text{-}C_{60}$ (72 basis states); **C.** $C_{60}\text{-}A\text{-}C_{60}^-$ (72 basis states). B and C are excited configurations where s- and d- states of the alkali atom can be occupied. The transitions A↔B and A↔B involve the corresponding electron transfer from C_{60} to the alkali atom and back.

FIGURE 2. The probabilities to find the system of two fullerenes and the cesium atom in the following configurations: 1. Two electrons occupy two fullerene molecules (C_{60}-C_{60}); 2. One of the electrons occupies one of the C_{60} molecules while the second one is in the s- state of Cs (C_{60}-$Cs(s)$); 3. One of the electrons occupies one of the C_{60} molecules while the second one is in the d- state of Cs (C_{60}-$Cs(s)$). Upper panel: 90° superexchange; Lower panel: 180° superexchange.

of Cs, and $P(C_{60}\text{-}Cs(d))$ for one electron on C_{60} and the other in the d- states of Cs. These probabilities were calculated from the eigenvector of the ground state (temperature T=0 K), Fig. 2. We find that for E_{t1u}=-5 eV there is still a significant weight of s- and d- states of Cs: $P(C60\text{-}Cs(s))$=0.089 and $P(C60\text{-}Cs(d))$=0.244. The calculation supports our initial idea [8] of importance of excited d- states in alkali atoms in these compounds. In Fig. 2 we observe two different regimes. One corresponds to the superexchange between two C_{60} molecules, while the other to valence bonds resonating around the alkali atom. The transition from one regime to another is abrupt for the 180° geometry, but it also occurs although in a smooth way for the 90° pathway.

Both the ground state and the gap between it and a first excited state are found to be very sensitive to the molecular orientations and the alkali position, as summarized in Table 1.

TABLE 1. The ground state and the first excited states in the orthorhombic structure. F stands for the "ferro"- (φ_2=φ_1=45°) and A for the "antiferro"- (φ_2=-φ_1=45°) orientations of two C_{60} molecules.

Superexchange pathway	C_{60}-C_{60} orientation	ground state	1st excited state	Gap ΔE, in K
1-a_1-2	F	triplet	singlet	2.2
1-a_1-2	A	triplet	singlet	2.4
1-a_2-3	F	singlet	triplet	43.1
1-a_1-4	F	triplet	singlet	0.5
1-a_1-4'	F	singlet	triplet	20.2
1-a_1-1'	F	singlet	triplet	0.5

ACKNOWLEDGMENTS

We acknowledge financial support from the Bijzonder Onderzoeksfonds, Universiteit Antwerpen (BOF-NOI) and the Fonds voor Wetenschappelijk Onderzoek, Vlaanderen.

REFERENCES

1. B. Simovič, D. Jérome, F. Rachdi, G. Baumgartner, and L. Forró, *Phys. Rev. Lett.* **82**, 2298 (1999).
2. B. Simovič, D. Jérome, and L. Forró, *Phys. Rev. B* **63**, 125410 (2001).
3. V. Brouet, H. Alloul, F. Quéré, G. Baumgartner, and L. Forró, *Phys. Rev. Lett.* **82**, 2131 (1999).
4. V. Brouet, H. Alloul, and L. Forró, *Phys. Rev. B* **66**, 155123 (2002).
5. P.W. Anderson, *Phys. Rev.* **79**, 350 (1950).
6. F. Keffer, and T. Oguchi, *Phys. Rev.* **115**, 1428 (1959).
7. A.V. Nikolaev and K.H. Michel, to be published (2004).
8. K.H. Michel and A.V. Nikolaev, *Phys. Rev. Lett.* **85**, 3197 (2000).

Phonon dispersion of graphite

J. Maultzsch*, S. Reich†, C. Thomsen*, H. Requardt** and P. Ordejón‡

*Institut für Festkörperphysik, Technische Universität Berlin, Hardenbergstr. 36, 10623 Berlin,
Germany
†Department of Engineering, University of Cambridge, Cambridge CB2 1PZ, United Kingdom
**European Synchrotron Radiation Facility (ESRF), B.P. 220, 38043 Grenoble, France
‡Institut de Ciència de Materials de Barcelona (CSIC), Campus de la Universitat Autònoma de
Barcelona, 08193 Bellaterra, Barcelona, Spain

Abstract. The phonon dispersion of graphite was determined by inelastic X-ray scattering along the Γ-K, K-M, and Γ-K direction. In contrast to many predictions, the dispersion of the transverse branch is large, leading to a minimum at the K point. We present *ab initio* calculations which agree very well with the experiment. The minimum of the transverse branch at the K point is due to a strong electron-phonon coupling for this phonon. This coupling dominates the scattering mechanism in both electronic transport and Raman scattering.

Many properties of carbon nanotubes are closely related to those of graphite. The simplest approximation is the zone-folding approach, where the Brillouin zone of a graphite sheet (graphene) is cut into lines that correspond to the allowed states of the nanotube. For the phonon spectrum of nanotubes, there are some fundamental limitations to this approximation, such as the presence of four acoustic modes in one-dimensional systems like nanotubes, and the finite Γ-point frequency of the radial breathing mode, which corresponds to an acoustic mode in graphene. Nevertheless, since it has not been possible to measure the phonon dispersion of carbon nanotubes directly, we need to understand the vibrational properties of graphite very well. Surprisingly, the in-plane graphite phonon dispersion has not been reported completely from experiments. Only the acoustic branches below $\approx 400 \, \text{cm}^{-1}$ were measured by inelastic neutron scattering [1]. The optical phonons between the Γ and K point and between the Γ and M point were covered by electron-energy loss spectroscopy [2, 3, 4], but experimental data are missing near the zone boundaries and in the third high-symmetry direction K-M.

In this paper we present the optical phonon branches in the entire in-plane graphite Brillouin zone determined by inelastic X-ray scattering [5]. Our results show that in particular around the K point the phonon dispersion was incorrectly predicted by force-constants as well as by *ab initio* calculations [6, 7, 8]. The branch of the transverse optical phonon (TO) is strongly softened at the K point, suggesting a large coupling to the electronic system. Our *ab initio* calculations agree very well with the experimental data if the long-range nature of the forces in a semimetal like graphite is included. We discuss the implications of the softened K-point phonons for graphite and carbon nanotubes with respect to electron-phonon coupling and the disorder-induced Raman spectrum.

CP723, *Electronic Properties of Synthetic Nanostructures*, edited by H. Kuzmany et al.
© 2004 American Institute of Physics 0-7354-0204-3/04/$22.00

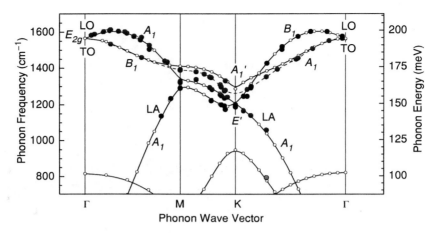

FIGURE 1. Phonon dispersion of graphite from inelastic X-ray scattering along the in-plane high-symmetry directions (full circles). The dashed line is a cubic-spline interpolation to the TO-derived branch. The open circles show an *ab initio* calculation for graphene, downscaled by 1%.

Inelastic X-ray measurements were performed at beamline ID28 at the European Synchrotron Radiation Facility. We used an incident beam of 17794 eV focused to a spot size of $30 \times 60 \mu m^2$. The energy resolution was 3.1 meV. The sample was a naturally grown graphite flake consisting of microcrystals with $\approx 100 \times 200 \mu m^2$ size.

In Fig. 1 the experimentally obtained phonon dispersion is shown along the Γ-M, Γ-K and K-M directions (full circles). The bands are labeled by their symmetry and by their displacements at the Γ point, LO (longitudinal optical), LA (longitudinal acoustic), and TO. We will use these labels for the entire branches, although the phonons are strictly transverse or longitudinal only at the Γ point and along Γ-M. The measured frequencies close to the Γ point agree with the Γ-point frequency known from Raman scattering. In the longitudinal optical branch, an overbending of $\approx 30 \, cm^{-1}$ is clearly seen. The TO and LO bands cross between the Γ and M point and between the Γ and K point. The greatest difference of the experimental data to many of the theoretical predictions is the shape of the TO branch near the K point. It has a minimum at the K point, invalidating force-constants calculations that predicted a local maximum [6], and its frequency is much closer to the LO/LA frequency than predicted from previous *ab initio* calculations [7, 8].

The open circles in Fig. 1 are an *ab initio* calculation within density-functional theory in the generalized gradient approximation, using the SIESTA package [9, 10]. The phonon dispersion is calculated in a supercell approach. Therefore in a semimetal with long-ranged forces, the calculated phonon frequencies are correct only at wave vectors commensurate with the chosen supercell. The agreement of our calculations with the experiment in Fig. 1 is very good; the softening at the K point is still a little underestimated for the TO branch. To obtain the good agreement, it was necessary to explicitly include the K point by taking an appropriate supercell (consisting of multiples of three of the unit cell). This indicates a strong electron-phonon coupling of the TO-

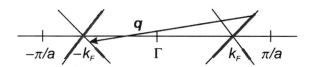

FIGURE 2. Schematic view of the linear electronic bands in an armchair tube. They cross at $k_F = 2/3\,\pi/a$ (a is the lattice constant of graphite), which corresponds to the graphite K point. A fully symmetric phonon with $q \approx 2k_F \equiv k_F$ can scatter the electrons near the Fermi level.

derived K-point phonon (A'_1 symmetry) with the electronic system, which is missed in calculations that do not include the K point. We calculated the electronic density of states in equilibrium and for atomic coordinates that were displaced according to the A'_1 phonon at the K point. In the case of the displaced atoms, a gap opens at the Fermi level, supporting the large electron-phonon coupling.

The K point in graphite and the K-point derived states in carbon nanotubes are special points in the Brillouin zone. At the K point, the conduction and valence bands cross at the Fermi level in a graphite sheet. A wave vector q connecting two equivalent K points is a reciprocal lattice vector, *i.e.*, $q = 0$; a vector connecting two non-equivalent K points is a K-point vector itself ($q = K$). Thus to couple electrons near the Fermi level, a B_1-symmetry phonon with $q \approx 0$ or a fully symmetric one with $q \approx K$ is required. The latter is given by the TO-derived mode (A'_1) at the K point. An analogous scattering process takes place in metallic carbon nanotubes. In Fig. 2 we show backscattering of electrons by this phonon in electronic transport. Transport measurements in carbon nanotubes indicate that indeed scattering by the phonons corresponding to the K-point TO in graphite dominates the backscattering mechanism [11, 12]. In these experiments, the current saturates after the electrons have gained an energy of about 160 meV, which is close to the energy of the TO phonon at the K point.

Similarly to the backscattering shown in Fig. 2, the excited electrons are scattered in the double-resonant Raman process that gives rise to the D mode in both graphite and carbon nanotubes [13, 14, 15]. Again, the D mode is constituted by the fully symmetric TO-phonons from near the K point [16]. We confirm this explicitly by calculating the D mode of graphite in the linear approximation [17], using the experimental phonon dispersion. In Fig. 3 we show the D-mode frequency as a function of excitation energy from a calculation with the TO branch (solid lines) and with the LO branch (dashed lines). It is clearly seen that the results of the TO branch agree very well with the experimental data (solid circles), whereas the LO-based calculations do not reproduce the experiments.

Finally, we want to illustrate what we expect for carbon nanotubes. In Fig. 4 we show the phonon dispersion of the $m = 0$ and $m = n$ bands in armchair and zig-zag tubes, obtained from the experimental data in Fig. 1 by zone folding. As mentioned above, zone folding contains systematic errors for the phonon dispersion; for example, the nanotube Γ-point

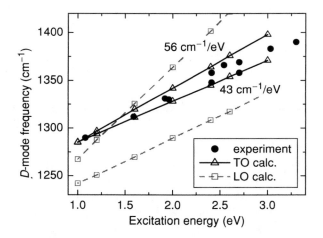

FIGURE 3. *D*-mode frequency in graphite as a function of laser energy calculated within the double-resonance model in the linear approximation. The triangles/solid lines (squares/dashed lines) were calculated using the TO (LO)-derived phonon branch. The calculation from the TO-derived branch matches the experimental values (full circles) very well. Experimental values are taken from Refs. [18, 19, 20].

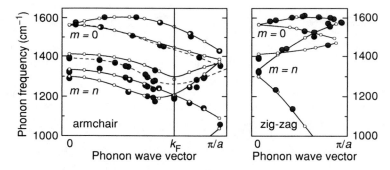

FIGURE 4. Zone-folding of the graphite dispersion; the symbols are the same as in Fig. 1. The folded branches of graphite form the $m = 0$ and $m = n$ branches in zig-zag tubes (Γ-M) and in armchair tubes (Γ-K-M direction).

modes should be split, and the $m = 0$ and $m = n$ bands should depend on the diameter. Nevertheless, we see that the graphite K point is transformed into the Fermi wave vector in metallic carbon nanotubes. As shown in Fig. 2, electrons at k_F and at $-k_F$ are coupled by phonons with $q = k_F$. Therefore, we predict a similar softening of the fully symmetric phonon band at k_F in carbon nanotubes as we observed in graphite.

In summary, we determined the phonon dispersion of graphite by inelastic X-ray scattering. The experimental results invalidate some of the existing theories, in particular, we found near the K point a much stronger dispersion of the TO-derived, fully symmetric branch than predicted. We presented *ab initio* calculations, which agree very well with the experiment. The softening of the A_1' mode at the K point is caused by a strong electron-phonon coupling, which is predicted for carbon nanotubes as well. This coupling leads to the double-resonant D mode in the Raman spectra of graphite and carbon nanotubes. We calculated the D mode using the experimental phonon dispersion and confirmed the symmetry-based prediction that the D mode comes from the TO-derived branch.

ACKNOWLEDGMENTS

We acknowledge support from the Deutsche Forschungsgemeinschaft under grant number Th 662/8-2 and from the ESRF. S.R. was supported by the Oppenheimer Fund and Newnham College. P.O. was supported by Spain's MCyT project BFM2003-03372-C03. We also acknowledge support from the Ministerio de Ciéncia y Tecnología (Spain) and the DAAD (Germany) for a Spanish-German Research action (HA2001-0065).

REFERENCES

1. Nicklow, R., Wakabayashi, N., and Smith, H. G., *Phys. Rev. B*, **5**, 4951 (1972).
2. Wilkes, J. L., Palmer, R. E., and Willis, R. F., *J. Electron Spectr. Rel. Phen.*, **44**, 355 (1987).
3. Oshima, C., Aizawa, T., Souda, R., Ishizawa, Y., and Sumiyoshi, Y., *Solid State Comm.*, **65**, 1601 (1988).
4. Siebentritt, S., Pues, R., and Rieder, K.-H., *Phys. Rev. B*, **55**, 7927 (1997).
5. Maultzsch, J., Reich, S., Thomsen, C., Requardt, H., and Ordejón, P., *Phys. Rev. Lett.*, **92**, 075501 (2004).
6. Jishi, R. A., and Dresselhaus, G., *Phys. Rev. B*, **26**, 4514 (1982).
7. Sánchez-Portal, D., Artacho, E., Soler, J. M., Rubio, A., and Ordejón, P., *Phys. Rev. B*, **59**, 12678 (1999).
8. Dubay, O., and Kresse, G., *Phys. Rev. B*, **67**, 035401 (2003).
9. Junquera, J., Paz, O., Sánchez-Portal, D., and Artacho, E., *Phys. Rev. B*, **64**, 235111 (2001).
10. Soler, J. M., Artacho, E., Gale, J. D., García, A., Junquera, J., Ordejón, P., and Sánchez-Portal, D., *J. Phys.: Cond. Mat.*, **14**, 2745 (2002).
11. Yao, Z., Kane, C. L., and Dekker, C., *Phys. Rev. Lett.*, **84**, 2941 (2000).
12. Bourlon, B., Glattli, D., Plaçais, B., Berroir, J., Miko, C., Forró, L., and Bachtold, A., *Phys. Rev. Lett.*, **92**, 026804 (2004).
13. Thomsen, C., and Reich, S., *Phys. Rev. Lett.*, **85**, 5214 (2000).
14. Maultzsch, J., Reich, S., and Thomsen, C., *Phys. Rev. B*, **64**, 121407(R) (2001).
15. Reich, S., Thomsen, C., and Maultzsch, J., *Carbon Nanotubes: Basic Concepts and Physical Properties*, Wiley-VCH, Berlin, 2004.
16. Mapelli, C., Castiglioni, C., Zerbi, G., and Müllen, K., *Phys. Rev. B*, **60**, 12710 (1999).
17. Maultzsch, J., Reich, S., and Thomsen, C. (2004), to be published.
18. Wang, Y., Alsmeyer, D. C., and McCreery, R. L., *Chem. Mater.*, **2**, 557 (1990).
19. Pócsik, I., Hundhausen, M., Koos, M., and Ley, L., *J. Non-Cryst. Sol.*, **227-230B**, 1083 (1998).
20. Matthews, M. J., Pimenta, M. A., Dresselhaus, G., Dresselhaus, M. S., and Endo, M., *Phys. Rev. B*, **59**, R6585 (1999).

Electron Interactions and Excitons in Carbon Nanotube Fluorescence Spectroscopy

C.L. Kane, E.J. Mele

Dept. of Physics and Astronomy, University of Pennsylvania, Philadelphia, PA 19104

Abstract. Recent fluorescence spectroscopy experiments on solutions of isolated single wall carbon nanotubes reveal substantial deviations in the dependence of absorption and emission energies from predictions of non interacting models of nanotube electronic structure. We study the effects of electron interactions in a theory of large radius tubes, which derives from the theory of two dimensional graphene. In graphene, the long range coulomb interaction leads to a logarithmic correction to the electronic self energy, which signifies marginal Fermi liquid behavior. Interactions on length scales larger than the tube circumference lead to both a strong excitonic binding and an additional contribution to the self energy. We show that these two one dimensional effects largely cancel one another, so that the subband and radius dependence of the optical transitions are dominated by the graphene self energy effect.

INTRODUCTION

The optical transition energies of semiconducting nanotubes, along with their dependence on the nanotube diameter and chiral angle have been studied in a recent series of fluorescence spectroscopy experiments [1, 2, 3]. It has become increasingly clear that electron interactions play an important role in determining the optical transition energies[4, 5, 6, 7]. As pointed out in early work by Ando[4], interactions lead to (1) an increase in the single particle energy gap and (2) binding of electrons and holes into excitons. More recently, Spataru et al.[6] have reached a similar conclusion by computing the optical spectra for selected small radius nanotubes. However, the systematic dependence of the transition energies on nanotube radius has not been addressed.

Here we examine the optical excitations of carbon nanotubes in the limit of large radius, R, where they inherit their electronic structure from that of an ideal sheet of two dimensional (2D) graphene. This permits a systematic study of the radius and subband dependence of the excitations. In this limit the electron interactions fall into two categories: (1) 1D interactions on scales longer than the tube circumference, and (2) 2D interactions on scales smaller than the tube circumference. We find that the 1D long range interaction leads to both an enhancement of the energy gap and an exciton binding energy which both scale as $1/R$. Although both effects are large they have opposite sign and ultimately lead to a moderate enhancement of the predicted optical transition energy. By contrast, we find the 2D interactions (2) lead to a $\log R/R$ correction to the bandgap renormalization. This singular behavior can be traced to the effect of a the Coulomb interaction on the dispersion of 2D graphene, which leads to marginal Fermi liquid behavior[8]. The presently available optical data indeed show this non linear scaling

CP723, *Electronic Properties of Synthetic Nanostructures*, edited by H. Kuzmany et al.
© 2004 American Institute of Physics 0-7354-0204-3/04/$22.00

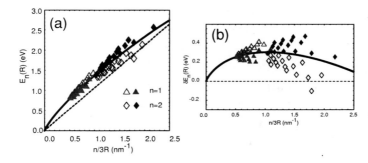

FIGURE 1. (a) Optical transition energies in the first two subbands for semiconducting nanotubes measured in Ref. 1 as a function of $n/3R$. The filled/open symbols correspond to $[p,q]$ nanotubes with chiral index $\nu = p - q \bmod 3 = +/-1$. The dashed line is non interacting prediction using $\hbar v_F = 5.3$ eVÅ from graphite. The solid line is Eq. (3), which incorporates the effect of the 2D Coulomb interaction. (b) The same data plotted as the difference from the non interacting prediction.

behavior and agrees favorably with the predictions of the large radius theory even for tubes with moderately small radii $R \sim 0.5$ nm.

SCALING BEHAVIOR

The simplest model of nanotube electronic structure, based on non interacting electrons in a linear graphene spectrum, predicts that the energy gaps of semiconducting nanotubes are

$$E_n^0(R) = 2n\hbar v_F/3R, \tag{1}$$

where R is the nanotube radius, $n = 1, 2, 4, 5$ describes the 1st, 2nd, 3rd and 4th subbands, and v_F is the graphene Fermi velocity. For a tight binding model on a honeycomb lattice with with lattice constant a and a nearest neighbor hopping amplitude γ_0, $\hbar v_F = \sqrt{3}\gamma_0 a/2$. The linearized model (1) is exact in the limit of large radius, and is the first term in an expansion in powers of $1/R$. Corrections due to curvature[9] and trigonal warping[10] are proportional to $\nu \sin 3\theta/R^2$, where θ is the chiral angle ($\theta = 0$ denotes an armchair wrapping) and $\nu = \pm 1$ is the chiral index. A central prediction of the non interacting model is thus that for large R the band gaps scale linearly with n/R - a fact that can be traced to the linear dispersion of graphene at low energies. The large R limit is most accurate for nearly armchair nanotubes for which the $\sin 3\theta$ corrections are smallest. For such tubes Eq. 1, describes the tight binding energy gaps to better than 1% for tubes with radii as small as 0.5 nm. The next term in the expansion at $O(1/R^3)$ is negligible. Here we focus exclusively on nearly armchair nanotubes, where large R scaling can be meaningfully applied. $\sin 3\theta$ corrections, when present in specific nanotubes, lead to deviations from the scaling predictions[6, 7].

The observed transition energies do not obey this linear scaling behavior. In Fig. 1 we plot the transition energies reported in Ref. 1 as a function of $n/3R$, where n is the subband index, and R is the tube radius deduced in Ref. 1 by exploiting the pattern of $\sin 3\theta/R^2$ corrections. We have used different symbols to represent the data with $v = \pm 1$. The separatrix between the data for the positive and negative v locates the data for the nearly armchair tubes with $\theta \sim 0$. At the separatrix the $\sin 3\theta/R^2$ corrections are absent, so that the radius dependence should be described by the large R limit to order $1/R^3$. It is clear, however, that even at the separatrix, the transition energies are blue shifted relative to (1) and the linear scaling relation is not satisfied. Nonetheless, it is striking that the data near the separatrix for the two subbands lie approximately on the same nonlinear curve. The simplest interpretation of this apparent scaling behavior is that these energies probe the dispersion of 2D graphene at a wavevector $q_n = n/3R$.

2D SELF ENERGY EFFECT

Gonzalez et al.[8] have shown that the Coulomb interaction in 2D graphene leads to a singular correction to the electron self energy. Consider the Hamiltonian

$$
H = \hbar v_F \int d^2 r \psi^\dagger \vec{\sigma} \cdot \frac{\vec{\nabla}}{i} \psi + \frac{e^2}{2} \int d^2 r d^2 r' \frac{n(\mathbf{r})n(\mathbf{r}')}{|\mathbf{r} - \mathbf{r}'|}, \tag{2}
$$

where ψ is a Dirac spinor with two copies for the K-K' degeneracy, and $n = \psi^\dagger \psi$. The Coulomb interaction is characterized by a dimensionless interaction strength $g = e^2/\hbar v_F$. In lowest order perturbation theory the electronic dispersion is

$$
E(q) = \hbar v_F q [1 + (g/4) \log(\Lambda/q)], \tag{3}
$$

where Λ is an ultraviolet cut off of order the inverse lattice constant. The nonlinear behavior as $q \to 0$ is a consequence of the long range singularity of the 2D Coulomb interaction $V(q) = 2\pi e^2/q$. The effect of screening is determined by the static polarizatibility $\Pi(q) = (1/4)q/v_F$. The linear dependence on q exactly cancels the $1/q$ singularity of $V(q)$, leading to a multiplicative renormalization of the interaction analogous to screening in a 3D dielectric. The $q \to 0$ logarithmic correction to $E(q)$ survives screening although its coefficient is renormalized. In a static screening approximation the renormalized interaction is $g_{\mathrm{scr}} = g/(1 + g\pi/2)$.

Though it is derived in for small g, this result has deeper implications, since it shows that the weak interaction limit is perturbatively stable. As emphasized by Gonzalez et al.[8], the form of (3) is *exact* for $q \to 0$, although the parameters v_F and g are renormalized, and scale with the cutoff Λ. This singular behavior is a signature of a marginal Fermi liquid. For an unscreened interaction we estimate $g \sim 1.1$ and $\hbar v_F \sim 12.9$ eVÅ at a cutoff scale $\Lambda \sim .5\mathrm{nm}^{-1}$. In a statically screened approximation we find $\hbar v_F = 7.2$ eVÅ, $g = 2.0$ and $g_{\mathrm{scr}} = 0.48$. The nonlinear scaling form of the separatrix in Fig. 1 is consistent with Eq. (3). Choosing the scale $\Lambda = 0.5\mathrm{nm}^{-1}$, the data is well fit with the parameters $v_F = 7.8$ eVÅ and $g = 0.74$. These parameters are in acceptable agreement with the statically screened theory described above, given the theory's simplicity.

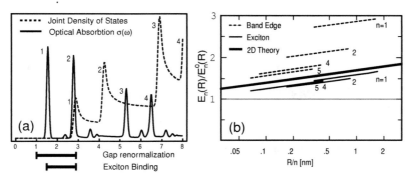

FIGURE 2. (a) Optical absorption with exciton peaks and joint particle-hole density of states with single particle van Hove singularities. (b) Single particle gaps (dashed line) and particle hole gaps (solid line) for the first four subbands of semiconducting $[p, p]$ nanotubes with phase shifted boundary conditions calculated for $5 < p < 25$. The thick line is the prediction of the 2D theory Eq. 4.

1D EXCITONIC EFFECTS

The agreement between the data and the interacting theory of 2D graphene is nevertheless surprising because the latter does not account for excitonic effects, which are due to 1D interactions on scales *larger* than R. In addition to exciton binding, these interactions lead to an increase of the single particle gap. To address this issue we have numerically calculated both the single particle and particle-hole gaps. We find that the two 1D interaction effects largely cancel one another, so the R dependence of the particle-hole gap is ultimately well described by the 2D theory.

We have computed the the single particle and particle hole energy gaps for nanotubes in a statically screened Hartree-Fock approximation. Our calculation is similar to that previously reported by Ando[4], though here we focus on the R dependence of the energy gaps. We use a π electron tight binding model, including all 1D subbands. To avoid the complications of the $v \sin 3\theta / R^2$ corrections we study semiconducting tubes by calculating excitations of armchair tubes with an appropriate phase shifted boundary condition to impose an energy gap. The single particle band gaps are computed by evaluating the exchange self energy using a statically screened Coulomb interaction. The particle-hole gap is determined by numerically diagonalizing the Schrodinger equation for the particle and the hole in the renormalized bands with the screened interaction.

In Fig. 2 we plot the single particle and particle hole gaps as a function of radius and subband index. To emphasize the corrections to linear scaling behavior we provide a log-linear plot of $E_n(R)/E_n^0(R)$ as a function of R/n, where E_n^0 is given by (1), and is proportional to n/R. The prediction based on the statically screened 2D theory of graphene given in (4) is shown for comparison. The single particle gaps are strongly enhanced relative to their non interacting values, while the particle-hole gaps are only moderately enhanced. Thus, most of the enhancement of the single particle band gap is cancelled by the electron hole interaction that binds the exciton. The particle-hole gaps for the different subbands lie nearly on a *single* straight line, close to the prediction of the 2D interacting theory. This is consistent with the scaling behavior in the experimental data in Fig. 1. In contrast, the single particle gaps are well above the predictions of the 2D theory, and do not obey scaling with subband index.

The essential features in Fig. (2) can be understood within a simpler model for the 1D interactions on scales larger than the tube radius. Consider a semiconducting nanotube with a bare energy gap 2Δ with an *infinite range* interaction $V(x) = V_0$. This is the constant interaction model, familiar from the theory of the Coulomb blockade. In this model the interaction energy is $V_0 N^2/2$, where N is the total number of electrons. The single particle energy gap is then simply $2\Delta + V_0$. The particle-hole energy gap, which determines the energy of optical transitions is 2Δ. Since the exciton is electrically neutral, its energy is unaffected by the infinite range interaction. For this model the exciton binding energy *exactly* cancels the enhancement of the single particle gap.

Because of the near cancellation between the exciton binding and the self energy, the effects of the 2D electronic interactions can be seen clearly in the experimental data. It is interesting that theory derived from the leading order contributions in $1/R$ to the excitation energies provides a good description of the data over the range of experimentally measured tube radii. The large single particle gaps shown in Fig. 2 are likely to be important for many nanotube-derived devices, but have yet to be measured directly in experiments. They are accessible in principle by measuring the activation energy for transport in a semiconducting tube, or by measuring the threshold for photoconductivity following optical excitation into the lowest subbands. Interpretation of the gaps measured in scanning tunneling spectroscopy are complicated by screening effects from the substrate, and make it difficult to extract the single particle gap of individual tubes.

ACKNOWLEDGMENTS

We thank Bruce Weisman for helpful discussions and the organizers of IWEP04 for their generous hospitality. This work was supported by the NSF under MRSEC grant DMR-00-79909 and the DOE under grant DE-FG02-ER-0145118.

REFERENCES

1. S.M. Bachilo, et al., Science **298**, 2361 (2002).
2. J. Lefebvre, Y. Homma and P. Finnie, Phys. Rev. Lett. **90** 217401 (2003).
3. A. Hagen and T. Hertel, Nano Lett. **3** 383 (2003).
4. T. Ando, J. Phys. Soc. Japan **66** 1066 (1996).
5. C.L. Kane and E.J. Mele, Phys. Rev. Lett. **90** 207401 (2003).
6. C.D. Spataru, S. Ismail-beigi, L. Benedict and S.G. Louie, Phys. Rev. Lett. **92**, 077402 (2004).
7. V. Perebeinos, J. Tersoff and P. Avouris, cond-mat/0402091 (2004).
8. J. Gonzalez, F. Guinea and M.A.H. Vozmediano, Phys. Rev. B. **59**, 2474 (1999).
9. C.L. Kane and E.J. Mele, Phys. Rev. Lett. **78**, 1932 (1997).
10. S. Reich and C. Thomsen, Phys. Rev. B **62** 4273 (2000).

Double resonance Raman spectroscopy and optical properties of single wall carbon nanotubes

R. Saito, A. Grüneis, J. Jiang*, A. Jorio, L. G. Cançado, C. Fantini,
M. A. Pimenta†, Ge. G. Samsonidze, G. Dresselhaus, M. S. Dresselhaus**
and A. G. Souza Filho‡

*Dept. of Phys. Tohoku Univ. and CREST JST, Sendai 980-8578 Japan
†Dept. de Física, Universidade Federal de Minas Gerais, Belo Horizonte-MG 30123-970, Brazil
**Massachusetts Institute of Technology, Cambridge, MA, 02139-4307, USA
‡Dept. de Física, Universidade Federal do Ceará, Fortaleza-CE, 60455-760, Brazil

Abstract. Recent progress on double resonance Raman spectroscopy and related optical properties of single wall carbon nanotubes (SWNTs) is presented. The electron-photon and electron-phonon interactions for graphite and SWNTs are calculated within the tight-binding scheme. Optical absorption matrix element and phonon relaxation time calculations are consistent with experimental observations.

INTRODUCTION

In carbon materials, we know about several weak Raman signals which often show dispersive behavior of the phonon frequency[1, 2]. That is, the phonon frequency changes with changing the laser excitation energy[3, 4, 5, 6, 7]. When optical absorption or emission occurs in the Raman process, the Raman intensity is enhanced significantly. This effect is known as the resonance Raman effect[1, 8]. Here the resonance condition means that emission or absorption of an electron takes place, respectively, from or to a real electronic state. Double resonance (DR) Raman theory works well for explaining the dispersive phonon modes in which the non-zone-center ($q \neq 0$) phonon mode and the second-order Raman process are relevant to these weak spectra[6, 7, 9].

In second-order Raman processes, the electron (1) absorbs a photon at a $\mathbf{k}$ state, (2) scatters to $\mathbf{k} + \mathbf{q}$ states, (3) scatters back to a $\mathbf{k}$ state, and (4) emits a photon by recombining with a hole at a $\mathbf{k}$ state. The two scattering processes consist of either elastic scattering by defects of the crystal or inelastic scattering by emitting a phonon. Thus (1) one-elastic and one-inelastic scattering event and (2) two-inelastic scattering events are relevant to Raman spectroscopy. Hereafter we call them, respectively, one-phonon and two-phonon double resonance Raman spectra[2].

In a DR Raman process, two resonance conditions for three intermediate states should be satisfied, in which the intermediate $\mathbf{k} + \mathbf{q}$ state is always a real electronic state and either the initial or the final $\mathbf{k}$ states is a real electronic state. This situation is similar to first-order Raman processes in which the incident and scattered light resonance

CP723, *Electronic Properties of Synthetic Nanostructures*, edited by H. Kuzmany et al.
© 2004 American Institute of Physics 0-7354-0204-3/04/$22.00

conditions give two different excitation laser energies. The difference for the DR Raman spectroscopy is that the two processes gives $\mathbf{q}$ and $-\mathbf{q}$ vectors for two-phonon DR Raman spectroscopy. As for one-phonon DR Raman spectroscopy, the elastic and the inelastic scattering processes give a different phonon $\mathbf{q}$ vector from an initial $\mathbf{k}$ state to satisfy the energy-momentum conservation. Thus for graphite, the D-band spectra appearing at $1350\,\mathrm{cm}^{-1}$ (one-phonon DR) can be fitted to two Lorentzians while the G'-band at $2700\,\mathrm{cm}^{-1}$ (two-phonon DR) can be fitted to one Lorentzian.

FORMULATION OF RAMAN INTENSITY

The first-order Raman intensity as a function of phonon energy, $\hbar\omega$, and of the incident laser energy, E_L is calculated by

$$I(\omega, E_L) = \sum_i \left| \sum_a \frac{M^d(\mathbf{k}-\mathbf{q}, b \to i)M^{ep}(-\mathbf{q}, a \to b)M^d(\mathbf{k}, i \to a)}{(E_L - E_{ai} - i\gamma)(E_L - E_{ai} - \hbar\omega - i\gamma)} \right|^2, \tag{1}$$

in which i, a and b denote, respectively, an initial state, the excited states, and the scattered state of an electron. An electron at wavevector $\mathbf{k}$ is (1) excited by an electronic dipole transition, $M^d(\mathbf{k}, i \to a)$, (2) then scattered by emitting a phonon with phonon wavevector $\mathbf{q}$ by an electron-phonon interaction, $M^{ep}(-\mathbf{q}, a \to b)$ and (3) finally emits a photon by an electronic dipole transition, $M^d(\mathbf{k}-\mathbf{q}, b \to i)$. For an energy separation between the i and a states, E_{ai}, the resonance conditions are either the incident resonance condition, $E_L = E_{ai}$, or the scattered resonance condition, $E_L = E_{ai} + \hbar\omega$. The sum in Eq.(1) is taken over possible intermediate states specified by a and initial states i. In order to take the sum for the intermediate states, we need the matrix elements of the electron-photon interaction, M^d, and electron-phonon, M^{ep}, interactions which are given below. In the scattering process, energy momentum conservation for an electron and phonon holds, which is not explicitly written in Eq.(1).

The second-order, two-phonon Raman intensity as a function of the sum of the two phonon energies $\omega = \omega_1 + \omega_2$ and E_L is given by a similar formula,

$$I(\omega, E_L) = \sum_i \left| \sum_{a,b,\omega_1,\omega_2} \frac{M^d(\mathbf{k}, c \to i)M^{ep}(-\mathbf{q}, b \to c)M^{ep}(\mathbf{q}, a \to b)M^d(\mathbf{k}, i \to a)}{(E_L - E_{ai} - i\gamma)(E_L - E_{bi} - \hbar\omega_1 - i\gamma)(E_L - E_{ai} - \hbar\omega_1 - \hbar\omega_2 - i\gamma)} \right|^2. \tag{2}$$

Now we have two phonon scattering processes with phonon wavevectors $\mathbf{q}$ and $-\mathbf{q}$. In order to get two resonance conditions at the same time, an intermediate electronic state E_{bi} is always resonant ($E_L = E_{bi} + \hbar\omega_1$), and either the incident resonance condition ($E_L = E_{ai}$) or scattered resonance condition ($E_L = E_{ai} + \hbar\omega_1 + \hbar\omega_2$) is satisfied. For a second-order process, the one-phonon Raman intensity is calculated by substituting one of the two phonon scattering processes into the elastic impurity scattering. However the elastic scattering matrix element depends on the nature of the defects. Thus we will not consider this matrix element in this paper. Another point to mention is the energy uncertainty of γ. Since the phonon scattering and photon emission processes are time-dependent processes which have life times on the order of 0.1 ps and 0.1 ns, respectively,

408

the resonance width of the energy is expected by the uncertainty relation, to be 10 meV and 0.01 meV, respectively. It is important to consider these values for reproducing the Raman spectral shape. However in this paper, we will not go in detail for simplicity.

Electron-photon matrix element

The absorption and spontaneous emission of light can be treated as an optical dipole transition matrix element at wavevector ($\mathbf{k}$) which has the form of an inner product of the polarization vector, $\mathbf{P}$ and the dipole vector, $\langle \Psi^c(\mathbf{k}) | \nabla | \Psi^v(\mathbf{k}) \rangle$ [10]

$$M(\mathbf{k}) = \mathbf{P} \cdot \langle \Psi^c(\mathbf{k}) | \nabla | \Psi^v(\mathbf{k}) \rangle, \tag{3}$$

in which $\Psi^i(\mathbf{k})$, $i = c, v$ are eigenfunctions for the conduction (valence) energy bands which can be expanded by Bloch orbitals at A and B sites of the graphite unit cell

$$\Psi^i(\mathbf{k}) = \sum_{j=A,B} C_j^i(\mathbf{k}) \Phi_j(\mathbf{k}, \mathbf{r}), \quad i = c, v. \tag{4}$$

Optical absorption or emission in graphite (and SWNTs) occurs at $\mathbf{k}$ points near the K point of the two dimensional (2D) Brillouin zone of graphite, since the valence and conduction energy bands touch each other at the K point. An important fact for the two coefficients of the Bloch wave-functions for A or B sites, $C_j^i(\mathbf{k})$ (j =A, B), is that the $C_j^i(\mathbf{k})$ are either constant or linearly proportional to the magnitude of $\mathbf{k}$, as measured from the K point. As a result, the dipole vector is proportional to $(-k_y, k_x, 0)$, which is a linear function of k_x (or k_y) in the case of graphite. This fact tells us that the optical absorption (or emission) matrix element has a node on an equi-energy circle around the K point and the node position is rotated by the polarization direction[10]. It is not easy to observe the existence of such a node by optical measurements since the optical absorption occurs for all $\mathbf{k}$ states on the equi-energy contour. Recently Cançado et al. observed this node by using a thin graphite (nanographite) ribbon[11]. Since the graphite ribbon has a small width in one direction, the wavevector becomes discrete in this direction. Thus by rotating the polarization relative to the ribbon direction, we can observe oscillations in the absorption intensity experimentally.

In the case of isolated SWNTs, optical absorption and emission are observed. The dipole vector of a SWNT is not similar to that for graphite since every carbon atom within the large unit cell of a SWNT has a different atomic dipole vector relative to the polarization vector. Nevertheless the analytic expression for the matrix element has a node, if we plot the matrix element as a function of $\mathbf{k}$ in the 2D Brillouin zone[12]. In Fig. 1, we plot the value of the matrix element as a function of the van Hove singularity (VHS) points for many SWNTs around the K point and for optical polarization parallel to the nanotube axis. Each point corresponds to a SWNT VHS point. From this figure, we can estimate the diameter and chirality dependence of the optical intensity.

The matrix element has a larger value in the three directions from K to the M points than from K to the Γ points. The matrix element values increase with increasing distance from the K point in k space, corresponding to the experimental fact that smaller diameter

FIGURE 1. Electron-photon matrix element as a function of the VHS k point of SWNTs. The numbers on the various contours are the values of matrix elements in arbitrary units.

SWNTs give stronger optical absorption and photoluminescence (PL). PL spectra are usually observed at the energy of the lowest VHS for isolated semiconducting SWNTs, E_{11}^S. It is known that the E_{11}^S energy position in k space is either along the K-M side or K-Γ side, respectively, depending on whether it is a type I [mod($2n+m$,3)=1] or type II [mod($2n+m$,3)=2] semiconducting SWNT[13]. Thus for a similar diameter SWNT, the PL intensity is stronger for type I than type II, which is consistent with the experimental results[14].

Electron-phonon matrix element

Next we consider the electron-phonon matrix element for an electron in the conduction band. Initially, the electron is in a $\mathbf{k}$ state on the energy-contour around the K (or K') point for photo-excited states. The scattered $\mathbf{k}+\mathbf{q}$ states are energy-momentum conserving electronic states obtained by emitting a phonon with wavevector $-\mathbf{q}$ and energy $\hbar\omega(-\mathbf{q})$. The matrix element is given by

$$M_j(\mathbf{k},\mathbf{k}+\mathbf{q}) = A_j \mathbf{u}_{-\mathbf{q}}{}^j \cdot \langle \Psi^c(\mathbf{k}+\mathbf{q})|\nabla V(r)|\Psi^c(\mathbf{k})\rangle, \qquad (5)$$

in which $\langle \Psi^c(\mathbf{k}+\mathbf{q})|\nabla V(r)|\Psi^c(\mathbf{k})\rangle$ is the deformation potential vector, and A_j and $\mathbf{u_q}^j$ are the amplitude and eigenvectors of the phonon. For a given initial state in graphite, there are six different phonon modes and two possible scattering paths, known as intra-valley and inter-valley scattering[9]. The final states are on circles around the K point whose size is slightly smaller by each phonon energy than the initial energy contour. In the case of a SWNT, since the wave vector in the circumferential direction becomes discrete, $6N$ phonon states are involved[15]. Each phonon mode is labeled by an irreducible representation of the C_N point group. Only A symmetry phonon modes contribute to the scattering within the 1D electronic energy band. The other E symmetry phonon modes contribute to interband transitions.

In Fig. 2(a) we plot the electron-phonon matrix element for graphite and for a zone center phonon with $q = 0$ in which the initial and final states can be approximated to be the same. Strictly speaking it is not possible by energy-momentum conservation to plot this case. However, since the electron energy from the Fermi energy (2 eV in this

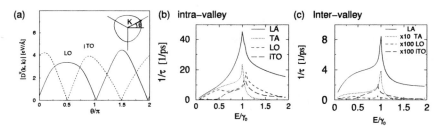

FIGURE 2. (a) Electron phonon coupling constant for graphite for $q = 0$ around the K point, where 0 degrees corresponds to the k_x axis. (b) Intra-valley and (c) Inter-valley contributions to the inverse of the electron relaxation time for four in-plane phonon modes. $\gamma_0 = 2.89$ eV.

case) is much larger than the phonon energy (0.2 eV at most), we can use this value for understanding the electron-phonon matrix element for the Raman intensity.

Because of the mirror symmetry of the graphite plane, the deformation potential vector is within the graphite plane and thus the electron-phonon coupling constant for out-of plane modes becomes zero. For $\mathbf{q} = 0$ phonons, the two in-plane acoustic phonon modes do not contribute to the electron-phonon coupling constant. Thus only the longitudinal optic (LO) and in-plane tangential optical (iTO) phonon mode contribute to the electron-phonon modes for ($\mathbf{q} = 0$), that gives the G-band Raman signals.

When the initial state rotates around the K points, the LO and iTO electron-phonon matrix elements oscillate alternatively as shown in Fig. 2(a). Thus, the relevant phonon for the G-band signal depends on the $\mathbf{k}$ position. In the case of graphite, however, we can not distinguish between the contributions from LO and iTO phonons to the Raman spectra. In the case of SWNTs, LO and iTO phonon modes give G_+ and G_- spectra whose relative intensity to one another depends on the chirality of the SWNT[16]. The present calculation of the electron-phonon matrix element is consistent with our previous calculation of the chirality dependence[16].

For general $\mathbf{q}$ phonon modes, we expect the electron-phonon coupling to contribute to the electron relaxation. In graphite and SWNTs, the relaxation rate of an electron for phonon emission is on the order of 0.1 ps, which is much faster than an optical transition (0.1 ns), but much slower than electron-electron scattering (plasmon emission) (1 fs) for the high energy electron states. Hertel and coworkers reported fast optics measurements for photo excited electrons, showing that the excited electron population at first decays very fast by e-e interaction and then decays more slowly by the e-p interaction by phonon emission [17].

Using the Fermi Golden Rule, we can calculate the lifetime for intra- and inter-valley electron-phonon scattering from a given k state using four in-plane phonon modes. In Figs. 2(b) and (c), we plot the inverse of the lifetime for four in-plane phonon modes as a function of electron energy. The fastest recombination is for LA phonon modes both for intra-valley scattering [Fig. 2(b)] and inter-valley scattering [Fig. 2(c)]. The electron-phonon coupling constant for the LA phonon mode increases quickly with increasing q values. The relaxation rate is singular around $E = 3$ eV, because of the singularity in the electronic density of states. The relaxation rate of an electron decreases with decreasing energy to the Fermi energy. It is a unique fact that in graphite the density of states $D(E)$ near the Fermi energy (E_F) is proportional to E relative to E_F. When the electron energy decreases to the Fermi energy, the number of possible scattered states decreases to zero. Thus the lifetime increases with decreasing energy and finally the lifetime for electron-phonon processes becomes longer than the lifetime for photoluminescence and at this energy PL occurs[12].

SUMMARY

In summary, we have calculated electron-photon and electron-phonon matrix elements for both graphite and SWNTs. The calculated results for the electron-photon matrix elements explain the diameter and type dependence of the PL for semiconducting SWNTs. The electron-phonon interaction near $q = 0$ for phonons is anisotropic around the K point, which is consistent with the observed chirality dependence of the relative intensity of the G_+ and G_- features. The special electronic structure around the K points gives rise to a divergence for the electron-phonon scattering, and is the source of strong emission of far-infrared electro-magnetic radiation for both graphite and SWNTs.

REFERENCES

1. M. Cardona, in *Light Scattering in Solids II: edited by M. Cardona and G. Güntherodt*, pages 19–176, (Springer-Verlag, Berlin, 1982), Vol. 50. Chapter 2, Topics in Applied Physics.
2. R. Saito, A. Grüneis, Ge. G. Samsonidze, V. W. Brar, G. Dresselhaus, M. S. Dresselhaus, A. Jorio, L. G. Cançado, C. Fantini, M. A. Pimenta, A. G. Souza Filho, New J. of Phys. **5**, 157 (2003).
3. M. S. Dresselhaus, G. Dresselhaus, A. Jorio, A. G. Souza Filho, R. Saito, Carbon **40**, 2043 (2002).
4. F. Tuinstra, J. L. Koenig, J. Phys. Chem. **53**, 1126 (1970).
5. M. S. Dresselhaus, P. C. Eklund, Advances in Physics **49**, 705 (2000).
6. S. Reich, C. Thomsen, J. Maultzsch, "Carbon Nanotubes", (Wiley-VCH, 2004); C. Thomsen and S. Reich, Phys. Rev. Lett. **85**, 5214 (2000).
7. J. Kürti, V. Zólyomi, A. Grüneis, H. Kuzmany, Phys. Rev. B **65**, 165433 (2002).
8. R. M. Martin, L. M. Falicov, in *Light Scattering in Solids I: edited by M. Cardona*, pages 79–145, (Springer-Verlag, Berlin, 1983), Vol. 8. Chapter 3, Topics in Applied Physics.
9. R. Saito, A. Jorio, A. G. Souza Filho, G. Dresselhaus, M. S. Dresselhaus, M. A. Pimenta, Phys. Rev. Lett. **88**, 027401 (2002).
10. A. Grüneis, R. Saito, Ge. G. Samsonidze, T. Kimura, M. A. Pimenta, A. Jorio, A. G. Souza Filho, G. Dresselhaus, M. S. Dresselhaus, Phys. Rev. B **67**, 165402 (2003).
11. Cançado, M. A. Pimenta, A. Jorio, R. A. Neves, G. Medeiros-Ribeiro, T. Enoki, Y. Kobayashi, K. Takai, K. Fukui, M. S. Dresselhaus, R. Saito, unpublished.
12. J. Jiang, R. Saito, A. Grüneis, G. Dresselhaus, M. S. Dresselhaus, unpublished.
13. R. Saito, G. Dresselhaus, M. S. Dresselhaus, Phys. Rev. B **61**, 2981 (2000).
14. Y. Miyauchi, S. Chiashi, Y. Murakami, Y. Hayashida, S. Maruy, Chem. Phys. Lett **387**, 198 (2004).
15. R. Saito, G. Dresselhaus, and M. S. Dresselhaus, *Physical Properties of Carbon Nanotubes* (Imperial College Press, London, 1998).
16. R. Saito, A. Jorio, J. H. Hafner, C. M. Lieber, M. Hunter, T. McClure, G. Dresselhaus, M. S. Dresselhaus, Phys. Rev. B **64**, 085312 (2001).
17. G. Moos, R. Fasel, T. Hertel, J. Nanosci. Nanotech. **3**, 145 (2003) related papers therein.

NEW MATERIALS AND BIOLOGICAL NANOSTRUCTURES

Boomerang-shaped VO$_X$ nanocrystallites

U. Schlecht*, L. Kienle*, V. Duppel*, M. Burghard* and K. Kern*

*Max-Planck-Institut für Festkörperforschung, Heisenbergstr. 1, D-70569 Stuttgart, Germany

Abstract. "L"-shaped VO$_X$ nanobelts were obtained by hydrothermal synthesis. These nanobelts represent the first example of nano-sized objects, containing well-defined kinks. The angle was found to be $96° \pm 3°$. Here we report on initial experiments with transmission electron microscopy (TEM) and selected area electron diffraction (SAED), which revealed twinning to be the origin of the kinked structure. The interesting boomerang-shaped nanocrystallites were compared with their more widely known counterpart, the V$_2$O$_5$ nanofibers. Furthermore, thin films with areas exceeding $10 \times 10\ \mu$m^2 have been found to be produced by the hydrothermal synthesis route. The SAED data revealed, that all three morphologies are based on a similar crystal structure.

INTRODUCTION

V$_2$O$_5$ nanofibers (see Figures 1 (a), 2 (a)) have attracted attention due to their possible use as electrostatic coatings and as building blocks in nanoelectronics [1, 2]. These nanofibers, with a width of 10 nm and a height of 2.5 nm (see inset of Figure 1 (a)), can be easily grown in aqueous solution. Their room-temperature synthesis, however, bears the drawback of a long synthesis time ($\approx$ 3 month), until they reach a length exceeding several μm.

In order to speed up the growth, a hydrothermal synthesis route was attempted to reduce the preparation time. These procedure yielded two new morphologies instead of regular nanofibers. The first morphology consists of broad bands (Figures 1 (b), 2 (b)) with a width of 500 nm, while their height matches that of the nanofibers, as displayed in the inset of Figure 1 (b). Most remarkably they appear in the shape of a boomerang, containing a kink with a reproducible angle of $96° \pm 3°$ (see Figure 1 (b)). The second new morphology, as displayed in Figure 2 (c), consists of thin films with dimensions exceeding $10 \times 10\ \mu$m^2.

EXPERIMENTAL

For the synthesis of V$_2$O$_5$ nanofibers 0.2 g ammonium(meta)vanadate (NH$_4$VO$_3$) and 2 g acid ion exchange resin (DOWEX 50WX8-100) were added to 40 ml of water. Due to a polycondensation process, fibers are formed within an orange colored gel. In order to obtain fibers with length exceeding 5 μm, the solution is kept at ambient conditions for more than 10 weeks.

In the hydrothermal approach – used for the synthesis of the boomerang-shaped material, as well as the thin film – 150 μl of vanadium-oxytriisopropoxide have been mixed with 15 ml of water. The mixture was stirred at 70-80 °C until the vanadium-precursor

CP723, *Electronic Properties of Synthetic Nanostructures*, edited by H. Kuzmany et al.

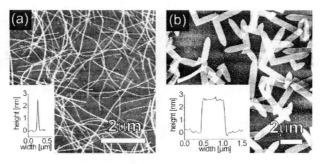

FIGURE 1. Atomic force microscopy images of V_2O_5 nanofibers (a) and boomerang-shaped VO_x nanocrystallites (b). The inset of the images display the cross-sections of the corresponding material

completely dissolved in the solvent. The solution was transferred in an autoclave cell and heated for 3.5 h at 180 °C.

The obtained vanadiumoxide dispersions were dip-coated on amino-silanized SiO_2-wafers and Formvar-coated copper-grids for AFM- and TEM-experiments, respectively. AFM-measurements have been performed on a DI Nanoscope IIIa, while TEM and SAED have been carried out on a Philips CM30/ST (300 kV). See Reference [3] for details.

RESULTS AND DISCUSSION

Figure 2 compares the morphology (a-c) and SAED pattern (d-f) of the three species. The [001]-diffraction pattern of an isolated L-shaped species (Figure 2 (e)) attests the monocrystallinity within an arm of the nanobelts. This pattern is in accordance to a pattern simulation based on the model of $V_2O_5 \cdot n\ H_2O$ xerogel [4], (assuming lattice-parameters a=12.55 Å, c=3.76 Å, $C2/m$). As the nanofibers of Figure 2 (a) are randomly oriented within the selected area used for the diffraction, a powder-like diffraction pattern is observed (Figure 2 (d)). The diffraction fringes of the fibers are in accordance to the diffraction spots of the boomerang-shaped nanocrystallites as indicated with the dashed-lines in both diffraction patterns. In addition, the diffraction pattern of the thin film (Figure 2 (f)) is in close agreement to that of the monocrystalline L-shaped material. However, one observes a double spot feature, where the two spots are related to each other by a small angle of rotation (~5°). This phenomenon can be attributed from the wrinkled structure of the foil and the resulting superposition of two layers whose planes are rotated relative to each other by the specified angle. Despite their different morphology, all three species have the same crystal structure along the [001] direction, which is in accordance to that reported previously by Petkov et al. [4]. It should be stressed, that supplementary experiments are needed in order to access the third dimension.

In order to elucidate the origin of the kinked morphology, we recorded a detailed series of SAED-patterns along an isolated boomerang-shaped nanobelt [3]. From this

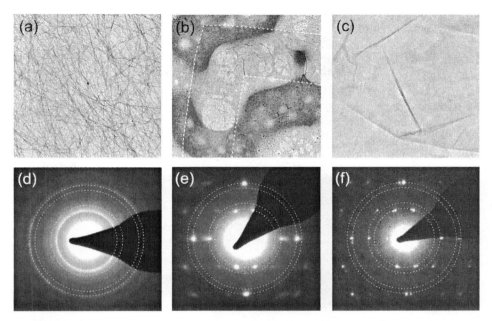

FIGURE 2. Transmission Electron Microscopy (TEM) Images of V_2O_5 nanofibers (a), boomerang-shaped VO_X nanocrystallites (b), and VO_X foil (c). The image size is 2.4 μm each. Selected Area Electron Diffraction (SAED) of an agglomerate of nanofibers (d), an isolated arm of a boomerang-shaped nanocrystallite (e), and a folded region within the foil (f)

analysis it became evident, that the SAED-pattern obtained in the center of the nanobelt is a superposition of the distinct patterns of the two arms [3]. It is therefore concluded, that twinning is the origin of the kinked structure. Similar twins have previously been found in V_2O_5-crystals [5].

The twinning-direction can be determined from the superimposed pattern of the center region (Figure 3 (a)). For comparison, Figure 3 (b) shows the simulated pattern of the central region, which is based on the model of Petkov et al. [4]. Whereas the spots of only one color (black or gray) are giving rise to the diffraction pattern of an isolated arm (as diplayed in Figure 2 (e)), the sum of both patterns is required to reproduce the superimposed pattern observed in the center of the kink (Figure 3 (a)).

The spots along the vertical axis belong to both patterns as indicated with the two-colored circles. Therefore this direction is the twinning-axis. Based on the $[010]*_{1,2}$ directions in reciprocal space, the twinning axis can be identified to be $[310]*$ in reciprocal space, corresponding to $[130]$ in direct space. The angle between the $[010]*_{1,2}$ directions in reciprocal space is found to be $84°$ and agrees well with the angle of $96°$ observed in direct space (Figures 1 (b), 2 (b)).

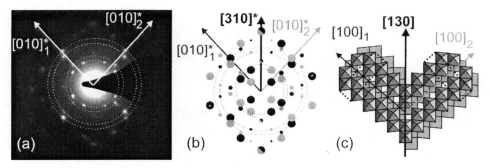

FIGURE 3. (a) SAED-pattern of the center position of a boomerang-shaped VO_x nanocrystallite. (b) Simulated SAED pattern of two V_2O_5 crystals, twinned along the [310]* orientation (vertical direction). (c) Model of the twin boundary region, the dashed box corresponds to the unit cell of each arm

CONCLUSION

Despite their different morphologies, V_2O_5 nanofibers, boomerang-shaped VO_x nano-crystallites and VO_x foils have a similar crystal structure. The kinked structure of the L-shaped nanobelts could be identified to originate from a twinning along the [130] direction. Based on the crystal-model of Petkov et al. [4], we propose a structural model of the kinked region [3]. The model as depicted in Figure 3 (c) is built up of two layers. Each layer consists of edge-sharing VO_5 pyramids. Slight distortions are needed in order to join together the two arms of the L-shaped nanobelt. The presence of distortions has also been observed in V_2O_5-crystals [5].

REFERENCES

1. Muster, J., Kim, G., Krstic, V., Park, J., Park, Y., Roth, S., and Burghard, M., *Adv. Mater.*, **12**, 420 (2000).
2. Kim, G., Muster, J., Krstic, V., Park, J., Park, Y., Roth, S., and Burghard, M., *Appl. Phys. Lett.*, **76**, 1875 (2000).
3. Schlecht, U., Knez, M., Duppel, V., , Kienle, L., and Burghard, M., *Appl. Phys. A*, **78**, 527 (2004).
4. Petkov, V., Trikalitis, P., Bozin, E., Billinge, S., Vogt, T., and Kanatzidis, M., *J. Am. Chem. Soc.*, **124**, 10157 (2002).
5. Hyde, B., and Tilley, R., *Phys. Stat. Sol. (a)*, **2**, 749 (1970).

Conducting Properties of Single Bundles of $Mo_6S_3I_6$ Nanowires

Marko Uplaznik [1], Ales Mrzel [1], Daniel Vrbanic [1,2], Peter Panjan [1], Bostjan Podobnik[3], Dragan Mihailovic [1,4]

[1] Jozef Stefan Institute, Jamova 39, 1000 Ljubljana, Slovenia; e-mail: marko.uplaznik@ijs.si
[2] Faculty of Chemistry and Chemical Technology, Askerceva 5, 1000 Ljubljana, Slovenia
[3] LPKF d.o.o., Planina 3, 4000 Kranj, Slovenia
[4] Mo6 d.o.o., Tehnoloski park, Teslova 30, 1000 Ljubljana

Abstract. The implementation of $Mo_6S_3I_6$ nanowires (MoSI) in microelectronics requires basic evaluation of their electronic properties. Low-ohmic contacts between the bundles and metallic guides is crucial. We describe resistivity measurements for bulk samples and for single bundles of MoSI nanowires. In the first case (Moly-paper), variable range hopping consistent with electrons traveling through the mesh of bundles was observed. Using e-beam lithography a device for longitudinal measurement of conductivity of single bundles was manufactured. A dependence on the different contact configurations is reported.

INTRODUCTION

MoSI nanowires are an alternative to carbon nanotubes in many applications. Both have unique electrical and mechanical properties and have been proposed as nanoscale building blocks, in particular for the construction of molecular electronic devices. Experiments on the basic properties of contacts between the MoSI nanowires and the metallic contact are crucial for making devices. The simplest proposal for a device is bulk material MoSI mesh connected to the electrodes. In that case the properties of individual bundles are not so important, and the interactions *between* them determine the properties of the device. Alternatively, using e-beam lithography, devices with single MoSI bundle can be produced. Their properties depend mainly on the contacts between the metallic contacts and the bundle and on the interactions between the nanowires inside each individual bundle[1]. This dependence is crucial, because it is possible to synthesize bundles with different diameters and thus to control the electrical performance of the devices.

We also present experimental results of temperature dependent resistivity for the bulk material. For single bundle measurement a comparison was made between the longitudinal resistivities for two layouts of the bundle: a) on-top and, b) under the metallic contacts.

$Mo_6S_3I_6$ is obtained by direct synthesis from the elements[2]. Using an ultrasonic bath, the bundles are dispersed in isopropanol and separated from other crystalline products. For the bulk sample, the dispersed bundles are dried , whereas for the single bundle measurement the sample is prepared using a drop of the dispersed liquid.

CP723, *Electronic Properties of Synthetic Nanostructures*, edited by H. Kuzmany et al.
© 2004 American Institute of Physics 0-7354-0204-3/04/$22.00

Bulk Measurement

The sample for the bulk measurement was prepared in the form of a round pressed pellet, with a diameter of 3 mm and a mass of about 50 mg. The sample was connected with silver paste to 4 golden wires (Fig.1a).

The sample was cooled down to 10 K in a cryostat. Four contact measurements were performed using two a Keithley 238 as the current source, and a Keithley 2000 for the measurement of the drop of voltage over the sample pellet. Because we performed four wire measurement, the contact resistivity did not influence the results.

a)

b)

FIGURE 1. (a) a schematic picture of pressed pellet and golden wires ahed to it; (b) SEM picture of a bundle mesh

In these measurements, the conducting is determined by a perculative path through a mesh of bundles as shown in Figure 1b). If we neglect the longitudinal resistivity of the bundles and assume random distances between links, we can describe the behavior of resistivity with Variable Range Hopping (VRH) theory expressed in Eq.1, and shown in Figure 2.

$$\rho = \rho_0 \exp\left[\left(\frac{T}{T_0}\right)^{-\frac{1}{4}}\right] \tag{1}$$

The diagram on Fig.2 a) shows the plot of the results for one sample; smaller graph shows saturation of the voltmeter, when the resistivity reaches the performance of the instrument. In order to verify the VRH formula for resistivity we plot the $T^{-1/4}$ dependence in logarithmic scale, because the shape of the curve is expected to be linear.

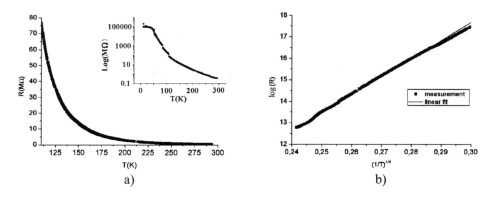

FIGURE 2. a) temperature dependence of the resistivity for cooling and heating of the sample; the inset shows whole temperature range b) temperature plot of $T^{-1/4}$ dependence in logarithmic scale

The diagram b) on Fig.2 shows the results plotted in described way and fitted with linear function. Due to the small discrepancy we can say, that the VRH law is confirmed.

Single Bundle Measurements

The single bundle devices were fabricated using standard electron lithography. A comparison was made between dielectrophoretical attachment of the bundle on prefabricated electrodes (Fig. 2a) and the direct metallization of contacts on top of the bundle (Fig. 2b). We compared the basic conducting properties of this two different types of contacts between the bundle and the electrode.

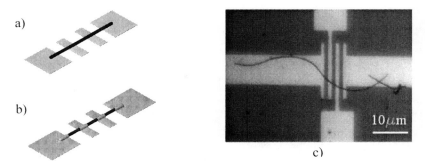

FIGURE 3. A scheme of a bundle over a) and under b) four contacts; c) A picture of a bundle under 4 contacts

The IV curves appear to be linear in both cases of bundles over and under the contacts. This implies metal like behavior. The resistivity in the first case was in range of several GΩ (Fig.3a) whereas in case of direct contacts was 1000 times smaller in range of MΩ (Fig.3b). We suggest an explanation as follows: when the bundle is introduced on top of the metal, a contact is created only with a very small number of nanowires under the bundle. If the metal is sputtered over the bundle, the number of nanowires contributing to the current increases, because there is a much larger contact area for the metal, than in the first case.

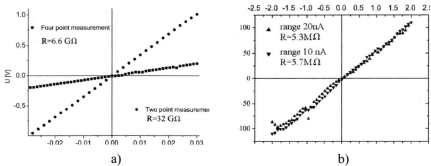

a)   b)

FIGURE 3. a) IV dependence for samples with bundle *on top* of contacts; b) IV dependence for samples with bundle *under* contacts

CONCLUSIONS

MoSI bulk measurements of conducting properties on Moly-paper indicate variable range hopping between single bundles in a bundle mesh. Such devices on Moly-paper, can be used as sensitive sensors for gases or pollutant in a liquid as the intercalation of molecules between the bundles can change the hopping rate and the conductivity of the device.

Single bundle devices show a strong dependence for different layouts of the bundle. The contact resistance for the on-top position is three orders of magnitude higher than in case of direct coating of a metal over the bundle. These results indicate that the procedure by which a circuit is made has a great impact on the basic properties of the system and they need to be taken into account in potential applications.

ACKNOWLEDGMENTS

We would like to acknowledge Nanotemp project and Ministry of Education, Science and Sport of Slovenia for financial support.

REFERENCES

1. Kis, A. et. al. , D., *Shear and Young's moduli of MoS$_2$ nanotube ropes,* Advanced Materials **15(9)**, 733-736, 2003
2. Vrbanić D. et *al., Nanotechnology* **15**, 635- 638 (2004)
3. Remskar, M. et. Al.., *Self -Assembly of Subnanometer - Diameter Single -Wall MoS2 Nanotubes,*Science, 292, p. 479 (2001)
4. Uplaznik, M., *Meritev vzdolžne električne prevodnosti nanožičk Mo6S4I4 z uporabo elektronske nanolitografije,* Thesis, Faculty of Mathematics and Physics, University of Ljubljana, 2003.
5. Mott, *Electronic processes in noncrystalline materials,* Clarendon press-Oxford, 1979.

$Mo_6S_3I_6$ nanowires

D. Vrbanic[1], A. Meden[1], B. Jancar[2], M. Ponikvar[2], B. Novosel[1],
P. Venturini[3], S. Pejovnik[1], D. Mihailovic[2,4]

[1] Faculty of Chemistry and Chemical Technology, University of Ljubljana, Askerceva 5, Ljubljana,
Slovenia
[2] Jozef Stefan Institute, Jamova 39, Ljubljana, Slovenia
[3] National Institute of Chemistry, Hajdrihova 19, Ljubljana, Slovenia
[4] Faculty of Mathematics & Physics, University of Ljubljana, Jadranska 19, Ljubljana, Slovenia

Abstract. We report on the synthesis and characterization of new nano-wire-like material with
a chemical formula $Mo_6S_3I_6$ or MoSIx-6 (x= 6); nomenclature: $Mo_6S_{9-x}I_x$. The distinguishing
features of the material are rapid one-step synthesis, easy isolation and dispersion into small-
diameter wires. Elemental analysis, X-ray diffraction, thermal gravimetric analysis, electron
microscopy and Raman spectroscopy were used to characterize the compound.

INTRODUCTION

Compounds in the molybdenum-sulphur-iodine system appear to form a variety of
different quasi-one-dimensional materials. MoS_2 nanotubes (NTs) [1] appear to be
just one of a large number of diverse transitional- metal chalcogenide (TMC)
nanostructured materials. After a systematic study designed to find alternate one-
dimensional transition-metal chalcogenide-based nanowires and nanotubes, we report
here the discovery of a new quasi-one-dimensional nano-structured material with the
formula $Mo_6S_3I_6$. The material can be synthesized in a single step and it is composed
of identical small-diameter nanowires, *weakly bound* in bundles, which can be
handled in the same way as CNTs, yet have the added advantage that they can be
dispersed in organic solvents. Particularly the reproducibility, ease of handling and its
functional properties open the way for widespread bulk applications, possessing
significant advantages over other inorganic nanowires and nanotubes.

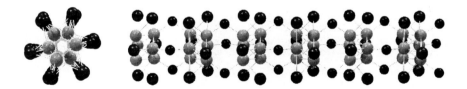

FIGURE 1. A proposed structure of $Mo_6S_3I_6$ nanowires.

CP723, *Electronic Properties of Synthetic Nanostructures*, edited by H. Kuzmany et al.
© 2004 American Institute of Physics 0-7354-0204-3/04/$22.00

EXPERIMENTAL

$Mo_6S_3I_6$ is obtained by direct synthesis from the elements [2]. The majority of the resulting material has a furry-like appearance, with individual needles having a diameter of about 50 nm to 1000 nm and a wide range of lengths, up to 5 mm. Figure 2 a) shows a scanning electron microscope (SEM) image of as-produced material. The material is not brittle and rather tough, and the individual needles are not easily broken on handling. It can be dispersed in organic solvents, such as methanol, ethanol, isopropanol, acetone, benzene or toluene using an ultrasonic bath, where the degree of dispersion depends on concentration and duration of sonification. Upon examination with a transmission electron microscopy (TEM) these nanoropes are seen to be composed of very fine nanowires, approximately 1 nm in diameter (Figure 2b).

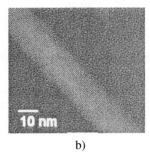

a)   b)

FIGURE 2. a) SEM image of as-grown $Mo_6S_3I_6$. The individual nanorods typically have a diameter of 50 nm to 1000 nm. b) TEM image of a 10 nm $Mo_6S_3I_6$ bundle.

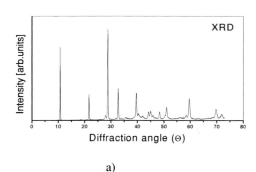
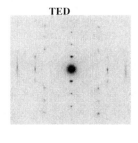

a)   b)

FIGURE 3. a) X-ray powder diffraction pattern of $Mo_6S_3I_6$. b) A transmission electron diffraction (TED) pattern obtained from the bundle.

To perform elemental analysis, we carefully selected nanowire material from the growth batch, excluding any crystals (a small amount of layered MoS_2 appears to be present as an impurity), and the wires were washed in acetone to remove traces of unreacted iodine. The chemical composition was determined to be (in mass %): Mo: 40.4%; S: 6.4%; I: 53.1%, with an uncertainty of ± 0.3 %. Compared with the calculated ratio for $Mo_6S_3I_6$, Mo: 40.2%; S: 6.7%; I: 53.1 %, we see that the deviations from exact stoichiometry are largest in the S content, but are within the uncertainty limits of the measurement. Thermogravimetry (TG) was performed in a gas mixture of argon: oxygen (80 : 20) with flow of 6 l/h and heating rate 10 ºC/min.

From the TG data, (Figure 4) a total weight loss of 100 % was derived when the material was heated up to 1400 °C. A weight loss of 40 %, with an exothermic effect occurring at temperatures between 300-500 °C corresponds to the decomposition and oxidation to molybdenum trioxide MoO_3. The second weight loss of 60 % with an endothermic effect occurs between 800-1150 °C, which can be explained by the total evaporation of molten MoO_3 (melting point 800 °C).

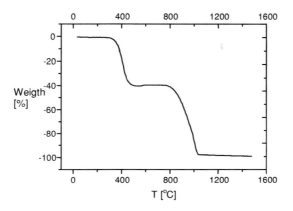

FIGURE 4. a) TG analysis of the $Mo_6S_3I_6$ in an argon/oxygen (80 : 20 %) atmosphere. Decomposition of $Mo_6S_3I_6$ to MoO_3 occurs above 300 °C.

RESULTS AND DISCUSSION

In this discussion we focus on the possible advantages of $Mo_6S_3I_6$ in terms of functional properties. The straightforward one-step synthesis and simple isolation procedure clearly means that scaling up can be performed, suggesting a great technological advantage over existing materials. We find that the trace impurities (such as MoS_2) can be removed by a simple dispersion and decanting and/or filtering procedure. Different degrees of dispersion yield different diameter nanowire bundles directly from the raw material, smaller diameters being obtained on greater dilution. TG measurements show that the material is stable in a mixture of argon and oxygen (80 : 20) up to 300 °C, and above 600 °C in an Ar atmosfere, making it useful for nanomechanical and electronic applications. The fact that the nanowires appear to be stable in air when dispersed to only a few nm confirms that the forces between the NTs are relatively weak. The obvious consequence of these weak inter-nanowire interactions is imperfect ordering of nanowires within the bundles. This gives very broad X-ray and streaked transmission electron diffraction peaks, with the drawback that an unambiguous determination of the detailed structure is rather difficult. The fact that the material is compatible with water, together with the known excellent tribological properties of this class of materials, solid lubrication might by another promising application for $Mo_6S_3I_6$.

We conclude by noting that the combination of easy synthesis and dispersion, together with many functional properties of the Mo-S-I family suggest $Mo_6S_3I_6$ to be a very commercially promising nano-material. In addition to the usual applications, $Mo_6S_3I_6$ might be useful in some areas where existing nano-materials are not suitable, such as in catalysis, as polyelectrolytes, as battery electrodes [3], field emission tips [4], or in tribology [5], where related materials have already been shown to have potential applications.

ACKNOWLEDGMENTS

The authors acknowledge prof. dr. Venceslav Kaucic from the National Chemical Institute, Ljubljana for making available the X-ray powder diffractometer and Ministry of Education, Science and Sport of Slovenia and European Community's Human Potential Program (NANOTEMP) for financial support.

REFERENCES

1. Remškar, M. et al., *Science* **292,** 479- 481 (2001)
2. Vrbanić D. et al., *Nanotechnology* **15,** 635- 638 (2004), Jesih A. et al., EU patent PCT/EP04/001870
3. Dominko R. et al., *Advanced Materials* **14,** 1531-1534 (2002)
4. Nemanič V. et al., *Appl. Phys. Lett.,* **82,** 4573- 4575 (2003)
5. Joly-Pottuz L. et al., *Tribology Lett.,* in press (2004)

Synthesis and Magnetic Characterization of $Cu(OH)_2$ Nanoribbons

P. Umek[*†], J. W. Seo[†], L. Fórró[†], P. Cevc[*], Z. Jagličič[**], M. Škarabot[*], A. Zorko[*] and D. Arčon[*‡]

[*]Institute "Jožef Stefan", Jamova 39, 1000 SI- Ljubljana, Slovenia
[†]Institute of Physics of Complex Matter, FSB, EPFL, CH-1015 Lausanne, Switzerland
[**]Institute of Mathematics and Physics, Jadranska 19, SI-Ljubljana, Slovenia
[‡]Faculty for Mathematics and Physics, Jadranska 19, SI-Ljubljana, Slovenia

Abstract. We report on the synthesis of $Cu(OH)_2$ nanoribbons in aqueous media with a typical diameter around 15 nm. The length of the nanoribbons depends on the synthesis route and can achieve up to several μm. We also demonstrate the possibility of transforming $Cu(OH)_2$ to CuO structures. The magnetic properties of $Cu(OH)_2$ nanoribbons are due to antiferromagnetic interactions between Cu^{2+} spins bridged by OH groups and show typical low-dimensional antiferromagnetic behaviour. Finite size effects are suggested to be responsible for the increase of the magnetic susceptibility below 20 K.

INTRODUCTION

One-dimensional semiconducting and metallic nanostructures [1] have attracted considerable attention because of their special properties such as a large surface-to-volume ratio or unique electronic and optical properties as compared to those of the bulk materials. Nanowire arrays were for instance grown from Co [2], In_2O_3 [3] or MoS_2 [4] layered material. However, to our surprise we found only few attempts to grow copper nanostructures described in the literature although one would expect that controlled growth of Cu nanostructures could find their place in the electronic devices almost immediately. A possible starting point for the design of copper-based nanostructures is copper(II) hydroxide ($Cu(OH)_2$) material. $Cu(OH)_2$ is a well-known layered material [5], and its orthorhombic crystal structure may be ideal for the growth of nanoribbons. The growth of $Cu(OH)_2$ is expected to be a result of the assembly of olated chains $>Cu(OH)_2Cu<$ and under some specific conditions nanoribbons of $Cu(OH)_2$ can be grown. The $Cu(OH)_2$ structure is rather unstable and can be transformed to black copper(II) oxide (CuO). For instance, strong electron beam irradiation transforms $Cu(OH)_2$ to CuO, at higher temperatures to Cu_2O and at even higher tempereatures to Cu [6]. If the nanoribbon morphology of the material is preserved during these transformations, then we have a simple recipe for the cheap synthesis of Cu-based nanostructures on a large scale.

Here we report on the synthesis of $Cu(OH)_2$ nanoribbons in aqueous media and the transformation of $Cu(OH)_2$ into CuO nanoribbons in organic solvents. It is also the purpose of this paper to shed some additional light on the structure, stability and magnetic properties of $Cu(OH)_2$ nanoribbons.

CP723, *Electronic Properties of Synthetic Nanostructures*, edited by H. Kuzmany et al.
© 2004 American Institute of Physics 0-7354-0204-3/04/$22.00

EXPERIMENTAL DETAILS

The procedure for the precipitation of $Cu(OH)_2$ nanoribbons consists of mixing of $CuSO_4 \times 5H_2O$ solution with [7] or without ammonium hydroxide before adding sodium hydroxide as the source of hydroxide anions. The samples prepared with ammonium hydroxide are denoted as $Cu(OH)_2 - 1$, while samples prepared without ammonium hydroxide are named as $Cu(OH)_2 - 2$. The entire synthesis is performed at room temperature. The precipitated particles of $Cu(OH)_2$ were then filtrated after 5 minutes of aging the solution under constant stirring at room temperature. The precipitated materials were finally washed on filter with distilled water until the pHs of the filtrates were not bellow 6 and then with ethanol. The products were dried on air at room temperature over the night and later stored at 4 °C.

Produced materials were characterized with transmission electron microscopy (TEM, Phillips, CM-20, 200kV) and scanning electron microscopy (SEM, Phillips, XL30 FEG, accelerating voltage was 15 kV). The magnetic susceptibility χ and the magnetization curves were measured with a Quantum Design SQUID magnetometer.

RESULTS AND DISCUSSION

Structural characterization and stability

We studied the morphology of the synthesized $Cu(OH)_2$ materials using SEM and TEM techniques. A typical SEM image of $Cu(OH)_2$ nanoribbons synthesized with addition of ammonia ($Cu(OH)_2$-1) is shown in Fig. 1a). These nanoribbons tend to form bundles, which can be even longer than 2 μm. From TEM images (not shown here) we estimated that the average diameter of observed nanoribbons is around 15 nm and that the material itself is polycrystalline.

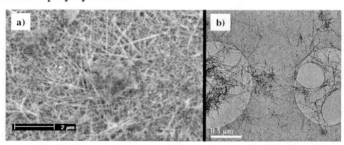

FIGURE 1. a) SEM image of $Cu(OH)_2$ nanoribbons synthesized with addition of ammonia, and b) TEM image of $Cu(OH)_2$ nanoribbons synthesized without addition of ammonia.

Nanoribbons of $Cu(OH)_2$ synthesized without any addition of ammonia ($Cu(OH)_2$-2) are shown on Fig. 1b). As can be concluded from the TEM image, the nanoribbons produced in this way are much shorter (up to 0.5 μm). The tendency to form bundles is in this case not so pronounced when compared to the $Cu(OH)_2$ nanoribbons synthesized following the first procedure.

can be slowed down when the samples are kept in organic solvents like methanol or ethanol. This is in particular true for $Cu(OH)_2 - 1$ nanoribbons whose structure seems to be rather stable. On the other hand $Cu(OH)_2 - 2$ material, slowly decomposes even in the organic solvent but the dehydrated material preserves the shape and dimensions of $Cu(OH)_2$ nanoribbons [8]. The exact origin of the difference in the stability of the two samples is not well understood at the moment but it can be due to different concentration of defects or even due to different amount of Na^+ ions intercalated into the nanoribbons.

Magnetic properties

The magnetic properties of hydroxobridged binuclear complexes are of particular interest because they can exhibit either ferromagnetic or antiferromagnetic character depending on the details of the crystal structure. A correlation between the magnetic coupling constant J and the Cu-O-Cu bridging angle as well as out-of-plane displacement of hydroxo groups has been found [9]. The $Cu(OH)_2$ structure (Fig. 2b) consists of assembly of Cu-OH-Cu chains where Cu-O-Cu angle is along the chain 96.7°. Hydrogen is out of plane for 5.5° [10]. These angles suggest predominantly antiferromagnetic $Cu^{2+} - Cu^{2+}$ interactions [9].

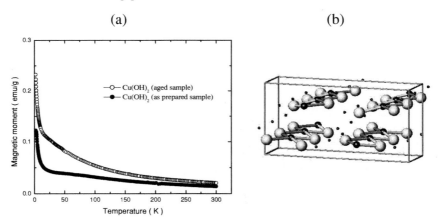

(a)   (b)

FIGURE 2. (a) The temperature dependence of the magnetic susceptibility of as-prepared (solid circles) and aged (open circles) $Cu(OH)_2$ samples. (b) The structure of the $Cu(OH)_2$. Here black circles represent Cu atoms, while large gray circles stand for oxygen. The smallest circles are at H positions.

The temperature dependence of the static magnetic susceptibility is shown on Fig. 2a for the freshly synthesized and aged $Cu(OH)_2$ nanoribbon sample. The two susceptibilities basically show the same type of behavior: at high temperatures the susceptibility follows a Curie-Weiss law down to about 60-70 K. At this temperature the susceptibil-

the value expected for a system of $S = 1/2$ spins. The dominant magnetic interactions between Cu^{2+} spins are antiferromagnetic as can be deduced from the temperature dependence of the inverse magnetic susceptibility. The Curie-Weiss temperature is in fresh samples -43(2) K while in aged sample this temperature decreases to -23(2) K.

Below 15 K the susceptibility suddenly strongly increases again. Such increase can be either an evidence for the magnetic ordering found for instance in some intercalated copper hydoxides [11] or due to the chain-end spins that are coupled to magnetic excitations of the Cu-OH-Cu chains. Additional high-field EPR and ^{1}H NMR measurements [12] should resolve between these two possibilities.

CONCLUSIONS

In conclusion, a simple synthetic approach is proposed for the synthesis of $Cu(OH)_2$ nanoribbons with an average diameter of about 15 nm and lengths up to 4 μm. The magnetic properties of $Cu(OH)_2$ nanoribbons are characteristic of low-dimensional antiferromagnets. The increase of the magnetic susceptibility at low-temperatures seems to reflect finite size effects and coupling of the "chain-end" spins to the Cu-OH-Cu chain magnetic excitations. With aging the length of the nanoribbons drastically decreases and the $Cu(OH)_2$ structure can be ultimately transformed to CuO-based nanoribbons.

ACKNOWLEDGMENTS

The authors gratefully acknowledge the financial support from the Slovenian Ministry for Education, Science and Sport, the European Commission (RTN Program, NANOCOMP network, RTN 1-1999-00013) and NATO (SfP 976913 grant).

REFERENCES

1. A. Agfeldt, M. Gretzel, *Chem. Rev.* **95** (1995) 49.
2. H. Q. Cao, Z. Xu, H. Sang, C. Y. Tie, *Adv. Mater.* **13** (2001) 121.
3. M. J. Zheng, L. D. Zheng, X. Y. Zhang, J. Zhang, G. H. Li, *Chem. Phys. Lett.* **334** (2001) 298.
4. M. Remškar, A. Mrzel, Z. Škraba, A. Jesih, M. Čeh, J. Demšar, P. Stadelmann, F. Levy, D. Mihailovič, *Science* **292** (2001) 479.
5. Y. Cudennec, A. Riou, A. Lecerf, Y. Gerault, *Eur. J. Solid State Inorg. Chem.* **32** (1995) 1013.
6. Z.L. Wang, X.Y. Kong, X. Wen, S. Yong, *J. Phys. Chem.* **B107**, 8275 (2003).
7. R. Rodriguez-Clemente, C. J. Serna, M. Ocana, E. Matijević, *J. Crystal Growth* **143** (1994) 277.
8. P. Umek et al., to be published.
9. A. Ruiz, P. Alemany, S. Alvarez, and J. Cano, *J. Am. Chem. Soc.* **119**, 1297 (1997).
10. H.R. Oswald et al., *Acta Cryst.* **C46**, 2279 (1990).
11. W. Fujita and K. Awaga, *J. Am. Chem. Soc.* **119**, 4563 (1997).
12. D. Arčon et al., to be published.

XPS Study of Carbyne-Like Carbon Films

T. Danno*, Y. Okada, J. Kawaguchi

Department of Lifestyle Design, Kochi Prefectural University
5-15 Eikokuji-Cho, Kochi 780-8515 JAPAN
*tdanno@cc.kochi-wu.ac.jp

Abstract. X-ray photoelectron spectra (XPS) of carbyne-like carbon (CLC) film obtained by the polymer reaction route were measured and analysed. Dehydrohalogenation reaction was carried out effectively regardless of the starting precursors. C1s core-level spectrum of CLC film shows the anomalous lowest peak maxima at 282.6 eV compared with the different carbon materials. An advanced model of carbyne chains concerning the conjugation length distribution is needed to analyse the electronic structure based on the valence-band XPS.

INTRODUCTION

Carbyne (polyyne) or carbyne-like carbon (CLC) has attracted the particular interest of researchers from the point of view of the simplest linear chains of carbon atoms and the next allotrope of carbon [1]. We obtained the CLC films by dehydrochlorination of solid poly(vinylidenechloride) using 1,8-diazabicyclo[5,4,0]undec-7-ene (DBU), which is the strongest organic base so far known, in polar solvents[2]. The film shape was kept by the two step treatments, where the amorphous and the crystalline regions are dehydrochlorinated stepwise.

We also demonstrated the clear dispersion effect for the stretch mode of carbon-carbon triple bonds ($\nu_{C\equiv C}$) on the resonance Raman spectra from the CLC film [3]. Comparing the Raman spectra with the quantum chemical calculations, the average number of the conjugated sp-carbon in the CLC film was estimated to be 10, which means 5 carbon-carbon triple bonds [3,4].

X-ray photoelectron spectroscopy (XPS) has emerged as a powerful tool to provide surface information of solids, on which the core-level electrons are excited by external photons and measured by an electron spectrometer. In molecular or solid compounds, the energy of a bonded electron in a specific element is subjected to a shift which conveys information about the chemical environment of the atoms under investigation. In addition to the core-level spectra, spectrum at the lower energy, so called the valence band spectrum, reflects the electronic structure of the material and has been widely investigated for the carbon materials.

In this paper we measured three types of XPS , *i.e.*, the wide scan spectrum (1200-0 eV), C1s core-level spectrum (295-275 eV), and valence-band spectrum (30--5 eV) for

CP723, *Electronic Properties of Synthetic Nanostructures*, edited by H. Kuzmany et al.
© 2004 American Institute of Physics 0-7354-0204-3/04/$22.00

the CLC films and discuss the structure from both chemical and physical aspects.

EXPERIMENT

CLC film was obtained from PVDC or PVDF spin-coated film on the Si substrate by the dehydrohalogenation treatments using DBU as described previously [2]. The specimen was stored under N_2 atmosphere before XPS measurements. XPS were obtained from a Shimadzu / Kratos AXIS-HS spectrometer with Mg $K\alpha$ (1253.6 eV). A flood type charge neutraliser attached to the spectrometer was used simultaneously during the measurement. Calibration of binding energy was carried out using the spectrum of Ag $3d_{5/2}$ (368.3 eV) before and after sample measurements.

RESULTS AND DISCUSSION

In the wide scan spectrum of CLC film obtained from PVDF, C(100), F(5), O(1), and N(2) were detected. The values in the parentheses represent the atomic ratios of the elements after correcting the raw peak intensities with the photoionization cross section. The above result suggests that the dehydrofluorination from the precursor material (PVDF in this case) was carried out effectively, which is consistent with the transmitted IR spectrum of CLC film [2]. Similar result was obtained for the CLC film from PVDC.

Figure 1 shows the C1s core-level spectra of CLC film and PVDF. Single peak at anomalous lower binding energy (282.6 eV) was observed in the spectrum of CLC, where the characteristic peaks of PVDF almost vanished. Similar single C1s spectrum was observed at 284.4 eV for SWCNT [5] and 283.7 eV for *tr*-polyacetylene [6], respectively. The peak shifts towards lower binding energy with increasing the bond order of the carbon atoms. The anomalous feature of the C1s spectrum of CLC was explained by Pesin as the different electrostatic charging of dielectric samples instead of the

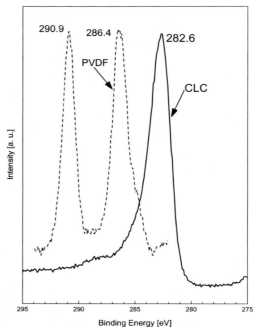

Figure 1 . C1s core-level XPS of CLC and PVDF.

intrinsic electronic structure of CLC [7]. Indeed the Raman spectrum of CLC showed the stronger intensity from amorphous carbon region than the carbyne region, this suggest that the dominant C1s peak could be assigned to the sp^2- or sp^3- carbon and the down shift of the peak. But the charging of samples in the spectrometer is observed normally as the up shift of the peaks with broadning of band width. Band width of CLC is comparable as the one for starting material. Thus the anomalous down shift of the C1s peak for CLC is still open to discuss with more precise experiments.

Figure 2. shows the valence-band XPS of CLCs obtained from PVDF and PVDC, respectively. Spectra were very noisy since the sample damaging by the X-ray exposure easily occurred. Thus the iteration of scanning was limited to obtain an unperturbed spectra. Increasing of the intensity around 30 eV observed in the spectrum of CLC from PVDF comes from the F_{2s} peak because of the larger photoionization cross section of fluorine. A couple of peaks around 8 eV and 18 eV for CLC from PVDF and 4 eV and 17 eV from PVDC were observed, respectively. Peak around 17 or 18 eV well corresponds to the Density of State (DOS) from σ electron compared with diamond and graphite [8]. Peak around 4 or 8 eV should correspond to the DOS from π electrons. Beitinger proposed a DOS model for an infinite α-carbyne chain as a sum of the partial contribution from each σ and π subbands [9]. Comparing the

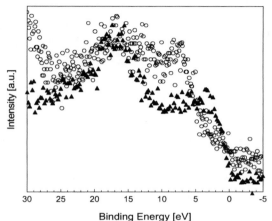

Figure 2. Valence-band XPS of CLC films. ○: CLC from PVDF, ▲: CLC from PVDC

experimental spectrum with the model, strength of the DOS from π subbands shows large difference than the one for σ subbands. The model treated a single infinite carbyne chain whereas the actual sample is an assembly of the chains of the different conjugaton length[4]. Thus the conjugation length distribution should be also considered for the model.

CONCLUSIONS

XPS of CLC films were measured and analysed as following :
(1) Dehydrohalogenation was carried out effectively regardless of the starting precursor.
(2) C1s core-level XPS of CLC film shows the anomalous lowest peak maximum at 282.6 eV compared with the different carbon materials.
(3) An advanced model of carbyne chains concerning the conjugation length distribution is needed to analyse the electronic structure based on the valence-band XPS.

ACKNOWLEDGEMENT

We thank Kochi Prefectural Industrial Technology Center for their kind permission to use the spectrometer. Support of this work by the Project for International Research Exchange of the Kochi Prefectural University is also acknowledged.

REFERENCES

[1] Kudryavtsev Yu.P., Heimann R.B., in *Carbyne and Carbynoid Structures*, eds. R.B.Heimann, S.E.Evsyukov, L.Kavan, Dordrecht/Boston/London, KLUWER ACADEMIC PUBLISHERS, 1999, Ch. 1, pp. 1-7.

[2] Danno T., Murakami K., and Ishikawa R., in *Electronic Properties of Novel Materials -Science and Technology of Molecular Nanostructures* (AIP Conference Proceedings 486), eds. H.Kuzmany, J.Fink, M.Mehring, and S.Roth, Melville, New York, American Institute of Physics, 1999, pp213-216.

[3] Danno T, Murakami K, Krause M, and Kuzmany K., in *Electronic Properties of Novel Materials -Science and Technology of Molecular Nanostructures*(AIP Conference Proceedings 544), eds. H.Kuzmany, J.Fink, M.Mehring, and S.Roth, Melville, New York, American Institute of Physics, 2000, pp473-476

[4] Kastner J., Kuzmany H., Kavan L., Dousek F.P., Kürti J., *Macromolecules*, **28(1)**, 344-353(1995).

[5] Chiu P.W., Dissertation for the technical university of München, 2003

[6] Salaneck W.R., Thomas H.R., Bigelow R.W., Duke C.,B., Plummer E.W., Heeger A.J., and MacDiarmid A.G., *J. Chem. Phys.* **72**, 3674 (1980).

[7] in Ref. [1] pp. 377.

[8] McFeely E.R., Kowalczyk S.P., Ley L, Cavell R.G., Pollak R.A., and Shirley D.A., *PRB*, **9(12)**, 5268-5277(1974).

[9] in Ref. [1] pp. 375.

Isolation, positioning and manipulation of $Mo_6S_3I_6$ by (di)electrophoresis

M.Ploscaru[1], A.Mrzel[1], D.Vrbanic[2], P.Umek[1], M.Uplaznik[1], B. Podobnik[4], D.Mihailovic[1,3], D.Vengust[1,3], V. Nemanic[1], M.Zumer[1], B. Zajec[1]

[1]Jozef Stefan Institute, Jamova 39, 1000 Ljubljana, Slovenia, e-mail: mihaela.ploscaru@ijs.si
[2]Faculty of Chemistry and Chemical Technology, Askerceva 5, 1000 Ljubljana, Slovenia
[3]Mo6 d.o.o. , Teslova 30, 1000 Ljubljana, Slovenia, www.Mo6.com
[4]LPKFd.o.o., Planina3, 4000 Kranj, Slovenia

Abstract. We report on the isolation, positioning and manipulation of nanowire bundles of $Mo_6S_3I_6$. Electrophoresis is found to be useful for separating bundles of different size. Dielectrophoresis was shown to be an effective way to self-assemble $Mo_6S_3I_6$ nanowires onto metal electrodes, which were used for field emission and for electrical measurements.

INTRODUCTION

Interest in nanomaterials has been growing rapidly in the past several years. New types of nanomaterial, such as are $Mo_6S_3I_6$ nanowires [1], are promising for a variety of potential applications. The distinguishing features of these materials are rapid one-step synthesis from the elements in bulk quantities. The material is composed of sub-nanometer diameter nanowires, weakly bound in bundles (Figure 1a) which can be easily dispersed in polar solvents like isopropanol or ethanol. After dispersion, a quick analysis of the material with a scanning electron microscope (SEM), shows that the solution contains $Mo_6S_3I_6$ nanowires of different diameters and also some impurities. In this contribution we present the results of electrophoresis experiments, where we managed to separate the $Mo_6S_3I_6$ nanowires from the impurities. We also present results of dielectrophoresis experiments for positioning $Mo_6S_3I_6$ nanowires onto contacts for different applications.

ELECTROPHORESIS FOR PURIFICATION

For these experiments we prepared 0.625 M TBE (trisbase, boric acid, EDTA) buffer with pH=8 and 1% agarose gel, in which, after cooling down, we made small pockets (Figure 2a). Two types of solution were placed in the gel pockets: one was made from 1 ml of DMF (dimethylformamide), 3 mg $Mo_6S_3I_6$, 0.5 ml glycerol and 0.5 ml deionized water, everything dispersed for 1 hour in an ultrasonic bath, and subsequently for 5 minutes with a sonification tip at 750 Watts power. The second solution was prepared in the same way, but DMF was replaced with isopropanol.

CP723, *Electronic Properties of Synthetic Nanostructures*, edited by H. Kuzmany et al.
© 2004 American Institute of Physics 0-7354-0204-3/04/$22.00

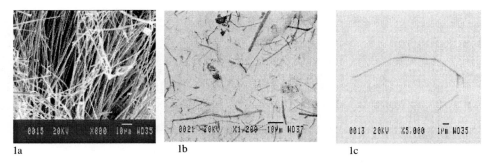

1a   1b   1c

FIGURE 1. SEM image of: 1a -$Mo_6S_3I_6$ nanowires as grown; 1b -DMF solution where it can be seen bundles of $Mo_6S_3I_6$ and also impurities; 1c -collected nanowires after electrophoresis experiments

We expect that $Mo_6S_3I_6$ nanowires cannot travel through the gel like carbon nanotubes [2], because their diameter is too big comparative with the diameter of the gel's pores. So, performing electrophoresis experiment, after 20 minutes under 300V bias, we could observe that part of the nanowires are deposited on the bottom of the pockets and another part is deposited on the vertical wall corresponding to the positive electrode (shown by arrow in Figure 2b). A SEM image of the material removed from the wall was compared with a SEM image of the solution put in the pockets at the beginning (the two are compared in Figures 1b and 1c). The result is that the impurities and those big bundles that could not be dispersed are deposited on the bottom of the pockets while the small bundles are deposited on the "positive wall". Still, the method can be used for separation of the nanowires from impurities and can be adopted to separate large quantities of material by DC electric field. More importantly, the experiments also show that the bundles are charged when are dispersed in polar liquids.

Nanowires deposited on the wall

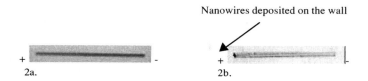

+   -   +   -
2a.   2b.

FIGURE 2. Image showing the diffused solution in the pocket at the beginning of the experiment (2a.). In the second one we can see that a fraction of the nanowires is deposited on the bottom of the pocket and the other fraction is deposited on the vertical wall, shown by the arrow (2b)

Dielectrophoresis for self-assembling $Mo_6S_3I_6$ onto contacts

Dielectrophoresis is the electro-kinetic motion of dielectrically polarized materials in non-uniform electric fields and is currently an active area of interest research for positioning and orientation or alignment of CNTs [5-7] and other nanomaterials.

In our experiments we used dielectrophoresis for positioning $Mo_6S_3I_6$ nanowires onto Ni wires covered with 99.95% pure Indium. The diameter of Nickel covered wires is = 100 µm.

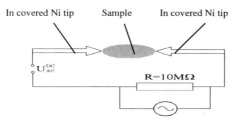

In covered Ni tip   Sample   In covered Ni tip

U_{AC}^{DC}

R=10MΩ

FIGURE 3 A schematic diagram of the circuit used for trapping single bundle of $Mo_6S_3I_6$

A suspension of $Mo_6S_3I_6$ was dispersed in isopropanol by sonification. The tips were very close together (d=100 µm) on a glass support and wired up with a series resistance of 10 MΩ as shown in Figure 3. The circuit is powered either by a DC or AC power supply. For the measurement of the electric current during the experiment, we used an oscilloscope which monitored the drop of voltage across a series resistance. After switching on the power supply, a drop of nanowires in DMF suspension (c=0.05 mg/ml) was applied onto the tips. The trapping we performed with an applied DC field of 20V and an applied AC field of 10 V at a frequency of 10 MHz. The process was run under a microscope where we were able to observe the experiment from the beginning, when the suspension was applied and how the nanowires were moving to the positive electrode, until the solution is dried (Figure 4a and 4b). Because of limited resolution of the optical microscope, we could not see well if something is trapped in between the contacts. A SEM picture was this subsequently taken after a contact was established. A single bundle of $Mo_6S_3I_6$ could then be observed on the tip (Figure 4c).

4a   4b   4c

FIGURE 4 Optical microscope image made immediately when the suspension was applied on the contacts (4a); microscope picture made when the solution is dried (4b); SEM picture showing a single bundle attached to Ni tip (4c)

RESULTS

During the experiment, the drop of voltage across the resistance was measured with an oscilloscope (Figure 5a). A jump of voltage can be observed in the graph which is attributed to the moment when the suspension is applied on contacts. Almost immediately after that, a bundle of nanowires is trapped on the emitter. The increase of the voltage, indicated in the graph by an arrow, correspond to the moment in which a bundle is trapped Once we have obtained a single bundle on a tip we can use it as an emitter for field emission measurements. The results of such measurements are presented in Figures 5b and 5c. An I-V curve is plotted together with a Fowler-Nordheim plot for the emitted current. Importantly, during the field emission measurements, the $Mo_6S_3I_6$ bundle didn't detach from the tip which suggests that the contact between the $Mo_6S_3I_6$ nanowire and emitter is quite strong.

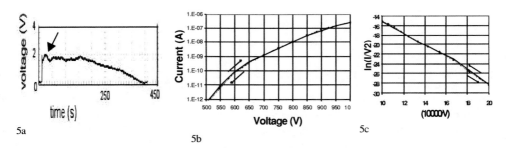

5a

5b

5c

FIGURE 5 Drop of voltage across the resistance (5a); I-V field emission curve (5b); F-N plots (5c)

CONCLUSIONS

Electrophoresis has been demonstrated to transport MoSIX nanowire bundles in solution. Bundles with small diameters travel to the positive electrode while impurities and larger bundles are deposited on the bottom of the chamber. This method demonstrates that the NWs carry charge in solution and move in a DC electric field, which can be successfully utilized for positioning or for separation of MoSIX material from impurities, as well as for separation of different diameter NWs. We have also shown that dielectrophoresis can be used for self-assembly of nanowires onto contacts. We have succeeded to obtain single bundles attached onto Indium covered Nickel tips which were used in field emission measurements.

REFERENCES

1. D. Vrbanic et al., Nanotechnology, 15,635-638(2004)
2. Umek P., D. Mihailovic, Separation of SWNTs by diffusion, Synthetic Metals, 121(1-3):1211-1212
3. H.A. Pohl, Dielectrophoresis (Cambridge University Press, Boston, MA, 1978)
4. M.P. Huges, AC Electrokinetics: application for nanotechnology, Nanotechnology 11(2000)124-132
5. Bubke K. Gnewuch, H., Hempstead M., Hammer J. and Green M. L. H., 1997 Appl. Phys. Lett. 71,1906
6. Krupke R., Hennrich F., Weber H. B., Beckmann D., Hampe O., Malik S., Kappes M.M. and Lohneysen H. V., 2003 Appl. Phys. Lett. 80,3826
7. Wakaya F., Nagai T., Gamo K., 2002 Microelectron. Eng. 63,27

Motions In Catenanes And Rotaxanes

Francesco Zerbetto

*Dipartimento di Chimica "G. Ciamician", Università di Bologna, Via F. Selmi 2, Bologna, Italy, email
francesco.zerbetto@unibo.it*

Abstract. Benzylic amide-based interlocked molecular architectures are becoming important players in the super-molecular field. They take the form of either catenanes or rotaxanes (which are not true supramolecular systems since the separation of the two independent molecules requires breaking of a chemical bond). Subsequent to their synthesis, much effort carried out by several groups has been aimed at measuring, understanding, controlling and directing the dynamics of unique degrees of freedom that arise from the formation of the mechanical bond between the two systems where either one ring is threaded through the cavity of another or the same ring is locked onto a dumbbell molecule. Here I present a review of the computer modeling performed in recent years on these systems with the final aim of discussing the recent unidirectional rotation that has been achieved for a very special catenane composed by three interlocked rings.

INTRODUCTION

Catenanes are molecules that owe their name to the Latin word *catena*, i.e., chain. They are formed by two rings, or macrocyles, that are mechanically interlocked. Rotaxanes are molecules that owe their name to the combination of the latin word *rota*, i.e., wheel, and the English word axis. They are formed by a ring locked onto a linear molecule, or thread, which is terminated by two bulky stoppers whose purpose is to avoid the falling off of the ring. When the interactions between the two independently stable molecules – namely the two rings, or the ring and the dumbbell species – are relatively small, the ring may undergo large amplitude motions that are by-and-large governed by the temperature or by possible external stimuli. Not all families of catenanes and rotaxanes show these motions. The reasons for the lack of inter-moiety dynamics may be manifold and are essentially related to the strength of their interactions, which, in turn tends to be large since it is directly related to the forces that leads to the assembly of the interlocked compounds. Amongst the most mobile ever reported, there are benzylic amide interlocked systems pioneered by the Leigh group, now at the University of Edinburgh. They share a common ring, shown in figure 1. Three types of large amplitude motions are usually present in these molecules. In catenanes, the spinning of the macrocycles inside each other's cavity is called circumrotation. In rotaxanes, the rotation of the ring about the axis of the thread is called pirouetting, while the motion of the ring along the axis takes the name of shuttling. In recent years, in my group, we have become interested in the computational modeling of the properties and the dynamics of a class of catenanes and rotaxanes based on the aforementioned benzylic amide motif. In these systems, the

CP723, *Electronic Properties of Synthetic Nanostructures*, edited by H. Kuzmany et al.
© 2004 American Institute of Physics 0-7354-0204-3/04/$22.00

special degrees of freedom resulting from the constrained motion of the two molecular units with respect to each other can have very different rates that cover nearly eight orders of magnitudes and range from the second to the sub-microsecond time regime. Of course these motions can also be blocked altogether mechanically by the presence of appropriate hindering group.

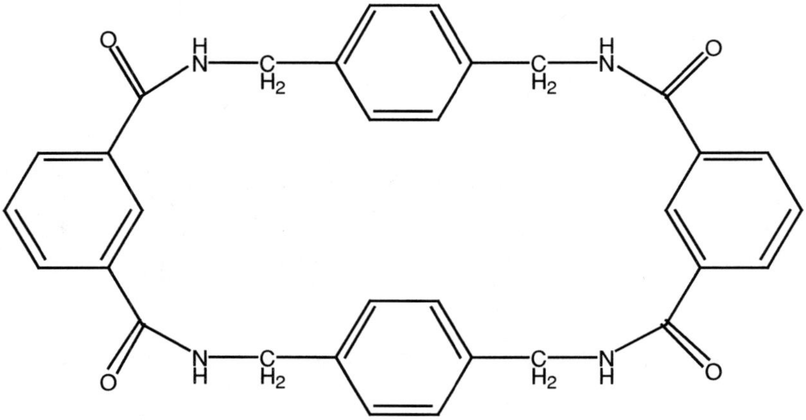

Figure 1. The benzylic amide catenanes and rotaxanes are characterized by the presence of a macrocyclic ring.

The origin of such wide spread of rates is the tunability of the amount of energetic inter-moiety interactions. In some of the benzylic amide systems, the barrier for the motion is just a few kcal mol^{-1}, while in others it is more than 15 kcal mol^{-1}. The whole range is thermally accessible, but it is important to remember that as a rule of thumb the change in activation barrier of 1.4 kcal mol^{-1} yields for a chemical process, at room temperature, a variation of its rate by an order of magnitude.

To the outsider, the question arises of why the large amplitude motions of interlocked molecules are so inspiring to many chemists. The short answer relies on the utilization of molecular-level motion in biology to govern a large number of phenomena that range from cell mobility to muscle activity and the possibility to exploit catenanes and rotaxanes to mimick Nature.

Here, I review part of the most recent systematic computational and theoretical effort to characterize the class of molecular architectures based on the benzylic amide unit. Both catenanes and rotaxanes are considered with the intent of explaining how our understanding of these molecules has developed over time and how it has been possible to achieve for the first time a pre-determined unidirectional motion.

DISCUSSION

The elucidation of the properties of molecular materials of the complexity of catenanes and rotaxanes demands a combined approach, where different kinds of spectroscopies and microscopies must contribute. Indeed for benzylic amide based interlocked molecules, experiments have been carried out using (i) infrared, Raman and inelastic neutron scattering spectroscopies to probe their vibrational dynamics, (ii) traditional uV-vis electronic spectroscopy in solution (both in absorption and in emission), together with electron energy loss and XPS spectroscopies in the solid or on surfaces, to understand their electronic states and their dynamics, (iii) electrochemistry to comprehend the phenomena that take place upon oxidation or reduction, (iv) atomic force and scanning tunneling microscopies to monitor their ordered assemblies on surfaces and their dynamics, (v) non linear optical measurements to gain further understanding of the electronic properties and the dynamics. This list of experimental techniques is by no means comprehensive and other approaches are utilized as the need arises.

Because of the complexity of the systems and the experiments, from the onset of the work it was deemed necessary to seek the additional assistance of various forms of computer modeling that ranged from molecular mechanics, to molecular dynamics, to quantum chemistry, to quantum mechanical models. The use of the simple-looking molecular mechanics may at times seem a necessary evil because of the size of the systems. It must, however, be emphasized that all of the intermolecular interactions in the interlocked species, and a substantial part of the intra-molecular ones, are based on the formation either of π-stacks or of hydrogen bonds. The description of these weak interactions contains dispersion forces that are usually poorly predicted by the Hartree-Fock self-consistent field quantum chemical model and are problematic also for the widely successful Density Functional Theory based methods.

From the start, molecular mechanics calculations were therefore used to describe the dynamics of benzylic amide catenanes and rotaxanes and, in particular, the pathway of rotation of one ring inside the cavity of the other, i.e. circumrotation, of three catenanes.[1,2] We were able to simulate the experimental energy barriers obtained by temperature dependent nuclear magnetic resonance, NMR, measurements that ranged from 0.3 to 0.7 eV with noteworthy accuracy. Analogously, rates of spinning going from the millisecond to the sub-microsecond time regime were reproduced. The success of the modeling afforded further insight into the microscopic dynamics of the circumrotation: we found that this does not occur as a uniform motion, rather it is composed by several, or even a large number, of intermediate steps that involve considerable molecular rearrangements. In practice, the internal hydrogen bond and π-stack patterns fight a losing war against the circumrotation (triggered by the thermal bath). Simply put, along the pathway, a hydrogen bond may be disrupted and become "replaced" by a π-stack or by the formation of a weak bifurcated hydrogen interaction. In the same spirit, a π-stack may turn into a slightly less binding arrangement. In short, the process is a continuous orchestrated dance whose purpose is to decrease the loss of energy associated with the circumrotation.

The same experimental and computational approach was used to investigate the dynamics of two rotaxanes.[3] The combination of Kerr experiments (carried out with an alternating electric field), NMR experiments and simulations showed that in an N-oxide and in a fumaramide rotaxane the macrocyclic ring rotates about the thread with rates that can be tuned by the external field. Experiments and simulations showed that the former presents only the pirouetting motion, while the latter gives a more complicated picture where the main motion is coupled to pivoting of the macrocycle against the thread.

441

The work described above was carried out in solution and it was felt that if practical applications are to come from these molecules, attention had to be re-routed to the solid. This does not imply any detrimental considerations to measurements performed in solution since many phenomena, for instance those of biological interest, have a wet component. Our initial solid state investigation was directed to the dynamics of a catenane on the mica surface.[4] Afterwards we attempted the prediction of which of these interlocked molecules would be more likely to display dynamics in a crystal.[5] As a rule, such a prediction is a nearly impossible task when molecules with different characteristics and properties are involved. The focus, however, was on a set of ~30 benzylic amide rotaxanes which share a common ring and for which the crystal structure was known. The simple reasoning behind the approach was that the molecules had several common features so that a prediction could be feasible. Extensive calculations were analyzed with techniques borrowed from pharmacology where small chemical modifications of a drug may enhance or decrease the biological activity. We finally narrowed down our choice to a few rotaxanes that indeed showed a tendency to display dynamics when smaller forces were used in atomic force microscopy experiments, AFM.[5] Although the experiments were not able to prove conclusively that the dynamics was due to ring motion, a reasonable cause could be made for it.

The AFM experiments gave a further result that is worth discussing. When the force of the tip was kept below the threshold of scratching, the surfaces where the most mobile rotaxanes were deposited spontaneously formed sets of dots that were equally spaced. The spacing between the dots was ultimately governed by the thickness of the film. This unforeseen phenomenon was exploited to write information with an extremely high density.[6] Simulations were instrumental to reach an understanding of the formation of the dots upon the tip stimulus, which was ultimately ascribed to nanocrystallization.

The interplay between experiments and theory made us extend our effort to something of general scientific interest, whose results would hopefully reach beyond the chemistry of these molecules. Inspired by the use of Nature of unidirectional motions to drive mobility in microorganisms or to achieve control of muscles, we investigated and reported the first example of stimuli-driven sequential and unidirectional rotation in molecules where the components were not connected by covalent bonds.[7] This process shares a degree of similarity with the well-known path trodden by myosin walking on actyn, where the two proteins that produce the muscle activity are not covalently linked.

The mechanically interlocked molecular assemblies consisted of one or two small rings that moved, in discrete steps, around a larger ring. The motion went from one binding site or 'station' to another, and was driven by changes in inter-component binding energies induced either by light, or heat, or chemical stimuli. The simpler system was made by two rings, that is a [2]catenane: the small ring moved about the larger one (with high positional integrity), but no control over the direction of the movement was achievable. In the more complicated [3]catenane, two rings moved about a third one in a pre-determined unidirectional sequence. Such larger ring has four stations, A, B, C, D each with a different affinity for the smaller rings.

The sequence of external stimuli that triggerred the unidirectional motion consisted of two electronic absorptions that caused trans to cis isomerization of a carbon-carbon double bond, followed by a final back isomerization by chemical means. Each stimulus modified the binding energy of the isomerized station and made one of the rings move to another site. True samples resulting from the isomerization and the subsequent ring displacement were isolated and characterized by the NMR spectroscopy.

In practice, the presence of a second ring blocks one of the pathways along which the other ring can move once it can no longer bind to an isomerized station (see figure 2).

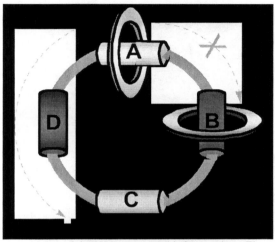

FIGURE 2. Schematic representation of the [3]Catenane and it four stations. The movement of a ring triggered by an external stimulus can only go in one direction because of the presence of another ring that blocks its way.

Initially the rings are located on stations A and B. If one labels the smaller rings as 1 and 2, a shorthand notation for the system is A(1)B(2), that is ring 1 sits on station A and ring 2 sits on station B. The first stimulus yields C(1)B(2), the second, C(1)D(2), while the third apparently restores the original situation but with the smaller rings exchange, i.e. it gives A(2)B(1). In this way a predetermined motion was achieved.

Simulations were used to study the local dynamics of the rings at the stations for each of the isolated samples. In figure 3, a snapshot of the molecular dynamics of one of the systems in solution can be appraised. The calculations were also instrumental to address a fundamental question that arose during our work. The unidirectional motion is due to the exchange or switch of the two smaller rings that, however, can also occur by random thermal motions. Such exchange is not uni-directional because is accompanied by all the other possible variations of the locations of the two rings.

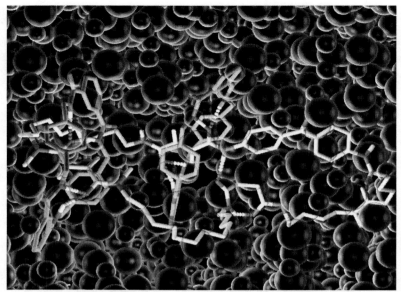

FIGURE 3. A snapshot of molecular dynamics simulation of a [3]Catenane in solution of 839 molecules of CH_2Cl_2. The spheres are atoms of the solvent molecules.

Integration of the relevant set of kinetic equations showed that the time required for the system to go from A(1)B(2) to A(2)B(1) using the thermal motions is about 8 hours, a time longer than the experiments. In turn, this implies that true unidirectionality was achieved.

In conclusion, in the simulation work discussed here on interlocked systems, computer modeling joins both the synthetic effort and various characterization techniques with the aim of understanding this class of materials. In a sense, simulations are becoming just one of the tools of the trade along with several experimental approaches that are trying to make the systematic achieving of predetermined unidirectional motion in chemical systems an ever closer goal.

REFERENCES

1. Leigh, D.A., Murphy, A., Smart, J.P., Deleuze, M.S., Zerbetto, F., *J. Am. Chem. Soc.* **120**, 6458-6467 (1998).
2. Deleuze, M.S., Leigh, D.A, Zerbetto, F., *J. Am. Chem. Soc.* **121**, 2364-2379 (1999).
3. Bermudez, V., Capron, N., Gase, T., Gatti, F.G., Kajzar, F., Leigh, D.A., Zerbetto, F., Zhang, S., *Nature* **406**, 608-611 (2000).
4. Cavallini, M., Lazzaroni, R., Zamboni, R., Biscarini, F., Timpel, D., Zerbetto, F., Clarkson, G.J., Leigh, D.A., *J. Phys. Chem. B* **105**, 10826-10830 (2001).
5. Biscarini, F., Cavallini, M., Leigh, D.A., León, S., Teat, S.J., Wong, J.K.Y., Zerbetto, F., *J. Am. Chem. Soc.* **124**, 225-233 (2002).
6. Cavallini, M., Biscarini, F., Léon, S., Zerbetto, F., Bottari, G., Leigh, D.A., *Science* **299**, 531 (2003).
7. Leigh, D.A., Wong, J. K. Y., Dehez, F., Zerbetto, F., *Nature* **424**, 174-179 (2003).

Synthesis of Silicon Nanowires

A. Colli[1], A. C. Ferrari[1a], S. Hofmann[1], J. A. Zapien[2], Y. Lifshitz[2], S. T. Lee[2], S. Piscanec[1], M. Cantoro[1], J. Robertson[1]

[1]Engineering Dept, Cambridge University, Cambridge CB2 1PZ, UK
[2]Department of Physics and Materials Science, City University of Hong Kong, Hong Kong, SAR, China.

Abstract. We present alternative routes for the production of silicon nanowires (SiNWs). We have successfully synthesized SiNWs by plasma enhanced chemical vapor deposition, high temperature annealing, and thermal evaporation. Selective growth is achieved by using a patterned catalyst. Bulk production of SiNWs can be achieved by thermal evaporation.

INTRODUCTION

Crystalline semiconducting nanostructures have become very popular in recent years both for fundamental physics and their potential applications in electronic and optoelectronic devices [1,2] One-dimensional silicon nanowires (SiNWs) are particularly attractive, due to the central role of the silicon semiconductor industry. Moreover, Si can become a direct band gap semiconductor at nanometer size due to quantum confinement [3] so it could be used in optoelectronics, unlike bulk Si.

Here we discuss three complementary approaches to SiNWs production. Plasma-enhanced chemical vapor deposition (PECVD) with silane as a source and gold as catalyst is used for low-temperature (<400 °C) selective growth of SiNWs [4]. Controlled growth can also be achieved by thermolysis of nanopatterned, multilayered Si/Au thin film precursors [5]. Bulk production of SiNWs is obtained by thermal-vapor-deposition from Si/SiO_2 powders, in a high temperature furnace (>1000 °C). In this case SiNWs grow either by condensing on a gold catalyst [5], or by the self-condensation of the vapor in a lower temperature region of the furnace [6].

RESULTS AND DISCUSSION

We demonstrated the growth of SiNWs by PECVD using SiH_4 as the Si source and Au as the catalyst, at a substrate temperature of 380 °C [4]. Fig. 1(a) shows as-grown SiNWs from a patterned 1 nm thick Au layer. Fig. 1(b) shows a high-resolution transmission electron micrograph (HRTEM) of a SiNW with a 7.4 nm thick crystalline core. The details of the deposition process are reported elsewhere [4]. Elemental mapping by electron energy loss spectroscopy (EELS) shows that they wires consist of a pure Si core surrounded by ~2 nm of SiO_x [4]. PECVD allows selective growth, whilst a growth temperature below 400°C can significantly expand the set of available substrates.

---

[a] Email:acf26@eng.cam.ac.uk

CP723, *Electronic Properties of Synthetic Nanostructures*, edited by H. Kuzmany et al.
© 2004 American Institute of Physics 0-7354-0204-3/04/$22.00

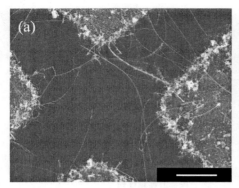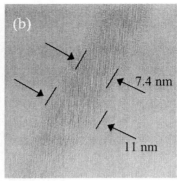

FIGURE 1. (a) SEM image of selectively grown SiNWs on Au patterned substrate. The scalebar is 4 μm. (b) HRTEM micrograph of a 10nm-thick crystalline SiNW [4].

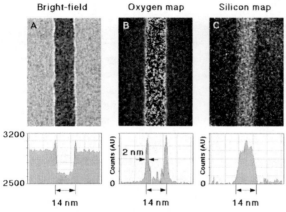

FIGURE 2. EELS mapping of a SiNW, showing the crystalline Si core and the oxide shell [4].

Bulk quantities of free-standing SiNWs can be deposited by a solid-vapor process [6,7]. Si/SiO$_2$ powders were placed in a quartz or alumina boat and heated at 1200°C in a high temperature tube furnace. The substrates were held in the low temperature zone of the tube in the downstream direction of an inert gas flow (Fig. 3). SiNWs nucleated both on the tube walls and on the substrate surface. Furnace SiNWs growth can be either catalyst-assisted or catalyst-free. The catalyst-free mechanism is often referred to as oxide-assisted growth, to stress the critical role of oxygen in providing one-dimensional nanostructure growth [7]. SiO$_x$ clusters condense preferentially at the wire tip, providing the desired one-dimensional growth. This vapor transport method

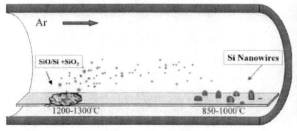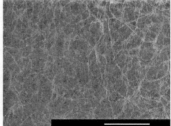

FIGURE 3. (left) Schematic of the oxide-assisted growth. (right) Bulk quantities of SiNWs are grown. Their average diameter is 10-20 nm. Scalebar is 1μm.

is applicable to a large class of materials, and does not require dangerous gases, such as SiH_4, as material sources. On the other hand, the high temperatures involved in the process strongly limit the substrate selection. Nanowires grown by this technique can then be dispersed in solution for further post-growth assembling [8].

Selective growth can also be achieved in a furnace, by annealing pre-patterned precursors. This synthesis approach is simpler than gas phase methods, and allows large-scale production of NWs [9]. We have synthesized Si/SiO_2 nanowires by

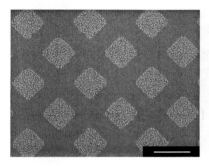

annealing pre-patterned SiO_2/Au vertical heterostructures [5]. Islands of different size were patterned on an oxidized Si wafer by means of shadow evaporation (Fig. 4) or lithography techniques. Evaporation through a 2000 mesh TEM copper grid resulted in a densely packed array of square islands with lateral size of 4-5 μm and periodicity of 10 μm (grid-patterned substrates, Fig. 4). 4x4 arrays of 3-μm-wide islands were obtained by UV photolithography. Patterns of 1μm dots were achieved by electron beam lithography (lithography-patterned substrates). SiO_2

FIGURE 4. SiO_2/Au precursor islands. The scalebar is 10μm.

(200nm-thick) was deposited by magnetron sputtering from a high purity SiO_2 target (99.999%), whilst gold (10nm-thick) was thermally evaporated in a resistive-boat metal evaporator. The base pressure was ~1 x 10^{-6} mbar in both cases. The samples were then placed in a quartz boat and annealed for 2 hours in an atmospheric pressure furnace at 1200°C under N_2 flux, resulting in the growth of Si/SiO_2 nanowires.

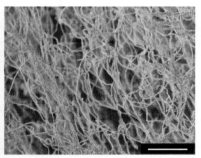

FIGURE 5. (a) SEM micrograph of a grid-patterned sample after annealing. (b) A close-up of the bottom-right part of Fig 5(a). Scalebars are 10μm.

The nanowire density and distribution depend on the initial size and periodicity of the patterned islands. Fig. 5(a) shows an SEM micrograph of a grid-patterned sample after annealing. Wire growth is present all over the patterned area, with an enhancement of wire density concentrated on the central part of the pattern (right side of Fig. 5a). Short wires emerge from the islands close to the margin of the array (left side of Fig. 5(a), showing a strongly position-dependent growth rate. The longest and thinnest wires are produced in the high growth rate zone. A close-up of the high-density area is shown if Fig. 5(b). Wires are tens of microns in length, with some branching and overlapping due to the very high density. Their worm-like morphology suggests the presence of amorphous structures.

Lithography-patterned substrates give different results. Fig. 6 shows an array of nanowire bushes originating from smaller and less dense precursor islands (~2μm in diameter). In this case, every set of wires stems from a specific location and grows independently of the neighbor islands. The wire diameter is between 200 and 400 nm. By further scaling the pattern features to less than 1μm, single nanowires per island can be obtained, in a similar fashion to what done for single carbon nanofiber growth. Decreasing the catalyst thickness will be an effective way to reduce the wires diameter. Further work is underway in order to optimize the growth.

 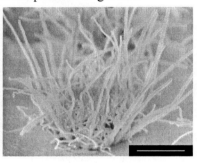

FIGURE 6. Nanowire bushes nucleating from lithography patterned substrates. Scalebars: 4μm.

Raman spectroscopy can become a preferred tool for the non-destructive characterization of SiNWs, like it is for carbon nanotubes. Recent papers found large shifts in the main Raman peak, varying with laser power and wavelength, which they attribute to resonant selection of wires of different diameter or to electronic Raman scattering [10,11]. We previously showed that these shifts are due to intense local heating by the laser [12]. These thermal effects are avoided at very low laser power and phonon confinement can be demonstrated. The high sensitivity of SiNWs to laser irradiation shows that thermal conductivity of SiNWs is much lower than bulk Si.

ACKNOWLEDGMENTS

This work was supported by the EU project CARDECOM. A. C. F. acknowledges funding from the Royal Society.

REFERENCES

1. S. J. Wind et al. *Appl. Phys. Lett.* **80**, 3817 (2002)
2. X. Duan, Y. Huang, R. Agarwal, and C. M. Lieber, *Nature* **421**, 241 (2003)
3. L. T. Canham, *Appl. Phys. Lett.* **57**, 1046 (1990);G. D. Sanders et al. *Phys. Rev. B* **45**, 9202 (1992)
4. S. Hofmann et al. *J. Appl. Phys.* **94**, 6005 (2003)
5. A. Colli et al. unpublished (2004)
6. D.D.D. Ma et al. Science **299**, 1874 (2003)
7. N. Wang, Y.H. Tang, Y.F. Zhang, C.S. Lee, and S. T. Lee, *Phys. Rev. B* **58**, R16024 (1998)
8. *see for example:* M.C. McAlpine et al. *Nano Lett.* **3**, 1531 (2003) *and references therein.*
9. S. Hofmann et al. *Adv. Mater.* **24**, 1821 (2002); T.C. Wong et al. *Appl. Phys. Lett.* **84**, 407 (2004)
10. S. L. Zhang et al. *Appl. Phys. Lett.* **81**, 4446 (2002)
11. R. Gupta et al. *Nano Lett.* **3**, 627 (2003)
12. S. Piscanec et al. *Phys. Rev. B* **68**, 241312(R) (2003)

Electrical Properties of InAs-Based Nanowires

C. Thelander*, M.T. Björk, T. Mårtensson, M.W. Larsson, A.E. Hansen,
K. Deppert, N. Sköld, L.R. Wallenberg, W. Seifert, and L. Samuelson

*Solid State Physics, Lund University, Box 118, 221 00 Lund, Sweden

Abstract. Semiconductor nanowires are grown using chemical beam epitaxy and metal organic vapor phase epitaxy from size-selected gold nanoparticles acting as catalysts. By changing materials during the growth it is possible to form heterostructures both along the length of the nanowires but also in a core-shell fashion. In particular, incorporation of pairs of InP tunnel barriers in InAs nanowires has been used to fabricate single-electron transistors and resonant tunneling diodes.

INTRODUCTION

In contrast to the hollow, layer-by-layer structure of carbon nanotubes, semiconductor nanowires are generally solid, and can be seen as cylinders cut out of a bulk semiconductor. The diameter of a nanowire is mainly determined by the size of the catalyst particle from which it was grown. Growth of wires with diameters down to only a few nanometers has been reported, whereas the lengths can be tens of micrometers.

Nanowires may play an important role in future electronics, both as passive interconnects and as devices, e.g. in the form of transistors and diodes. In addition, semiconductor nanowires have very interesting optoelectronic properties – devices such as photodetectors, light emitting diodes and even lasers have already been fabricated in nanowire systems.[1] Many different types of techniques can be used to grow wires, such as chemical beam epitaxy (CBE), metal-organic vapor phase epitaxy (MOVPE), chemical vapor deposition and laser ablation techniques.

NANOWIRE GROWTH

Charged and size-selected gold aerosols that will later act as catalysts for the nanowire growth are deposited onto III-V substrates by means of an electric field. The particles are produced in the gas phase and deposited randomly onto the growth substrate. Compared to wet deposition techniques, the dry aerosol deposition method produces little to no contamination and also gives rise to evenly spread out gold particles. Catalyst definition can also be made by lithographic techniques, such as electron beam lithography and nano-imprint lithography, as shown in Refs. 2-3.

CP723, *Electronic Properties of Synthetic Nanostructures*, edited by H. Kuzmany et al.
© 2004 American Institute of Physics 0-7354-0204-3/04/$22.00

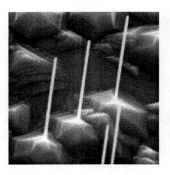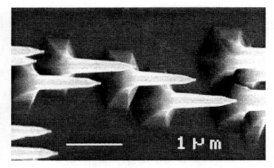

FIGURE 1. (A) InP nanowires grown with CBE. (B) GaAs/AlGaAs core-shell nanowires grown with MOVPE.

The sample is transferred to the growth chamber of the epitaxy machine (CBE or MOVPE), where the temperature is raised to deoxidize the surface and to create a eutectic alloy in the catalyst by incorporation of group III material from the sample. The growth species are then introduced into the growth chamber and the material forms a super-saturation in the catalyst and a wire starts to grow at the metal-semiconductor interface. The diameter of a wire is thus determined by the size of catalyst from which it is grown. Fig. 1(a) shows 2 μm long InP nanowires grown with CBE.

CBE is a vacuum technique where the growth species are transported to the sample as a beam. The growth rate is low, on the order of one monolayer per second, but the advantage is the ability to grow atomically abrupt interfaces in heterostructures. The MOVPE has a back-ground pressure of the growth species, which flow in a carrier gas (H$_2$). This leads to much higher growth rates, around 10-50 ML/s. Available sources for growth are In, Ga, As, and P for both systems, and also Al for the MOVPE.

As shown in Fig. 1(b) the MOVPE can be used to grow core-shell nanowires. Here the GaAs core is formed at a temperature where epitaxial growth is only possible via the catalyst; whereas the shell is formed at a higher temperature where bulk growth is possible. Such capping layers are important not only for passivation of wires, but perhaps also for future modulation doping of nanowires.

ELECTRICAL CHARACTERIZATION

For electrical measurements of individual wires, nanowires are removed mechanically from the growth substrate and transferred to a Si sample with an insulating SiO$_2$ top layer where the Si is used as a back-gate. The position of nanowires is recorded, and contacts are fabricated by electron beam lithography. For InAs nanowires the contact areas are etched in an NH$_4$S$_X$ solution prior to metal evaporation (Ni/Au).

InAs wires produced with CBE generally show a low resistance if the diameter of the wires is more than 35 nm, with a strong n-type conductance. Wires with diameters less than 35 nm generally become depleted from conduction electrons at 4.2 K unless a positive gate voltage is applied to fill the wire with electrons. The relatively large diameter required for conduction in InAs nanowires can partly be explained by the low effective electron mass in InAs (0.023 m_0) which gives rise to strong quantum confinement effects.

By changing the growth species during wire growth it is possible to grow semiconductor heterostructures not only radially but also along the length of a wire. Due to the slow growth rate of the CBE the interfaces can be made with a sharpness on the atomic scale, and sections as short as 1-2 monolayers can be grown. Ref. 4 shows this in detail for InAs/InP heterostructure nanowires and an example of an nanowire InAs/InP heterostructure is given in Fig. 2(a).

The conduction band offset for InAs-InP is estimated to be around 600 meV, which also is verified by temperature dependent electrical measurements on wires with thick barriers. If the InP barriers are grown with a thickness of less than 10 nm electron tunneling becomes possible. By incorporating pairs of tunnel barriers the 1D constriction of the nanowires will generate 0D dots between the barriers. If such dots are grown to a length of 100 nm quantum confinement effects are not strong enough to give a level spacing large enough to be visible compared to the charging energy, and such wires thus behave as single-electron transistors (SETs).[5] Here the capacitance of the tunnel junctions mainly determine the charging energy of the dot, which is on the order of 5 meV for these structures, and the Coulomb blockade is visible for temperatures up to around 12 K.

The fact that the semiconductor nanowire SET contains much fewer electrons than a corresponding metallic system becomes evident for negative gate voltages, where the wire is eventually pinched off and breaks up into a series of islands separated by tunnel barriers. A multiple island SET is revealed by a complicated gate dependence of the conductance and the different Coulomb blockades of the islands add up to form a large total blockade for the wire.

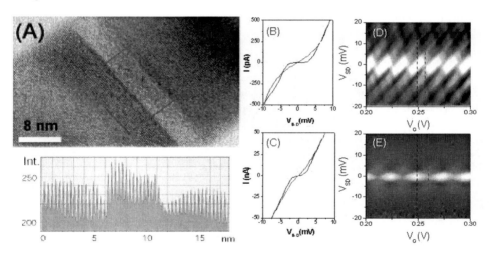

FIGURE 2. (A) InP barrier inside an InAs nanowire. (B-C) *I-V* measurements for nanowire SETs with different dimensions. (D-E) Grayscale differential conductance plots for nanowire SETs, where the lines indicates where the *I-V* curves in Figs. B-C have been obtained.

451

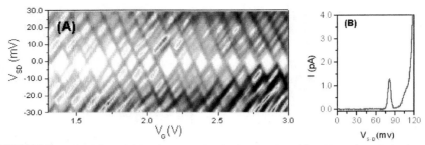

FIGURE 3. (A) Differential conductance plot for a nanowire with a 30 nm long InAs dot at 4.2 K. (B) *I-V* curve for a nanowire with 15 nm long inAs dot at 4.2 K..

The random deposition of the catalysts makes some wires considerably shorter than the average. This is explained by a limitation of material in the CBE growth, which means that wires grown from larger catalysts or wires formed close to other wires grow slower. The spread in wire length has a profound effect of the transistor characteristics of the nanowire SETs. A wire grown from a larger catalyst particle will have a larger area for the tunnel junctions which gives rise to a lower charging energy. The barriers also grow thinner, which is the main reason for the decrease in resistance by a factor of a 100 as shown in Fig. 2(b-c). The wire was probably nucleated by a merger between two gold particles. In addition, the InAs dot length is shorter, which is apparent from a comparison between Fig 2(d-e), where fewer Coulomb blockade diamonds are visible for the thicker wire for the same range in gate voltage.

In a different growth the dot length was designed to have a length of 30 nm. Now the z-confinement along the growth direction is greater than the x-y confinement defined by the diameter, which was around 60 nm. The quantum confinement energies in the z-direction is now comparable to the total charging energy of the device, and some electron configurations therefore show an increased stability, identified by larger diamonds in the stability plots [Fig. 3(a)].

If the dot length is reduced even further, down to 15 nm, the quantum confinement grows even stronger, and the electron states are considerably pushed up in energy.[6] Such a structure can be used as a resonant tunneling transistor as illustrated in Fig. 3(b). For a particular bias voltage there will be an energy level aligned with the Fermi level of the emitter and the conductance increases. Eventually the energy level falls below the lowest occupied state of the emitter and the conductance drop. The width of the conductance peak thus depends on the Fermi level position compared to the lowest occupied 1D state of the emitter.

REFERENCES

1. X. Duan, Y. Huang, R. Agarwal, and C.M. Lieber, Nature **421**, 241 (2003).
2. T. Mårtensson, M. Borgström, W. Seifert, B.J. Ohlsson, and L. Samuelson, Nanotechnology **14**, 1255 (2003).
3. T. Mårtensson, *et al.* Nano Lett. 4, 699 (2004)
4. M.T. Björk *et al.* Appl. Phys. Lett. **80** 1058 (2002).
5. C. Thelander *et al.* Appl. Phys. Lett. **83**, 2052 (2003)
6. M.T. Björk *et al.* Appl. Phys. Lett. **81**, 4458 (2002)

NANOCOMPOSITES

CNF RE-INFORCED POLYMER COMPOSITES

Max L. Lake, Gary G. Tibbetts, and D. Gerald Glasgow

Applied Sciences, Inc
Cedarville, Ohio 45314
mllake@apsci.com

Abstract. In properties of physical size, performance improvement, and production cost, carbon nanofiber (CNF) lies in a spectrum of materials bounded by carbon black, fullerenes, and single wall to multi-wall carbon nanotubes on one end and continuous carbon fiber on the other. Results show promise for use of CNF for modified electrical conductivity of polymer composites. Current compounding efforts focus on techniques for nanofiber dispersion designed to retain nanofiber length, including de-bulking methods and low shear melt processing. Heat treatment of CNF as a postproduction process has also been evaluated for its influence on electrical properties of CNF-reinforced polymer composites.

INTRODUCTION

Carbon nanofibers (CNF) synthesized by the floating catalyst method have a diameter of about 100 nm and a length in excess of 50 micrometers. CNF has been observed to have a high degree of graphitization, exceptional mechanical and transport properties, and excellent potential as an engineering material. There are a variety of applications into which the nanofibers can be inserted with little additional development. These include addition of the nanofibers to polymers to add electrical conductivity, add mechanical modulus and strength, or act as a CTE moderator. An overriding factor influencing use of such reinforcements is the performance/cost ratio for the additive. Since carbon nanofibers are currently produced in relatively small quantities and consequent higher cost compared to substitute materials such as carbon black, glass fiber, or PAN-based carbon fiber, the current applications are essentially limited to those applications where significant improvements in properties can be achieved with low loadings of CNF as an additive. Electrical conductivity is one such property. To improve the performance of electrical conductivity additives, one would choose an additive with high intrinsic electrical conductivity, and would choose, in contrast to spherical particles, particles which are long and thin, as are represented by carbon nanofibers. This paper describes work directed towards achieving desired benefit of electrical conductivity through the use of carbon nanofibers.

CP723, *Electronic Properties of Synthetic Nanostructures*, edited by H. Kuzmany et al.
© 2004 American Institute of Physics 0-7354-0204-3/04/$22.00

EXPERIMENTAL

2.1 Materials

The polypropylene (PP) used in this study was Montel Pro-Fax 6301 in flake form.

Several varieties of fibers are produced by Applied Sciences, Inc., in Cedarville, Ohio (apsci.com) with different gas space velocities using various mixtures of feed gas. PR-11 fibers are grown in a high velocity flow of methane, air, ammonia and the catalyst constituents hydrogen sulfide and iron pentacarbonyl.

PR-19 fibers are a newer variety of CNF produced at high gas space velocity with a natural gas feedstock. PR-19 PS fibers have been heated in an inert gas in a post treatment to remove any hydrocarbons deposited during growth. PR19-HT fibers are heat treated to over 3000°C in bulk after growth to graphitize the fibers; this treatment presumably makes the fibers stiffer. They appear to be more robust during processing after this treatment.

PR-24 fibers are produced in a manner so as to reduce deposited turbostratic carbon on the graphitic catalytic core of the CNF, so that the diameter is approximately 100 nm, and the CNF has a higher index of graphitization.

Because the as-grown fibers exit the reactor in large clumps, it is necessary to debulk the material to facilitate polymer infiltration. Fibers were de-bulked in a slurry using low speed stirring. We also experimented with a more vigorous processing in a Littleford Day mixer (model FM-130-D) which reduced both clump size and fiber length even more.

Average fiber length was determined in the scanning electron microscope.

2.2 Heat Treatments

The fibers were loaded into graphite crucibles. The crucibles were opened when placed in the heat treatment furnace and were purged by cyclic evacuation and backfilling with argon. The crucibles containing the sample fibers were heated to the target temperatures at a rate of 5 °C/min, and were held at the target temperature, determined by optical pyrometry, for one hour. The samples were then allowed to cool to room temperature over a 20-hour period.

2.3 Composite Processing

A bench-top CS-183 MiniMax Molder (Custom Scientific Instruments, Inc., Cedar Knolls, NJ, U.S.A.) was used for fabricating PYROGRAF/polymer composite specimens [1]. The injection molder is equipped with a cylindrical mixing cup (12.7 mm in diameter and 25.4 mm height) and a rotating and vertically moving rotor for mixing and injection. The cup temperature was 230°C for polypropylene, while the specimen mold was held at room temperature. Initially polymer pellets were loaded into the mixing cup and melted. The carbon nanofibers were then gradually loaded

456

into the mixing cup filled with polymer melt and mixed by rotational and vertical motions of rotor. Ultimately, the composite mixture was injected into the mold by opening the valve connecting the mixing cup and the mold while vigorously pushing the lever attached to the rotor downward. The mold was immediately submerged in cold water for at least 5 minutes to minimize crystal growth during solidification.

2.4 Electrical Resistivity Measurements

A rectangular bar (12.7 x 70 x 0.33 mm) obtained by injection molding was lightly sanded in order to smooth the surface and expose the fibers for measuring the electrical resistivity. The volume resistivity was measured using a Keithley 2000 electrometer (Keithley Instruments, Inc., Cleveland, OH, U.S.A.) at room temperature by connecting electrodes with silver paint to the surfaces of the specimens. For resistivities over 10^4 00°C in Ohm cm, the two point measurement was accurate, while a four-point measurement was performed for lower resistivities.

The intrinsic electrical resistivity of the fibers could also be measured by putting the neat fibers in a cylindrical die. A stainless steel pin of diameter 12.94 mm compresses the fibers in a mating Teflon cylinder so that electrical resistivity measurements may be made while the fibers are compressed with a maximum of 1000 lb. force. The volume fraction calculations utilize the intrinsic density of the fibers, which we have previously measured to be 2.02 g/cc [2].

RESULTS AND DISCUSSION

The effect of fiber aspect ratio on electrical resistivity is shown in Fig. 1.

Figure 1. Effect of Debulking on Electrical Resistivity of Carbon Nanofiber Based Composites.

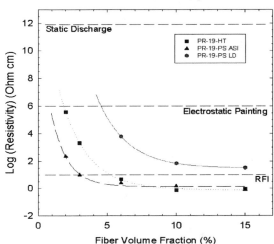

The carbon nanofibers produced by ASI can be debulked to various bulk densities. The process used to do the debulking operation results in generation of small fiber bundles and some reduction of fiber aspect ratio. The aspect ratio impacts the resultant electrical resistivity and mechanical properties in a composite. The process that exposes the fiber to the least amount of shear forces, with the least reduction in fiber aspect ratio, will produce the composite with the highest level of electrical conductivity. In contrast, mechanical properties benefit from the use of processes that reduce the fiber aspect ratio during fabrication, resulting in improved resin permeation and lower void content. The ASI and LD designations refer to two different debulking methods. The ASI method produces fiber with higher aspect ratio than the LD method. These results show that while graphitization of the fiber having equal aspect ratio yields composites with improved conductivity, the same values of conductivity may be recovered using un-heat treated fiber which has a high aspect ratio. In addition the process used to incorporate fiber into a specific matrix resin also affects the fiber aspect ratio. Both these effects must be considered when choosing a method for incorporating carbon nanofibers into composites.

A study was conducted on the effect of heat treatment temperature (HTT) on the electrical properties of both the fiber and corresponding composites. Single batches of both the PR-19 and PR-24 type fibers were used so that production and processing variables would be eliminated. These fibers were heat treated at four different temperatures in an inert atmosphere using a constant time at temperature. The resistivity data generated for the fibers as well as corresponding polypropylene based composites containing 24 wt.% fiber are shown in Fig. 2.

Figure 2. Effect of Heat Treatment Temperature on the Electrical
Resistivity of Bulk Fiber and Polypropylene Composites

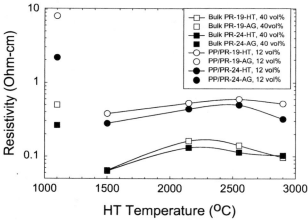

As can be seen the data for both fibers as well as the composites produced the same trend with respect to the effect of HTT on resistivity. This trend may be accounted for by an improvement lattice coherence length and consequent increase in electrical conductivity and mechanical strength of the nanofiber. Such observation is similar to that of Heremans et al., who measured electrical resistivity of VGCF as a function of temperature [3]. Heremans identified two declines in resistivity, the first of which plateaus near 1500 C, and a second decline above 2000 °C. Heremans attributes both drops to an increase in lattice coherence length (crystallite size) with temperature that increases the electron mean free path and hence decreases resistivity. For small diameter fibers, increases in crystallite size below 30 nm may not be as effective in decreasing resistance, as this approaches the characteristic wall thickness of the fiber. Hence, no further reduction in resistivity is observed above 2000 °C.

CONCLUSIONS

Methods of carbon nanofiber production and postproduction processing have been developed which improve the ability to successfully compound such reinforcements in selected polymers. Heat treatment temperature is a candidate method that may improve transport and mechanical properties; however, the value of heat treatment above 1500 °C is not evident, particularly if low shear processing, retaining fiber aspect ratio, is used to disperse the nanofibers in the polymer. Electrical properties of nanofiber-reinforced compounds even at loadings in the range of 1 to 2 % by volume appear suitable for static dissipation, electrostatic painting, and EMI shielding.

ACKNOWLEDGEMENT

Thanks to R. Alig, D. Burton, and C. Kwag at ASI for synthesis of fiber and composite samples, R. Jacobsen for electrical conductivity measurements, and K. Lafdi of the University of Dayton Research Institute for heat treatments. This work was supported in part by NIST ATP under Cooperative Agreement Award No. 70NANB5H1173, *Vapor-Grown Carbon Fiber Composites for Automotive Applications.*

REFERENCES

1.   Tibbetts GG, and McHugh JJ, *J. Mater. Res.* **14**:2871 (1999)
2.   Tibbetts GG, Doll GL, Gorkiewicz DW, Moleski JJ, Perry TA, Dasch CJ, and Balogh MP, *Carbon* **31**:1039 (1993)
3.   Heremans, J, *Carbon* **23**: 432 (1985)

Carbon Nanotubes as Backbones for Composite Electrodes of Supercapacitors

*[1]F. Béguin, [1,2]K. Szostak, [3]M. Lillo-Rodenas, [2]E. Frackowiak

[1]CRMD, CNRS-University, 1B rue de la Férollerie, 45071 Orléans, France
[2]ICTE, Poznan University of Technology, ul. Piotrowo 3, 60-965 Poznan, Poland
[3]GMCMA, Dept. Quimica Inorganica, Apartado 99, 03080 Alicante, Spain

Abstract. Carbon nanotubes have been used as a backbone for preparing a new type of C/C composite by one-step carbonisation of CNTs/PAN blends. Although their specific surface area is very low, the composites demonstrate high values of capacitance in two-electrode supercapacitors. A template effect of nanotubes on the mesoporous character of the composite is observed. The exceptional capacitance values are due to the combination of the open network of mesopores which favors the access of ions to the active surface and a nitrogen functionality which provides pseudocapacitance properties.

INTRODUCTION

The most common electrical energy storage systems are supercapacitors and batteries. Rechargeable batteries generally provide a high energy and a rather low specific power. On the contrary, the electrochemical capacitors (often called supercapacitors or ultracapacitors) [1,2] store less energy but can be operated at substantially greater specific power than most batteries, while presenting a much higher cycleability. They are basically classified in two different categories depending on the charge storage mechanism, i.e. the electric double layer capacitors (EDLCs) generally based on activated carbon materials and the so-called pseudocapacitors based on electronically conducting polymers (ECPs) [3] or metal oxides [4]. Due to the particular interest of supercapacitors, an intense research effort is devoted to design new electrode materials or new systems in order to enhance the power and the energy stored.

The power is reversibly proportional to the internal resistance of the supercapacitor, i.e. commonly the equivalent series resistance (ESR) R_S. The ESR of the entire device is the sum of the resistances of all the materials between the external contacts, *i.e.* substrate, carbon, binder, separator and electrolyte. Mesopores are essential to lower R_S, allowing a quick transportation of ions in the bulk of the electrodes. Carbon nanotubes (CNTs) are very promising for this function of electrochemical capacitors, because owing to their entanglement they form a well-developed network of open mesopores almost impossible to obtain with activated carbons [5]. Taking into account that their intrinsic specific capacitance is rather low, they can be used profitably as a support for materials with pseudo-capacitance properties, e.g. ECPs [6].

CP723, *Electronic Properties of Synthetic Nanostructures*, edited by H. Kuzmany et al.
© 2004 American Institute of Physics 0-7354-0204-3/04/$22.00

In this paper, a new type of C/C composite obtained by direct pyrolysis of a CNTs/polyacrylonitrile (PAN) blend, without any additional activation step, has been used for building two electrode capacitors, demonstrating interesting pseudo-capacitance properties related with the presence of nitrogen functionalities.

EXPERIMENTAL

We used multiwalled carbon nanotubes prepared by catalytic decomposition of acetylene at 600°C on a solid solution $Co_xMg_{(1-x)}O$ catalyst precursor [7]. Polyacrylonitrile (PAN) from Aldrich has been used as received. CNTs with a content ranging from 15-70 wt% and PAN are mixed in an excess of acetone to obtain a slurry. Then acetone is evaporated and the composite mixture is pressed at 1-2 tons/cm² to form precursor electrodes. The pellets are carbonised at 700-900°C for 30-420 min under nitrogen giving rise to C/C composites. The pure components of the composites, i.e. carbon nanotubes and PAN, and the C/C composites were observed by scanning electron microscopy (Hitachi S 4200) and transmission electron microscopy (TEM Philips CM20). The chemical composition of the composites has been determined by elemental analysis and by X-ray photoelectron spectroscopy (XPS, Escalab 250, VG Scientific). The porous texture has been characterized by nitrogen adsorption at 77K on an Autosorb 6 (Quanta Chrome) after degassing 15 hours at 200°C.

Two electrode Teflon Swagelok® supercapacitors were built from the self-standing C/C composite pellets directly after carbonisation without any additional binder. For a comparison, the catalytic multi-walled carbon nanotubes and PAN after carbonisation at 700°C under nitrogen flow were also investigated. In this case, the electrodes were prepared by pressing a mixture of 85 wt% of active material (nanotubes or carbonised PAN) with 10 wt% of polyvinylidene fluoride (PVDF-Kynarflex, Atochem) and 5 wt% of acetylene black. The current collectors were from gold, and the electrodes were separated by a thin glassy paper separator, in 1M H_2SO_4 aqueous solution as electrolyte. The values of capacitance were obtained from cyclic voltammetry (scan rate of potential from 2 to 10 mV) and galvanostatic charge/discharge (VMP-Biologic-France).

RESULTS AND DISCUSSION

The atomic percentages determined by elemental analysis and XPS for the C/C composite formed by carbonisation of the CNTs/PAN (30/70 wt%) blend at 700°C under nitrogen are given in Table 1. The values are fairly comparable in the bulk and on the surface. The final C/C composite is still quite rich in nitrogen, particularly in the bulk (9.2 at%), demonstrating that PAN is efficient nitrogen carrier. The amount of

TABLE 1. Elemental composition of the C/C composite obtained from the CNTs/PAN (30/70 wt%) blend carbonized at 700°C under nitrogen.

	C (at%)	N (at%)	O (at%)
Elemental analysis	86.7	9.2	4.1
XPS	89.5	7.4	3.1

oxygen in the composite is quite high, probably due to its incorporation by addition on the dangling bonds when the C/C composite is exposed to air after its formation.

The different nitrogen functionalities in carbon materials are presented in Figure 1 [8]. The N1s core level spectrum of this composite is fitted by four peaks at 398.5 eV (3.9 at%; pyridinic (N-6) type nitrogen); 400 eV (1.5 at%; nitrogen in a 5-membered pyrrolic ring and/or pyridone); 401 eV (1.5 at%; quaternary (N-Q) nitrogen groups); 401.9 eV (0.4 at%; oxidized nitrogen (N-X) form).

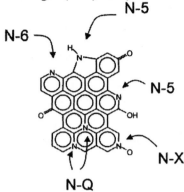

FIGURE 1. Possible nitrogen functionalities in a carbon-type material.

SEM images of the CNTs/PAN (30/70 wt%) blend carbonised at 700°C under nitrogen are presented in Figure 2. This C/C composite is quite homogeneous and similar to the carbonised pure PAN, at least at the scale of this investigation technique. It is possible to distinguish zones where individual or interconnected aggregates from PAN carbonisation are present (Figure 2a). At such a high content of PAN (70%), most of the nanotubes are probably embedded, therefore they are not easily visible in the SEM pictures.

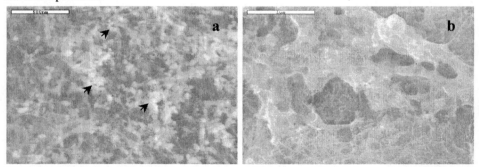

FIGURE 2. SEM images at two different magnifications of the C/C composite obtained by carbonisation of the CNTs/PAN (30/70 wt%) blend at 700°C under nitrogen during 180 min. The arrows indicate aggregates of carbonised PAN.

During pyrolysis of the CNTs/PAN mixture, while becoming rigid, the pellets keep their original shape without appearance of any noticeable cracks or defects, indicating that CNTs act as a backbone preventing from dimensional changes and shrinkage during the C/C composite formation. However, when the CNTs content is less than 15 wt% in the original composite, the weight loss increases dramatically during carbonisation and shrinkage appears.

The electrochemical properties of the pure components and of the composites were determined by cyclic voltammetry and galvanostatic charge/discharge. All the composites give a typical "box-like" shape voltammogram characteristic of an ideal capacitor; the example of the CNTs/PAN (30/70 wt%) blend carbonized at 700°C is shown in figure 3.

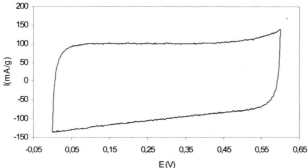

FIGURE 3. Voltammogram of a supercapacitor based on the CNTs/PAN (30/70 wt%) blend carbonised at 700°C under nitrogen during 180 min. Scan rate 2 mV/s; electrolyte 1M H_2SO_4.

The values of capacitance, pores volumes and specific surface area for the pristine components and for different composites prepared at 700°C are presented in Table 2. As it could be expected, a negligible value of capacitance is determined for the microporous carbon from PAN with a very low specific surface area of 29 m^2/g. For all the other materials, including the pristine nanotubes and the C/C composites, the isotherms are of type IV, with S_{BET} ranging from 130 to 230 m^2/g. A template effect of the nanotubes is at the origin of the mesoporous character of the composites. Interestingly, the capacitance values of the C/C composites are always higher than for the pristine CNTs, even if their specific surface area is lower than for nanotubes. It clearly shows that there is a profitable contribution of the carbon moiety from PAN. Considering the three composites obtained after 180 minutes of pyrolysis, but with different proportions of CNTs and PAN, there is not any real relationship between the capacitance values and the specific surface area. For the CNTs/PAN (15/85 wt%) composite, the proportion of nanotubes is not high enough to keep a good backbone

TABLE 2. Capacitance, specific surface area and pore volume for C/C composites prepared at 700°C from different CNTs/PAN blends and various pyrolysis time.

CNTs (%)	PAN (%)	Time (min)	S_{BET} (m^2/g)	V_{micro} (cm^3/g)	V_{meso} (cm^3/g)	Capacitance (F/g)
100	-	-	220	0.0004	1.00	18
-	100	180	29	-	-	Negligible
30	70	90	130	0.011	0.062	76
30	70	180	208	0.012	0.099	100
15	85	180	164	0.01	0.08	58
50	50	180	233	0.01	0.011	57

preventing from shrinkage. The specific surface area is rather low (164 m^2/g), giving rise to a moderate value of capacitance (58 F/g). Although the specific surface area of the composite from CNTs/PAN (30/70 wt%) is lower than that of the composite from CNTs/PAN (50/50 wt%), the former material gives a noticeably higher value of capacitance (100 F/g), because it certainly contains more residual nitrogen contributing to pseudocapacitance properties.

In order to confirm the effect of nitrogen on the capacitance properties, two composites have been prepared from the CNTs/PAN (30/70 wt%) blend at carbonisation temperatures of 700°C and 900°C. The specific capacitance decreases from 100 F/g for the 700°C composite to 48 F/g for the 900°C one, whereas the nitrogen values determined by XPS are 7.3 at % and 3.8 at %, respectively.

Taking into account the presented results, we can assume that the obtaining of an optimal performance is a compromise between: 1/ an enough high amount of PAN to favour a large gas evolution able to create pores; 2/ a sufficient amount of CNTs, that prevents shrinkage of the composite during the carbonisation of PAN and consequently assists the pores formation; 3/ an important proportion of PAN in order to get the largest amount of residual nitrogen in the C/C composites, and to enhance the contribution of the pseudocapacitance effect.

CONCLUSION

Self-standing supercapacitor electrodes have been prepared by one-step carbonisation of CNTs/PAN blends in nitrogen atmosphere. The strongly entangled network of nanotubes enhances the conductivity of the electrodes, prevents an important shrinkage during the carbonisation and facilitates the composite matrix auto-activation by the evolving decomposition gases. The remarkable capacitance properties of these new composites are due to the combination of the template effect of nanotubes on the mesoporous character and to the pseudocapacitance properties of carbonised PAN.

ACKNOWLEDGEMENTS

This research has been supported by the European Research Training Network NANOCOMP under the contract number HPRN-CT-2000-00037

REFERENCES

1. Conway, B.E., *Electrochemical Supercapacitors*, Kluwer Academic/Plenum Publishers, New York, 1999.
2. Frackowiak, E., Béguin, F., *Carbon* 39, 937-50 (2001).
3. Mastragostino, M., Arbizanni, C., Soavi, F., *J Power Sources* **97-98**, 812-5 (2001)
4. Wu, N.L., *Mater Chem Phys* **75**, 6-11 (2002).
5. Niu, C., Sichel, E.K., Hoch, R., Moy, D, Tennet, H., *Appl Phys Lett* **70**, 1480-2 (1997)
6. Frackowiak, E., Béguin, F., *Carbon* **40**, 1775- 87 (2002)
7. Delpeux, S., Szostak, K., Frackowiak, E., Bonnamy, S., Béguin, F., *J Nanosc Nanotech* **2**, 481-4 (2002)
8. Jurewicz, K., Babel, K., Ziolkowski, A., Wachowska, H., *Electrochim Acta* **48**, 1491-1498 (2003).

Synthesis And Characterization of Carbon Nanotubes/Amylose Composites

Pierre Bonnet[1], David Albertini[1], Christine Godon[1], Mickael Paris[1], Herve Bizot[2], Joelle Davy[2], Alain Buleon[2] and Olivier Chauvet[1]

[1]. Institut des Materiaux Jean Rouxel, Universite de Nantes, 44322 Nantes cedex 3, France
[2]. INRA, BP 71627 - 44316 Nantes cedex3 , France

Abstract. Wrapping single-walled carbon nanotubes (SWNT) by an helical polymer may be an interesting way to solubilize them in water or organic solvents or to prepare biocomposites with multifunctionnal properties. Here we report the synthesis in aqueous solution of amylose/SWNT composites where amylose is a biological polymer. Upon reacting amylose with SWNT, a stable suspension is formed which is probed by optical absorption and Raman spectroscopies. While nicely resolved absorption spectra suggest that isolation of SWNTs likely related to a complex formation is achieved in the suspension, this is not confirmed by the fluorescence behavior.

INTRODUCTION

A large research effort is now devoted on forming nanocomposites with single walled carbon nanotubes (SWNT),. One of the problems which is encountered is the dispersion of the nanotubes which tends to aggregate to form bundles or ropes. Another point is the non solubility of SWNT in aqueous solution. Addition of surfactants[1] or polymers[2] may be a way to face these problems. Another alternative is to use non covalent functionalization through wrapping a water soluble polymer around them. Amylose is a biological polymer which forms helical complex in aqueous solution. It is thus a good candidate to achieve such a goal and it was suggested that amylose could wrap SWNT [3]. Here, we investigate the ability of amylose to disperse the SWNT in aqueous solution and to compete with a surfactant. We show that a good dispersion is maintained even when the surfactant is removed.

EXPERIMENTAL

Amylose is one of the starch constituent (~20%) with amylopectin (~80%). The amylose macromolecule is typically built from 200-2000 anhydroglucose units [4]. Aqueous solutions of amylose are very unstable due to intermolecular attraction and association of neighboring amylose chains. Modification of amylose with hydroxypropyl group inhibits the auto-association process [4]. Here we use homemade hydroxypropyl amylose (AmH)

Raw nanotubes (HiPCo nanotubes from Carbon Nanotechnologies Inc., average diameter 0.8 nm) are dispersed in 20 ml (0.5 mg/cm^3) of aqueous (D$_2$O) sodium

CP723, *Electronic Properties of Synthetic Nanostructures*, edited by H. Kuzmany et al.
© 2004 American Institute of Physics 0-7354-0204-3/04/$22.00

dodecyl benzenesulfonate (NaDDBS) (1%wt) followed by magnetic agitation (1 hour) and sonication (10 hours). The dispersion is then centrifuged (2 hours at 12 000 rpm) and the supernatant is recovered. This first step is supposed to isolate the SWNT in NaDDBS micelles[4]. This starting point sample is referred as sample LA.

Competitive wrapping of amylose has been tested following two different routes. In the first one, AmH (10 mg/ml) is directly added to the LA solution followed then by sonication and centrifugation (sample LB1). Next, the NaDDBS surfactant is eliminated with a ion retardation resin (from BioRad) (sample LB2). In the second route, the two steps are reversed with first the removal of NaDDBS with the resin (sample LC1), then mixing with AmH (sample LC2).

The samples are characterized by optical absorption spectroscopy (Cary 5G spectrophotometer in the 4000-50000 cm^{-1} range) and Raman spectroscopy (RFS 100 Bruker FT Raman spectrometer, excitation line 1064 nm).

SPECTROSCOPIC CHARACTERIZATION

O'Connel et al. [5] have recently shown that efficient tools to characterize the dispersion of SWNT are optical absorption and luminescence spectroscopies. Well dispersed samples result in nicely resolved absorption spectra reflecting the optical transitions between the Van Hove singularities of the SWNT density of states. Aggregations into bundles lead to a redshift and a broadening of the spectrum.

The optical absorption and Raman spectra of AmH in {NaDDBS + D_2O} solution are given in figure 1 for the sake of comparison. The absorption spectrum is essentially featureless in the NIR-vis region. This is also the case of the Raman spectrum except for the large D_2O contribution close to 2500 cm^{-1}. It will not contribute to the spectra as will be seen later.

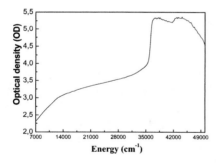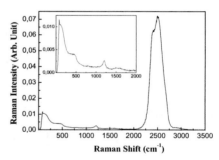

FIGURE 1. (left panel) NIR-vis-UV optical absorption of AmH in {NaDDBS+D_2O}. (right panel) Corresponding Raman spectrum at 1064 nm. Inset: zoom on spectrum between 0 and 2000 cm^{-1}.

The absorption spectra of AmH/SWNT solutions obtained following the LB (left panel) or LC (right panel) routes are shown in figure 2 after subtraction of the plasmon baseline. The spectrum of sample LA (without amylose) is shown as trace 1 in both panel. Except for sample LC1, all the spectra are pretty well resolved. The region below 11000 cm^{-1} corresponds to the E_{11} transitions involving the first Van Hove

466

singularities while the second ones for semiconducting tubes and the for metallic tubes appears at higher energy.

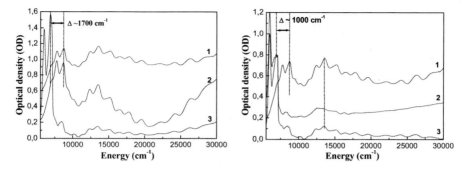

FIGURE 2. Optical absorption spectra of AmH/SWNT solutions. (left panel): Traces 1, 2, 3 corresponds to samples LA, LB1, LB2 respectively. (right panel)): Traces 1, 2, 3 corresponds to samples LA, LC1, LC2 respectively.

The most interesting fact is that even when the surfactant has been removed but in presence of amylose (samples LB2 and LC2), the resolution is maintained. Still, a redshift is observed which suggests that a partial aggregation occurs. Another interesting point is given by the results of the LC route (right panel). Starting from SWNT/NaDDBS micelles (sample LA), the resolution is lost when NaDDBS is removed (sample LC1). But it is at least partly recovered when AmH is introduced (sample LC2). Indeed it suggests that amylose is really able to wrap SWNT which thus maintains a good degree of dispersion. Whether amylose is able to compete with NaDDBS or not is more difficult to answer. Comparison of samples LB1 and LB2 may indicate that either amylose was already wrapping the SWNT in sample LB1 and, NaDDBS removal does not change significantly the results, or it may imply a process similar to the LC route where amylose waits for NaDDBS removal before wrapping.

The Raman spectra (at 1064 nm) of samples LA, LB1 and LB2 are shown in figure 3. The spectra of samples LC1 and LC2 (not shown) are quite similar to the one of LB2. The spectrum is dominated by the SWNT one with a RBM contribution at 266 cm^{-1}, the G band at 1591 cm^{-1} and the D' band at 2550 cm^{-1}. The D band at 1282 cm^{-1} is very weak. The solution features shown in figure 1 do not contribute to this spectrum. Addition of amylose does not seem to affect the Raman spectrum, especially in the RBM region. Still, three important fluorescence peaks at 390 cm^{-1}, 1000cm^{-1} and 1550cm^{-1} can be seen in LA and LB1 spectra with a typical width of 300 cm^{-1}. By comparison with the LA sample, these peaks are weakened when AmH is introduced (LB1 sample) and they disappear when NaDDBS is removed (LB2 sample). As shown in the right panel of the figure, these peaks are in correspondence with the E$_{11}$ peaks observed in absorption. It suggests that they are due to isolated SWNT fluorescence emission. SWNT fluorescence peaks observed in a Raman scattering experiment has already been reported [1] with a visible line excitation. Here since we use a 1064 nm excitation, the attribution of these peaks to SWNT fluorescence means that we have to deal with a resonant fluorescence process. Further studies are in progress to clarify this point. Note that the absorption peak at 8750 cm^{-1} is not observed in emission while the 8560 and 9000 cm^{-1} are observed both in absorption and emission. It our assignment is correct and following Bachilo et al.[6], it means that the (9,2) SWNT does not contribute to the emission in our samples.

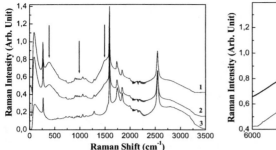 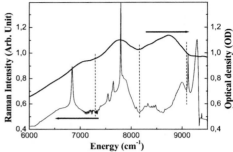

Figure 3. (left panel) Raman spectra at 1064 nm of AmH/SWNT solutions. Traces 1, 2 and 3 correspond to sample LA (without amylose), sample LB1 and sample LB2 respectively. Arrows indicate fluorescence peaks (right panel): Comparison between the absorption (dashed line) and the Raman (solid line) spectra for sample LA.

As seen in figure 3, adding AmH to the SWNT/NaDDBS micelles supension (sample LB1) decreases the fluorescence. It suggests that amylose does compete with NaDDBS. Removing the remaining NaDDBS (sample LB2) totally quenches the fluorescence. This result is quite disappointing. It means either that amylose does not hinder a partial aggregation of the SWNT or that it does not wrap totally the SWNT allowing Van der Waals contacts between some uncovered parts of the tubes which quenches the fluorescence.

CONCLUSION

In this work, we have tested the efficiency of amylose to disperse SWNT as seen from optical characteristics. While absorption spectroscopy results are quite good and suggests that amylose can compete with NaDDBS, fluorescence behavior seems to indicate that amylose is less efficient than NaDDBS. Further optical investigations are under way in order to clarify this point. Investigations of the composite morphology by AFM and TEM are also in progress.

REFERENCES

1. Moore V. C. et al., *Nano. Letters* **3**, 1379 (2003).
2. O'Connell M.J et al., *Chem. Phys. Letters* **342**, 265 (2001).
3. Star A et al., *Angew Chem.***114**, 2618 (2002) ; Lii C.Y. et al., *Carbohydrate Polymers* **51**, 93 (2003)
4. Wulff G et al., *Carbonhydrate Research* **307**, 19-31 (1998).
5. O'Connell M.J et al., *Science* **297**, 593 (2002).
6. Bachilo S et al., *Science* **298**, 2361 (2002).

ROUTE FOR SINGLE-WALLED NANOTUBE-POLYMER COMPOSITES

Michael Holzinger[a], Johannes Steinmetz[a, b], Damien Samaille[a],
Patrick Bernier[a], Vickie Aboutanos[a] and Marianne Glerup[a, c]

[a]GDPC(UMR5581), Université Montpellier II, 34095 Montpellier, France,
[b] Max-Planck-Institut für Festkörperforschung, Heisenbergstr. 1, 70569 Stuttgart, Germany
[c]Grenoble High Magnetic Field Laboratory, MPI-FKF/CNRS, 25, 38042 Grenoble, Cedex 9, France
Corresponding author; E-mail: holzinger@gdpc.univ-montp2.fr

Abstract. The main issue in the research field of NT composites is the load transfer and homogeneity of the composites. Our investigations open a route to reach this homogeneity by functionalizing these nanotubes. We have tested several functionalisation methods and here, we want to focus on the addition of amines to the oxidized nanotube sidewalls. Using bi-functionalities, a chemical link between the nanotubes and polymer should result in more stable and homogeneously dispersed composites. The chemistry behind the modification of the nanotubes and the process in lab scale will be discussed in detail.

INTRODUCTION

The interest in the use of single-walled carbon nanotubes incorporated in polymer matrices for making composites quickly sprang up in the last years because of the possibility to produce mechanically advanced and conductive high performance materials. To reach this goal, a lot of efforts have been carried out.[1] In most cases, it is not possible to disperse the nanotubes sufficiently in the polymer by using ultrasound or compounder[2-4]. One possibility to overcome this problem is to polymerize monomers in the presence of nanotubes and another way is to functionalize the nanotubes to make them dispersable in the polymer. Extensive and systematic studies would be necessary for the second method. Our route is here the functionalisation of oxidised nanotubes by forming amides with amines. It has already been shown that using diamines oxidised nanotubes can be linked to each other.[5] The idea is to use these diamines as reactant and polycarbonate as polymer. With cyclohexylcarbodiimide (DCC) as coupling reactant we have the possibility to link covalently the nanotubes with the polymer matrix by a trans-esterification-like reaction.

EXPERIMENTAL, RESULTS AND DISCUSSION

Nitric acid oxidized SWCNTs were dispersed in dimethyleforamide (DMF) together with 20 fold-excess of polycarbonate. By estimating the concentration of the carboxylic acid groups, 10^{-3} mg of diaminohexane was added with an equimolar

CP723, *Electronic Properties of Synthetic Nanostructures*, edited by H. Kuzmany et al.
© 2004 American Institute of Physics 0-7354-0204-3/04/$22.00

amount of cyclohexylcarbodiimide. The mixture was stirred at ambient temperature for 24h. After the reaction, the mixture was filtered and washed with DMF and THF to remove by-products and non-reacted polymers. After drying, we obtained a highly fragile composite containing around 40% nanotubes referred to the amount of SWCNT starting material.

There are several reactions that can occur using these conditions. Fig. 1 shows a sketch of the important reaction products. There is a probability that one carboxylic acid group of one nanotube reacts with another carboxylic acid group of another nanotube and connects them. The excess of diamine, however, lowers the probability of this reaction. The more likely possibility is that the nanotube gets connected to the polymer. The polymer chains can also be linked by this reaction (not shown in this Fig.).

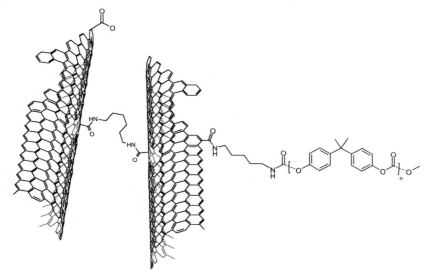

Figure 1: possible products after reaction with diaminohxane.

The IR spectrum of the resulting product was compared with the spectra of the starting materials. The most interesting changes can be seen in the aliphatic (Fig.2a) and in the carbonyl region (Fig. 2b). The aliphatic area of the polycarbonate (Fig. 2a top) shows three minima at 2973 cm^{-1}, 2933 cm^{-1} and at 2876 cm^{-1}. In the spectrum of the linked composite (Fig. 2a middle), the peaks are placed at 2959 cm^{-1}, 2916 cm^{-1} and at 2850 cm^{-1}. These clear differences of these IR features indicates that there must be an influence of the diaminohexane which serves as linker. The spectrum of the nanotube starting material shows no features in this region (Fig. 2a bottom). The spectrum of polycarbonate (Fig. 2a, top) shows three significant peaks at 1775 cm^{-1}, 1633 cm^{-1} and 1505 cm^{-1}. The IR-spectrum of the oxidized nanotubes (Fig. 2b, bottom) shows a weak broad peak at 1726 cm^{-1}, a stronger signal at 1584 cm^{-1} and a small shoulder at 1630 cm^{-1}. The IR-spectrum of the linked composite (Fig. 2b, middle) has features which have similarities to both spectra of the starting materials. One can see that the peak at 1775 cm^{-1} (Fig. 2b top) is slightly shifted to to 1764 cm^{-1}and the peak at 1726 cm^{-1} in Fig. 2a bottom to 1736 cm^{-1}. The more obvious indication of a possible linking between the SWCNTs and the polymer matrix is the change in peak intensities. Because of the high number of different possible reactions, a clear assignment of the peaks cannot be done.

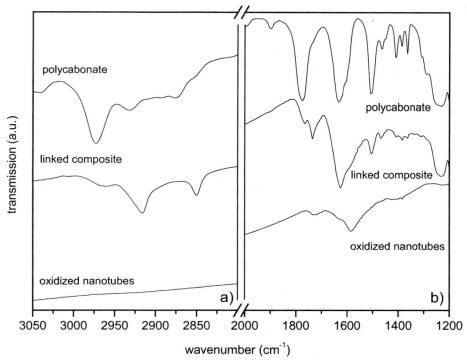

Figure 2: IR spectra of the linked composite (a and b middle) in comparison with the starting materials (top: polycarbonate, bottom: oxidized SWCNTs).

In Fig. 3a the SEM image shows nanotubes sticking out of the polycarbonate matrix. The density of the freestanding nanotubes is quite low in respect to the 40 weight percent of SWCNTs in the sample. The reason for the small amount of visible nanotubes is not presently clear. One possible explanation could be that the surface is covered by a thin layer of polycarbonate which was not removed in the washing process. Fig. 3b shows a SEM image of the sample with a higher density of freestanding nanotubes. The only difference in these two images is that the SEM image in Fig. 3a was recorded at the surface of the as filtered sample and the SEM image in Fig. 3b was recorded at a fraction edge.

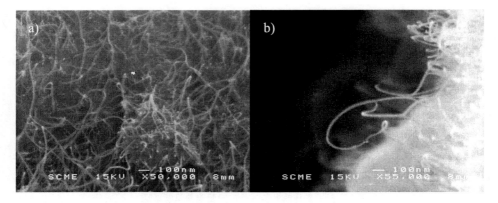

Figure 3: SEM images of the linked NT-polycarbonate composite. a): recorded surface of the as filtered sample. b): recorded at a fraction edge.

Conclusion and Outlook

We have presented a new route for the formation of SWCNT-polymer composites via functionalisation. Spectroscopic and microscopic investigations cannot give a clear characterization of the product. Unfortunately, it was up to now not possible to measure reliably the mechanical and electronic properties because of the high fragility of the sample. Therefore, further experiments have to be done to increase the stability of the sample and to reach a more defined and tuneable nanotube-polymer ratio.

References

[1]   R. Andrews, M. C. Weisenberger, *Current Opinion in Solid State and Materials Science* **2003**, *In Press, Corrected Proof,* Available online 6 December 2003.

[2]   P. Pötschke, A. R. Bhattacharyya, A. Janke, *European Polymer Journal* **2004**, *40*, 137.

[3]   P. M. Ajayan, L. S. Schadler, C. Giannaris, A. Rubio, *Adv. Mater. (Weinheim, Ger.)* **2000**, *12*, 750.

[4]   J. M. Benoit, B. Corraze, S. Lefrant, W. J. Blau, P. Bernier, O. Chauvet, *Synth. Met.* **2001**, *121*, 1215.

[5]   U. Dettlaff-Weglikowska, J.-M. Benoit, P.-W. Chiu, R. Graupner, S. Lebedkin, S. Roth, *Current Applied Physics* **2002**, *2*, 497–501.

Investigations on Polycarbonate-Nanotube Composites

B. Hornbostel[1a], M. Dubosc[1a], P. Pötschke[2] and S. Roth[1]

[1]Max Planck Institute for Solid State Research, Heisenbergstr. 1, 70569 Stuttgart, Germany
[2]Institut für Polymerforschung Dresden e.V., 01069 Dresden, Germany

Abstract. In this work we have been studying electrical properties of polycarbonate-single-walled-nanotube composites for purified and non-purified material. The raw material was produced by standard nanotube arc discharge or supplied by CNI (HiPco process). The SWNT were mixed by hand or by a micro compounder. From the electric conductivity as function of the amount of nanotubes in the polymer matrix the percolation threshold has been determined. This threshold is different for purified and non-purified material and depends on the purification method. For mechanical characteristics we determined the elastic modulus and the yield strength. Surprisingly, we found that the purified ArcD material revealed worse results when worked up into the polymer matrix in comparison to the unpurified raw material. However, the good electrical conductivity of bucky paper proofed the success of the purification method.

INTRODUCTION

Thermoplastics have developed into key structural materials for the electrical engineering sector. By progressing miniaturization of electrical equipment together with the use of digital assemblies and the growing proportion of sensitive electronic components employed in the systems, measures are required to restrict mutual electromagnetic interference. An additional requirement is the removal of heat generated by the electronic circuits. Polycarbonate-nanotube-composites may be a solution for these tasks. The interest is mainly based on the unique characteristics of carbon nanotubes and their high aspect ratio [2]. This facilitates among others a lower electrical percolation threshold and better mechanic properties in comparison to globular particles like black carbon in polymers [3][4][5].

For our experiments we used standard nanotube arc discharge (ArcD) material from Yangtze, Shanghai, and material synthesized in our lab. HiPco material from CNI, Houston, was used for comparison. The ArcD material was homogenized and purified. Purification was accomplished by different methods. Some of the material was additionally doped by $SOCl_2$, because we found a slightly higher conductivity in previous experiments with PMMA. The raw material and its purified derivative were characterized by Raman spectroscopy and Xray diffraction but the information on the quality has been taken from the electrical conductivity of the bucky paper.

---

[a] both authors contributed equally

CP723, *Electronic Properties of Synthetic Nanostructures*, edited by H. Kuzmany et al.
© 2004 American Institute of Physics 0-7354-0204-3/04/$22.00

EXPERIMENTAL

Purification and Pre-Treatment of Raw SWNT Material

The pristine SWNT (single-walled nanotube) ArcD material was collected separately according to the defined regions in the Krätschmer reactor [1]. The material was stirred in ethanol to attain a macroscopically homogenous starting material. Some parts of the homogenized SWNT material was purified or purified and additionally doped, respectively. We applied three different purification techniques: 1) ArcD material was purified by dry oxidation in 355°C hot air. After 2 x 200 minutes a weight loss of ~15wt.% was attained.. 2) Same procedure as 1) but at 370°C and for 13h. We measured a weight loss of 20-30wt.%. 3) As 2) but additionally centrifuged at 5000 rpm for 1h. For this treatment the material was suspended in SDS with ultra sonic agitation. Two runs of centrifugation were applied where after each run the lighter material was separated and taken for the next step. After the last centrifugation the suspension was filtrated and rinsed by distilled water. We also tried to improve the conductivity by doping the SWNT material. For this the material was put in a flask with $SOCl_2$ at room temperature overnight.

Determination of Raw Material Quality

For the determination of the raw material quality Xray, Raman, TGA and TEM were used but most reliable we found measurements of the electric conductivity of the bucky paper. According to Table 1 a significant improvement has been achieved with the applied purification methods.

TABLE 1. Average conductivity of the starting material in bucky paper.

Sample	Conductivity
Unpurified ArcD material	55 S/cm
Purified ArcD material (355°C, 2x200')	108 S/cm
Purified ArcD material (370°C, 13h)	107 S/cm
Purified ArcD material (370°C, 13h, 2x centrifuged)	200 S/cm
HiPco material (bucky pearls)	155 S/cm

Mixing Methods of Composites

For the hand-mixing method we used Poly-(Bisphenol-A-)Carbonate from Aldrich (43,512-0) which was dissolved in chloroform before adding a defined amount of SWNT material. After intense stirring the mixture was put onto a holder to evaporate the organic solvent. The residue on the holder after evaporation was a composite film which was than used for characterization. This method is termed *solvent casting*.

A second hand-mixing method was tested to let material coagulate in a mixture of organic solvent and polycarbonate. By this method a very homogeneous distribution of SWNT in the polymer matrix has been attained. Therefor a solution of 0.25 mg SWNT material per 1 ml DMF was prepared by ultra sonic agitation for 2 minutes. In parallel to this step polycarbonate was dissolved in chloroform (100 mg / 1 ml). After the complete dissolution of the PC-CHCl$_3$-dissection both mixtures were mixed together, vigorously shaked and sonicated for a couple of seconds. The mixture was then dripped into an ethanol bath, where the polycarbonate precipitated immediately due to its insolubility. The ratio of ethanol to SWNT-PC mixture was 1 to 5. The precipitating polycarbonate chains entrap the SWNT and prevent SWNT from intense bundling. By filtration the raw composite material was collected, and for comprehensive testing processed in a hot press to obtain a pellet.

For the automated melt mixing method the composite was prepared in a DACA Micro Compounder (DACA Instruments, Goleta, CA, USA) at 280°C and 50 rpm. The mixing process lasted 5 or 15 minutes, respectively. The amount of the polycarbonate (Iupilon E2000, Mitsubishi) and SWNT pre-mixture was 4.2 g at each run. SWNT material was added to the PC prior injecting it into the automated compounder in two different conditions: 1) Direct into the PC-powder without any pre-treatment. 2) Pre-treatment by slurryfication (s) in acetone using ultra sonic agitation. The second condition was additionally applied with the HiPco material. The strand-like product was ejected by the machine. From this strand 5 cm were cut off and put into a hot press (pressing at 260°C). Finally an approximately 350μm thick circular plate was obtained. The diameter was >60mm.

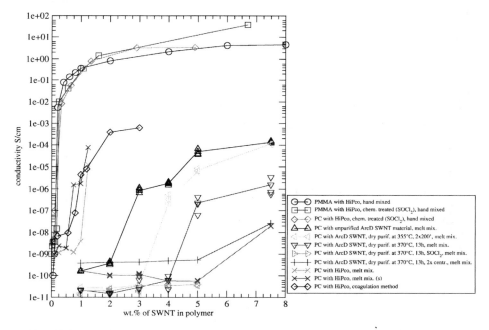

FIGURE 1. Conductivity versus nanotube concentration

Measurement of Electrical Conductivity

The resistance was measured (bucky paper and clean strips of the composite) by using the four-lead-method and a Keithley 197 AR Microvolt meter or a Knick Teraohm-

Meter, respectively. All outer dimensions of the cut rectangular samples were determined by caliper tools.

RESULTS AND DISCUSSION

Fig. 1 depicts the electrical conductivity over wt.% SWNT material. Contrary to our expectations, the various purification processes which have resulted in an increased electrical conductivity in the bucky paper (Table 1), have not much influenced the characteristic of the composite (Table 2). One possible explanation is the introduction of defects by the purification. Additional SWNT breaking takes place while SWNT and polymer are melt-mixed due to shear forces at high temperatures. The HiPco material shows a better conductivity and a lower percolation threshold in comparison to the purified ArcD SWNT.

TABLE 2. Electrical parameters of polymer composites.

Sample	Mixing	Percolation	Conductivity[1] S/cm
ArcD (pur.) in Epoxy	Hand	<1wt.%	0.1
HiPco in PC + SOCl2	Hand	0.17 wt.%	3
HiPco in PC	Coagul.	>0.5 wt.%	6.1e-4
HiPco in PC	Melt mix.	>1.0 wt.%	-
HiPco in PC (s)	Melt mix.	0.75 wt.%	1.18e-5
ArcD (unpur.) in PC	Melt mix.	2-3 wt.%	1.4e-4
ArcD (pur., 355°C, 2*200') in PC	Melt mix.	3-4 wt.%	1.1e-4
ArcD (pur., 370°C, 13h) in PC	Melt mix.	4-5 wt.%	1.6e-6
ArcD (pur., 370°C, 13h) in PC + SOCl2	Melt mix.	-	-
ArcD (pur., 370°C, 13h, 2xcentr.) in PC	Melt mix.	4-5 wt.%	2.6e-8

SUMMARY

Purification of ArcD material enhances the electrical conductivity of SWNT material. However, the expected high improvement in electrical properties was not observed in the composite. HiPco material reveals higher conductivity and lower percolation threshold than purified and not purified ArcD material. The HiPco material mixed with the polymer by the coagulation method exposes a reproducible process for homogenous composite making.

ACKNOWLEDGMENTS

The project was funded by the European Union CARDECOM project.

---

[1] Highest measured conductivity of the composite. Saturation, leading to end-conductivity, was not inevitably reached.

REFERENCES

1. C. Journet et al., *Nature* **388**, 1997, pp. 756.
2. R. Saito, G. Dresselhaus, M.S. Dresselhaus, *Physical Properties of Carbon Nanotubes*, Imperial College Press (1998).
3. R. Zallen, *The Physics of Amorphous Solids*, John Wiley & Sons, New York (1983).
4. Wagner et al., *Appl. Phys, Lett.* **72**, 1998, pp. 188.
5. R. Ruoff et al., *International Conference: Nanotechnology in Carbon and Related Materials*, Brighton (UK) (1999).
6. P. Pötschke et al, « Melt mixing as method to disperse carbon nanotubes into thermoplastic polymers», *Fullerenes, Nanotubes and Carbon Nanostructures,* 2004 (accepted).

Dispersion of Carbon Nanotubes into Thermoplastic Polymers using Melt Mixing

P. Pötschke[1], A.R. Bhattacharyya[1+], I. Alig[2], S.M. Dudkin[2], A. Leonhardt[3], C. Täschner[3], M. Ritschel[3], S. Roth[4], B. Hornbostel[4] and J. Cech[4]

[1]*Institute of Polymer Research Dresden, Hohe Str. 6, 01069 Dresden, Germany*
[2]*Deutsches Kunststoff-Institut Darmstadt, Schlossgartenstr. 6, 64289 Darmstadt, Germany*
[3]*Leibniz-Institute for Solid State and Materials Research Dresden, Helmholtzstr. 20, 01069 Dresden, Germany*
[4]*Max-Planck-Institute for Solid State Research, Heisenbergstr. 1, 70569 Stuttgart, Germany*
[+] *Present address: Department of Metallurgical Engineering & Materials Science, Indian Institute of Technology Bombay, Powai, Mumbai 400076, India*

Abstract. This paper presents melt mixed composites where two ways of introducing nanotubes in polymer matrices were used. In the first case, commercially available masterbatches of nanotube/polymer composites are used as the starting material which are diluted by the pure polymer in a subsequent melt mixing process (masterbatch dilution method) while in the other case nanotubes are directly incorporated into the polymer matrix. As an example of the masterbatch dilution method, composites of polycarbonate with MWNT are presented which are produced using different melt mixing equipments. The lowest percolation threshold was found at about 0.5 wt% MWNT using a Brabender PL-19 single screw extruder. The nanotube dispersion as observed by TEM investigations is quite homogeneous. The direct incorporation method is discussed in composites of polycarbonate with MWNT and SWNT. The nanotube addition significantly changes the stress-strain behavior of the composites: modulus and stress are increased; however, elongation is reduced especially above the percolation concentration.

INTRODUCTION

Effective utilization of nanotubes in composites with polymers with regard to enhancement of mechanical properties and electrical conductivity depends primarily on the ability to disperse the nanotubes homogeneously in the polymer matrix. However, homogeneous dispersion of nanotubes is difficult due to the intermolecular van der Waals interactions between the nanotubes, thus resulting in the formation of aggregates. This problem presents a major challenge irrespective of the method of composite preparation. In context with industrial applications of nanotube/polymer systems, melt mixing is the preferred method of composite preparation. It is expected that aggregates can be minimized by appropriate application of shear in melt mixing [1-5].

This contribution presents melt mixed composites where in the first case commercially available masterbatches of nanotube/polymer composites are used as the base material which are diluted by the pure polymer in a subsequent melt mixing process (masterbatch dilution method) while in the other case nanotubes are directly incorporated into the polymer. The investigations are focused on the percolation composition as detected by electrical resistivity measurements and the stress-strain behavior of the composites.

CP723, *Electronic Properties of Synthetic Nanostructures*, edited by H. Kuzmany et al.
© 2004 American Institute of Physics 0-7354-0204-3/04/$22.00

EXPERIMENTAL

A masterbatch of 15 wt% MWNT1 in polycarbonate (provided by Hyperion Catalysis International Inc, Cambridge, USA) was diluted with different kinds of polycarbonate. MWNT1 are vapor grown and typical diameters range from 10-15 nm while lengths are between 1 and 10 μm. [6,7]. They are produced as agglomerates and exist as curved intertwined entanglements [8-10]. The polycarbonates used for these studies were PC Iupilon E2000 (PC1, powder, Mitsubishi Engineering Plastics, Japan) with a zero shear viscosity at 260°C ($\eta_{0,260}$) of 6800 Pa-s, PC Lexan 121 (PC2, granules, GE Europe, $\eta_{0,260}$ of 3500 Pa-s), and the PC used for the masterbatch production (PC3, powder, supplied by Hyperion Cat., $\eta_{0,260}$ of 1000 Pa-s). Melt compounding was carried out between 240 and 280°C using a Haake twin screw extruder [3], a small scale DACA Micro Compounder [11, 15], or a Brabender PL-19 single screw extruder [16]. The materials were dried at 120°C in vacuum for at least 4 h and fed as granular premixture.

Direct incorporation of MWNT and SWNT was performed using the small scale (4.5 cm^3 material) DACA Micro Compounder. After drying of polymer and nanotubes (100°C, 2 h vacuum) the materials were premixed and fed to the running compounder. For direct incorporation of the MWNT2 (purity >95%, diameters 20-50 nm, length up to more than 100 μm, produced at the Leibniz-Institute for Solid State and Materials Research Dresden by thermal catalytic chemical vapor deposition) were incorporated into PC2 at 260°C, 50 rpm, and 5 min. SWNT (unpurified arc-discharge material, bundled, single tube diameters of 1.0-1.3 nm, prepared at the Max-Planck-Institute for Solid State Research Stuttgart) were incorporated into PC1 powder at 280°C, 50 rpm, and 15 min.

Electrical volume resistivity was measured on compression molded sheets (diameter > 60 mm, thickness ~ 0.35 mm). A Keithley electrometer Model 6517 equipped with an 8009 Resistivity Test Fixture was used to measure high resistivity samples. Lower resistivity composites were measured by a four point test fixture combined with a Keithley electrometer Model 2000 using strips (20 mm x 3 mm) cut from the sheets. In addition, dielectric measurements were performed at room temperature in a frequency range from 10^{-3} to 10^7 Hz using a frequency response analysis system consisting of Solartron SI1260 Impedance/Gain Phase Analyzer and Novocontrol broadband dielectric converter with the BDC active sample cell [12]. Discs (diameter of 20 mm) were cut from the sheets and gold layers were sputtered onto both sides as electrodes.

Scanning electron microscopy (SEM) was performed using a LEO VP 435 scanning electron microscope (Leo Elektronenmikroskopie, Germany) on raw materials or fractured samples. Mechanical testing of miniature dogbones (length 20 mm, parallel length 6 mm, gauge width 2 mm) punched from the pressed sheets was performed similar to ISO 527-2 on a Zwick Z010 tensile tester at an extension rate of 5 mm/min. The values are mean values of 10 tests.

RESULTS & DISCUSSION

Using the masterbatch dilution method, the mixing task is to expand the already existing percolated MWNT-polymer structure gradually by incorporating the diluting polymer between the individual tubes by maintaining the "tube percolation". The PC based masterbatch with 15 wt% MWNT consists of highly interconnected tubes in the

PC3 matrix which show a good wetting of the tubes with polymer (see discussion in [3]). If the dilution is properly performed, the MWNT should be homogeneously distributed and dispersed in the matrix. Resistivity vs. composition dependencies can be used to detect the percolation composition unequivocally. In addition, the resistivity values can give information about the state of dispersion obtained at a given composition [7, 13].

Figure 1 summarizes the electrical resistivity values obtained by using the different PC and different mixing equipments.

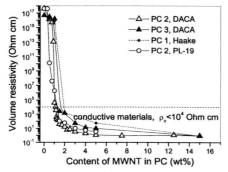

FIGURE 1. Electrical volume resistivity vs. MWNT content in composites prepared by masterbatch dilution method

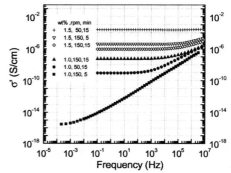

FIGURE 2. AC conductivity σ' vs. frequency for PC3-MWNT1 composites processed in the DACA Micro Compounder at different processing conditions, legend contents amount of MWNT in wt%, rotation speed in rpm, and mixing time in min.

Depending on the PC and mixing equipment, percolation was reached between 0.5 and 1.5 wt% MWNT1 content. With PC1 as diluting material and processing in a Haake twin screw extruder, electrical percolation was found between 1 and 2 wt% [3], whereas dilution using the lower viscosity PC3 using the DACA Micro Compounder led to percolation between 1.0 and 1.5 wt% MWNT1 [11]. Using PC2 and processing in the DACA Micro Compounder, percolation was reached between 0.875 and 1.0 wt% MWNT1 [14]. For PC2 as diluting polymer and processing in the single-screw extruder the percolation was found to occur between 0.25 and 0.5 wt% MWNT1 in PC. In all the cases, composites with contents starting at 1.5 wt% MWNT1 can be regarded as electrically conductive (volume resistivity $<10^4$ Ohm-cm). For the example of masterbatch dilution with PC3 using the DACA Micro Compounder the influence of processing conditions at compositions near the percolation was investigated more in detail using dielectric spectroscopy. Figure 2 shows the AC conductivity for composites containing 1.0 and 1.5 wt% MWNT1. It is interesting to note that in the case of 1.0 wt% variations in the processing conditions (increasing mixing time) can transform a nonpercolated structure in a percolated one. The relatively low percolation thresholds indicates a quite good distribution and dispersion within the PC matrices as it was shown by TEM investigations in [15].

Using the direct incorporation method, next to a good distribution and dispersion of the tubes also a wetting of the nanotube surface with polymer has to be achieved which depends on the CNT surface characteristics and amount, the interfacial tension between nanotube surface and polymer melt, and the polymer melt viscosity.

MWNT2 used for this study (Fig. 3) are relatively long, less tangled and quite straight. It is assumed that some of the very long tubes break during melt mixing with high viscous polymers. The cryofracture shown in Fig. 4 illustrates a homogeneous distribution and dispersion without agglomeration of the nanotubes. In this system, percolation started at 3 wt% MWNT2 addition.

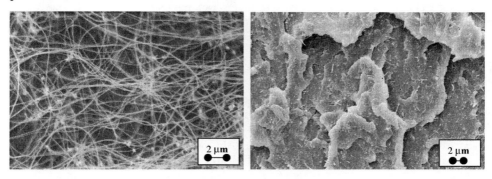

FIGURE 3. SEM micrographs of MWNT2(left) and a composite of PC2 with 1.5 wt% MWNT2

The incorporation of nanotubes significantly changes the stress-strain behavior of the composites as it is shown in Fig. 4. PC2 exhibits a pronounced yielding behavior followed by cold drawing and shows elongation at break of about 30% [14]. After adding MWNT2, next to an increase in Young's modulus (9% at 1wt% and 16% at 5wt% MWNT2) the elongation at break is reduced dramatically starting at 1.5 wt% MWNT2 since the nanotubes start to form a network. Starting at 2 wt% MWNT2, the pronounced yielding and cold drawing behavior is no longer visible and the break occurs before reaching the elongation (and stress) typical for the yield point of the pure PC. Thus, tensile strength as the maximum stress in the stress-strain curves is only increased at low MWNT contents (9% at 1.5 wt% MWNT2).

In the composites of PC1 with SWNT, percolation was reached between 2 and 3 wt% SWNT addition [16]. The changes in the stress-strain behavior (Fig. 4) are similar to those ones described before.

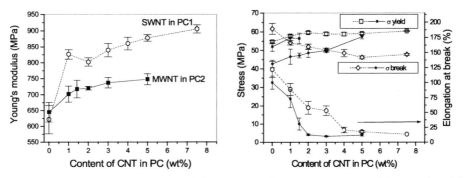

FIGURE 4. Properties of PC2/MWNT2 (full symbols) and PC1/SWNT (open symbols) composites; left: Young's modulus; right: stress at yield point (σ_{yield}), stress at break (σ_{break}), and elongation at break

The modulus shows an increase with SWNT addition (46% at 7.5 wt% SWNT) which is higher than that one achieved for PC2/MWNT2 composites. The pronounced cold drawing behavior after yielding is observable up to 3 wt% SWNT. The stress at yield is increased by about 7 MPa as well as the stress level beyond the yield point [16]. At higher SWNT contents the break occurs just after the yield point leading to significantly decreased elongations at break. This behavior can be understood by the percolated network of the SWNT which hinders the polycarbonate in its typical deformation behavior especially in developing a cold drawing process [14].

CONCLUSIONS

Melt mixing using the masterbatch technique is a method which is applicable and easily accessible for small and medium-sized enterprises. The results obtained for polycarbonate based materials show that the percolation only slightly depends on the PC used for dilution, the mixing equipment, and the processing conditions. In composites based on PC percolation was always reached at 1.5wt% MWNT using a masterbatch provided by Hyperion Catalysis International. Using a suitable equipment, in our case a Brabender PL-19 extruder, percolation could be reached already at 0.5 wt% MWNT.

It could be shown that also the direct incorporation of CNT using melt mixing leads to a good dispersion of CNT in PC. For the selected materials used, percolation occurred in the range between 2 and 3 wt% CNT. The incorporation of nanotubes significantly changes the stress-strain behavior of the composites: modulus and stress are enhanced; however, the elongation is reduced especially above the percolation concentration. Further optimization of melt processing conditions and nanotube pretreatment (purification, predispersion in solvents, functionalization) seem to be promising tasks.

REFERENCES

[1] R. Haggenmueller. H.H. Gommans, A. Rinzler, J.E. Fischer and K.I. Winey, *Chem. Phys. Lett.* **330**, 219-225 (2000).
[2] J.R. Hagerstrom and S.L. Greene. Electrostatic dissipating composites containing Hyperion fibril nanotubes. Commercialization ofNanostructured Materials, Miami, USA, April 7, 2000.
[3] P. Pötschke, T.D. Fornes, D.R. Paul *Polymer* **43**, 3247-3255 (2002).
[4] R. Andrews, D. Jacques, M. Minot, T. Rantell *Macromol. Mat. Eng.* **287**, 395-403 (2002).
[5] M. Sennett, E. Welsh, J.B. Wright, W.Z. Li, J.G. Wen, Z.F. Ren. *Appl. Phys.* A **76**, 111-113 (2003).
[6] *Plastics Additives & Compounding* **3**(2001) 9, 20-22.
[7] D.W. Ferguson, E.W.S. Bryant, H.C. Fowler, ESD thermoplastic product offers advantage for demanding electronic applications, ANTEC'98, 1998. p.1219-1222.
[8] M.S.P. Shaffer, X. Fan, A.H. Windle, *Carbon,* **36**, 1603-1612 (1998).
[9] J. Sandler, M.S.P. Shaffer, T. Prasse, W. Bauhofer, K. Schulte, A.H. Windle, *Polymer* **40**, 5967 (1999).
[10] M.S.P. Shaffer, A.H. Windle. *Adv Mat.* **11**, 937-941 (1999).
[11] P. Pötschke, A.R. Bhattacharyya, A. Janke, H. Goering, *Comp. Interf.* **10**, 389-404 (2003).
[12] P. Pötschke, S.M. Dudkin, I. Alig, *Polymer* **44**, 5023-5030 (2003).
[13] P. Pötschke, A.R. Bhattacharyya, A. Janke, *Carbon* **42**, 965-969 (2004).
[14] P. Pötschke, Percolation and stress strain behaviour of polycarbonate- multiwalled carbon nanotube composites, *Polymer*, 2004 (submitted).
[15] P. Pötschke, A.R. Bhattacharyya, A. Janke, *European Polymer Journal* **40**, 137-148 (2004).
[16] P. Pötschke., A.R. Bhattacharyya, A. Janke, S. Pegel, A. Leonhardt, C. Täschner, M. Ritschel, S. Roth, B. Hornbostel, J. Cech, Melt Mixing as Method to Disperse Carbon Nanotubes into Thermoplastic Polymers, *Fullerenes, Nanotubes, and Carbon Nanostructures*, 2004 (submitted)

APPLICATIONS

Field emission from individual thin carbon nanotubes

Niels de Jonge*, Myriam Allioux*[†], Maya Doytcheva*, Monja Kaiser*,
Kenneth B. K. Teo[¶], Rodrigo G. Lacerda[¶] and William I. Milne[¶]

*Philips Research Laboratories, Prof. Holstlaan 4, 5656 AA Eindhoven, The Netherlands
[†]Present address: ESPCI, Paris, 10 Rue Vauquelin 75005 Paris, France
[¶]Department of Engineering, University of Cambridge, Trumpington Street, Cambridge CB2 1PZ, UK

Abstract. Individual thin carbon nanotubes with closed ends were mounted on tungsten support tips and their surfaces were thoroughly cleaned. We show these carbon nanotube electron emitters have Fowler-Nordheim behavior and we observed no deviation from this model for thin nanotubes. The measurements were consistent with numerical simulation.

INTRODUCTION

Carbon nanotubes emit electrons at low turn-on voltages [1] and carbon nanotubes have several advantages over other materials used for electron emitters, such as tungsten, molybdenum, diamond and amorphous carbon. Most important, a nanotube is not a metal, but a structure build by covalent bonds. As a result, the nanotube can withstand the extremely strong electric fields (several V/nm) needed for field emission without changing its shape. In addition, carbon nanotubes have a high conductance, an extremely large Young's modulus, are chemically highly inert [2] and carbon has one of the lowest sputter coefficients. Electron emission from carbon nanotubes has experienced large interest for several applications, like field emission displays [3], cathode ray lamps [4], X-ray sources [5] and electron microscopes [6].

But, despite the large amount of reports on field emission from carbon nanotubes, several issues remain unclear. One of the main questions is, whether or not a Fowler-Nordheim model applies, or whether correction factors are needed. The difficulty in addressing this question lies in the small size of the carbon nanotube. It is almost impossible to obtain a sample containing one single nanotube with properties that are exactly known, and the surface of this nanotube has to be perfectly clean. We have developed a method to mount an individual carbon nanotube on a tungsten support tip, using a well-characterized sample of thin carbon nanotubes. We master a reliable cleaning procedure and we manage to obtain closed nanotubes. This enables us to perform experiments under the desired conditions.

CP723, Electronic Properties of Synthetic Nanostructures, edited by H. Kuzmany et al.

FOWLER-NORDHEIM THEORY

Often it is assumed that the electron emission process of carbon nanotubes in the presence of a strong electric field can be described by field emission. The Fowler-Nordheim theory [7] describes the field emission process in terms of a tunneling current through a potential barrier between a conducting surface and vacuum, with a current density J [8]:

$$J = \frac{e^3 F^2}{8\pi h \phi\, t^2(y)} \exp\left\{ -\frac{8\pi\sqrt{2m}\,\phi^{3/2}}{3heF} v(y) \right\} \tag{1}$$

with workfunction ϕ, electron mass m, electric field F, Planck's constant h, the electron charge e and the functions $t(y)$ and $v(y)$. A plot of $\log(J/F^2)$ versus $1/F$, the so-called Fowler Nordheim plot, is an almost linear curve. The functions $t(y)$ and $v(y)$ were calculated by Good and Mueller [9] and can be approximated by [8], $t(y) = 1 + 0.1107y^{1.33}$ and $v(y) = 1 - y^{1.69}$. The function y is expressed as:

$$y = \frac{1}{\phi}\sqrt{\frac{e^3 F}{4\pi\varepsilon_0}} \tag{2}$$

with the permittivity of free space ε_0. The Fowler-Nordheim equation is valid for low temperatures only and a correction factor is needed to include a temperature effect [8]:

$$J_T = J \frac{\pi k_b T / d}{\sin(\pi k_b T / d)} \tag{3}$$

The total current I is a function of J and the emitting surface A, which is often taken as a half sphere with radius of curvature R.

$$I = AJ = 2\pi R^2 \tag{4}$$

The field at the nanotube end is:

$$F = \beta U \tag{5}$$

with the extraction voltage U and the field enhancement factor β. The field enhancement factor of a carbon nanotube can be calculated numerically, see Fig. 1a. The potential was calculated for a spherically symmetric emitter inside a much larger casing (10 mm) using Munro's software. The axial potential was differentiated to obtain the electric field strength as function of the axial position. The maximal field strength was obtained directly at the nanotube end and represents the field enhancement factor.

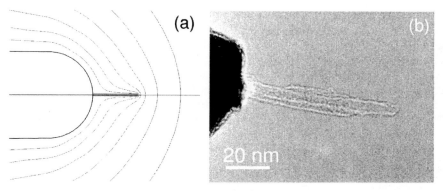

FIGURE 1. Carbon nanotube electron source on a support tip. (a) Numerical calculations on the field enhancement factor for a nanotube with a length of 100 nm and a radius of 2 nm, giving a field enhancement factor of 1.7×10^7 m^{-1}. Equipotential lines every 50 mV are also shown. (b) Bright field transmission electron microscopy image of a carbon nanotube with a closed end on a tungsten support tip. The image was taken deliberately out-of-focus to enhance the contrast of the nanotube.

EXPERIMENTS

Mounting and cleaning of the carbon nanotubes

Thin carbon nanotubes (1-4 walls) were grown on an oxidised silicon substrate by thermal chemical vapour deposition[10]. The substrate was prepared with an Al/Fe/Mo (10/1/0.2 nm) triple layer, heated to 950°C in a He (1000 sccm flow, 15 mbar pressure) ambient and subjected to a rapid 5 seconds burst of C_2H_2 (250 sccm flow, 2.5 mbar pressure). Individual carbon nanotubes were mounted on tungsten tips in a highly controlled manner [11], using a scanning electron microscope (SEM, Philips), equipped with a piezo driven nano-manipulator (Omicron). TEM images of 6 nanotube electron sources revealed that always a short and thin nanotube had been mounted, with lengths of 25 – 110 nm, radii of 2 – 4.5 nm and mostly a closed end.

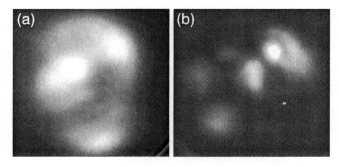

FIGURE 2. Emission patterns of individual carbon nanotube electron emitters. (a) Pattern of closed tube, nanotube No. 1. (b) Pattern of an open tube.

The field emission measurements were performed in an ultra-high vacuum system with a base pressure of 2×10^{-10} Torr. A fresh nanotube was always thoroughly cleaned by heating it to a temperature of about 700° C in vacuum for 10 minutes to remove adsorbed species and impurities from the tube. The emission pattern was recorded with a micro-channel plate and phosphor screen (Hamamatsu) to check the cleaning procedure. Figure 2a shows the emission pattern of nanotube No. 1 with a closed cap, which was highly stable with time. In contrast, the patterns of nanotubes that were not sufficiently cleaned showed one or more spots that fluctuated with time. A few nanotubes displayed an emission pattern with separate spots, fluctuating with time (Fig. 2b), even after repeated heating to the carbonization temperature. These patterns were assigned to the emission patterns of nanotubes with an open end.

Field emission

The emitted current of nanotube No. 1 was measured as function of the extraction voltage, see Fig. 3a. The measurement was performed at room temperature. Fitting the Fowler-Nordheim equation, assuming a workfunction of 5 eV, to this data gave a value of the field enhancement factor $\beta = 8.0\times10^6$ m^{-1}. The tube radius was calculated from the simultaneously obtained value of the emitting area and amounted to 4.9 nm. The numerically calculated value of the field enhancement factor is 8.4×10^6 m^{-1} for a tube with a radius of 5 nm and a length of 25 nm. This value is close to the typical values of our emitters.

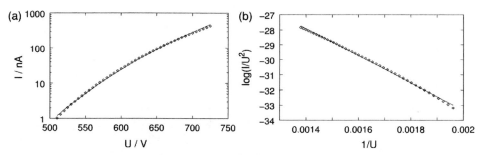

FIGURE 3. Emission measurements of nanotube No. 1. (a) The emitted current as function of the extraction voltage measured at room temperature and a fit of the Fowler-Nordheim Theory. (b) Fowler Nordheim plot with a slope of -9.0×10^3 and a linear fit.

The data follows a straight line in the Fowler Nordheim plot (Fig. 3b), which is a clear indication of a field emission process. At low currents a few data points do not fit, on account of a small leakage current in the measure system. Another nanotube revealed similar behavior and the values $\beta = 1.6\times10^7$ m^{-1} and $r = 1.6$ nm. Concerning carbon nanotubes, the question arises whether the Fowler-Nordheim theory is still valid, as it was derived for a surface that appears flat as 'seen' from the electrons, which might not be the case for carbon nanotubes on account of their strong curvature at the tube end [12]. Other effects may take place, since the density of states is not energy independent around the Fermi level as in 'real' metals [2]. However, we did not observe deviations from the Fowler-Nordheim mechanism and the numerical calculated value of the field enhancement factor agrees well with the experimentally obtained value.

If the measurement had not been taken at room temperature, but for example, at a temperature of 900K, the total current level would have been larger. The resulting fit of R would have been 17% smaller. Yet, the effect of the increased temperature on the slope of the Fowler-Nordheim is only minor, both the slope and the field enhancement factor would differ less than 4%. Therefore, it is allowed to neglect the temperature correction for our range of parameters. At high temperatures also other emission mechanisms may play a role, such as Schottky emission and thermionic emission. The occurrence of such mechanisms was not observed.

CONCLUSIONS

We show that carbon nanotube electron sources show Fowler-Nordheim behaviour and we observed no indication of inconsistency from this behaviour. The measurements are consistent with numerical calculations. Our data demonstrates the need for measurements on samples of individual carbon nanotubes with closed ends and the importance of a good cleaning procedure.

ACKNOWLEDGEMENT

We thank G. Amaratunga, T. van Rooij, A. Teh and M. Yangs for experimental help. This work was supported by FEI Company, the Dutch Ministry of Economic Affairs, the EPSRC and the EC

REFERENCES

[1]   Rinzler A. G., Hafner J. H., Nikolaev P., Lou L., Kim S. G., Tomanek D., Nordlander P., Colbert D. T. and Smalley R. E., *Science* **269**, 1550-1553 (1995).
[2]   Saito, R., Dresselhaus, G. and Dresselhaus, M. S., *Physical properties of carbon nanotubes*, London: Imperial college press, 1998.
[3]   Choi W. B., Chung D. S., Kang J. H., Kim H. Y., Jin Y. W., Tan I. T., Lee Y. H., Jung J. E., Lee N. S., Park G. S. and Kim J. M., *Appl. Phys. Lett.* **75**, 3129-3131 (1999).
[4]   Saito Y. and Uemura S., *Carbon (UK)* **38**, 169-182 (2000).
[5]   Sugie H., Tanemure M., Filip V., Iwata K., Takahashi K. and Okuyama F., *Appl. Phys. Lett.* **78**, 2578-2580 (2001).
[6]   de Jonge N., Lamy Y., Schoots K. and Oosterkamp T. H., *Nature* **420**, 393-395 (2002).
[7]   Fowler R. H. and Nordheim L., *Proc. Roy. Soc. London A* **119**, 173-181 (1928).
[8]   Hawkes, P. W. and Kasper, E., *Principles of electron optics II: Applied geometrical optics*, London: Academic Press, 1996.
[9]   Good, R. H. and Mueller, E. W., "Field Emission" in *Handbuch der Physik, XXI*, edited by Fluegge, S., Berlin: Springer verlag, 1956, pp. 176-231.
[10]   Lacerda R. G., Teh A. S., Yang M. H., Teo K. B. K., Rupesinghe N. L., Dalal S. H., Koziol K. K. K., Roy D., Amaratunga G. A. J., Milne W. I., Chowalla M., Hasko D. G., Wyczisk F. and Legagneux P., *Appl. Phys. Lett.* **84**, 269-271 (2004).
[11]   de Jonge N., Lamy Y. and Kaiser M., *Nano Letters* **3**, 1621-1624 (2003).
[12]   Edgcombe C. J. and Johansen A. M., *J. Vac. Sci. Technol. B* **21**, 1-5 (2003).

Fundamental Aspects and Applications of Low-Field Electron Emission from Nano-Carbons

A.N. Obraztsov[1], A.P. Volkov[1], Al.A. Zakhidov[1], D.A. Lyashenko[1], Yu.V. Petrushenko[2], O.P. Satanovskaya[2]

[1] Department of Physic, Moscow State University, Moscow119992, Russia
[2] SRPC "Platan", Fryasino 141120, Russia

Abstract. Low-field electron emission (LFEE) from various carbon related materials has high attractiveness for application to replace metal or semiconductor micro-tip arrays in vacuum electronic devices and electron beam technologies such as flat panel displays, microwave tubes, miniature X-ray sources, mass-spectrometers, electron-beam lithography etc. Evaluation of yhe carbon cold cathode applicability requires understanding of the LFEE mechanisms. In our work we have performed experimental and theoretical study of the electron emission from nano-carbon thin film materials grown by CVD. An empirical model of the LFEE sites was proposed on base of comprehensive study of structural and electronic characteristics of the carbon cathodes. The modifications of usual Fowler-Nordheim theory were proposed for adequate qualitative and quantitative description of the experimental LFEE observations. The prototypes of highly efficient light sources based on the nano-carbon cathodes were designed and tested. The lamps brightness of 200000 cd/m^2 and record energy efficiency were demonstrated.

INTRODUCTION

Field emission is a quantum-mechanical phenomenon in which electrons tunnel through a potential barrier at the surface of a solid as a result of the application of a large electric field. Field emission is distinct from thermionic emission and photoemission in which electrons acquire sufficient energy via heating or energy exchange with photons, respectively, to overcome the potential barrier. Field emission might occur, correspondingly, without any power consumption that is very attractive for many applications. In field emission from metals and semiconductors external electric fields on the order of 10^9 V/m are required for appreciable electron currents. Numerous experimental studies have shown also "unusually" low-field electron emission (LFEE) from various carbon materials. While LFEE mechanism is not understood well their different applications are very attractive because allow significant device simplification.

In our recent publications (see for example [1]) we found experimentally evidences of general nature of LFEE from different carbon materials. In accordance with the proposed empirical model the particular property of carbon materials to emit electrons at low fields is a result of well-ordered structures of nano-sized carbon species like carbon nanotubes (CNT) and nano-crystallites of graphite (NCG). At the same time

CP723, *Electronic Properties of Synthetic Nanostructures*, edited by H. Kuzmany et al.
© 2004 American Institute of Physics 0-7354-0204-3/04/$22.00

these nanostructured carbon (nC) materials must emit electrons under fields corresponding to normal Fowler-Nordheim FE like usual metals. In this paper we present results of low-field electron emission study for nC thin film material grown by chemical vapor deposition (CVD) in dc discharge plasma activated gas mixture of hydrogen and methane. The fundamental results are illustrated by the prototypes of cathodoluminescent lamps with nC cold cathodes.

STRUCTURE, MORPHOLOGY AND FIELD EMISSION FROM NANOCARBON CVD FILMS

The film nC material fabrication was performed using CVD in a hydrogen-methane gas mixture activated by a direct current discharge. The original facility of the dc CVD system is described in detail elsewhere [2]. In brief, the nC films were deposited on Si or Ni substrates usually. The maximal size of the substrates is 50 mm. A static pressure of approximately 100 Torr of a H_2/CH_4 gas mixture was maintained during the deposition. The dc discharge was activated in the CVD reactor between the substrate, located on a water cooled anode, and a tungsten cathode. The substrate temperature was maintained at approximately 1000°C. The dc voltage applied between the electrodes was approximately 800V and the discharge current density was in range of 0,25-0,75 A/cm^2. The methane concentration in the gas mixture was about 6%.

The nC film morphology and composition were tested by using electron microscopy, Raman spectroscopy and other methods. Raman spectra of the CVD films contain lines at 1350 cm^{-1} and in the vicinity of 1580 cm^{-1} (between 1550 and 1620 cm^{-1}) attributed to graphite and different forms of disordered graphite. Similar 1580 cm^{-1} line is typical for multiwall carbon nanotubes [3]. The typical morphology of our nC CVD films consisting of NCG is shown in SEM image on Fig.1. The characteristic sizes of the sharp edges of NCG are in range of 2 to 20 nm. These nC species form on substrate surface homogeneous and rather porous layers with thickness of about 1 to 2 μm. The aspect ratios of the nC species (ratios of the species height to the edge size – H/r) are in range of 50 to 500 as estimated from the electron microscopy observations. The field emission (FE) tests were performed in a vacuum diode configuration with a flat parallel anode and cathode. The anode was a glass plate coated with a transparent, conducting indium tin oxide (ITO) film. The conducting anode film was covered with a phosphor layer allowing an image of FE site distribution over the cathode surface to be obtained. While the macroscopic value of field (E) is in range of 1 to 10 V/μm in the test measurements the local field value (F) on FE site surface may be estimated as $F=\beta E$ where factor β is the same order as the aspect ratios H/r. Some reduction of the field enhancement is possible due to field screening effect because of rather dense growth of NCGs in the nC layer. The opposite effect of the local field enhancement factor increase is possible in the case where the actual size of the FE sites is less than the geometric shape, as evaluated from electron microscopy observations. The typical example of current-voltage (I-V) dependence measured for those nC cathodes is shown in Fig.2. The emission was quite stable at repeatable measurements in vacuum of about 5×10^{-6} Torr during a few hours. The distinctive features of the FE characteristics are: (i) strong linearity of the *I-V* curve in Fowler-Nordheim (FN) coordinates, and in a wide range of voltages and currents; (ii) substantial curvature of the FN *I-V* plot at the lowest voltages and currents; (iii) homogeneity of the FE site distribution over the film surface.

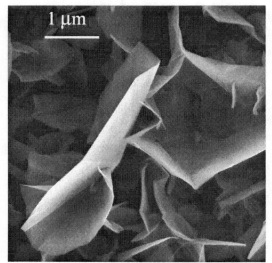

FIGURE 1. Typical SEM image of nC CVD film surface.

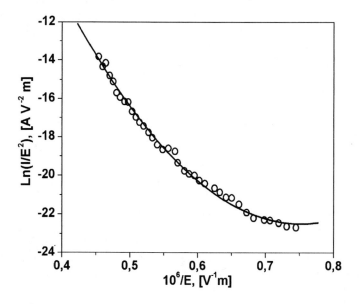

FIGURE 2. Typical experimental I-V curve for electron emission from nC cathode plotted in FN coordinates (dots) and fitted curve calculating using formula (1) (line).

STATISTICAL ANALYSIS OF ELECTRON EMISSION FROM NANOCARBON CATHODES

The linear character of the FN plot corresponds to the classical electron tunneling mechanism of FE [4] while its substantial curvature may be explained by statistical distribution of the emitting sites' geometric, structural and electronic characteristics [5]. Although FE site characteristics have complex physical origins, a normal (Gaussian) distribution can be a good first approximation to the actual distribution. The FE image homogeneity at moderate electric fields [1] is evidence of a narrow width of the normal distribution. Believing, for simplicity, that the subject for the statistical distribution is an effective size (radius r) of emission sites only and assuming that emission current occurs from the entire surface area of each emission site $S=\pi r^2$, we obtain for total emission current I formula:

$$I = CNE^2 \exp\left[-\frac{Dr_0}{E} + \frac{\sigma^2}{2}\left(\frac{D}{E}\right)^2\right]$$

$$\tag{1}$$

where constants $C = \frac{A\pi H^2}{\varphi}\exp[10.1/\sqrt{\varphi}]$; $D = 0.95B\frac{\varphi^{3/2}}{H}$; $A = 1.5414 \times 10^{-6}$ A eV V^{-2}; $B = 6.8309 \times 10^9$ eV$^{-3/2}$ V m^{-1}; φ is the work function, and σ is the standard statistical deviation [6].

Formula (1) for *I-V* dependence differs from the classical expression deduced from FN theory for FE of metals by the second term in the exponent. Fig. 2 illustrates the good agreement of experimental results with the calculation performed by using equation (1). The second term causes excess current and upward curvature in the FN plot. This term vanishes when all emission sites have the same geometrical characteristics since $\sigma=0$. However, the term increases at lower values of macroscopic field (or applied voltage) if $\sigma\neq0$. The inputs of both terms in (1) become comparable at σ in range of 0.5Å to 2Å when we assume $r=5$nm, $H=1\mu$m, $\varphi=5$eV and E in range of 1V/μm to 10V/μm, correspondingly. This numerical estimation is in good agreement with our initial assumption about narrow width of the normal distribution with a small standard deviation, σ. The FN plot curvature increases with σ, but for the linear part of the plot we can believe σ and the second term in (1) equal to zero.

The experimental *I-V* curve allows estimation of the number of emission sites, *N*. This estimation may be made for the linear part of the FN plot from an intercept by assuming that all emission sites have the same surface area. Thus for the particular case of FN plot in Fig. 1 we have $N\approx10^6$ cm^{-2} which is in good agreements with our observations by using a phosphor screen [1,2,6]. But, for the low-voltage range curved part of the plot our estimation gives us the value $N\approx10^{13}$ cm^{-2}. The emission site density decrease with voltage increase may be explained by the screening effect leading to suppression of emission from the sites having smaller field enhancement factors. However, even in the case of the screening the number $N\approx10^{13}$ cm^{-2} is possible, if emission sites have atomically small sizes of their top edges only. This implies that the main physical assumptions made in common FN theory of FE are not valid for these emitters. Correspondingly, the results of this theory cannot be used for any numerical estimation but only for qualitative speculations. In such speculations we should take into account the atomic structure of the nano-carbon emitters in contrast to that of the free electron metal.

DOUBLE BARRIER MODEL FOR LFEE SITE

The alternative explanation consists of the notion that emission occurs from an area exceeding the top edge of the nano-carbon species and including some part of the lateral surface, as well. In this case estimated number of emission sites will be decreased by two or three orders without extreme decrease of the characteristic sizes of emitting carbon species. This explanation agrees with our previous publications [1,2] as well as with other observations showing origination of field emission from CNT lateral walls [7,8]. A mechanism providing such lateral emission is the nonmetallic behavior of nano-carbon emitters and electron tunneling through the potential barrier reduced if in the vicinity of defects of the sp^2 coordinated graphite-like carbon atoms network [1,2]. These defects have sp^3 hybridization in well-ordered carbon materials and provide partly localized electron states on the emitter surface.

Figure 3 illustrates this approach schematically, showing the energy barrier on a cathode-vacuum interface for normal graphite and for nC graphite-like material with an sp^3 cluster on the surface (figures 1(a) and (b), respectively). The quantum wells related to the carbon atoms' positions are shown in the figure instead of the usual energy bands. The upper electron levels lie near the top of the wells in the case of graphite, providing zero band gap and semi-metallic properties. The barrier height on the surface is equal to the work function of about 5 eV for graphite. The non-graphite (sp^3) carbon cluster is represented in the energy diagram by two bands separated by a gap between the highest occupied states (HOS) and the lowest unoccupied states (LUS). The LUS-HOS gap width is assumed to be similar to that in defective diamonds and according to our cathodoluminescent studies, it can be taken to be slightly larger than 4 eV.

It may also be assumed that the vacuum level lies near the LUS, similar to that of diamond with sp^3 atomic configuration. The internal potential barrier between two carbon phases should be smaller than the work function of graphite and we assume it to be equal to 4.5 eV. The thickness of this interface region should be in the order of interatomic distances in graphite or diamond and in our estimations we assume it equals 4 Å. The outer potential barrier on the cathode surface is determined by the position of the Fermi level and work function for the layer of diamond-like material. It is natural to suppose that the Fermi level is fixed near the LUS-HOS midgap and the work function is about 4.5 eV for the untreated surface of CVD diamond films. Therefore, it would be a good estimation to consider the outer barrier seen by electrons on Fermi level of the cathode to be about 4.5 eV. Application of an external field to the cathode surface leads to deformation of this two-barrier structure. In the simplest approximation, the first barrier exhibits a rectangular shape and the second one becomes triangular, as sketched in Fig. 3.

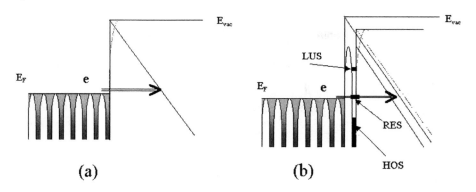

FIGURE 3. Energy band diagram of vacuum-cathode interface for normal graphite (a) and for nC graphite-like material (b).

The theoretical calculations of electron tunneling probability show that significant difference may be obtained for the double barrier system in case if additional level exist in between LUS and HOS bands. It this case of resonance tunneling we got finally equation for emission current density (see [9]):

$$ j = \frac{e}{8\pi h} \left[\frac{\varphi^{1/2}}{eF} - \frac{w}{\sqrt{\chi}} \right]^{-2} \exp\left(-\frac{4\pi\sqrt{2m}}{h} \left(\frac{2(\varphi)^{3/2}}{3eF} - \sqrt{\chi}w \right) \right) $$

(2)

In this equation V is barrier height and w is barrier width; $\varphi = E_{vac} - E_F$ and $\chi = V - E_F$, h is Plank constant; e and m are charge and mass of electron. This formula is different from the usual Fowler-Nordheim law [4] by the presence of a second additional term in the exponent. This new term ($\chi^{1/2}w$) may lead to significant increase in the current density. Using for estimation $\chi = 4.5$eV and $w = 4$ Å, as it was proposed above, we will have current density increase on 4 orders for the same fields in comparison with values predicted by FN for metal emitters.

As it follows from our analysis, the presence of the resonance electronic states (RES) in the quantum well corresponding to the surface layer is an important feature which provides low-field and intensive electron emission. The efficient electron tunneling requires the special energetic position of these RES near Fermi level and slightly above it to explain the existence of nonzero FE thresholds. When an external field is applied to the cathode surface, these resonance electron states are shifted down due to partial voltage drop in the surface carbon layer (corresponding electron potentials are shown by solid lines in Fig. 3b). Various carbon materials including CNT, NCG, nanodiamonds etc. show very similar threshold values of macroscopic electric field (E) in range of 1 to 1.5 V/μm that may be considered as an evidence of the emission mechanisms similarity.

The difference between macroscopic (E) and local (F) values of electric field is determined by: (i) configuration of electrodes during FE measurements, (ii) geometrical shape of emission sites and (iii) emission site density on the cathode surface. As it was mentioned above emission site density (or total emission surface) on surface of nC cathodes increases with decreasing applied field. Together with increase of the electron tunneling efficiency it allows to obtain total emission current from such cathodes at macroscopic fields having threshold values which are much lower than predicted by usual FN approach and correspond to moderate geometrical field enhancement factors less than 500. These conclusions are important for evaluation of the nC cathode applicability and, in particular, the most efficient use of these cathode may be in vacuum electronic devices requiring moderate values of current densities generating from large surfaces by using low voltages. The high current density applications of the carbon cathodes seems to be less appropriate because of very probable destruction of specific material structure and corresponding double barrier configuration on the surface.

PROTOTYPES OF LAMPS WITH NANOCARBON CATHODES

One of the most attractive areas for application of the LFEE cathodes relate to flat panel displays where electron beams emitted from the cathode may be used to produce light emission from phosphor layer deposited on the display screen. The abilities of our nC cathodes for this purpose were demonstrated already in prototypes of flat vacuum luminescent lamps [10]. The moderate light intensities required for the flat panel displays at the level of about 1000 cd/m^2 are easily achievable with use of the low voltages. But higher light intensities and, that is more important, higher efficacy of the light generation require use of much higher voltages to accelerate electrons emitted from the cathode. This requirement of high accelerating voltage is due to intrinsic property of the phosphor material and phenomenon of cathodoluminescence. An evident technical solution to provide higher electron energy is introduction of an additional electrode into vacuum lamp between cathode and anode. In this case electrons may be emitted using relatively low voltages applied between the nC cathode and this additional electrode and than accelerated by application higher voltage to anode. Such triode schema of the lamps was tested and show abilities to get much higher light intensities and total device efficacy [11].

An alternative schema providing high electron energy and a correspondingly high efficiency of light emission exploits a cylindrical diode configuration [12,13]. An important advantage of the cylindrical diode configuration in comparison with a planar one is that for the same applied voltage U, the macroscopic electric field $E=U/[r\ ln(R/r)]$, as compared to $E=U/d$ for two planar electrodes with an interelectrode distance d. Practically, this means that the macroscopic field at the cathode surface $E(r)$ will be much higher in the cylindrical case for identical $d=R-r$ and U. This makes possible the realization of a field emission diode working at suitable voltages applied between electrodes separated by rather large distance. An essential improvement of the lamp parameters was achieved by using a scheme with reflecting Al layer deposited on the glass tube instead of the ITO [14]. This schematic and actual lamp images are shown in Fig. 4. A lamp brightness up to 2×105 cd/m^2 was achieved. The power efficiency of the lamps with reflecting Al anode exceeds 30%. Such power efficiency is an absolute record for light emission devices and is obtained due to combination of nC cold cathode excellence and the lamp configuration providing more optimal conditions for excitation of the phosphor by electrons in comparison with other similar devices. The other advantage of our lamp providing their high efficiency is direct output of light generated in the phosphor layer in contrast to the scheme with a transparent anode where CL light must penetrate through the phosphor layer in order to exit the lamp.

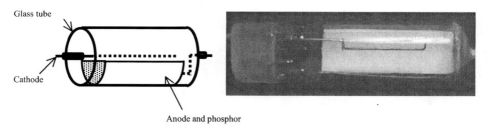

Glass tube

Cathode

Anode and phosphor

FIGURE 4. Schema and Photograph of cylindrical diode with nC wire cathode and cylindrical light reflective anode.

ACKNOWLEDGMENTS

This work was supported in part by INTAS grant No. 01-0254 and by CRDF award No. RE2-5008-MO-03.

REFERENCES

1. Obraztsov A.N., Volkov A.P., Nagovitsyn K.S., et al., J. Phys. D: Appl. Phys. 35, 357-362 (2002).
2. Obraztsov A.N., Zolotukhin A.A., Ustinov A.O., Volkov A.P., Svirko Yu.P., Carbon 41, 836-839 (2003).
3. Carbon Nanotubes, 2001, edited by M.S. Dresselhaus, G. Dresselhaus, P. Avouris, Springer, Heidelberg.
4. Gomer R., Field Emission and Field Ionization, AIP, NY, 1993.
5. Levine J.D., J.Vac. Sci. Technol. B 13, 553-557 (1995).
6. Obraztsov A.N., Zakhidov Al.A., Volkov A.P., Lyashenko D.A., Diamond and Related Mat. 12, 446-449 (2003).
7. Wang Z.L., Gao R.P., De Heer W.A., Poncharal P., Appl. Phys. Lett. 80, 256-258 (2002).
8. Chen Y., Shaw D.T., Guo L., Appl. Phys. Lett. 76, 2469-2470 (2000).
9. Obraztsov A.N., Zakhidov Al.A., Diamond and Related Mat., (2004) in press.
10. Obraztsov A.N., Pavlovsky I.Yu., Volkov A.P. et al., Electronic Properties of Novel Materials – Science and Technology of Molecular Nanostructures, edited by H. Kuzmany, et al., AIP Conference Proceedings 486, American Institute of Physics, New York, 1999, 651-654.
11. Obraztsov A.N., Volkov A.P., Petrushenko Yu.V., Satanovskaya O.P., Tech. Digest of 15th Int. Vacuum Microelectronics Conf. & 48th Int. Field Emission Symp., July 70-11, 2002, Lyon, France, Abs. OB4.07.
12. Silzars A.K., Springer R.W., Fluorescent Lamp, PCT Application PCT/US96/13091, 1995.
13. Bonard J.-M., Stoeckli Th., Noury O., Chatelain A., Appl. Phys. Lett. 78, 2775-2777 (2001).
14. Obraztsov A.N., Cathodoluminescent light source, PCT Application PCT/RU02/00175, 2002.

Light-Driven Molecular Motors

Richard A. van Delden and Ben L. Feringa

*Department of Organic Chemistry, Stratingh Institute, University of Groningen,
Nijenborgh 4, 9747AG Groningen, The Netherlands, feringa@chem.rug.nl.*

Abstract. Molecular motors can be defined as molecules that are able to convert any type of energy input (a fuel) into controlled motion. These systems can be categorized into linear and rotary motors, depending on the motion induced. This brief account will discuss the state of affairs of the research on light-driven rotary molecular motors.

MOLECULAR ROTARY MOTION

Inspired by the unidirectional rotary motion found in F_1-ATPase[1], current pursuits towards nanomachines and synthetic molecular motors focus on systems that allow controlled molecular rotation and translation. Intramolecular rotary motion itself is trivial as, *e.g.* most single bonds in organic molecules freely rotate under ambient conditions. Already in the late sixties, Akkerman *et al.*[2] tried to exert control over the rotation of a carbon-carbon single bond, a concept that was later exploited by Mislow *et al.*[3]. In their molecular gear systems, two intramolecular rotations are coupled due to steric effects. However, no control over the direction is exerted and the motion is merely an oscillation as is the case for most reported examples of forced rotary motion, like *e.g.* in catenane systems[4], double-decker metal complexes[5] and molecular turnstiles[6]. An important design feature of unidirectionally rotating systems, which might find application in nanotechnological machinery, is the presence of an asymmetry in the molecular system. In an approach towards a chemically driven molecular motor, Kelly *et al.* demonstrated a unidirectional 120° rotation for a helical chiral molecule[7]. Recently, Leigh *et al.* developed a catenane system in which, due to four distinct sites on one ring, a unidirectional stepwise rotation (translation along a circular trajectory) around the other ring can be induced[8].

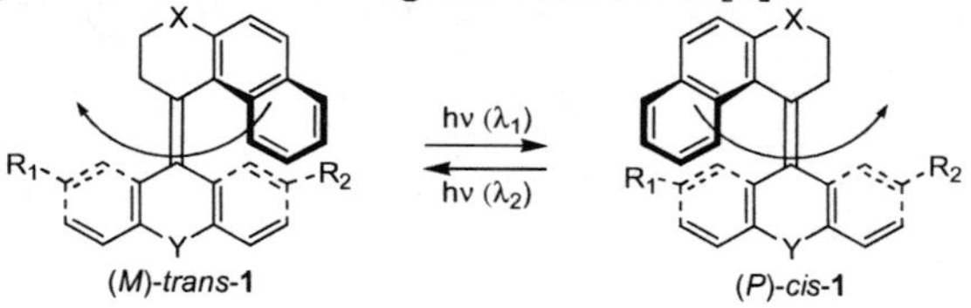

FIGURE 1. Unidirectional rotary motion in a chiroptical molecular switch **1**.

The first systems we designed that employed the controlled motion of a chiral molecule are molecular switches based on overcrowded alkenes[9]. These alkenes for steric reasons adopt a helical structure and the olefinic bond allows photochemical

CP723, *Electronic Properties of Synthetic Nanostructures*, edited by H. Kuzmany et al.
© 2004 American Institute of Physics 0-7354-0204-3/04/$22.00

isomerization. For asymmetrically substituted compounds the state of the system, the (*cis* or *trans*) configuration and the helical (*M* or *P*) chirality, could be controlled by the wavelength of light employed (Figure 1). Selective chiroptical switching at the molecular level was demonstrated for a range of compounds, both in solution and in liquid crystal matrices. Next to the potential exploitation of these systems in optical data storage applications, it is demonstrated with these systems that controlled rotary motion around an olefinic bond, serving as an axis, is possible. Upon excitation, depending on the initial state of the system and the wavelength of the light employed, either a clockwise or a counterclockwise rotation of one half of the molecule relative to the other is induced. The step from partial to full repetitive unidirectional rotation (a prerequisite for the development of molecular motors) was realized by combining two photoisomerization steps with two irreversible thermal helix inversion steps.

LIGHT-DRIVEN UNIDIRECTIONAL ROTATION

First Light-Driven Molecular Motor

Biphenanthrylidene **2** is an overcrowded alkene where the intrinsic helical chirality is combined with two stereocenters of (*R*)-configuration (Figure 2)[10]. The combination of two chiral entities gives rise to four distinct stereoisomers, differing in photochemical properties and thermal stability. In the stable (*cis* or *trans*) configuration, the molecule adopts a (*P*)-helix and the methyl-substituents are allowed to adopt the sterically less hindered axial orientation.

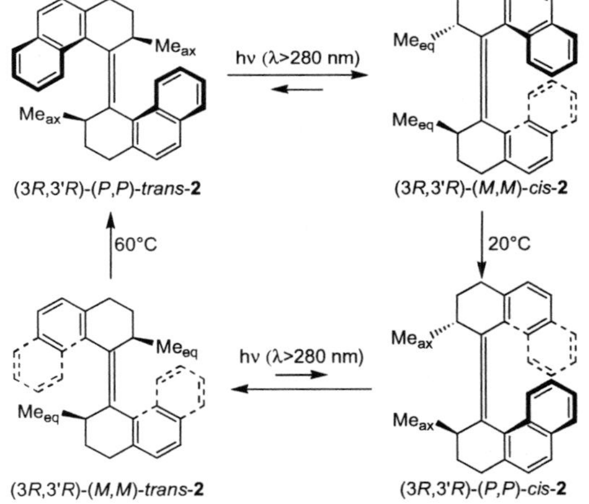

FIGURE 2. The first light-driven molecular motor **2**.

Irradiation of the molecules in their stable form results in an energetically uphill *cis-trans* isomerization, which reverts the helicity of the system and forces the methyl substituents into an unfavorable equatorial orientation. Upon heating, an irreversible energetically downhill helix inversion takes place, allowing the methyl groups to adopt an axial orientation again. Combining the photochemical and thermal steps, by continuous irradiation at elevated temperature, a four step light-driven clockwise unidirectional rotation is induced of one half of the molecule relative to the other around a central axis. Although the irreversible helix inversion steps ensure unidirectional rotation, the photochemical *cis-trans* isomerizations are extremely selective. Already upon irradiation with polychromic light, high selectivities were found for both photoequilibria. Using monochromic light of different wavelengths, it was shown that the selectivity of the second photoequilibrium can fully be controlled and the system can function as a perfect chiroptical molecular switch[11].

A next step toward application of this molecular motor would be to perform actual work, using the unidirectional rotation to drive another process. For this purpose, the molecule was doped into a liquid crystalline matrix[12]. Here, the molecular chirality is amplified to macroscopic chirality that can visually be detected. Doping of an aligned liquid crystal film with pure (*P,P*)-*trans*-**2** resulted in a violet color, which upon irradiation gradually changed to red. This is the first demonstration of (albeit primitive) work performed by a molecular motor, where in this particular case molecular rotation drives a pitch elongation of the liquid crystalline host assembly.

Second-Generation Molecular Motor

The major drawback of the original system is the requirement of elevated temperatures to achieve continuous rotation. In order to tune the properties and lower the barriers for helix inversion (the rate determining step for unidirectional rotation), a second-generation motor was developed. The molecular skeleton combines one *rotor* half of the original motor with a *stator* half of the molecular switches (**3**, Figure 3).

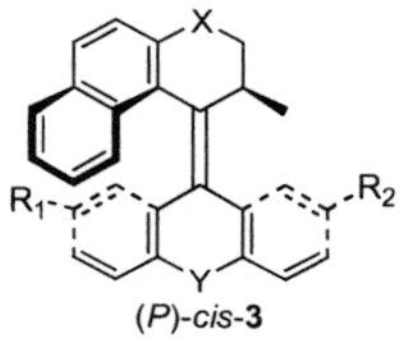

(*P*)-*cis*-**3**

FIGURE 3. The second-generation light-driven molecular motor **3**.

The prototype second-generation motor (X,Y=S; R_1=OMe; R_2=H) was shown to function in essentially the same way as the original system, combining two energetically uphill photoisomerization steps with two thermal helix inversion steps to allow, in a four step process, unidirectional rotation of the rotor part of the molecule with respect to the stator part[13]. The key feature here is that the unidirectional rotation is fully controlled by a single stereogenic center. The design should allow lowering of the energy barrier for the thermal steps by decreasing the steric hindrance for helix inversion. This can be achieved by changing the bridging atoms (X,Y) whereas the spectral properties can be tuned by changing the substitution pattern, mainly in the lower half of the molecule (*e.g.* R_1,R_2). Furthermore, the distinct upper and lower half might allow anchoring of the sterically overcrowded alkene to surfaces, an important step toward future application.

500

It was demonstrated for a large range of molecules that indeed tuning of the thermal barriers and as a consequence the speed of rotation is possible[14]. Whereas the original second-generation motor had a half-life for helix inversion of 215 h at room temperature, for the fastest analogue (X=CH$_2$; Y=O; R$_1$,R$_2$=H) this value is only 40 min. The introduction of donor and acceptor substituents in the lower half (R$_1$=NMe$_2$; R$_2$=NO$_2$) resulted in a bathochromic shift of the absorption bands, allowing unidirectional rotation driven by visible light[15]. These examples demonstrate the versatility of the design and current efforts towards more elaborate systems focus on geared molecular motor systems and mounting these molecular motors onto surfaces.

Exploring the Boundaries

Although the fastest second-generation motor readily functions at room temperature, half-lives are still too long for any practical application, for which Brownian motions have to be overcome. In our pursuit for faster motors we explored the boundaries of our motor design in three ways. First, the methyl substituent adjacent to the central olefinic bond, which causes steric hindrance in the helix inversion steps to some extent, was shifted one position away from the axis (**4**, Figure 4)[16]. In this homologous overcrowded system, the thermal helix inversion steps were proven to be reversible due to a decreased energy difference between stable and unstable isomers. The rotary motion is no longer unidirectional under the conditions employed, although there still is a preferred direction of rotation. Remarkably, the barrier for thermal helix inversion was slightly increased. Rather than speeding up the motor system, the rotary motion was slowed down.

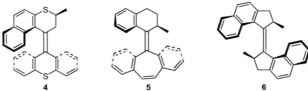

FIGURE 4. Exploring the boundaries of a molecular motor design.

In a second attempt to lower the barriers for thermal helix inversion, the upper half of the molecule was changed from a dihydrophenanthrene to a dihydronaphthalene unit (**5**, Figure 4)[17]. Although this system functions as a unidirectional motor, also here, the barrier for thermal helix inversion was increased with respect to the second-generation counterpart. Both examples show that the unidirectional rotary process relies on a delicate balance of ground state and excited state parameters. In a successful approach toward faster molecular motors, the initial design was reexamined. Although all parts of the molecular system add up to the steric hindrance in the thermal helix inversion steps, an important factor is the conformation of the two rings directly adjacent to the rotation axis. These six-membered rings impose substantial repulsion in the molecule. When the ring size is decreased, as is the case in five-membered rings, this repulsion should be lowered. A pertinent question is whether such a five-membered ring compound (**6**, Figure 4), where the difference between axial and equatorial orientation of the methyl substituents are less pronounced, is still capable of performing a light-driven unidirectional rotation[18]. Analogous to the six-membered ring systems, this molecular motor performs a four-step light-driven unidirectional rotation with the distinct advantage that it occurs readily at room temperature. This last system offers tremendous possibilities for the design of faster systems but the key step to be taken in the research on molecular motors is the incorporation of this molecular component into more complex supramolecular systems and demonstrate the ability of the presented photoactive alkenes to perform useful work as part of nanomachines.

REFERENCES

1. D. S. Goodsell, *Our Molecular Nature: The Body's Motors, Machines and Messages*, Springer, New York, 1996.
2. a) O. S. Akkerman and J. Coops, *Rec. Trav. Chim. Pays-Bas* **86**, 755-761 (1967); b) O. S. Akkerman, *Rec. Trav. Chim. Pays-Bas* **89**, 673-679 (1970).
3. F. Cozzi, A. Gueni, C. A. Johnson, K. Mislow, W. D. Hounshell and J. F. Blount, *J. Am. Chem. Soc.* **103**, 957-958 (1981).
4. a) F. M. Raymo and J.F. Stoddart, "Switchable Catenanes and Molecular Shuttles" in *Molecular Switches*, edited by B. L. Feringa, Wiley-VCH, Weinheim, 2001, pp. 219-248; b) *Molecular Machines and Motors*, edited by J. -P. Sauvage and V. Amendola, Structure and Bonding Vol.99, Springer, Berlin, 2001.
5. a) M. F. Hawthorne, J. I. Zink, J. M. Skleton, M. J. Bayer, C. Liu, E. Livshits, R. Baer and D. Neuhauser, *Science* **303**, 1849-1851 (2004); b) K. Tashiro, K. Konishi and T. Aida, *J. Am. Chem. Soc.* **122**, 7921-7926 (2000).
6 T.C. Bedard and J. S. Moore, *J. Am. Chem. Soc.* **117**, 10662-10671 (1995).
7. T. R. Kelly, H. de Silva and R. A. Silva, *Nature* **401**, 150-152 (1999).
8. D. A. Leigh, J. K. Y. Wong, F. Dehez and F. Zerbetto, *Nature* **424**, 174-179 (2003).
9. a) B. L. Feringa, R. A. van Delden and M. K. J. ter Wiel, "Chiroptical Molecular Switches" in *Molecular Switches*, edited by B. L. Feringa, Wiley-VCH, Weinheim, 2001, pp. 123-163; b) B. L. Feringa, R. A. van Delden, N. Koumura and E. M. Geertsema, *Chem. Rev.* **100**, 1789-1816 (2000).
10. N. Koumura, R. W. J. Zijlstra, R. A. van Delden, N. Harada and B. L. Feringa, *Nature* **401**, 152-155 (1999).
11. R. A. van Delden, M. K. J. ter Wiel and B. L. Feringa, *Chem. Commun.*, 200-201 (2004).
12. R. A. van Delden, N. Koumura, N. Harada and B. L. Feringa, *Proc. Nat. Acad. Sci.* **99**, 4945-4949 (2002).
13. N. Koumura, E. M. Geertsema, A. Meetsma and B. L. Feringa, *J. Am. Chem. Soc.* **122**, 12005-12006 (2000).
14. N. Koumura, E. M. Geertsema, M. B. van Gelder, A. Meetsma and B. L. Feringa, *J. Am. Chem. Soc.* **124**, 5037-5051 (2002).
15. R. A. van Delden, N. Koumura, A. M. Schoevaars, A. Meetsma and B. L. Feringa, *Org. Biomol. Chem.* **1**, 33-35 (2003).
16. R. A. van Delden, M. K. J. ter Wiel, H. de Jong, A. Meetsma and B. L. Feringa, *Org. Biol. Chem.* **2**, 1531-1541 (2004).
17. E. M. Geertsema, N. Koumura, M. K. J. ter Wiel, A. Meetsma and B. L. Feringa, *Chem. Commun.*, 2962-2963 (2002).
18. M. K. J. ter Wiel, R. A. van Delden, A. Meetsma and B. L. Feringa *J. Am. Chem. Soc.* **125**, 15076-15086 (2003).

Progress Towards a Rotary Molecular Motor

Gwénaël Rapenne*, Alexandre Carella*, Romuald Poteau[†], Joël Jaud*, and Jean-Pierre Launay*

*NanoSciences Group, CEMES-CNRS, 29 rue Jeanne Marvig, BP 94347,
F-31055 Toulouse Cedex 4, France
† Laboratoire de Physique Quantique, IRSAMC, Université Paul Sabatier, 118 route de Narbonne,
F-31062 Toulouse Cedex 4, France

Abstract. We present here our strategy to build a molecular motor following the bottom-up approach. This motor is designed to operate as a single molecule which is supposed to convert an electric current into a directionally-controlled rotary motion.

Introduction

Artificial molecular motors have recently emerged as a new field of chemistry related to a newly explored dimension of molecular sciences : controlled movement at the molecular scale.[1] Since multistep chemical synthesis allows chemists to prepare tailor-made molecules with predetermined shape and programmed movement, the bottom-up approach will permit technology to reach the ultimate miniaturization of electronic and mechanic devices.

A motor is a machine which consumes energy to create work continuously via a unidirectional and controlled movement. The inspiration of such motors comes both from Nature with the fascinating machinery of ATP synthase,[2] and in the macroscopic world where rotary motors are very common. Some examples of molecular rotary motors have been described by a few groups[3] and a rotor[4] has been proposed, but the molecule needs its neighbours in a supramolecular bearing to act as a rotor. The major difficulties concern the control of the directionality of the rotation, the continuous response to the stimulus and the manipulation of a single molecule. In our strategy, the source of energy will be electric.

Our Original Concept

Our target molecule symbolized in Figure 1 comprises a stator, *i.e.* one part fixed between two electrodes 3-4 nm apart (*e.g.* nanojunction), and on this stator is connected a rotor which should transform a current of electrons into a unidirectional rotation motion. The rotor is a rigid aromatic platform constructed around a cyclopentadienyl ligand with five linear and rigid arms, each terminated by an electroactive group. As electroactive group, ferrocene has been selected because it exhibits reversible oxidation in various solvents.[5] The stator is a hydrotris(indazolyl) borate ligand of the family of scorpionates developed by Trofimenko[6] with a piano

CP723, *Electronic Properties of Synthetic Nanostructures*, edited by H. Kuzmany et al.
© 2004 American Institute of Physics 0-7354-0204-3/04/$22.00

stool shape. The joint between the rotor and the stator is a ruthenium(II) ion to obtain a stable molecule bearing zero net charge, both criteria being essential for surface deposition and hence for building single molecule nanomotors. The upper part should be free to turn whilst the basis should stay still, anchored on the surface between the two electrodes of the addressing system.

The concept is shown in Figure 1. The electroactive group (EG) closest to the anode would be oxidized (oxidized form EG⁺) and pushed back by electrostatic repulsion like it has been shown for a [60]-fullerene between two electrodes.[7] This motion corresponds to a fifth of a turn. As a result, the oxidized electroactive group would approach the cathode and subsequently be reduced. At the same time, a second electroactive group would come close to the anode and a second cycle would occur. A complete 360° turn would be achieved after five cycles, corresponding to the transport of five electrons from the cathode to the anode. This would correspond to the conversion of an electron flow into a movement of rotation, *i.e.* a redox-triggered molecular rotary motor. In order for the rotation to be directional, the molecule should be placed in a dissymmetrical environment. This could be achieved either by its disposition in the nanojunction, or by a secondary electric field perpendicular to the nanojunction.

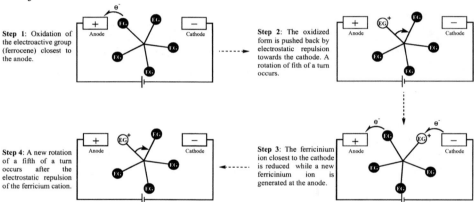

FIGURE 1. Concept of a single molecular motor placed between two electrodes of a nanojunction. The molecule has a C5 symmetry with five arms ended with an electroactive group. The oxidized form is in white, the reduced one in black.

Synthesis of a Molecular Motor

The synthesis is presented in Figure 2. Our strategy starts with the bromination of 1,2,3,4,5-pentaphenylcyclopentadiene (**1**) followed by reaction with $Ru_3(CO)_{12}$. Coordination of the hydrotris(indazolyl)borate stator yielded the key intermediate of our strategy, the piano stool complex **4** described as an organometallic molecular turnstile.[8] The presence of the five aryl bromide groups allows the simultaneous connection of the five electroactive groups to the rotor, leading to our designed molecular motor after a quintuple coupling reaction of ethynylferrocene provided by the powerful chemistry of the Pd-catalyzed carbon-carbon bond formation.[9]

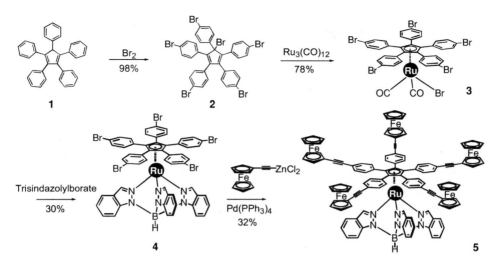

FIGURE 2. Synthetic pathway leading to a molecular motor in 5 steps starting from commercially available compounds. The rotor (upper ligand) is connected to the stator (lower ligand) by the means of a ruthenium(II) center (in black).

Study of the Rotation in Solution

The rotation has been studied both experimentally and theoretically on the intermediate compound **4**. In this complex, an additional steric interference exists with respect to the case of [TIB Ru Cp].[10] This is due to the interaction between the bromophenyl units and the legs, so that one could anticipate a slow or restricted rotation. But experimentally, we observe a rapid rotation on the NMR timescale, even at $-90°C$, meaning that the energy barrier for rotation is certainly very small.

The symmetry of the two parts of the molecule suggests that the potential energy curve should have a high periodicity. If we start from a conformation where a « paddle » sits just above a « leg », it is clear that a rotation by 24° (1/15th of a turn) makes another coincidence happen (Figure 3). Hence we predict that the potential energy curve should present closely spaced extrema. To systematize the analysis, we define « eclipsed » and « staggered » conformations from the value of the smallest dihedral angle between a paddle and a leg, i.e. 0° and 12° respectively (Figure 3). Simple steric considerations suggest that the staggered conformation should be an energy minimum, while the eclipsed one should be a transition state.

The X-ray crystal structure shows however a geometry close to the « staggered » one (see Figure 4, c). But this is not reproduced by calculations, using DFT with the B3PW91 functional. At this level, it is found that, surprisingly, the most stable conformation is the « eclipsed » one (Fig 4, a). The paddle which sits exactly above a leg is tilted with respect to the Cp ring plane by 31° only, i.e. much less than the others (53 to 85°).

FIGURE 3. Sketch of the lower part, with three « legs » (dotted), and the upper part, with five « paddles » (plain). Definition of the « eclipsed » (left), and « staggered » (right) conformations.

A transition state, corresponding to an intermediate during the rotation process, has been found by DFT at 4.5 kcal.mol^{-1} above the minimum. It resembles the « staggered » geometry (Figure 4, b), but one paddle (*) starts to interact with a leg, prefigurating the nearby minimum geometry. The differences between the X-Ray structure and the computed minimum and transition state structures are subtle and also involve differences in orientation of the phenyl rings with respect to the Cp ring.

Closer inspection of the structure strongly suggests that the rotation must involve some « gearing effect », i.e. a correlated motion of the upper part with respect to the lower one (Figure 4, motion (1)), **and** of a « paddle » around the Cp-phenyl single bond (Figure 4, motion (2)). This is reminiscent of the « Fosbury flop ».[11]

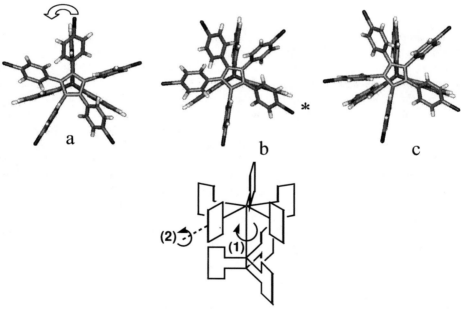

FIGURE 4. Upper part : different conformations, a : minimum found by DFT ; b : transition state by DFT ; c : X-Ray crystal structure. Lower part : the gearing effect (Fosbury flop).

Finally, the comparison of the X-ray crystal structures of [TIB Ru Cp],[10] and **4** reveals a small increase in bond lengths in the latter : +0.02 Å for the Ru-C bonds, and +0.02 Å for the Ru-N bonds. This trend is reproduced by DFT calculations on both systems (+0.05 Å and +0.01 Å for the Ru-C and Ru-N bonds respectively). These increases are due to the steric crowding in **4**, but the magnitude of the effect is small. Thus we can conclude that the upper part remains strongly linked to the lower part during the rotation process, while being almost free to rotate.

In conclusion, a precursor of a molecular motor has been prepared. Functionalization of the hydrotris(indazolyl)borate ligand (stator) is now in progress to ensure its deposition on insulating surfaces. Observation of the conversion of an electric current into a controlled rotary movement will then be attempted, using suitable methods such as analysis of the time dependence of the current and scanning probe microscopes. Work is currently underway to achieve this goal.

Acknowledgments

We thank the CNRS, the European Union and the University of Toulouse for financial support and the French Ministry of National Education for a fellowship to AC. One of us (RP) thanks the Centre Informatique National de l'Enseignement Supérieur (CINES) for allocation of computational resources.

References

1. Special Issue on Molecular Machines, *Acc. Chem. Res.* **34**, 409-522 (2001), and references therein.
2. Walker, J. F. *Angew. Chem. Int. Ed.* **37**, 2308-2319 (1998).
3. (a) Kelly, T. R., De Silva, H., Silva, R. A. *Nature* **401**, 150-152 (1999). (b) Kelly, T. R., Silva, R. A., De Silva, H., Jasmin, S., Zhao, Y. *J. Am. Chem. Soc.* **122**, 6935-6949 (2000). (c) Koumura, N., Zijlstra, R. W. J., van Delden, R. A., Harada, N., Feringa, B. L. *Nature* **401**, 152-155 (1999). (d) ter Wiel, M. K. J., van Delden, R. A., Meetsma, A., Feringa, B. L., *J. Am. Chem. Soc.* **125**, 15076-15086 (2003). (e) Tashiro, K., Konishi, K., Aida, T. *J. Am. Chem. Soc.* **122**, 7921-7926 (2000). (f) Ikeda, M., Takeuchi, M., Shinkai, S., Tani, F., Naruta, Y., Sakamoto, S., Yamaguchi, K. *Chem. Eur. J.* **8**, 5541-5550 (2002). (g) Leigh, D. A., Wong, J. K. Y., Dehez, F., Zerbetto, F. *Nature* **424**, 174-179 (2003). (h) Horinek, D., Michl, J. *J. Am. Chem. Soc.* **125**, 11900-11910 (2003). (i) Jian, H., Tour, J. M. *J. Org. Chem.* **68**, 5091-5103 (2003).
4. Gimzewski, J. K., Joachim, C., Schlitter, R., Langlais, V., Tang, H. *Science* **281**, 531-533 (1998).
5. (a) Connelly, N. G., Geiger, W. E., *Chem. Rev.* **96**, 877-910 (1996). (b) Astruc, D. *Acc. Chem. Res.* **33**, 287-298 (2000).
6. (a) Trofimenko, S., *Scorpionates, The Coordination Chemistry of Polypyrazolylborate Ligands*, London, Imperial College Press, 1999. (b) Rheingold, A. L., Haggerty, B. S., Yap, G. P. A., Trofimenko, S. *Inorg. Chem.* **36**, 5097-5103 (1997).
7. Park, H., Park, J., Lim, A. K. L., Anderson, E. H., Alivisatos, A. P., McEuen, P. L. *Nature* **407**, 57-60 (2000).
8. Carella, A., Jaud, J., Rapenne, G., Launay, J.-P. *Chem. Commun.* 2434-2435 (2003).
9. Carella, A., Rapenne, G., Launay, J.-P. submitted for publication.
10. [TIB Ru Cp] : Carella, A., Jaud, J., Rapenne, G., Launay, J.-P, unpublished results.
11. From the name of Dick Fosbury who revolutionized high jump in 1968, by introducing the technique of leaping backward over the bar, and rolling around it.

Catalytic CVD of SWCNTs at Low Temperatures and SWCNT Devices

Robert Seidel*[+], Maik Liebau*, Eugen Unger*, Andrew P. Graham*,
Georg S. Duesberg*, Franz Kreupl*, Wolfgang Hoenlein*,
and Wolfgang Pompe[+]

* Infineon Technologies AG, Corporate Research, 81730 Munich, Germany
[+] Institut für Werkstoffwissenschaft, Technische Universität Dresden, 01062 Dresden, Germany

Abstract. New results on the planar growth of single-walled carbon nanotubes (SWCNTs) by catalytic chemical vapor deposition (CVD) at low temperatures will be reported. Optimizing catalyst, catalyst support, and growth parameters yields SWCNTs at temperatures as low as 600 °C. Growth at such low temperatures largely affects the diameter distribution since coalescence of the catalyst is suppressed. A phenomenological growth model will be suggested for CVD growth at low temperatures. The model takes into account surface diffusion and is an alternative to the bulk diffusion based vapor-liquid-solid (VLS) model. Furthermore, carbon nanotubes field effect transistors based on substrate grown SWCNTs will be presented. In these devices good contact resistances could be achieved by electroless metal deposition or metal evaporation of the contacts.

INTRODUCTION

The integration of SWCNTs into future electronic devices requires the controlled placement of a large number of SWCNTs on substrates. Catalytic CVD of SWCNTs directly on a substrate [1] might be one promising alternative to solution-based deposition methods [2]. To achieve a high degree of compatibility with standard semiconductor processes it is important that the growth temperature of SWCNTs is low while maintaining a high yield and good quality. Further the CVD growth mechanisms have to be well understood from an experimental point of view. This may lead to a better controlled and more selective growth of SWCNTs. In this paper, results on the CVD synthesis of SWCNTs at 600 °C will be presented together with a simple phenomenological growth model. Further, a high-current SWCNT field effect transistor (SWCNT-FET) based on a randomly grown SWCNT network and SWCNT-FETs with electroless deposited Pd contacts will be discussed.

RESULTS AND DISCUSSION

Systematic experiments and variations of the catalyst were performed in order to reduce the growth temperatures. Figure 1 shows that the growth of SWCNT is possible at temperatures around 600 °C [3, 4] if the catalyst particles have the right

CP723, *Electronic Properties of Synthetic Nanostructures*, edited by H. Kuzmany et al.
© 2004 American Institute of Physics 0-7354-0204-3/04/$22.00

size and the carbon supply is sufficient. The size of the catalyst particle is primarily a result of the coalescence, which depends on the thickness of the catalyst layer, the interactions with the catalyst support (e.g. wetting, surface roughness, and surface diffusion coefficients), the heat-up conditions, and the pretreatment time. The SWCNTs in Figures 1a) and 1d) appear as thick bundles due to their dense growth.

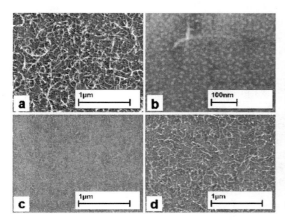

FIGURE 1. SEM images showing the influence of the duration of the hydrogen pretreatment on the SWCNT growth at 600 °C (Catalyst: 0.2 nm Ni on either plain SiO_2 (1a) and 1b)), or on SiO_2 with an additional layer of 3 nm Al (1c) and 1d))). Samples in 1a) and 1c) were pretreated in H_2 for 10 minutes but samples in 1b) and 1d) each for 60 minutes (growth time: 10 minutes, CH_4 pressure: 375 Torr).

Based on experimental observation it is suggested that the CVD growth of SWCNTs at low temperatures is essentially the result of surface diffusion along the catalyst support or along the walls of the growing SWCNTs [3]. The vapor-liquid-solid model might only hold for the initial steps of the growth if the formation of a carbon cap around a supersaturated catalyst particle is necessary to initiate growth. It appears unlikely, however, that the catalyst particles have enough free surface area during growth to catalytically decompose the carbon source and dissolve the carbon before precipitation since they are covered on one side by the growing SWCNT and on the other side by the substrate. The surface diffusion based model is supported by considering the four different growth situations illustrated in Figure 2.

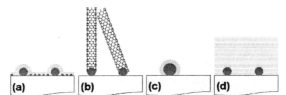

FIGURE 2. Possible scenarios during the catalytic CVD of SWCNTs: (a) surface poisoned by catalyst atoms (interaction between catalyst and support too strong, temperature too low, H_2 pretreatment time too short); (b) good SWCNT growth – catalyst particles have the right size and carbon supply is optimal; (c) catalyst particles too large (interaction between catalyst and support too weak, temperature too high, H_2 pretreatment time too long); (d) amorphous carbon or graphite layer forms at the high temperature or high pressure limit of the process.

These situations have all been experimentally observed. In particular the deposition of thick carbon layers at higher temperatures and/or high partial pressures of the carbon feedstock (Figure 2d)) shows that the pyrolytic decomposition of the carbon feedstock occurs on certain surfaces (SiO$_2$, Al$_2$O$_3$) at temperatures well below those at which the actual autopyrolysis happens. This indicates that at slightly lower temperatures and pressures a high concentration of adsorbed carbon species on the catalyst support can exist and can deliver carbon to the growing SWCNTs.

Based on the optimization of the growth at low temperatures a high-current SWCNT-FET was developed that allows the switching of currents as high as 2 mA at $V_{ds} \sim 1V$. With such devices it could be demonstrated, for the first time, that carbon nanotubes based FETs can switch macroscopic devices such as LEDs and electromotors [5]. The high-current SWCNT-FETs were obtained after growing a random network of SWCNTs with a rather low density on Si with a 50 nm Al$_2$O$_3$ layer. Source and drain contacts were defined by e-beam lithography and subsequent e-beam evaporation (gate length: 200 µm, gate width: 90 – 300 nm). The Si substrate was used as a backgate. The on/off ratio of those FETs was increased to about 3 orders of magnitude by applying bias pulses together with a positive gate voltage (+ 20 V). The gate voltage was applied in order to switch the semiconducting SWCNTs off and allow the preferential electric breakdown of the metallic SWCNTs [6].

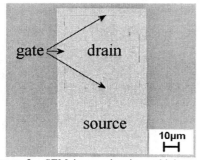 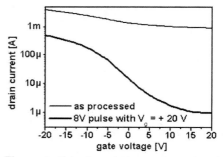

Figure 3. SEM image showing a high-current SWCNT FET. (< 0.2 nm Ni catalyst on Si with 50 nm Al$_2$O$_3$, growth @ 650°C with CH$_4$, gate length ~ 240 nm, gate width 200 µm).

Figure 4. Gate dependent measurements using the Si substrate as backgate. The thin curve shows initial behavior, the thick curve the improved performance after application of an 8 V bias pulse with a simultaneous gate voltage of + 20V.

Finally, electroless Pd deposition was developed in order to improve the contacts of SWCNTs grown between Ta/Co electrodes. This process is similar to the electroless Ni deposition presented previously [1, 7]. The advantage of electroless deposition is that it is a self-aligned process that allows good contact formation just by dipping the sample into a solution followed by a short annealing process. For electroless Pd deposition a tetraamminepalladium (II) chloride solution with a reducing agent was used. The annealing was performed at 400 °C in N$_2$. After electroless Pd deposition the semiconducting SWCNTs showed on-resistances from 200 kΩ to about 1 MΩ. The resistances of metallic SWCNTs ranged between 50 kΩ and 100 kΩ. The resistances are, therefore, similar to those reported after electroless Ni deposition [1] and to those obtained by e-beam evaporation of Pd onto SWCNTs with small diameters [4]. The main advantage of the Pd contacts is the higher corrosion resistance compared to electroless Ni contacts, which show degradation after several days in air.

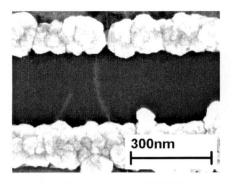

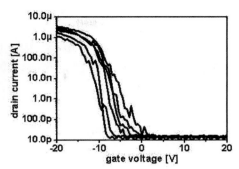

Figure 5. SEM images of a SWCNT contacted by electroless Pd deposition.

Figure 6. Gate dependent measurements of several SWCNTs contacted by electroless Pd deposition and using the Si substrate as a backgate (on-resistances between 200 kΩ and 1 MΩ).

CONCLUSION

The results presented here on the growth of SWCNTs at low temperatures offer a method that allows the integration of SWCNTs at temperatures compatible with microelectronic manufacturing. The high-current SWCNT-FET illustrates that carbon nanotubes might well be exploited for high power devices. Furthermore, electroless Pd deposition has been demonstrated as a simple and self-aligned process to achieve good contacts to SWCNTs grown or deposited between metal electrodes.

ACKNOWLEDGMENTS

This work has been supported by the German Ministry of Science and Technology (BMBF) under Contract No. 13N8402.

REFERENCES

1. R. Seidel, M. Liebau, G. S. Duesberg, F. Kreupl, E. Unger, A. P. Graham, W. Hoenlein, and W. Pompe, *Nano Lett.* **3**, 965-968 (2003).
2. R. Krupke, F. Hennrich, H. B. Weber, M. M. Kappes, and H. v. Lohneysen, *Nano Lett.* **3**, 1019-1023 (2003).
3. R. Seidel, G. S. Duesberg, E. Unger, A. P. Graham, M. Liebau, and J. Kreupl, *J. Phys. Chem. B* **108**, 1888-1893 (2004).
4. Y. Li, D. Mann, M. Rolandi, W. Kim, A. Ural, S. Hung, A. Javey, J. Cao, D. Wang, E. Yenilmez, Q. Wang, J. F. Gibbons, Y. Nishi, and H. Dai, *Nano Lett.* **4**, 317-321 (2004).
5. R. Seidel, A. P. Graham, E. Unger, G. S. Duesberg, M. Liebau, W. Steinhoegl, F. Kreupl, W. Hoenlein, and W. Pompe, *Nano Lett.* in press.
6. P. G. Collins, M. S. Arnold, and Ph. Avouris, *Science* **292**, 706-709 (2002).
7. M. Liebau, E. Unger, G. S. Duesberg, A. P. Graham, R. Seidel, F. Kreupl, and W. Hoenlein, *Appl. Phys. A.* **77**, 731-734 (2003).

Engineering Nanomotor Components from Multi-Walled Carbon Nanotubes via Reactive Ion Etching

T. D. Yuzvinsky, A. M. Fennimore and A. Zettl

Physics Department, University of California, Berkeley, CA 94720 USA
Materials Sciences Division, Lawrence Berkeley National Laboratory, Berkeley CA 94720 USA

Abstract. It has been shown that a multi-walled carbon nanotube (MWCNT) can be used as the rotation enabling element for nanoelectromechanical systems. Modification of the MWCNT to create a rotational bearing in previous devices has concentrated on smaller diameter tubes and mechanical methods to create the bearing. We here investigate reactive ion etching of a MWCNT as a means to engineer nanomotor components, including the rotational bearing.

INTRODUCTION

We recently reported on the fabrication of a rotational actuator mounted on a multiwall carbon nanotube (MWCNT) bearing [1]. In our attempts to better understand the nature of this bearing, we have tried several approaches (see, for example, the report on electrically driven vaporization, also in these proceedings). Here we present one approach, the use of reactive ion etching to partially remove the outer walls of a MWCNT.

A model of our previously produced devices is shown in Figure 1. A MWCNT is suspended between two anchor electrodes. A rotor plate is attached to the middle of the MWCNT, and voltages applied to three stators (two side electrodes and the conductive back gate) are used to control the position and velocity of the rotor. The chief obstacle in the fabrication of these devices is the modification of the MWCNT to create the bearing and axle assembly on which the rotor plate rotates. In our original report, we used very large electrostatic fields to torque the rotor and twist the outer walls of the MWCNT beyond the elastic limit, ultimately breaking those walls and freeing the rotor to rotate.

Although torsional stress did prove to be effective in freeing the nanotube for rotational movement, it also presented several limitations. We found it to be effective only on thinner MWCNTs, of diameter less than approximately 20 nm. For thicker tubes, the electrostatic fields required to rotate the device exceeded the breakdown strength of the surface dielectric. Furthermore, the electrostatic fields required to free each of the thinner MWCNTs varied, depending on tube's diameter and the rotor's exact dimensions, necessitating that each device be freed individually. Finally, the

CP723, *Electronic Properties of Synthetic Nanostructures*, edited by H. Kuzmany et al.
© 2004 American Institute of Physics 0-7354-0204-3/04/$22.00

FIGURE 1. A model of the rotational actuator. A rotor paddle (R) is attached to a suspended MWCNT. Two anchors (A1, A2) provide mechanical and electrical contact. Two stator electrodes (S1, S2) and the conductive back gate (S3) provide control over the position and velocity of the rotor.

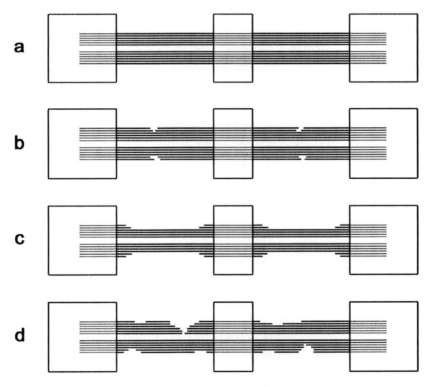

FIGURE 2. A model of bearing creation by various methods: **(a)** An actuator in its pristine state, with all walls intact. Anchors hold the MWCNT on each end, with a rotor paddle suspended in the middle. **(b)** The presumed result of torsional shearing. Though free to rotate, the rotor cannot slide along the nanotube. **(c)** An alternate geometry, in which the rotor can both spin and shuttle along the inner core. **(d)** The apparent outcome of RIE, with localized etch pits causing extensive damage to the inner core before rotational freedom can be achieved.

presumed mechanism of rotational liberation does not allow us to fully investigate the rotational bearing nature of the MWCNT. Torsional stress should damage the outer shell of the MWCNT locally, at one point on either side of the rotor, without affecting the rest of the shell: as the torsional stress is increased, and the shell begins to weaken at some point along its length, it will twist more at that point, and further damage will occur there. Therefore the position of the rotor along the axis of the nanotube is fixed, since the remaining sections of outer shell prevent it from sliding (see Figure 2b). This, in turn, prevents us from testing the full bearing nature of the devices, which we could do by shuttling the rotor along the MWCNT, testing both its linear and rotational freedom. This would also allow us to investigate chiral mismatch and the possibility of the axle bearing assembly undergoing a screw-like effect as it is translated rotationally [2,3].

In hopes of finding a method more suitable for use with all diameters of tubes, and that would at the same time allow the rotor axial freedom, we have investigated the effects of reactive ion etching (RIE).

METHODS

Reactive Ion Etching (RIE) is a technique that is commonly used in semiconductor fabrication to selectively remove material in a controlled and uniform manner. This motivated us to try to reactively etch a MWCNT, hoping to controllably remove the entire outer wall irrespective of its diameter. Given the common use of oxidation to purify raw MWCNT soot we chose to use O_2 as our etching gas. Samples containing suspended MWCNT devices (with rotors and stators already in place) with varying nanotube diameters were placed in a Plasma-Therm PK-12 Parallel Plate Plasma Etcher. The samples were etched in O_2 plasmas at a pressure of 100 mTorr and a power of 15 W for up to 40 seconds.

After etching, the devices were placed in a scanning electron microscope (SEM) for *in situ* analysis of their torsional behavior. We found that devices subjected to identical plasma treatment had their torsional strength weakened to varying degrees. Some devices even appeared to be free, requiring only low applied voltages for large angular displacements, and several times the rotors were already in the vertical position (rotated 90°) when first loaded into the SEM. Meanwhile, other devices, while weakened, still had appreciable torsional strength, and could only be slightly deflected with large applied voltages. We also found several devices that appeared to be stuck to the underlying silicon oxide surface. The MWCNTs were slightly extended, to the point where the rotor could not rotate without touching the substrate. We have seen both from this and from other experiments that once a rotor has touched the surface it is very hard to overcome the van der Waals attraction and pull it out of contact.

Unfortunately, the devices that appeared to be free were prone to failure, unlike the torsionally sheared devices reported earlier. After several rotations the MWCNT would snap somewhere along its length (see Figure 3). Previous work on O_2 etching of graphite surfaces provides insight as to the mechanism of these failures [4]. It was found that instead of etching the surface uniformly, layer by layer, deep etch pits develop at defect locations (see Figure 2d). The presence of etch pits would explain the sudden failure of the MWCNTs. This also provides an explanation for why the effect of the same plasma dose could vary greatly from tube to tube.

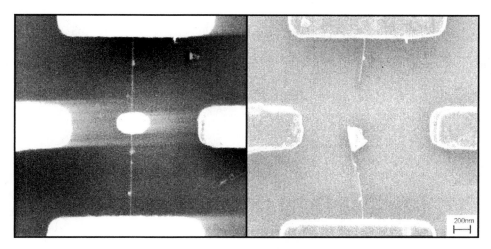

FIGURE 3. SEM micrographs of a MWCNT actuator. The image on the left shows it in its pristine state. The device was then subjected to RIE, after which the torsional spring constant was greatly reduced. Shortly after testing began, however, the device failed, shown in the image at right. The scale bar is 200 nm.

While RIE with O_2 plasma was successful in rotationally freeing the tubes, the lack of uniformity in rotational freedom from tube to tube, their abrupt snapping failure, and the common problem of adhesion to the surface demonstrate that more work on refining this technique is in order.

ACKNOWLEDGMENTS

This research was supported by the Office of Science of the U.S. Department of Energy. We thank Noah Bodzin for assistance with graphics.

REFERENCES

1. Fennimore, A.M. et al, Nature 424, 408-410 (2003)
2. Saito, R. et al, Chem. Phys. Lett. 348, 187-193 (2001)
3. Lozovik, Y.E., Mingoin, A. V., Popov, A.M., Phys. Lett. A 313, 112-121 (2003)
4. Paredes, J. I., Martinez-Alonso, A., Tascon, J. M. D. Carbon 38, 1183-1197 (2000)

Contact Resistance between Individual Single Walled Carbon Nanotubes and Metal Electrodes

Yunsung Woo[1], Maik Liebau[2], Georg S. Duesberg[2], and Siegmar Roth[1]

[1]*Max-Planck Institute for Solid State Research, Heisenbergstra. 1, 70569 Stuttgart, Germany*
[2]*Infineon Technologies AG, Corporate Research, 81730 Munich, Germany*

Abstract We report that ohmic contact between individual single walled carbon nanotubes (SWNTs) and palladium (Pd) was formed by using the electric-current-induced joule heating in Pd electrode. The SWNTs are deposited onto Pd electrode patterned on SiO$_2$/Si substrate, through which the electrical currents flow for microseconds. As a result, transport measurement exhibited a decrease in contact resistance between individual SWNT and Pd electrode from 3.3 MΩ to 176 kΩ at room temperature after joule heating induced by the electrical current. Atomic force microscopy also showed that a surface of Pd metal is smoothened in roughness after electrical current flowing, indicating metallic melting by joule heat. Therefore, we suggest the way to achieve a low contact resistance between SWNT and metal in field effect transistor at room temperature, instead of thermal annealing method at high temperature.

INRTODUCTION

Carbon nanotubes are attractive for the application in nanoscale electronic devices owing to their one-dimensional molecular structure with metallic or semiconducting properties. In particular, field effect transistors (FET) have been involved promisingly using a single walled carbon nanotube (SWNT) as a channel of electron transport [1]. It has been reported, however, that there is a substantial Schottky barrier at the contact between SWNT (channel) and metal electrode (source/drain), which could determine the transistor behavior dominantly [2]. The contact between SWNT and metal electrode, therefore, becomes a critical issue to explain the characteristics of carbon nanotube field effect transistor (CNTFET). In addition, much effort has been done to improve the contact resistance by thermal annealing [3], metal deposition on nanotubes [4], and soldering by electron beam [5].

The drawback of thermal annealing is that heat treatment will destroy delicate semiconductor circuits in the substrate. Therefore we have developed a method of locally confined thermal annealing where only the close vicinity of the nanotube-to-surface contact is heated. This local heating is achieved by passing short current pulse through the metal leads.

Individual SWNTs were deposited onto the Pd metal electrodes patterned on a SiO$_2$/Si substrate. Electrical current pulses were applied through the Pd electrode strips to generate Joule heating. Then, current-voltage characteristics of individual SWNT were measured and compared to the characteristics of untreated contacts.

CP723, *Electronic Properties of Synthetic Nanostructures*, edited by H. Kuzmany et al.

EXPERIMENTAL

The SWNTs used in this experiment are synthesized by HiPco at CNI. These SWNTs are suspended in a 1% aqueous solution of sodium dodecyl sulfate and dispersed onto the Pd electrodes patterned on highly doped silicon substrate with a 200 nm thermally grown silicon oxide layer. The Pd electrode was deposited with a width of 1.1 μm, a length of 3.4 μm, and a thickness of 200 nm by electron beam lithography and thermal evaporation. Then, microsecond voltage pulses were applied to the Pd electrodes, and Joule heating is produced by the electrical current, as described in Fig. 1. The distance between the two Pd electrodes was approximately 1 μm.

The transport measurements were carried out on the individual SWNT deposited onto Pd electrodes, and compared before and after pulse annealing. Atomic force microscopy (AFM) was also used to see the effect of pulse annealing on the surface morphology of Pd electrodes.

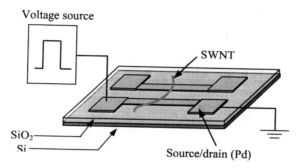

FIGURE 1. Schematic apparatus of the electric-current-induced joule heating method. The pulsed voltages of 6-7 V were applied for microseconds to Pd electrode with a length of 3.4 μm, a width of 1.1 μm, and a thickness of 200 nm.

RESULTS AND DISCUSSION

Figure 2 shows the variations of the current-voltage characteristics of SWNTs deposited onto Pd electrode with an increase of the pulsed voltage applied to the Pd electrode. Here, three SWNTs are observed between the two Pd electrodes. Initially the resistance of this SWNTs device is measured to be about 1.20 MΩ. As the pulses of 6.5 and 6.75 V are applied to the Pd electrode for 3 μs, the resistance decreases dramatically to 588.6 and 92.1 kΩ, respectively. However, pulses of 6 V do not change the resistance of this device.

The field effect transistor (FET) behavior of individual SWNT was also studied before and after pulse annealing, as shown in Fig. 3(a). Figure 4 shows an AFM image of the nanotube investigated in Fig. 3. The tube is 1.1 μm long and has a diameter of 0.80 nm. The resistance measured from current-voltage characteristics decreases from 3.35 MΩ initially to 176.50 kΩ after applying pulses of 7 V for 2 μs. As seen in Fig. 3(a), at

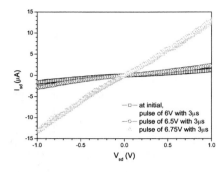

FIGURE 2. I –V$_{sd}$ characteristics obtained from three SWNTs between two Pd electrodes with the application of the pulsed voltage to the Pd electrode. The pulse of 6 V with 3 µs does not change the I-V characteristics, but the source/drain current increases dramatically after applying the pulse of 6.75 V with 3 µs.

first, the I-V curves show a non-linear behavior at high bias revealing a high Schottky barrier at the contact. However, with an increase of pulse voltage and duration, the I-V curves become linear, and furthermore show a saturation behavior at high source/drain bias. This indicates that the Schottky barrier between SWNT and Pd electrode is suppressed by applying pulse to the Pd electrode, and finally the ohmic contact between SWNT and Pd electrode is formed. Figure 3(b) shows the conductance dependence of an individual SWNT on the gate voltage at different temperatures after applying the pulses of 7V for 2µs. Interestingly, the SWNT exhibits ambipolar behavior, acting as both p- and n-channel in gate dependence. From this, we conclude that the absorbed oxygen on SWNT, which is known to cause p-type behavior in carbon nanotube FET (CNT-FET), is removed by the pulse annealing.

The surface morphology of Pd electrode before and after pulse annealing was

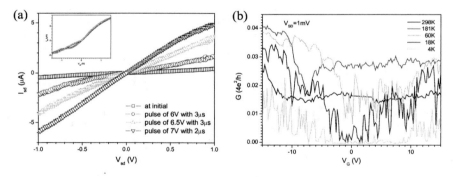

FIGURE 3. (a) I-V$_{sd}$ characteristics measured after applying the pulsed voltage on the Pd electrode with various volts and duration times. Inset shows the saturation behavior in I-V$_{sd}$ curve at high bias after applying a pulse of 7 V with 2 µs. (b) I-V$_g$ curves at different temperature measured at 1 mV of source/drain bias after applying a pulse of 7 V with 2 µs.

FIGURE 4. AFM images of individual SWNT contacted between the two Pd electrodes.

examined using AFM, and the results are summarized in Table 1. The mean value of surface roughness decreases from 0.461 to 0.359 nm as the pulses of 7 V for 2 μs were applied to the Pd electrode. This indicates that the electrical current induced by the pulses melts the Pd electrode temporarily. From this, it is seen that good wetting of Pd to SWNT wall can be formed.

SUMMARY

We have presented that the formation of ohmic contact between individual SWNT and Pd electrode was achieved by applying the pulsed voltage of several volts for microseconds in the Pd electrode. AFM observations imply that the Pd electrode melted temporarily by the electric-current-induced joule heating, resulting in a good contact between SWNT and Pd. This method has an advantage of not destroying other structures patterned on Si substrate because only the metal electrode is heated due to its resistance.

REFERENCES

1. Sander J. Tans, Alwin R. M. Verschueren, and Cees Dekker, Nature 393, 49 (1998)
2. S. Heinze, J. Tersoff, R. Martel, V. Derycke, J. Appenzeller, and Ph. Avouris, Phys. Rev. Lett. 89, 106801 (2002)
3. J. O. Lee, C. Park, Ju-Jin Kim, Jinhee Kim, Jong Wan Park, and Kyoung-Hwa Yoo, J. Phys. D, **33**, 1953, **2000**
4. M. Liebau, E. Unger, G. S. Duesberg, A. P. Graham, R. Seidel, F. Kreupl, and W. Hoenlein, Appl. Phys. A, **77**, 731, **2003**
5. D. N. Madsen, K. Mølhave, R. Mateiu, A. M. Rasmussen, M. Brorson, C. J. H. Jacobsen, and P. Bøggild, Nano Letters, **3**, 47, **2003**

Gate-Field-Induced Schottky Barrier Lowering in a Nanotube Field-Effect Transistor

T. Brintlinger, B.M. Kim, E. Cobas, and M. S. Fuhrer

Department of Physics and Center for Superconductivity Research, University of Maryland,
College Park, MD 20742-4111, USA

Abstract. We propose that in nanotube field effect transistors (FETs) with small effective dielectric thickness the vertical potential drop across the nanotube diameter at finite gate bias can lower or eliminate the Schottky barrier at the electrode. This effect is demonstrated in single-walled carbon nanotube FETs fabricated on top of ultra-high-κ dielectric constant $SrTiO_3/Si$ substrates. These FETs show transconductances normalized by channel width of 8900 S/m, one of the highest values to date. This transconductance cannot be explained within the conventional FET or Schottky-barrier models.

INTRODUCTION

The possibility of using semiconducting single-walled carbon nanotubes (CNT) as a replacement for, or complement to, silicon within a field effect transistor (FET) architecture has led to an intense research effort into both the NTs themselves and the NTs interfaced with technologically relevant materials. The intrinsic nanometer scale and pristine quality of the carbon lattice in SWNTs initially inspired great hope for a semiconducting, ballistic, easily manufacturable nanoscale material for use in FETs [1]. However, the Schottky barrier (SB) at the CNT-metal interface limited the improved performance with traditional vertical scaling (thinner dielectrics and higher dielectric constants) [2]. Subsequent efforts focused on either avoiding the SBs through engineering solutions or by achieving Ohmic contact with high work function metals and large diameter, smaller band gap SWNTs [3-5]. Continuing this research effort, we report here the integration of high-κ $SrTiO_3$ on Si substrates (STO/Si) with NT-FETs. The high transconductance per channel width (8900 $\mu S/\mu m$), limited number of lithography steps, and use of small-diameter NTs indicates an alternate avenue for high-performance CNT-FET fabrication. We explain the scaling of the transconductance as the emergence of a new vertical length regime within the Schottky barrier model, which causes a gate-induced Ohmic contact.

DEVICE FABRICATION

Our starting substrates were epitaxial STO/Si ($\kappa \approx 175$). Details on the growth and characterization of this STO/Si have been reported elsewhere [6]. CNTs were grown

CP723, *Electronic Properties of Synthetic Nanostructures*, edited by H. Kuzmany et al.

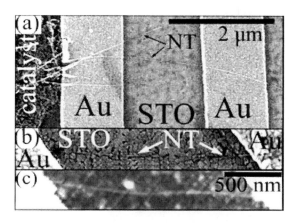

FIGURE 1. Images of CNT-FETs grown on SrTiO₃ (STO) substrates by chemical vapor deposition. Part (a) shows patterned catalyst (left) on STO, as well as several nanotubes extending from the catalyst island. One nanotube has been contacted by two Cr/Au electrodes. (b) Field-emission scanning electron micrograph of a semiconducting nanotube on STO bridging two Cr/Au contacts with 1.8 um separation. (c) AFM image of the nanotube in (b), giving a NT diameter of 1 nm.

by chemical vapor deposition (CVD), adapting from procedures for synthesis of CNTs on SiO_2. The resulting CNT-FETs are shown in Figure 1. Parts (a) and (b) show scanning electron microscope (SEM) images, employing a technique originally developed with SiO_2/Si substrates [7]. Part (c) illustrates a typical atomic force micrograph, used to determine NT diameter (1 nm in (c) and for measurements in this paper). Briefly, an alumina-supported Fe/Mo catalyst was patterned in islands on the substrate by electron-beam lithography. CVD synthesis was carried out in a 1 in. diameter tube furnace for 11 min at 900 °C using a methane/hydrogen co-flow. To ensure the STO/Si remained intact after growth and to verify the STO thickness, we performed transmission electron microscopy (TEM) and electron diffraction, shown elsewhere [8]. Transport results are shown in Fig. 2 below, with discussion following.

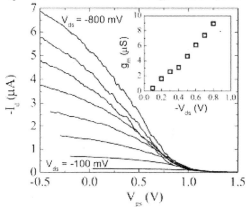

FIGURE 2. Drain current (I_d) vs. gate voltage (V_{gs}) at different bias voltages (V_{ds}) for a 1 nm CNT-FET. Inset shows peak transconductance (g_m) vs. V_{ds} through numerical differentiation of the curves in the main graph. The highest g_m at 800 mV bias is one of the highest to date.

RESULTS AND DISCUSSION

We have measured the electrical properties of CNT-FETs on STO using the Si substrate as a global back gate. Figure 2 shows the behavior of a typical device with the drain current, I_d, changing as a function of gate voltage, V_{gs}, for different bias voltages, V_{ds}. The inset shows peak transconductances, $g_m = dI_d/dV_{gs}$, determined by numerically differentiating the curves in the main graph, versus V_{gs}. The peak transconductance at 800 mV is 8.9 μS. To compare to other CNT-FETs, we divided by the channel width, the NT diameter, to obtain 8,900 μS/μm, one of the highest values to date (see Table 1). We also note that this value is most likely not in the saturation regime, and thus g_m may go even higher with increasing V_{ds}. In comparing this device with others, we first attempted to model its behavior within the 1D diffusive FET model in which the transconductance in the saturation region is given by $g_m \approx \mu c_g V_{ds}/L$, where μ is mobility, c_g is gate capacitance per length and L is channel length. Thus for a given material system (given μ) the transconductance can be increased through increasing c_g or V_{ds}, or decreasing L. In practice, however, the product $\mu V_d/L$ is expected to saturate in v_s, the saturation carrier velocity, and the maximum transconductance is $g_{m,max} = c_g v_s$. Thus increasing c_g becomes the goal for obtaining higher transconductance. In comparing our g_m (or g_m/d) with other published values of g_m, we see that it is 10-15 times larger than other values while the (quantum capacitance-limited) total gate capacitance, c_g, is only 3-5 times larger. This is not surprising as other workers have shown that SBs at the NT-metal interface greatly influence CNT-FET performance [9]. In Ref. 9, the vertical scaling of the SB effect on CNT-FET is investigated, and it is shown that $g_m \sim t^{-1/2}$, where t is the oxide thickness, and this scaling occurs *independent of dielectric constant*. Again comparing our results with Refs. 1 and 2 in Table 1, we see that the oxide thickness of our CNT-FETs is greater than or equal to the others while g_m is 10-15 times larger, an inadequate explanation of the behavior of our CNT-FETs. A third possible explanation of the high transconductance is Ohmic contact at the NT-metal interface, as shown in Ref. 4, but small diameter ($d < 2$ nm) NTs and unannealed Cr/Au contacts are shown to make poor/non-Ohmic contacts in Ref. 4 and Ref. 10 respectively. So we consider this possibility unlikely.

TABLE 1. Device parameters for CNT-FETs in this and other works.

Author	Dielectric (κ)	t (nm)	d (nm)	c_g (pF/cm)	g_m (μS)	V_{ds} (V)	g_m/d (μS/μm)
Bachtold (Ref. 1)	Al$_2$O$_3$ (5)	2-5	1	0.7-1.0	0.3	-1.3	300
Appenzeller (Ref. 2)	HfO$_2$ (11)	20	1-2	1.1	0.6	-1.5	300-600
This work	SrTiO$_3$ (175)	20	1.0	3.4	8.9	-0.8	8900
Javey* (Ref. 3)	ZrO$_2$ (25)	8	2	2.3	12	-1.2	6000
Nihey* (Ref. 3)	TiO$_2$ (40-90)	2-3	1.5	3.0	8.7	-1	5800
Javey*† (Ref. 4)	HfO$_2$ (20)	8	2.3	1.7	20	-0.5	10000
Rosenblatt‡ (Ref. 5)	Electrolyte (80)	~1	3	3.8	20	-0.8	6700

The columns display the dielectric material and dielectric constant κ, dielectric thickness t, nanotube diameter d, total gate capacitance c_g, transconductance g_m, source-drain bias V_{ds} and transconductance per width, g_m/d. The symbol * denotes local top gating, † Ohmic contacts, and ‡ electrolytic gating.

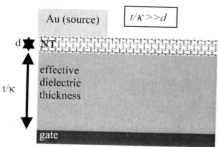
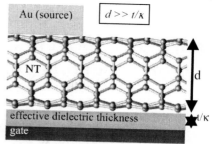

FIGURE 3. Visual model of different vertical length regimes within the Schottky barrier model. The left image shows a picture of the length regime described in Ref. 10, while the right image describes the length regime in this work.

Another possibility is that vertical scaling has a more pronounced effect on the Schottky barriers. The model of Ref. 9 may be inadequate for ignoring charge in the nanotube channel [11] and for treating the NT as infinitely thin. We expect the electric field at the contacts will be substantially modified when the effective thickness of the dielectric $t' = t/\kappa$ becomes significantly less than the CNT diameter d (where $\kappa_{NT} \sim 1$ [12]). In our devices, $t'/d \approx 0.1$ is significantly less than the values of $t'/d \approx 0.4$-1 and 1-2 in Refs. 1 and 2 respectively (see Fig. 3). For $t'/d \ll 1$, the potential drop *across* the CNT diameter becomes a large fraction of V_{gs}. When the potential drop across the radius of the nanotube is equal to the SB height, population of the valence band with carriers should become energetically favorable, allowing Ohmic contact with the channel. Stated another way, at moderate V_{gs}, the potential of the NT relative to the electrode can be greater than the SB height, eliminating the barrier. In our devices, this would occur at a V_{gs} of a few hundred mV from threshold. This model also offers an alternate explanation for the observation of high transconductances (even in small diameter CNTs) in FETs with an electrolyte dielectric ($t'/d \approx 0.01$) [5].

This research was supported by ARDA and ONR (Grant No. N000140110995), the NSF (Grant No. DMR-0102950), and the DCI Fellowship. The authors thank F. Gac.

REFERENCES

1. A. Bachtold *et al* Science **294**, 1317 (2001).
2. J. Appenzeller, *et al* Phys. Rev. Lett. **89**, 126801 (2002).
3. A. Javey *et al* Nature Materials **1**, 241 (2002) and F. Nihey *et al* J. J. Appl. Phys. **42**, L1288 (2003).
4. A. Javey *et al* Nature **424**, 654 (2003) and A. Javey *et al* Nano Letters **4**, 447 (2004).
5. S. Rosenblatt *et al* Nano Letters **2**, 869 (2002).
6. K. Eisenbeiser *et al* Appl. Phys. Lett. **76**, 1324 (2000) and McKee *et al* PRL **81**, 3014 (1998).
7. T. Brintlinger *et al* Appl. Phys. Lett. **81**, 2454 (2002).
8. B.M. Kim *et al* Appl. Phys. Lett. **84**, 1946 (2004).
9. S. Heinze *et al* PRL **89**, 106801 (2002) and S. Heinze *et al* Phys. Rev. B **68**, 235418 (2003).
10. Y. Yaish *et al* Phys. Rev. Lett. **92**, 046401 (2004).
11. J. Guo *et al* IEEE Trans. Electron Devices **51**, 172 (2004).
12. F. Leonard and J. Tersoff, Appl. Phys. Lett. **81**, 4835 (2002).

Fabrication of Field Effect Transistors based on carbon nanotubes made by LASER ablation

Marina Dipasquale[1], Pietro Repetto[2], Flavio Gatti[2], Davide Ricci[1], Ermanno Di Zitti[1]

[1]*Department of Biophysical and Electronic Engineering, University of Genoa , 16145 Genova, Italy*
[2]*Department of Physics, University of Genoa , 16145 Genova, Italy*

Abstract. We report results obtained performing the fabrication of a back gate Field Effect Transistor that uses a rope of single wall nanotubes as channel. Nanotubes were made using the LASER ablation technique and they were characterized using micro-RAMAN spectroscopy, SEM and AFM imaging. A self-assembly method based on a high frequency electric field was used to direct the placement of nanotube ropes between source and drain gold electrodes made by low cost standard microlithographic techniques. Measurements of the electrical transport in the nanotube ropes depending on the applied gate voltage are reported.

INTRODUCTION

Thanks to their noteworthy physical and electronic properties, such as high aspect ratio, metallic or semiconducting behaviour and low noise, carbon nanotubes are of great interest for electronic applications [1, 2]. Among the variety of possible devices, their most common exploitation is within either top gate [3] or back gate [4] field effect transistors (FETs), where the carbon nanotubes play the role of channel. The road towards nanoscale integrated carbon nanotube devices is laid, but up to now efforts are generally devoted to the comprehension of the behaviour of the tubes within hybrid nano-microscale devices. We contribute to this research by describing the results obtained in such a device that can be fabricated using simple and low cost methods.

NANOTUBE PRODUCTION, CHARACTERIZATION AND PURIFICATION

We have chosen the LASER ablation method for single wall nanotube (SWNT) preparation, as it can yield a high quality product [5, 6] and could be set-up with reasonable costs using equipment present in the laboratory. A 16 mm diameter target was made from Carbon powder doped with Nichel and Cobalt (90% C, 5% Ni and 5% Co in weight) and placed into a quartz tube (4 cm in diameter and 100 cm in length) inserted into a Fuji Electric furnace kept at a temperature of $1200°C \pm 10°C$ and under an 4 l/min argon flow at 600 mbar of pressure.

CP723, *Electronic Properties of Synthetic Nanostructures*, edited by H. Kuzmany et al.

The nanotubes were formed by vaporizing the target with light pulses (0.5 Joule, 10 ns) from a ND:YAG LASER (wavelength 1064 nm) and deposited onto a water-cooled copper collector.

The raw material was purified using a sequence of steps involving both chemical and physical treatments [7, 8]. The first step was applied by refluxing 5mg of raw material into 50ml of 2.6M HNO_3 for 48h in order to dissolve the metal catalyst particles. Hence the acid was removed from the solution by centrifugation, washing with 30ml ethanol and further centrifugation. Nanotubes were then suspended in a surfactant solution of 1ml Triton X-100 in 10ml of ethanol and contaminant particles were filtered using two different pore size membranes, respectively 5 μm and 0.2 μm.

After production and during the purification steps the nanotubes were characterized by Scanning Electron Microscopy (SEM), Transmission Electron Microscopy (TEM) and Micro-Raman spectroscopy. Figure 1a shows a SEM image of raw nanotubes, while Fig. 1b shows a SEM image of diluted purified nanotubes directly placed onto an aluminum sample holed. One can notice an individual bundle of nanotubes and also a nanotube ring. Figure 1c reports a high resolution TEM image taken after purification that allows to identify single nanotubes within a bundle. At last, in Fig. 1d a Raman spectra of the purified material (LASER wavelength 633 nm) is reported, showing typical features of single wall nanotubes.

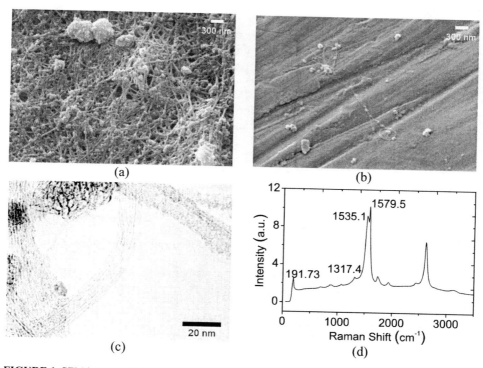

(a)

(b)

(c)

(d)

FIGURE 1. SEM images of (a) raw material and (b) purified nanotubes from diluted solution. TEM image (c) at high resolution of nanotube bundle. RAMAN spectra (d) of purified nanotubes.

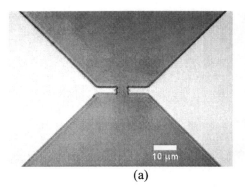 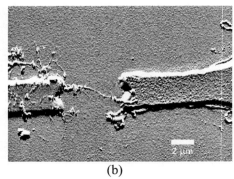

(a)   (b)

FIGURE 2. Optical microscope image of a contact pair (a). AFM image (b) showing the nanotube rope deposited between source and drain electrodes.

CARBON NANOTUBE FET FABRICATION

A silicon wafer was processed by means of two steps of UV lithography and subsequent evaporation of thin films, using a titanium film to build the back gate and a titanium-gold deposition for the surface metallic contacts, typically 100-200 nm thick.

The device geometry determines arrays of electrodes 3 μm wide and 100 μm long (used as source and drain), separated by a 5 μm wide gap.

Electric fields have been exploited to orient nanotubes between the contact pair shown in Fig. 2a. A small drop (about 5μl) of purified nanotubes ethanol solution was deposited between a contact pair while an ac bias was superimposed to a dc voltage before ethanol evaporation. Good positioning on the structures was typically obtained using a 5 Vpp ac bias @ 1 MHz and a 2 V dc bias for 30-180 s, as proven by AFM imaging (see an example in Fig. 2b).

ELECTRICAL MEASUREMENTS

An automated system consisting of a HP DC source monitor 4142B interfaced by GPIB to a PC has been used to perform the measurements of input and output characteristics of the different FETs at room temperature.

Source and drain electrodes were addressed by a set of spiked tips in an ad hoc designed sample holder. Care was taken to ensure that the gate electrode were electrically isolated from surface contacts. The typical measured leakage currents are in the range 0.1 pA – 10 nA.

Two main issues can be inferred from the measured output characteristics. A first class of curves (see Fig. 3a) exhibits a gate voltage dependent current with a low off-current, showing the effect of semiconducting nanotube connecting the two contact leads. In another set of curves (see Fig. 3b) the effect of metallic nanotubes is dominant, giving rise to a slight nonlinear characteristic with a very weak dependence on the gate voltage.

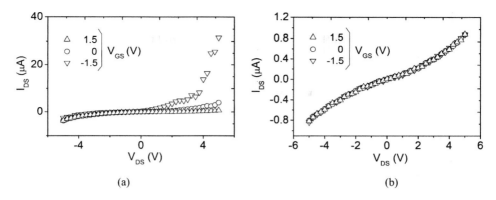

FIGURE 3. Examples of I_d vs. V_{ds} characteristics after nanotube deposition showing higher percent of semiconducting (a) or metallic (b) nanotubes in the rope between contacts.

CONCLUDING REMARKS

Prototypes of FET using ropes of nanotubes as channel have been fabricated using affordable technology, enabling engineering and device optimization for a variety of applications. The obtained electrical characteristics allow the design of exciting nonlinear devices once a better control of the process is achieved.

ACKNOWLEDGMENTS

Work supported by the Italian Ministry of University and Scientific and Technological Research (National Research Program "*Carbon nanotubes for electronics: synthesis, characterization and manipulation*") and by the National Research Council (Grant "*Nanoarchitetture organiche per l'elettronica e paradigmi computazionali*")

REFERENCES

1. M. S. Dresselhaus, G. Dresselhaus, and Ph. Avouris, *Carbon Nanotubes: synthesis, structures, properties and applications,* Springer, Springer, 2001, pp. 1-420.
2. S., Iijima, *Physica B* **323**, 1–5 (2002).
3. S.J. Wind, J. Appenzeller, R., Martel, V. Derycke, and Ph. Avouris, *Appl. Phys. Letters* **80**, 3817-3819 (2002).
4. V. Derycke, R. Martel, J. Appenzeller, and Ph. Avouris, *Nano Letters* **1**, 453-456 (2001).
5. A. Thess, R. Lee, P. Nikolaev, R. E. Smalley et al., *Science* **273**, 483-487 (1996).
6. W. F. Maser et al., *Nanotechnology* **12**, 147-151 (2001).
7. Z., Shi, Y., Lian, F., Liao, X., Zhou, Z., Gu, Y., Zhang, S., Iijima, *Solid State Communication* **112**, 35 – 37 (1999).
8. J., Liu, A. G., Rinzler, R. E., Smalley, et al., *Science* **280**, 1253-1255 (1998).

Emission Characteristics of CNT-Based Cathodes

G.S. Bocharov*, A.V. Eletskii**, A.F. Pal, A.G. Pernbaum**, V.V. Pichugin

State Research Center of Russian Federation "TRINITI". Troitsk. Moscow region. Russia
Moscow Power Engineering Institute. Moscow. Russia
**RRC "Kurchatov Institute". Moscow. Russia*

Abstract. There have been measured the current-voltage characteristics (CVC) of electron field emission cathodes fabricated on the basis of single walled carbon nanotubes (SWNT). SWNT's of $1.2 - 1.5$ nm in diameter were produced by the standard arc discharge method using Ni-Cr alloy foil as a catalyst. At relatively high electrical field strength the CVC are well agreed with the known Fowler-Nordheim dependence (FND). A notable deviation of those from FND at low fields has been observed. This deviation is due presumably to a spread in geometry of SWNT, which promotes even a larger spread in their emission properties owing to the electrical field amplification phenomenon. A model approach to description of the electron field emission characteristics of a CNT-based cathode with taking into account a spread in the geometry of individual nanotubes has been developed. Supposing a normal distribution in the electrical field amplification factor γ of individual CNT's, the generalized expression for CVC of a CNT-based cathode has been derived. This expression transforms to the FND in the limiting case of zero dispersion of the amplification factor. Close agreement of measured CVC and calculated through the generalized expression is reached at $\Delta\gamma/\gamma = 0.304$.

INTRODUCTION

Extraordinary emission properties of carbon nanotubes (CNT) that are due to their good electrical conductivity and high aspect ratio make them very attractive material for field emission cathodes [1-3]. Development of this (as well as others) direction of usage of CNT is hindered mainly by the lack in a regular non expensive procedure of large scale production of CNT and as a consequence by their high production cost. Another problem, retarding the advancement of applied usage of CNT, concerns to a statistical spread in parameters of individual nanotubes, for which reason operation characteristics of a CNT-based device show a dependence on such a spread. This dependence is studied experimentally and theoretically in the present work by the example of the field emission cathode, containing SWNT's as emitters.

PREPARATION OF SAMPLES

The procedure of CNT synthesis has been described in detail elsewhere [4]. Samples of the SWNT containing soot were produced in a cylindrical gas discharge

CP723, *Electronic Properties of Synthetic Nanostructures*, edited by H. Kuzmany et al.
© 2004 American Institute of Physics 0-7354-0204-3/04/$22.00

camera with water cooled walls of 7 l in volume. A cylindrical graphite rod of 6 mm in diameter with a sharpened end was used as a cathode. A "sandwich" structure was used as an expendable anode. It was consisted of two graphite rods of rectangular cross section (7x3,5 mm^2), between of which a thin Ni/Cr foil of weight ratio 80/20 used as a catalyst was inserted. The foil of 0.2 mm accounted about 10% (mass) from the anode material. Such an anode structure has provided a homogeneous delivery of a catalyst into the plasma region. The DC arc was burnt at the He pressure of 700 Torr, current of 60 A, voltage of 29 - 30 V and inter-electrode gap of 4 mm. The arc voltage and gap were stabilized automatically during arc burning by moving the anode rod [5]. The rate of thermal sputtering the anode material was about 1 g/h.

The material containing SWNT was found both on the discharge camera walls and on the cathode surface. The material covering the camera walls has a layered cloth-like structure with the porosity of about 90%. It could be easily separated from the wall and hold its stability under a light mechanical action. The rate of growth of this layer is estimated as about 0.1 mm/h. This material was subjected to investigation by Raman spectrometry, TEM microscopy and TGA analysis. Besides of that, the sample was purified by acid treatment. The Raman spectroscopy and TEM observations show that the samples contain single walled nanotubes up to 5 μm in length and diameter ranging within 1.2 – 1.5 nm with maximum in distribution at about 1.24 nm. The nanotubes are bound into bundles of about 10 nm in diameter, containing up to 100 individual SWNT's. The material is highly contaminated with carbon nano-particles as well as catalyst particles, which can be mostly removed through the thermal treatment in air (oxygen) atmosphere [4].

MEASUREMENT OF THE ELECTRON FIELD EMISSION

Samples of the soot containing SWNT's were used for manufacturing field emission cathodes. In so doing the soot annealed in air at 600° C or/and nitride acid treated was powdered in an agate mortar. Then the finely dispersed powder produced was used for preparing the CCl$_4$–based suspension. This suspension was applied onto the surface of a silicon or copper substrate of 1x1 cm^2 in size. Then the substrates were dried at the room temperature. The adhesion of SWNT films produced in such a manner is quite sufficient for performing emission measurements.

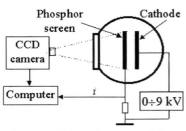

FIGURE 1. Broad-aperture diode-type measurement device

The emission characteristics of cathodes were measured with use of the broad-aperture diode-type device shown schematically in Fig.1. A phosphor coated glass substrate with the Indium-Tin-Oxide (ITO) interlayer was used as an anode, and a substrate coated with CNT film was used as a cathode. The experiments were carried out at the inter-electrode gap of 500 μm and the rest gas pressure lower than 10^{-6} Torr. The voltage up to 9 kV was applied to the gap in pulsed mode with pulse repetition rate of 50 Hz and pulse duration of 100 μs. Emission images of the cathode were recorded by CCD camera and stored in PC for processing. Usually we recorded the emission images of one cathode taken at different values of the applied field. The character of the emission shows a high sensitivity of the emission pattern to the thickness of the soot layer on the substrate. Thus a soot layer of 0.3 mm in thickness does not show any emission.

RESULTS AND DISCUSSION

A typical emission current voltage characteristic (CVC) measured at vacuum conditions (< 10^{-6} Torr), is given on Fig.2 in both convention and Fowler-Nordheim coordinates. As is seen, at low fields (E ≤ 10 V/μm) the emission current exceeds considerably (2-3 orders of magnitude) the Fowler-Nordheim value. This is caused by the statistical spread in parameters of CNT's, for which reason at low fields the threshold conditions for the emission are reached for few nanotubes, characterized by a maximum magnitude in the field amplification coefficient. The number of such nanotubes increases with rise in the electrical field, which results in an additional enhancement in the emission current. This effect can be seen from the comparison of the images of spatial distribution of the emission current density, obtained at various magnitudes of the applied voltage (Fig.3). As is seen the rise in the electrical field strength causes the corresponding increase in not only the brightness of phosphor radiation but also in area of the radiating surface.

Another type of the emitter studied in the work was a grid manufactured from a metal wire of ~ 0.1 mm in diameter. SWNT containing soot was applied to the metal surface by rubbing technique. In this case the emission was observed at the electrical field exceeding 3 V/μm, and the maximum emission current accounted 1 A/cm².

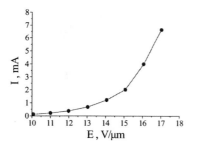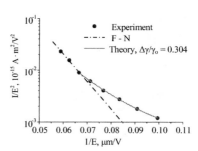

FIGURE 2. A typical example of the emission current-voltage characteristic of an SWNT-based cathode, presented in conventional (left) and Fowler-Nordheim coordinates (right)

The influence of the statistical spread in parameters of CNT on the character of CVC was studied on the basis of a theoretical approach [5], providing an interconnection between the shape of CVC and the degree of the spread. In accordance with this approach, the field amplification factor γ of individual CNT's is characterized by the assumed normal distribution, which results in the following relation for the current-voltage characteristic:

10.4 V/μm, 0.85 mA   12.2 V/μm, 1.2 mA   .13 V/μm, 2.4 mA

FIGURE 3. Images of the surface distribution of the phosphor radiation intensity, obtained at various magnitude of the electrical field strength.

$$J = C_1 E_o^2 \gamma_o^2 \exp\left[-\frac{C_2}{\gamma_o E_o} + \frac{C_2^2 \Delta\gamma^2}{4\gamma_o^4 E_o^2} \right].$$ (1)

Here J is the emission current density, E_o is the average magnitude of the electrical field strength in the gap, γ_o is the magnitude of the electrical field strength amplification factor, averaged over the massive of CNT's, $\Delta\gamma$ is the statistical dispersion of that, C_1 and C_2 are coefficients, dependent on the emission area, electron work function etc. The expression (1) can be considered as a generalization of the Fowler-Nordheim relation with taking into account the statistical spread in the emission characteristics of individual emitters and transforms into the latter at $\Delta\gamma = 0$. The CVC calculated by the relation (1) for $\Delta\gamma/\gamma_o = 0.304$ is compared on Fig. 2 (right picture) with the Fowler-Nordheim dependence and that observed in the present experiment. As is seen, the measured CVC can be used as the basis for estimating the degree of homogeneity of CNT-based field emission cathode.

ACKNOWLEGEMENTS

The work is supported by ISTC in frame of the project No.2484 and CRDF in frame of REC "Plasma".

REFERENCES

1. Gulyaev Yu.V. et al. Proc. 7th Int. Vacuum Microel. Conf. Grenoble 1994. P.322; *Vacuum Sci. & Tech.* (B13) 1995 P.234; Gulyaev Yu.V. et al. *Microelectronics* **26** (2) 84 (1997) (In Russian); Chernozatonskii L.A. et al., 8th Int. Vacuum Microel. Conf. Portland, Oregon, Technical Digest. 1995. P.363; *Chem. Phys. Lett.* **233** 63 (1995)
2. De Heer W.A., Chatelain A., Ugarte D. *Science* **270** 1179 (1995)
3. Eletskii A.V. *Physics-Uspechi* **45** 369 (2002)
4. Bezmelnitsyn V.N. et al. *Physics of the Solid State* **44** 630 (2002)
5. Bocharov G.S., Eletskii A.V., Korshakov A.V. *Rev. Adv. Mater. Sci.* **5** 371 (2003)

Application Of Metal Coated Carbon Nanotubes To Direct Methanol Fuel Cells And For The Formation Of Nanowires

*E. Frackowiak, G. Lota, K. Lota, [1]F. Béguin

Poznan University of Technolog y, Piotrowo 3, 60-965 Poznan, Poland
[1]CRMD, CNRS-University, 1b rue de la Férollerie, Orléans, France

Abstract. Homogeneously distributed particles of Pt, Ru or their alloys were electrochemically deposited on catalytic and template multiwalled carbon nanotubes (CNTs). The catalysts supported on the entangled or straight CNTs network with open mesopores have been tested for the reaction of methanol oxidation (1M) in acidic medium (1M H_2SO_4) by potentiodynamic and galvanostatic methods. The high current densities (500 mA/g) for methanol oxidation proved that such a composite enables a good accessibility of the reagent molecules to the electrode interface.
Contrarily, continuous layers of such metals as nickel, copper and silver were chemically and/or electrochemically deposited on carbon nanotubes. These novel materials of nanometer size can be adapted for some electronic application, e.g. nanowires.

INTRODUCTION

Carbon nanotubes (CNTs) due to their unusual electrical, microtextural, mechanical properties can find many potential applications, e.g. as a support for metallic particles. Carbon nanotubes coated or filled by dispersed metallic particles can be used in heterogeneous catalysis [1-5], particularly in fuel cells. Among the different kinds of fuel cells, the Direct Methanol Fuel Cells (DMFC) have recently attracted much attention for their potential usage as power sources due to the high theoretical energy, the low cost of methanol and its availability. Methanol oxidation in acidic medium requires the application of noble metals, generally platinum and/or their alloys. The efficiency of the oxidation reaction strongly depends on the catalyst dispersion that is quite often determined by the support, therefore, carbon nanotubes have been used for this target [2]. Indeed, due to the open mesoporous network of CNTs, such a support allows an easy diffusion of reagents and reaction products. Our objective is to test the efficiency of catalysts, i.e. single components Pt, Ru and their alloys supported on nanotubes and to compare this support with other carbon materials.

For special electronic applications, metallic nanowires or nanorods are of great interest. Therefore we have also used carbon nanotubes for a continuous coating by such metals as: silver, nickel and copper.

CP723, *Electronic Properties of Synthetic Nanostructures*, edited by H. Kuzmany et al.
© 2004 American Institute of Physics 0-7354-0204-3/04/$22.00

EXPERIMENTAL

Catalytic and template multiwalled carbon nanotubes (CNTs) were tested as a catalyst support. Platinum, ruthenium and Pt-Ru alloys were used as a catalyst for methanol oxidation. The catalytic particles were deposited galvanostatically in acidic medium. The electrolytic solutions were composed of 0.01 M $RuCl_3 \cdot xH_2O$ and/or 0.01 M H_2PtCl_6. Methanol oxidation was investigated in acidic medium (1M CH_3OH + 1M H_2SO_4) by voltammetry method at a scan rate of 5 mV/s and galvanostatic discharging. The counter electrode was from platinum whereas the reference electrode was the mercury/mercurous sulfate system (Hg/Hg_2SO_4). For the galvanostatic and cyclic voltammetry measurements, a potentiostat- galvanostat (AUTOLAB-ECOCHEMIE) was used.

Deposition of a continuous metallic coating on the CNTs was realized by chemical and electrochemical methods [6]. CNTs were preliminary functionalised by oxidation in a solution of nitric acid. Then, CNTs were specially treated with a tin chloride solution. Deposition of silver has been performed from silver nitrate with ammonia, nickel plating from nickel chloride/sulfate, whereas copper was deposited from its sulfate. The dispersion of catalyst and metallic coating were studied by SEM (Hitashi S 4200) and TEM (Philips CM20) techniques.

RESULTS AND DISCUSSION

Carbon Nanotubes – Catalytic Support for DMFC

Different types of carbon nanotubes/catalyst composite with Pt, Ru, Pt-Ru were observed by SEM and TEM techniques and characterized by EDX analysis. Examples of CNTs/Pt-Ru images are shown in Fig. 1a, b. Although the particle size of the catalyst clusters has a high dispersion, a few nm particles could be easily found.

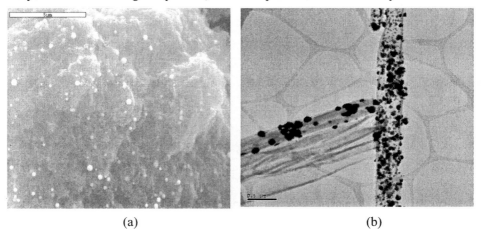

(a)   (b)

FIGURE 1. (a) SEM image of catalytically grown CNTs and (b) TEM of template CNTs with electrodeposited Pt-Ru particles.

The CNTs/catalyst composites were tested for the methanol oxidation reaction in 1M H$_2$SO$_4$. The voltammetry experiments allow to estimate the methanol oxidation current for the Pt catalyst (Fig. 2) and Pt-Ru alloy (Fig. 3). The most characteristic feature is a very useful shift of the oxidation peak to a more negative value in the case of the Pt-Ru alloy and also a rest potential from -0.184 V to -0.34 V vs Hg/Hg$_2$SO$_4$. From the galvanostatic experiments, even one order of magnitude higher current load could be reached for the nanotubular supported catalyst in comparison to catalysts based on other carbon supports.

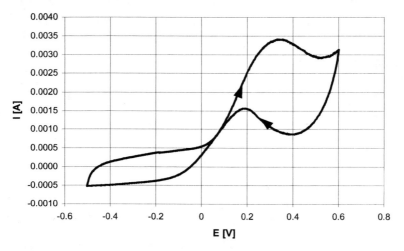

FIGURE 2. Voltammetry characteristics of methanol oxidation (1M) in sulfuric acid on catalytic CNTs with Pt electrodeposited particles. Scan rate of potential 5 mV/s.

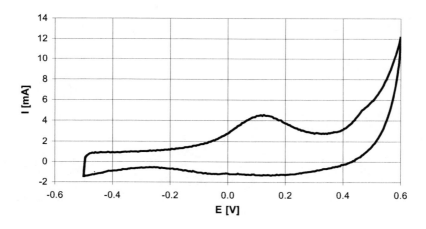

FIGURE 3. Voltammetry characteristics of methanol oxidation (1M) in sulfuric acid on catalytic CNTs with Pt-Ru electrodeposited particles. Scan rate of potential 5 mV/s.

Metallic Coating of Carbon Nanotubes

The formation of nanowires is another interesting domain where nanotubes play a great role as support. CNTs could be easily coated by metallic films such as silver, copper and nickel after a special treatment, i.e. tin sensitization [6]. In the case of nickel and copper, irregular deposits were obtained by electroless plating, however, silver deposition supplied a quite homogeneous coating (Fig. 4). It is noteworthy that the catalytically grown nanotubes become straight after metallization. Electrochemical coating, especially copper plating needs further experiments.

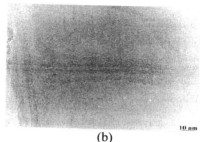

(a)   (b)

FIGURE 4. (a) SEM and (b) TEM images of catalytic CNTs chemically coated by silver.

After a thermal treatment in oxidizing atmosphere, the nickel coated CNTs can be transformed into nanowires. However, during removal of carbon, the metallic deposit has a tendency to agglomerate (Fig. 5).

FIGURE 5. TEM image of nickel nanowires.

REFERENCES

1. Serp, P., *Appl.Cat. A: General* **253**, 337-358 (2003).
2. Che, G., Lakshmi, B.B., Martin, C.R., Fisher, E.R., *Langmuir* **15**, 346-349 (1999).
3. Li, W.Z., Liang, C.H., Qiu, J.S., Zhou, W.J., Han, H.M., Wei, Z.B., Sun, G., Xin, Q. *Carbon* **40**, 791-794 (2002).
4. Li, W., Liang, C., Zhou, W., Qiu, J., Li, H., Sun, G., Xin, J. *Carbon* **42**, 436-439 (2004).
5. Sun, X., Li, R., Villers, D., Dodelet, J.P., Desilets, S. *Chem. Phys. Lett.* **379**, 99-104 (2003).
6. *Modern Electroplating* ed. M. Schleisinger, M. Paunovic, John Wiley & Sons Inc. New York (2000).

Electrical Interconnects Made of Carbon Nanotubes

Maik Liebau*, Andrew P. Graham, Zvonimir Gabric, Robert Seidel,
Eugen Unger, Georg S. Duesberg, and Franz Kreupl

Infineon Technologies AG, Corporate Research, 81730 Munich, Germany

Abstract. The unique properties of carbon nanotubes (CNTs) make them promising candidates for electrical conductors in microelectronic devices. The parallel integration of CNTs into processes that are compatible with the requirements of the microelectronic industry will be important for their future application in chip devices. We present lithography-based processes to create vertical electrical interconnects made of CNTs. The approach involves catalyst and dielectric (insulator) deposition, lithography, standard etch processes, CVD growth of the CNTs, and the structured deposition of metallic top contacts. This paper will discuss the electrical properties of these CNT vertical interconnects and compare them with the requirements of the ITRS roadmap.

1. INTRODUCTION

An alternative approach to meet the future requirements for electrical interconnects in microelectronics is the application of CNTs.[1] Interconnects based on CNTs have a high potential to overcome the limitations of Cu interconnects in particular because of the higher current densities they can sustain.[2] Electromigration of Cu becomes a critical element with the aggressive development toward 65 nm and 45 nm technologies where the feature size of copper interconnects is also in the nanoscale range. Current densities of 3×10^6 A/cm^2 that are anticipated by 2009 are most probably not practical using copper wires. That is why the semiconductor industry is currently looking for 'radical solutions' which might include the application of new materials such as CNTs.[3]

This paper gives an example for the formation of CNT based vertical interconnects (vias). Since vias are generally the most critical structures with respect to electromigration, this is a practical structure to evaluate alternative materials. A buried stacked catalyst approach is presented that allows the reliable growth of individual multi-walled CNTs in lithographically defined holes. Using this approach we were able to perform transport measurements on individual vertically aligned CNTs. We will also address some technological relevant problems that have to be solved for the integration of CNTs as wires in microelectronics.

CP723, *Electronic Properties of Synthetic Nanostructures,* edited by H. Kuzmany et al.
© 2004 American Institute of Physics 0-7354-0204-3/04/$22.00

2. DEVICE FABRICATION

A key issue when considering new processes in the microelectronics industry is the compatibility of the new process parameters, e.g. temperature, with the existing technology. The process described here is based on the CVD of CNTs that is considered to be a technology which is compatible with these requirements, although the growth temperature of 700°C is on the upper bound.

The process to create CNT based vias involves five steps (Fig. 1). The first step is the preparation of a buried stack catalyst involving the deposition of 10 nm iron and 20 nm tantalum on a substrate using an ion beam coater. Tantalum was used as bottom contact since preliminary studies showed that CVD growth of CNTs using tantalum as a substrate gives reasonable yield.[4] A 150 nm thick silicon oxide layer and a 25 nm thick SiN hardmask were deposited on the catalyst multilayer by standard plasma deposition processes.

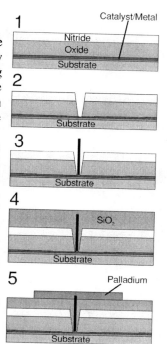

Figure 1: Schematic view of the CNT via processing

Second, the device pattern is written using electron beam lithography using a PMMA resist. This yields holes in the PMMA layer with diameters from 25 nm to 50 nm (Fig. 2a). This is followed by a two stage plasma etch process to break through the nitride hard mask layer and selectively etch the oxide layer. These processeses yield holes (vias) with diameters between 50 and 100 nm and aspect ratios (height to width) of 3 that stop at the catalyst/metal layer (Fig 2b). The hole diameter is much larger than defined in the PMMA resist so it is clear that more emphasis must be put on the etching technology to reduce the enlargement of the holes during etching.

The next process step is the growth of multi-walled CNTs from the catalyst in the predefined holes. An optimized chemical vapor deposition (CVD) growth at 700°C was used, as described elsewhere.[5] In this study acetylene was used as carbon source and the growth was performed at 700°C. Preheating for 10 min at the same temperature under hydrogen was found to enhance significantly the yield of the CNT synthesis. We attribute this to the time that is required for the formation of stable and catalytically active particles in the holes. The growth yield (number of holes that are filled with individual CNTs) was about 80%. Due to a templating effect we do expect higher yields if the hole size exactly matches the size of the catalytic particles or the diameter of the CNTs.

In the fourth step the CNTs were embedded in an insulating SiO_2 film that was spun on to the samples (OCD Tokyo OHKA) and baked for 5 min at 400°C. The oxide thickness was estimated using ellipsometry to be 70 ± 5 nm. The SiO_2 deposition was followed by a back etch step using hydrofluoric acid to partially uncover the CNTs. This was found to be necessary to avoid short circuits between the bottom and the top contacts. As shown Figure 2d, CNTs that were grown as long as several micrometers are coiled on top of the holes and are to some extent embedded in the oxide.

Finally, the individual CNTs are contacted with 20 nm palladium layer structured using e-beam lithography and lift-off.[6]

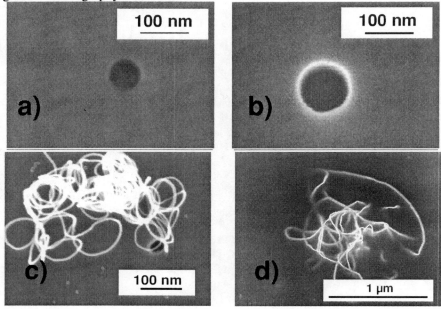

Figure 2: SEM image of a hole in resist (a) after electron beam lithography and widened holes after etching into SiO$_2$ using a hardmask of Si$_3$N$_4$ (b). Image 2c) shows an individual pristine CNT grown from a hole and the right image (2d) shows a CNT that is partially embedded in SiO2. Parts of the CNT that are overhanging from SiO2 (after HF etch) appear very bright, whereas embedded parts are more diffuse in the SEM image.

3. ELECTRICAL MEASUREMENTS

The electrical resistances were determined by two-point probe measurements on ten devices using a HP analyzer connected to a Süss Microprober. We obtained an average resistance of 30 ± 22 kΩ. The relatively high resistances are attributed to tunneling between the substrate and the nanotube at the bottom contact. A maximum current density for one of these devices was measured to be as high as 6 x 10^7 A/cm^2. A major problem of the via approach is the leakage current. To determine the leakage, reference structures without holes, respectively CNTs, were measured by two-point probe measurements. Here we obtained an average resistance of 216 ± 22 kΩ. The leakage originates either from impurities on the surface or from defects in the silicon oxide. These defects maybe generated by thermal stress during the growth process. However, the leakage current is about one order of magnitude lower than the current measured on the CNT vias.

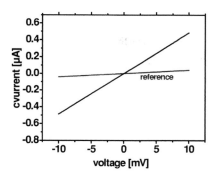 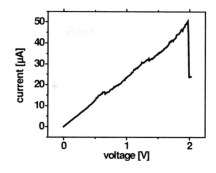

Figure 3: I-V characteristics of on an individual CNT compared to an identical stack without CNT (reference). The graph on the right shows an I-V curve at high bias. The current density for this particular sample was estimated to be as high as 6.10^7 A/cm^2.

4. CONCLUSIONS

A process flow for the parallel production of multi-walled carbon nanotube based vertical interconnects (vias) within a microelectronics compatible technology has been presented. Electrical measurements on the vias yielded resistances of around 30 kΩ. We attribute these high values to poor contact resistances and anticipate that, with an optimized combination of materials this can be significantly reduced.

5. REFERENCES

1. Kong P.G. Collins, P. Avouris: Sci. Am. **62**, (December 2000); H. Dai, E. W. Wong, C. M. Lieber: Science, **272**, 523 (1996)
2. Z. Yao, C.L. Kane, , C. Dekker: Phys. Rev. Lett. **84**, 2941 (2000); B.Q. Wei, R. Vajtai, P.M. Ajayan: Appl. Phys. Lett., **79**, 1173 (2001); Forro, A. Bachtold: Phys. Rev. Lett. **92**, 268041 (2004)
3. http://public.itrs.net/
4. G. S. Duesberg, F. Kreupl, A. Graham, M. Liebau, E. Unger, Z. Gabric, and W. Hönlein, in AIP Proceedings 633, Eds H. Kuzmany, J. Fink, M. Mehring, S. Roth, Kirchberg, 157 (2002)
5. F. Kreupl, A. P. Graham, G. S. Duesberg, W. Steinhögl, M. Liebau, E. Unger, W. Hönlein: Microelec. Eng., **64**, 399 (2002); G. S. Duesberg, A. P. Graham, M. Liebau, R. Seidel, E. Unger, F. Kreupl, W. Hönlein: Nano Lett. **3**, 237 (2003); G.S. Duesberg et al.: Diamond Relat. Mater., **13**, 354 (2004)
6. D. Mann, A. Javey, J. Kong, Q. Wang, H. Dai: Nano Lett. **3**, 1541 (2003); A. Javey, J. Guo, Q. Wang, M. Lundstrom, H. Dai: Nature **424**, 654 (2003)

Freestanding nanostructures for TEM-combined investigations of nanotubes

Jannik C. Meyer*, Dirk Obergfell*, Matthieu Paillet[†], Georg S. Duesberg**
and Siegmar Roth*

*Max-Planck Institute for solid state research, Stuttgart, Germany
[†]Universite de Montpellier II, Groupe de Dynamique des Phases Condensees, Montpellier, France
**Infineon Technologies CPR NP, Munich, Germany

Abstract. We present a method which allows to design almost arbitrary freestanding nanostructures by lithography in such a way that TEM investigations are possible in combination with various other measurements on the same carbon nanotube.

INTRODUCTION

Whenever investigations are carried out on nanometer-sized objects, it is difficult to know what the object of investigation really is. In the case of carbon nanotubes, it is often difficult to tell the difference between a bundle and a single tube, to estimate the diameter, or to know whether the tube is filled. A powerful tool to answer these questions is the Transmission Electron Microscope (TEM). However TEM investigations are not possible on objects on a bulk substrate, thus TEM and many other investigations on single nanotubes are mutually exclusive.

We present a simple approach to create well-defined nanostructures (including carbon nanotubes) which can be investigated by TEM. The idea is to create the structure close to a cleaved edge of the substrate and underetch the structure sideways. The resulting free-standing structure, on the edge or corner of the substrate, can be investigated by TEM.

The advantage of this approach is that potentially any measurment which requires a substrate can be performed before the etching step, anything that requires a free-standing tube can be done after the etching step, and a TEM investigation is possible on the very same tubes afterwards.

EXPERIMENTAL

Single-walled carbon nanotubes are absorbed from an SDS suspension or grown by CVD [1, 2] on Si substrates with a 100nm or 200nm oxide layer. A metal structure is created by e-beam lithography on top of the nanotubes. It consists of 3nm Cr and 100nm Au. In the examples shown it is a grid structure which is created on top of randomly located carbon nanotubes. The sample is cleaved through or close to the structure (Fig.

CP723, *Electronic Properties of Synthetic Nanostructures*, edited by H. Kuzmany et al.
© 2004 American Institute of Physics 0-7354-0204-3/04/$22.00

FIGURE 1. (a) Optical microscope image of a grid structure patterned onto randomly located carbon nanotubes on the corner of the substrate. (b) A freestanding grid structure after etching and critical point drying (different sample than (a)). Scale bar 10μm.

1a). It is etched in 30% aqueous KOH solution at 60°C for 7 hours. Since the Si bulk is etched much faster than the SiO_2 layer, the structure is undercut mainly from the side. As a result the grid sticks out over the edge of the substrate (Fig. 1b).

The freestanding part of the structure is investigated by TEM using a Philips CM200 microscope. The example shown in Fig. 2 contains carbon nanotubes grown on the substrate by CVD.

DISCUSSION

The described method is a simple way to create almost arbitrary freestanding structures with carbon nanotubes in such a way that they can be investigated by TEM. It results in well-defined TEM samples with all nanotubes orthogonal to the beam. It is possible to bring nanotubes that were grown on the substrate into the view of the TEM. We find that for the tubes grown by CVD on the substrate, all nanotubes are grown as individual tubes. Bundling occurs only if the density of the tubes is so high that tubes fall together. Nanotubes absorbed from an SDS suspension tend to bundle.

The aim of this work is to combine TEM with various other investigations. Work is in progress to combine TEM and transport on a single nanotube. This is especially useful for nanotubes filled with fullerenes. Preliminary results show that transport investigations are possible before the etching step, so that the bulk of the substrate can be used as back-gate. Investigating a nanotube in a transistor configuration (i.e., using the gate to change its potential) provides crucial information about its electronic structure. After the transport measurement the substrate can be etched away for the TEM investigation. Previous approaches for combining TEM investigations with transport [3, 4] could not provide a gate dependence. Using structures on thin membranes [5] one could in principle define a side-gate. However this has not been achieved yet to our knowledge, and it is a much more complex process than our approach. Furthermore, the membranes limit the resolution of the TEM image as they generate contrast on their own.

Tests show that it is also possible to access a single freestanding tube by AFM to measure its mechanical properties in correlation with e.g. its diameter. Another possibility is

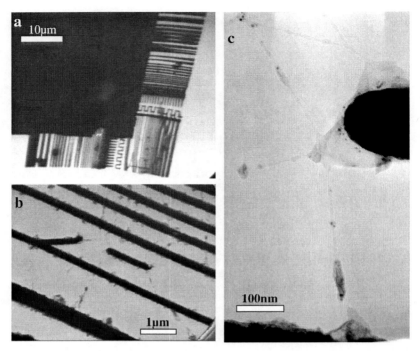

FIGURE 2. TEM images of a freestanding structure: Low-magnification overview (a), close-up on a metal particle which is suspended between just a few SWNTs (b). At higher magnification, the CVD-grown nanotubes along with catalyst particles and amorphous carbon are visible (c).

to combine Raman spectroscopy with TEM information on the same nanotube.

CONCLUSIONS AND OUTLOOK

We have developed a new and simple method which enables many types of investigations to be combined with TEM investigations. But we also envisage new types of in-situ investigations in the TEM, using well-defined structures created by lithographic methods. One example would be a TEM version of the nanoscale rotational actuator [6].

ACKNOWLEDGEMENTS

We acknowledge financial support by the BMBF project INKONAMI.

REFERENCES

1. M. Paillet and et al. 2004. to be published.
2. R. Seidel, G. S. Duesberg, E. Unger, A. P. Graham, M. Liebau, and F. Kreupl. Chemical vapor deposition growth of single-walled carbon nanotubes at 600°C and a simple growth model. *J. Phys. Chem. B*, 108, 2004.
3. A. Yu. Kasumov, R. Deblock, M. Kociak, B. Reulet, H. Bouchiat, I. I. Khodos, Yu. B. Gorbatov, V. T. Volkov, C. Journet, and M. Burghard. Supercurrents through single-walled carbon nanotubes. *Science*, 284, 1999.
4. M. Kociak, K. Suenaga, K. Hirahara, Y. Saito, T. Nakahira, and S. Iijima. Linking chiral indices and transport properties of double-walled carbon nanotubes. *Phys. Rev. Lett*, 89, 2002.
5. M. Sagnes, J-M. Broto, B. Raquet, C. Vieu, V. Conedera, P. Dubreuil, T. Ondarcuhu, Ch. Laurent, and E. Flahaut. Nanodevices for correlated electrical transport and structural investigation of individual carbon nanotubes. *Microelectronic Engineering, In Press*, 2004.
6. A. M. Fennimore, T. D. Yuzvinsky, Wei-Quiang Han, M. S. Fuhrer, J. Cumings, and A. Zettl. Rotational actuators based on carbon nanotubes. *Nature*, 424, 2003.

Controlling The Position And Morphology of Nanotubes For Device Fabrication

Emer Lahiff, Rory Leahy, Andrew I. Minett, Werner J. Blau.

Molecular Electronics and Nanotechnology Group, Department of Physics, Trinity College Dublin, Dublin 2, Ireland.

Abstract. In producing nanotube based devices as diverse as composite materials and sensing platforms, the *in-situ* growth of carbon nanotubes is most advantageous. Obtaining growth from organo-metallic catalysts pre-patterned on silicon wafers, precise structured nanotube patterns have then easily been incorporated into flexible stand-alone composites. In an alternative approach, aligned and sometimes ultra-long (>40μm) nanotubes have been obtained from catalytic growth in porous alumina membranes. Three-way (T) and now four-way (X) interconnects have been observed during the growth process, which can be incorporated into nanoscale electronic devices. Current approaches are for use as on-chip interconnects and single tube devices that can be used as the transducer in small scale bio- and chemical-sensors. In both these approaches, the density, morphology and position of the nanotubes can be controlled. This provides more precise placement of conduction channels in composites or devices, resulting in more efficient fabrication over conventional device formation.

INTRODUCTION

It is well documented that carbon nanotubes have a huge variety of potential applications. Nanotubes are of tremendous interest both in their original form and also as part of a composite material. Embedded nanotubes enhance the properties of an insulating polymer by increasing mechanical strength and conductivity [1-3]. Possible applications for nanotubes and their composites include; flat panel displays, sensors, flexible electronic devices, and actuators.

Before realizing their full potential, the issues of controllable and economic production of nanotubes and their subsequent composites, must be overcome. We report a method of nanotube production whereby we can control the morphology of the nanotubes. This production method additionally allows us to grow nanotubes with three- and four-way junctions. We also report an efficient and cost effective method of incorporating these carbon nanotube arrays into a poly(dimethylsiloxane), PDMS, polymer matrix.The presence of nanotube conduction channels through the composite was confirmed by both scanning electron microscopy and electron force microscopy.

EXPERIMENTAL DETAILS AND RESULTS

Carbon nanotubes were grown by the Chemical Vapor Deposition (CVD) method, using acetylene as the carbon source. The substrates used for nanotube growth were

CP723, *Electronic Properties of Synthetic Nanostructures,* edited by H. Kuzmany et al.
© 2004 American Institute of Physics 0-7354-0204-3/04/$22.00

prepared by soft lithography patterning as described elsewhere [4-5]. Using this method, micro-pattern feature sizes are dictated by the dimensions of the stamp and feature heights are easily controlled by varying the concentration of the catalyst solution. The catalyst used was poly(styrene-vinylferrocene). This was anionically synthesized in house. Characterization of the catalyst showed it to have an iron content of 2.1%.

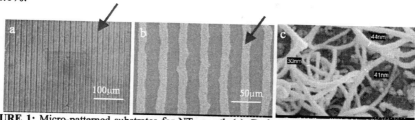

FIGURE 1: Micro-patterned substrates for NT growth (a). Darker areas, shown by the black arrow, correspond to regions of high catalyst concentration and paler areas represent the silicon oxide substrate. After CVD, carbon nanotube growth occurred only on the micro-patterned areas of the substrate (b), shown by the black arrow. (c) a higher magnification image of the nanotubes.

(Reprinted with permission from Nano Lett.; (Communication); 2003; 3(10); 1333-1337 by Lahiff, E.; Ryu, C. Y.; Curran, S.; Minett, A. I.; Blau, W. J.; Ajayan, P. M. Copyright 2003 American Chemical Society.)

By altering the CVD conditions such as temperature and deposition time we were able to control the length and diameter of tubes grown (FESEM images not shown). Also, by controlling the catalyst type and concentration, we were able to control the density of nanotube arrays produced (FESEM images not shown). By modifying the CVD conditions, structures including three- and four-way nanotube junctions were reproducibly observed.

FIGURE 2: Three-way junctions are most commonly formed however a number of 4-way junctions were also observed. Recently, through tailoring of the CVD conditions, up to 20% branched nanotubes were obtained on specially prepared substrates.

The as-grown nanotube arrays were incorporated into a poly(dimethylsiloxane), PDMS, matrix by spin coating a viscous mixture of base/curing agent onto the nanotubes. The spin speed can be used to control the thickness of the composite film. The PDMS mixture flows into areas between the nanotubes in the arrays to form a thin film composite. After curing, the PDMS composite can be easily peeled away from the substrate producing a flexi le freestanding film containing controlled nanotub morphologies.

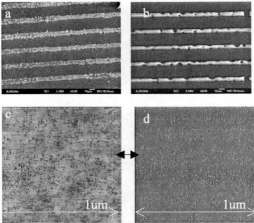

FIGURE 2: Nanotube patterns incorporated into a PDMS matrix showing (a) controlled nanotube growth on a SiO₂ substrate (b) nanotubes integrated into a polymer composite (c) an AFM topography image of the polymer composite surface and (d) an EFM image on the corresponding area shown in (c). Features are 0-10nm for AFM image (c), and 0-300mVolts for EFM image (d).

(Reprinted with permission from Nano Lett.; (Communication); 2003; 3(10); 1333-1337 by Lahiff, E.;

Ryu, C.Y.; Curran, S.; Minett, A.I.; Blau, W.J.; Ajayan, P.M. Copyright 2003, Am. Chem. Soc.)

A combination of atomic force microscopy, AFM, and electron force microscopy, EFM, confirm that the presence of nanotubes provide conduction channels through our composite film. AFM was used to obtain a height profile of the composite surface. EFM revealed the position of conduction channels through the film. It is clear that the height profile of the composite matches exactly the position of the conduction channels. The uneven height of the composite surface is due to the nanotubes protruding from the PDMS surface. Hence, the position of the nanotubes in the AFM scan matches the location of conduction paths in the EFM image.

DISCUSSION AND CONCLUSIONS

Good wetting at the polymer/nanotube interface was observed by FESEM (images not shown). This allows for better load transfer from the polymer matrix to the nanotubes. Good wetting was expected due to the hydrophobic nature of both components of the composite. Nanotubes protruding from the polymer matrix appeared well coated in PDMS, which indicates a favorable interaction between the two materials.

In conclusion, we report a scalable inexpensive technique to fabricate flexible carbon nanotube thin film composites [6]. Carbon nanotubes create conduction channels through the insulating polymer matrix. Soft lithography patterning allows for selective positioning of these conduction channels. It is the location and density of embedded nanotube networks, which determine the electrical and thermal conductivity properties of the composite material. The composite can then be used for applications, which require conductive channels within a bendable matrix. These applications include flexible electronics, electromagnetic shields and sensors to name but a few possibilities.

ACKNOWLEDGMENTS

The authors acknowledge the Irish Higher Education Authority (HEA) Enterprise Ireland and Intel Ireland for financial support.

REFERENCES

1. E.T. Thostenson, Z. Ren, T. W. Chou, "Advances in the science and technology of carbon nanotubes and their composites: a review", Compos. Sci. Technol., 61, 1899-1912, (2001).
2. P. Fournet, J.N. Coleman, B. Lahr, A. Drury, W.J. Blau, "Enhanced brightness in organic light-emitting diodes using a carbon nanotubes composite as an electron-transport layer", J. Appl. Phys., 90 (2), 969, (2001).
3. S. Curran, P.M. Ajayan, W.J. Blau, D.L. Carroll, J.N. Coleman, A.B. Dalton, A.P. Davey, A. Drury, B. McCarthy, S. Maier, A. Strevens, "A Composite from Poly(m-phenylenevinylene-co-2,5-dioctoxy-p-phenylenevinylene) and Carbon Nanotubes: A Novel Material for Molecular Optoelectronics", Adv. Mater., 10 (14), 1091, (1998).
4. Y. Xia, G.M. Whitesides, Angew. Chem. Int. Ed., 37, 550-575, (1998).
5. H. Kind, J.M. Bonard, C. Emmenegger, L. Nilsson, K. Hernadi, E. Maillard-Schaller, L. Schlapbach, L. Forro, K. Kern, Adv. Mater., 11 (15), 1285, (1999).
6. E. Lahiff, C.Y. Ryu, S. Curran, A.I. Minett, W.J. Blau, P.M. Ajayan, "Selective Positioning and Density Control of Nanotubes within a Polymer Thin Film", Nano Lett. (Communication), 3(10), 1333-1337 (2003).

Conjugated Polymeric Donor – Fullerene Type Acceptor Systems For Photoelectrochemical Energy Conversion

A. Gusenbauer[1], A. Cravino[1], G. Possamai[2], M. Maggini[2], H. Neugebauer[*1], N. S. Sariciftci[1]

[1]Linz Institute for Organic Solar Cells (LIOS),
Physical Chemistry, Johannes Kepler University Linz, 4040 Linz, Austria
[2]Department of Chemistry and ITM-CNR,
University of Padova, 35131 Padova, Italy

Abstract. The usage of fullerenes as electron mediator in photoelectrochemical cells with a conjugated polymer as electron donor and a dissolved redox couple as acceptor is described. To prevent solubility problems, the fullerenes are covalently linked to the conjugated polymer polybithiophene ("double cable"). Copolymers with different amount of fullerenes are studied in respect to light absorption and photovoltaic device properties. Compared with polybithiophene, the introduction of fullerenes in the devices enhances the photovoltaic response significantly.

INTRODUCTION

Photoelectrochemical solar cells with inorganic semiconductors, e.g. TiO_2, have been investigated since many years [1]. In the last decade, the versatility of organic compounds, in respect to enhanced light absorption and spectral sensitivity, was utilized for the construction of both inorganic/organic devices (e.g. dye sensitized inorganic semiconductor solar cells [2]) and polymeric organic semiconductor solar cells [3]. In the devices, the basic process is a charge transfer from the excited light absorber to a redox couple in an electrolyte solution. It has been shown, that the introduction of compounds with LUMO levels intermediate between the LUMO level of the absorber and the level of the redox system, e.g. fullerenes, acts as an electron mediator and may enhance the efficiency of the device [4]. In this paper, the usage of fullerenes covalently linked to a polymeric donor backbone ("double cable") is described. The energy of the LUMO level of the fullerene is inbetween the valence band edge of polybithiophene, which is used as the polymeric donor, and the energy level of a redox system (I_3^-/I^-) in the electrolyte solution. The covalent linkage prevents possible dissolution of the fullerene in the electrolyte solution, which may occur under illumination, leading to rapid deterioration of the device properties. To increase the absorption in the visible part of the spectrum, copolymers with different amounts of the fullerenic part are used. Compared with polybithiophene, the introduction of fullerenes in the devices enhances the photovoltaic response significantly.

CP723, *Electronic Properties of Synthetic Nanostructures*, edited by H. Kuzmany et al.
© 2004 American Institute of Physics 0-7354-0204-3/04/$22.00

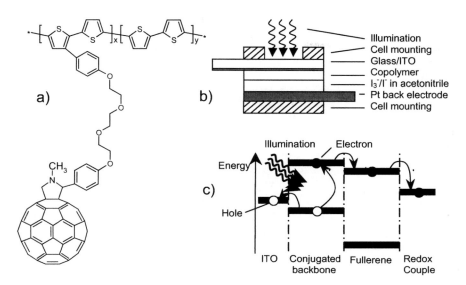

FIGURE 1. a) Structure of the copolymer of ITB-C60 and bithiophene; b) Scheme of the photoelectrochemical cell; c) Sketch of the energetics of the electron transfer.

EXPERIMENTAL

The synthesis the monomer ITB-C60 (x-component of the copolymer structure shown in Fig. 1a) has been described in a previous paper [5]. For the formation of the copolymers, the monomers ITB-C60 and bithiophene in different concentrations were dissolved in electrolyte solution (0.1 M $(C_4H_9)_4NPF_6$ in acetonitrile/toluene (30:70, v:v)), electro-copolymerized on ITO/glass substrate at 1.25 V vs. Ag/AgCl quasireference electrode and reduced to the neutral form in 0.1 M $(C_4H_9)_4NPF_6$ in acetonitrile. UV-vis spectra were recorded with an HP 8453 spectrometer. The photoelectrochemical properties were studied in a cell shown in Fig. 1b with illumination from a solar simulator (80 mW/cm^2).

RESULTS AND DISCUSSION

Figure 2 shows absorption spectra of films of polybithiophene, poly-ITB-C60, and copolymers obtained from solutions with mass concentration ratios bithiophene:ITB-C60 of 3:1, 2:1 and 1:1. As can be seen, the absorption of pure poly-ITB-C60 in the visible part of the spectrum is low. By increasing the bithiophene content in the copolymers, the relative absorption around 500 nm, which is important for solar cell applications by better matching the solar emission spectrum, increases.

In Fig. 3 current/voltage curves of devices made from polybithiophene, poly-ITB-C60 and different copolymers under solar simulator illumination are compared, and the values for the open circuit voltage (V_{oc}), short circuit current (j_{sc}), fill factor and

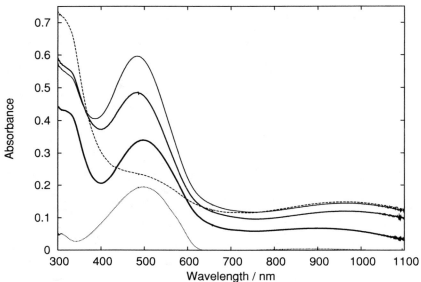

FIGURE 2. Absorption spectra of films of polybithiophene (dotted curve), poly-ITB-C60 (dashed curve), and copolymers obtained from solutions with mass concentration ratios bithiophene:ITB-C60 of 3:1 (thin solid curve), 2:1 (medium solid curve) and 1:1 (thick solid curve).

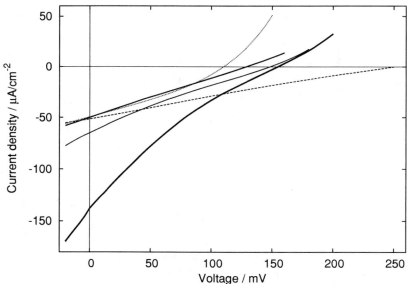

FIGURE 3. I/V characteristics under illumination of photoelectrochemical cells made with polybithiophene (dotted curve), poly-ITB-C60 (dashed curve), and copolymers obtained from solutions with mass concentration ratios bithiophene:ITB-C60 of 3:1 (thin solid curve), 2:1 (medium solid curve) and 1:1 (thick solid curve).

TABLE 1. Photovoltaic I/V characteristics

Sample (ratio)	V_{oc}/mV	j_{sc}/μAcm^{-2}	Fill factor	Efficiency %
Poly-Bithiophene	110	-49.5	0.32	0.0022
PBiTh/ITB-C60 (3 : 1)	148	-64.9	0.22	0.0026
PBiTh/ITB-C60 (2 : 1)	129	-49.1	0.25	0.0020
PBiTh/ITB-C60 (1 : 1)	155	-137.8	0.19	0.0051
Poly-ITB-C60	250	-51.1	0.23	0.0037

efficiency are compared in Table 1. Devices with pure polybithiophene show the best diode behaviour (dotted curve in Fig. 3), however with low values for V_{oc} and j_{sc}. With devices containing poly-ITB-C60 the fill factor decreases, probably related to poor morphology of the fullerene containing films. However, both the open circuit voltage and the short circuit current are generally higher. The highest open circuit voltage is reached with pure poly-ITB-C60 (dashed curve in Fig. 3). Fill factor and short circuit current are not improved compared with pure polybithiophene, which gives only a moderate improvement of the efficiency. The highest improvement is found in devices using copolymers obtained from solutions with mass concentration ratio bithiophene:ITB-C60 of 1:1 (thick solid line in Fig. 3). Especially the short circuit current is significantly improved. Even with the lowest fill factor obtained in the series measured (due to the "wrong" curvature of the I/V curve), the overall efficiency is more than doubled compared with pure polybithiophene.

The results show, that the incorporation of fullerene containing units into photoelectrochemical devices using conjugated polymers as absorbers and electron donors improve the photovoltaic properties significantly. However, the overall efficiencies in all devices are found to be only moderate. Improvements can be expected by influencing the morphology and by the usage of other redox couples with more negative redox potentials, which should give higher values of the open circuit voltage and therefore also higher efficiencies.

ACKNOWLEDGMENTS

The work was supported by the Austrian Ministry of Economic Affairs within the Christian Doppler Society's dedicated laboratory of Plastic Solar cells. M.M. wish to thank MIUR (GR. No. 2002032171) and The University of Padova (GR. No. CPDA012428) for financial support.

REFERENCES

1. Gerischer, H., and Tributsch, H., *Ber. Bunsenges. Phys. Chem.* **72**, 437 (1968).
2. O'Regan, B., and Grätzel, M., *Nature* **353**, 737 (1991).
3. Yohannes, T., Solomon, T., and Inganäs, O., *Synth. Met.* **82**, 215 (1996).
4. Kamat, P. V., Barazzouk, S., Thomasand, K. G., and Hotchandani, S., *J. Phys. Chem. B* **104**, 4014 (2000).
5. Cravino, A., Zerza, G., Neugebauer, H., Maggini, M., Bucella, S., Menna, E., Svensson, M., Andersson, M. R., Brabec, C. J., and Sariciftci, N. S., *J. Phys. Chem. B* **106**, 70 (2002).

Integration of Carbon Nanotubes with Semiconductor Technology by Epitaxial Encapsulation

J. Nygård*, A. Jensen*†, J.R. Hauptmann*, J. Sadowski** and P.E. Lindelof*

*Niels Bohr Institute and Nano-Science Center, University of Copenhagen, Universitetsparken 5, DK-2100 Copenhagen, Denmark
†Department of Physics, Technical University of Denmark
**MAX-lab, Lund University, Sweden

Abstract. Single-wall nanotubes are encapsulated in a semiconductor structure by Molecular Beam Epitaxy (MBE) of GaAlAs and GaMnAs. Microprocessing is used to form a field-effect transistor geometry, where the nanotube channel is supported on a bandgap engineered heterostructure and contacted by semiconductor electrodes. These devices behave as single-electron transistors (SET) at low temperature. The encapsulation technique can be extended to other semiconductor systems and shows promise for the integration of nanotubes in existing high-performance electronic and optoelectronic circuits.

Carbon nanotubes have appeared as promising candidates for incorporation of individual molecules in nanoscale electronics [1]. Semiconducting carbon nanotubes can act as high-performance field-effect transistors (FET) while wires from metallic tubes are capable of carrying very high currents with densities around 10^9 A/cm^2. In spite of their exceptional properties it is more likely that carbon nanotubes will be used in electronic applications if they can be smoothly integrated with existing technologies such as silicon metal oxide semiconductors (MOS) and III-V heterostructures. Surprisingly, only few experiments have addressed the aspects of integrating nanotubes in traditional micro-electronics. Recently, Tseng et al. combined single-wall nanotubes with a silicon MOSFET switching circuit by growing the tubes onto the circuit at predefined locations [2]. In most electron transport experiments nanotubes were lying on insulating gate oxides and contacted by polycrystalline thin film metal electrodes evaporated onto the tubes. In a few cases nanotubes were overgrown epitaxially by high-κ dielectrics using atomic layer deposition (ALD) [3]. To our knowledge nanotubes have not yet been encapsulated fully in epitaxially grown (semi)conducting materials.

We present here a new route for nanotube electronics, where single-wall carbon nanotubes are supported by a bandgap engineered gate barrier and contacted by epitaxially grown semiconductor electrodes [4]. We show that nanotubes can be contacted by GaAs based heterostructures and that our particular devices form single-electron transistors at low temperature. It should be noticed that devices based on GaAs play an important role in specialized high-speed electronics and optoelectronic applications. Our technique should also work for other important semiconductors which can be grown epitaxially, e.g. silicon.

CP723, *Electronic Properties of Synthetic Nanostructures*, edited by H. Kuzmany et al.
© 2004 American Institute of Physics 0-7354-0204-3/04/$22.00

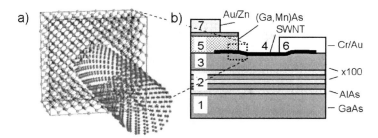

FIGURE 1. (a) The central concept of the device; a carbon nanotube is incorporated in a semiconductor crystal (zincblende structure). (b) Architecture of the fabricated devices, comprising (1) Si doped n-type GaAs backgate, (2) insulating barrier consisting of a 100 period superlattice of GaAs (2 nm) and AlAs (2 nm), (3) GaAs layer, (4) single-wall carbon nanotube, (5) GaMnAs source contact (40 nm), (6) Cr/Au drain contact (5 nm/20 nm), and (7) annealed p-type Ohmic contacts from Au/Zn.

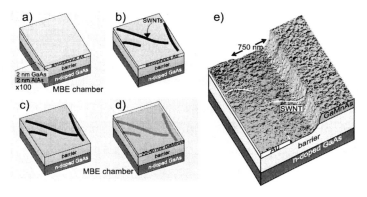

FIGURE 2. Fabrication sequence: (a) MBE growth of substrate, (b) deposition of nanotubes, (c) desorption of As cap, (d) epitaxial overgrowth, and (e) AFM image of final device where a nanotube is seen to bridge between the Au (left) and GaMnAs (right) contacts.

We utilize Molecular Beam Epitaxy to grow the nanotubes into semiconductor structures as illustrated in Fig. 1a. The device architecture is shown in Fig. 1b. The n-type GaAs substrate acts as a backgate, separated from the nanotubes by an insulating GaAs/AlAs superlattice barrier. The nanotubes are connecting two electrically separated islands made from GaMnAs and Cr/Au.

The fabrication sequence is illustrated in Fig. 2 (for details, see Ref. [4]). Firstly a superlattice barrier structure is grown by MBE on top of the backgate substrate, Fig. 2a. The superlattice is capped by a layer of amorphous As. Bundles of single-wall nanotubes are deposited under ambient conditions from a suspension in dichloroethane, Fig. 2b. The sample is then reloaded into an MBE system, where the amorphous As layer is desorbed at 400°C. Since the tubes stick to the substrate by van der Waals interactions throughout the process this leaves the tubes on the underlying clean and atomically flat GaAs surface, Fig. 2c. The substrate is now overgrown epitaxially by 40 nm of

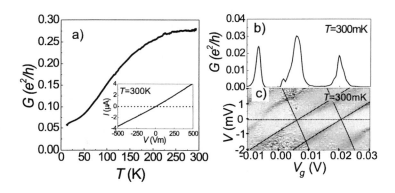

FIGURE 3. (a) Temperature dependence of the two-terminal conductance G for nanotube device with GaMnAs source and Cr/Au drain contacts. Inset: $I - V$ curves for nanotube with GaMnAs source and drain contacts (solid) and for a control sample with no nanotube (dotted) (b) G vs gate voltage V_g at 300 mK. (c) Diff. conductance dI/dV vs. V_g and bias V (dark=high dI/dV).

$Ga_{0.95}Mn_{0.05}As$ at 250°C, Fig. 2d [5]. The overgrowth is epitaxial as confirmed by reflection high energy electron diffraction. Standard UV and electron beam lithography is used to define the final device structure shown in Fig. 1b. 60 nm deep trenches in the substrate are etched by wet etch while metal contacts are applied by thermal evaporation. The process leaves the tubes as interconnects between semiconductor and metal islands which are separated by 0.5-1 μm. In the atomic force micrograph (AFM) shown in Fig. 2e a nanotube structure is seen in the trench between the contacts. The tube imprint is also visible in the surface of the evaporated polycrystalline Cr/Au contact as is usual for nanotube devices. Contrary, no imprint is seen on the GaMnAs surface. These facts confirm that the nanotubes are incorporated epitaxially by this technique without being harmed structurally. The effects of the oxidizing agents in the etchant on the conducting properties of the nanotubes remains to be investigated in detail.

In order to characterize the electrical properties of the devices we have performed transport measurements between room and sub-kelvin temperatures. At room T we find linear $I - V$ characteristics with resistances down to around 100 kΩ when applying a bias V up to 500 mV, see the $I - V$ characteristics in the inset to Fig. 3a. Control samples without nanotubes between source and drain leads carry comparatively insignificant leak currents ($I < 1$ nA). Upon cooling we observe that the conductance G is reduced as is usual for nanotube devices with tunnel contacts [6], see Fig. 3a. At the lowest temperatures G fluctuates strongly as a function of the gate voltage V_G. In some regions we find periodic Coulomb oscillations as seen in Fig. 3b, showing that the device acts as a single-electron transistor [7]. This is confirmed by the diamond patterns observed in the 2D map of the differential conductance vs. V and V_g, Fig. 3c. One can read off the single-electron charging energy $U \approx 1.5$ meV, while discrete quantum states cannot be discerned in this plot. The spacing between the contacts for this device is around 750 nm. Devices of this length would normally yield larger Coulomb and quantization energies

[6, 7]. The gate capacitance for the nanotube on GaAs device is three times greater than for nanotubes on SiO_2 due to the higher dielectric constant ($\kappa = 12$ for GaAS vs $\kappa = 4$ for SiO_2). Moreover, due to the epitaxial overgrowth the conducting part of the nanotube may extend underneath the GaMnAs contact, yielding a greater contact capacitance. An increase in the total capacitance results in a lower U.

We have shown that nanotubes can indeed be incorporated in semiconductor structures. Our technique shows promise for novel investigations and applications of nanotubes. A few examples will be given in the following. Since the tubes in our devices are lying on a (semi)conducting substrate it will be possible to perform scanning tunneling microscopy (STM) and spectroscopy on the surface and thus probe the conducting tubes in situ. This has not been possible in previous devices where the tubes were lying on insulating gate oxides, except when using specialized conducting-tip AFM instruments with poorer resolution than STM [8].

Carbon nanotubes can withstand elevated temperatures and the overgrowth is not limited to the low-T MBE employed here. Epitaxial encapsulation can be performed using a number of material systems and growth techniques, e.g. chemical vapour deposition. Moreover, a UHV compatible nanotube deposition technique was recently suggested [9]; such techniques would allow for cleaner nanotube depositions in situ without the need to remove the substrate from the semiconductor growth chamber. By overgrowing nanotubes with different semiconductor materials it will be possible to connect tubes to contacts with controlled doping levels, band gap engineered structures or materials designed in other ways. In the present study the chosen contact material, GaMnAs, is in fact a ferromagnetic semiconductor at low T. Compatibility with epitaxial growth allows for incorporation of nanotubes in more complex systems such as quantum wells, superlattices and optical resonators, where the nanotubes can act as active elements or passive interconnects. Monolithic integrated circuits can be formed by repeating the nanotube encapsulation in subsequently grown semiconductor layers. In conclusion we find reasons to expect that epitaxial encapsulation will expand the range for possible applications of nanotubes in electronics and optoelectronics.

We acknowledge fruitful discussions at the Kirchberg mini-workshops with P.M. Albrecht and C. Schönenberger. The MBE growth was carried out at III-V Nanolab, Copenhagen and at MAX-Lab, Lund. The work was supported by NEDO Spintronics and the Danish Research Councils (STVF, SNF).

REFERENCES

1. P.L. McEuen, M. Fuhrer, and H. Park, *IEEE Trans. Nanotec.* **1**, 78 (2002).
2. Y.-C. Tseng et al., *Nano Lett.* **4**, 123 (2004).
3. A. Javey et al., *Nature Mat.* **1**, 241 (2002), M. Biercuk et al., *Appl. Phys. Lett.* **83**, 2405 (2003).
4. A. Jensen et al., *Nano Lett.* **4**, 349 (2004).
5. J. Sadowski et al., *J. Vac. Sci. Tech. B* **18**, 1697 (2000).
6. J. Nygård et al., *Appl. Phys. A* **69**, 297 (1999).
7. M. Bockrath et al., *Science* **275**, 1922 (1997).
8. See, e.g., M. Freitag et al., *Phys. Rev. B* **62**, R2307 (2000).
9. P.M. Albrecht and J.W. Lyding, *Appl. Phys. Lett.* **83**, 5029 (2003).

Electrical Transport in Dy Metallofullerene Peapods

Dirk Obergfell[1], Jannik C. Meyer[1], Po-Wen Chiu[1], Shihe Yang[2], Shangfeng Yang[2], Siegmar Roth[1]

[1]*Max Planck Institute for Solid State Research, Stuttgart, Germany*
[2]*Department of Chemistry, The Hong Kong University of Science and Technology, Clear Water Bay, Kowloon, China*

Abstract. If endohedral metallofullerenes are inserted into a single-walled carbon nanotube (SWNT), a linear chain of metallofullerenes will form in the interior of the tube, the resulting structure is named *metallofullerene peapod*. C_{82} molecules each containing a single Dysprosium atom have been filled into single-walled carbon nanotubes. The success of filling was verified by transmission electron microscopy (TEM). Transport measurements in field-effect transistor configuration were performed on $(Dy@C_{82})_n@SWNT$ metallofullerene peapods. Several $(Dy@C_{82})_n@SWNT$ structures exhibit so-called *ambipolar $I_{sd}(V_g)$ characteristics* at 4.2 K.

INTRODUCTION

Numerous physical phenomena and technical aspects of electrical transport in carbon nanotubes have been investigated throughout the years since the early transport publications [1,2]. Concerning electrical properties of carbon nanotubes it was found that carbon nanotubes are ballistic conductors [3], that single-electron effects as known from quantum dot physics can appear [4,5] and that current carrying capabilities of carbon nanotubes are very high [6] in comparison to materials known so far. From the technical point of view carbon nanotubes are an attractive material since they can serve as channels of field effect transistors (FETs) with a high on-off ratio of more than 10^5 [1,8]. It has been shown that logical electrical elements can be realized by using single-walled carbon nanotube field effect transistors [7,8]. Very recently monolithic integration of SWNT-FETs with standard silicon technology was reported [9].

Meanwhile more complicated hybrid systems based on carbon nanotubes attract attention, for instance variations of electronic properties are expected when SWNTs are filled. Within this study SWNTs were filled with C_{82} fullerenes each encaging a single dysprosium atom. These Dy metallofullerenes arrange in a linear fashion in the interior of the SWNTs, the arising structure is named dysprosium metallofullerene peapod $[(Dy@C_{82})_n@SWNT]$ (Fig. 1). After the filling procedure transport measurements were carried out at 4.2 K.

CP723, *Electronic Properties of Synthetic Nanostructures*, edited by H. Kuzmany et al.
© 2004 American Institute of Physics 0-7354-0204-3/04/$22.00

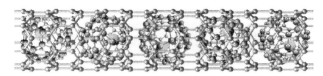

FIGURE 1. Schematic of an individual $(Dy@C_{82})_n@$ SWNT metallofullerene peapod. The SWNT is filled with a linear array of C_{82} fullerenes each encapsulating a single Dy atom.

EXPERIMENTAL

Laser-ablation SWNTs were purified by oxidisation and subsequent HCl treatment. This procedure was carried out several times. The $(Dy@C_{82})$ endohedral metallofullerenes were produced by the arc-discharge method and separated by high-performance liquid chromatography [10].

The endohedral metallofullerenes were dissolved in toluene and dropped onto a grain of purified SWNT material. After drying, the grain comprising SWNT host tubes and endohedral metallofullerenes was put into a quartz tube, which was evacuated to 10^{-6} Torr, sealed and annealed for 48 hours at 510 °C. This temperature is sufficient to sublimate the metallofullerenes which enter the nanotubes from the gas phase through opened SWNT end caps and/or defects along the SWNT walls.

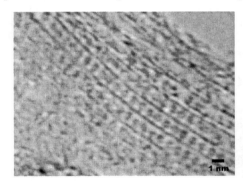

FIGURE 2. TEM micrograph of the SWNT material after the filling with $(Dy@C_{82})$ endohedral metallofullerenes (TEM: Philips CM 200, electron energy: 200 keV).

To verify the success of the filling procedure TEM micrographs of the produced $(Dy@C_{82})_n@$SWNT material were taken with a Philips CM 200 at 200 keV. In the center of Figure 2 one can see at least 3 nicely filled single-walled carbon nanotubes, the upper right and lower left corner of the TEM image are blurred due to amorphous carbon on top or below the nanotubes. Figure 2 indicates that the filling procedure generally was successful, but one has to keep in mind that TEM is a very local probe and that therefore not all the SWNTs of the sample have to be filled with endohedral metallofullerenes.

Highly doped silicon chips ($\rho \leq 6$ mΩ cm) covered with 200 nm thermally grown SiO_2 were used as substrates for the transport measurements. The bottom sides of the substrates had been metalized in order to achieve good electrical contact between the bulk Si and the bottom sides of the chips. The chips were provided with marker systems [11] by e-beam lithography so that one is able to detect the positions of the metallofullerene peapods after their deposition later on. Prior to the adsorption of the peapod materials the Si dioxide surface was silanized.

After the filling process the Dy metallofullerene material was dispersed in 1% aqueous solution of sodium dodecyl sulfate (SDS) and adsorbed on the substrates described above. The nanotubes were located by AFM making use of the mentioned marker system. Source and drain contacts were put on each selected tube by standard electron beam lithography. The spacing between source and drain contacts is approximately 300 nm, they are made from AuPd with a height of about 30 nm. The schematic in Figure 3 displays a contacted single metallofullerene peapod in field-effect transistor (FET) configuration on top of a substrate.

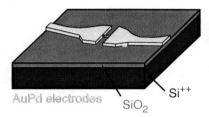

FIGURE 3. Schematic drawing of a metallofullerene peapod in field-effect transistor (FET) configuration. The metallofullerene peapod is contacted with two metal electrodes, namely the source and drain contact, each made from AuPd (height approx. 30 nm). Source and drain are approximately 300 nm apart. The highly doped silicon functions as back gate.

To determine the electronic properties of a contacted metallofullerene peapod, the dependences of the current I_{sd} through the peapod on the bias voltage V_{sd} and the gate voltage V_g are measured. Thereby only one voltage is swept at a time, the other being held constant. The gate voltage was swept within the limits of ± 20 V. The electronic measurements were carried out at 4.2 K.

RESULTS

Figure 4 displays the output characteristics (Fig. 4a) and the transfer characteristcs (Fig. 4b) of a $(Dy@C_{82})_n@SWNT$ metallofullerene peapod at 4.2 K. We estimated the diameter of the metallofullerene peapod by AFM to be ≈ 1.5 nm. The dependence of the current I_{sd} on the bias voltage V_{sd} is non-linear (Fig. 4a, inset), for small values of V_{sd} a differential resistance of approximately 850 kΩ is retrieved ($V_g = 0$ V). For large absolute values of the gate voltage V_g, e.g. $V_g = -20$ V or $V_g = +20$ V conductance is substantially higher than for $V_g = 0$ V (Fig. 4a). The sample exhibits *ambipolar transport behaviour* (Fig. 4b), i.e., there is a regime of intermediate values of the gate voltage, in which the current is suppressed, whereas the current I_{sd} through the sample is large for large absolute values of the gate voltage V_g. That means that by varying the gate voltage one can shift the band structure of the sample in such a way, that both hole and electron conduction is possible. Consequently the interval of suppressed current can be taken as the band gap of the sample. Qualitatively similar results have been received with several other $(Dy@C_{82})_n@SWNT$ samples.

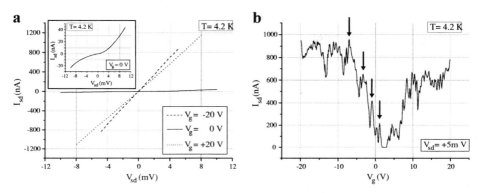

FIGURE 4. Transport data of a potentially single (Dy@C$_{82}$)$_n$@SWNT metallofullerene peapod taken at T = 4.2K. **(a)** Output characteristics I$_{sd}$(V$_{sd}$) for V$_g$ = -20 V, V$_g$ = 0 V and V$_g$ = +20 V. Inset: enlargement of the I$_{sd}$(V$_{sd}$) graph for V$_g$ = 0 V. **(b)** Transfer characteristic I$_{sd}$(V$_g$) for V$_{sd}$ = +5 mV and -20 V ≤ V$_g$ ≤ 20 V. The humps indicated by the arrows could correspond to the van Hove singularities in the density of states (DOS) of a carbon nanotube as a 1D system [14].

DISCUSSION

Lee et al. performed Scanning Tunneling Spectroscopy (STS) on a single (Gd@C$_{82}$)$_n$@SWNT metallofullerene peapod [12] and found that the bandgap is narrowed at the sites where the (Gd@C$_{82}$) metallofullerenes are located. The bandgap narrowing in metallofullerene peapods makes ambipolar transport behaviour likely, as reported in [13]. This bandgap modulation, which is due to insertion of metallofullerenes, can be the reason for the ambipolar transport behaviour observed within our work.

A further reason for the observation of ambipolar transport behaviour, i.e. for being able to get access to both electron and hole conduction by sweeping the gate voltage to moderate values, can be a large gate coupling factor α. The gate coupling factor α is defined as $\alpha = C_g/C_\Sigma$, where C_g is the capacity between the nanotube and the gate electrode and C_Σ the total capacity of the carbon nanotube with respect to gate, source and drain electrode. Thus α depends on a number of parameters, e.g. the gate oxide thickness, the dielectric constant of the gate oxide, the nanotube diameter and the coupling of the nanotube to source and drain contacts, which can be different from sample to sample. Knowing α, the bandgap E$_g$ can be obtained by E$_g$ = α ΔV$_g$ e, where ΔV$_g$ is the width of the gate voltage interval, in which the current is suppressed. There are publications reporting on ambipolar transport in samples of unfilled SWNTs with a large value of α ($\alpha \approx 0.2$) [14]. We could not determine a value for α within this study since Coulomb blockade behaviour was not observed for the samples investigated and α usually is retrieved from the Coulomb blockade diamond diagram.

Another explanation for the ambipolar transport behaviour can be that the SWNTs investigated are "nearly metallic" SWNTs with curvature-induced small bandgaps [15,16]. Also, the nanotube-contact interfaces may play an important role [17,18].

CONCLUSION

Several nanotubes, potentially filled by ($Dy@C_{82}$) fullerenes, exhibited ambipolar transport behaviour at 4.2 K. The origin of the ambipolar behaviour is assumed to be bandgap narrowing due to the insertion of the metallofullerenes, but there are other possibilities which have to be considered. In order to evaluate the transport behaviour of the metallofullerene peapod samples with higher precision, TEM micrographs of the very same peapod samples would be highly desirable [19]. These would allow to judge whether samples consist of a small bundle of SWNTs or of single SWNTs and to check whether the SWNTs are filled with metallofullerenes indeed. Currently we aim at performing TEM and transport measurements including transfer characteristics on the very same metallofullerene peapod.

ACKNOWLEDGEMENTS

The authors acknowledge financial support by the BMBF project INKONAMI. Furthermore we thank B. Maile (xlith GmbH) for producing the marker systems.

REFERENCES

1. Martel, R., Schmidt, T., Shea, H. R., Hertel, T., and Avouris, Ph., *Appl. Phys. Letters* **73**, 2447-2449 (1998).
2. Tans, S. J., Verschueren, A. R. M., and Dekker, C., *Nature* **393**, 49-52 (1998).
3. Frank, S., Poncharal, P., Wang, Z. L., and de Heer, W. A., *Science* **280**, 1744-1746 (1998).
4. Park, J., and McEuen, P. L., *Appl. Phys. Letters* **79**, 1363-1365 (2001).
5. Nygård, J., Cobden, D. H., Bockrath, M., McEuen, P. L., and Lindelof, P. E., *Appl. Phys. A* **69**, 297-304 (1999).
6. Wei, B. Q., Vajtai, R., and Ajayan, P. M., *Appl. Phys. Letters* **79**, 1172-1174 (2001).
7. Derycke, V., Martel, R., Appenzeller, J., and Avouris, Ph., *Nano Letters* **1**, 453-456 (2001).
8. Bachtold, A., Hadley, P., Nakanishi, T., and Dekker, C., *Science* **294**, 1317-1320 (2001).
9. Tseng, Y.-C., Xuan, P., Javey, A., Malloy, R., Wang, Q., Bokor, J., and Dai, H., *Nano Letters* **4**, 123-127 (2004).
10. Huang, H., Yang, Shi., and Zhang, X., *J. Phys. Chem. B* **104**, 1473-1482 (2000).
11. Kim, G.-T., Waizmann, U., and Roth, S., *Appl. Phys. Letters* **79**, 3497-3499 (2001).
12. Lee, J., Kim, H., Kahng, S.-J., Kim, G., Son, Y.-W., Ihm, J., Kato, H., Wang, Z. W., Okazaki, T., Shinohara, H., and Kuk, Y., *Nature* **415**, 1005-1008 (2002).
13. Shimada, T., Okazaki, T., Taniguchi, R., Sugai, T., Shinohara, H., Suenaga, K., Ohno, Y., Mizuno, S., Kishimoto, S., and Mizutani, T., *Appl. Phys. Letters* **81**, 4067-4069 (2002).
14. Babić, B., Iqbal, M., and Schoenenberger, C., *Nanotechnology* **14**, 327-331 (2003).
15. Kane, C. L., and Mele, E. J., *Phys. Rev. Letters* **78**, 1932-1935 (1997).
16. Zhou, C., Kong, J., and Dai, H., *Phys. Rev. Letters* **84**, 5604-5607 (2000).
17. Martel, R., Derycke, V., Lavoie, C., Appenzeller, J., Chan, K. K., Tersoff, J., and Avouris, Ph., *Phys. Rev. Letters* **87**, 256805 (2001).
18. Heinze, S., Tersoff, J., Martel, R., Derycke, V., Appenzeller, J., and Avouris, Ph., *Phys. Rev. Letters* **89**, 106801 (2002).
19. Meyer, J. C., Obergfell, D., Paillet, M., Duesberg, G. S., and Roth, S., "Freestanding nanostructures for TEM-combined investigations of nanotubes" in *Electronic Properties of Synthetic Nanostructures*, edited by H. Kuzmany, J. Fink, M. Mehring and S. Roth, AIP conference proceedings, 2004, in press.

Simultaneous Deposition of Individual Single-Walled Carbon Nanotubes onto Microelectrodes via AC-Dielectrophoresis

M. Oron[1,2], R. Krupke[1], F. Hennrich[1], H. B. Weber[1], D. Beckmann[1],
H. v. Löhneysen[2,4] and M. M. Kappes[1,3]

[1] Forschungszentrum Karlsruhe, Institut für Nanotechnology, D-76021 Karlsruhe, Germany
[2] Physikalisches Institut, Universität Karlsruhe, D-76128 Karlsruhe, Germany
[3] Institut für Physikalische Chemie, Universität Karlsruhe, D-76128 Karlsruhe, Germany
[4] Forschungszentrum Karlsruhe, Institut für Festkörperphysik, D-76021 Karlsruhe, Germany

Abstract. Based on AC-dielectrophoresis, we succeeded to deposit simultaneously individual single-walled carbon nanotubes (SWNTs). Using only two bond-wires, the individual tubes are site-selectively deposited onto many pairs of sub-micron electrodes from a surfactant stabilized tube suspension. The presented method goes well beyond previous efforts, in which single bundles of SWNTs were deposited [1-2], and is therefore a step towards a fast deposition and characterization of individual SWNTs. Atomic force microscopy and scanning electron microscopy techniques were used to demonstrate the presence of single tubes trapped between electrode pairs.

INTRODUCTION

The remarkable electrical properties of carbon nanotubes make them ideal candidates for molecular electronic devices and thus attractive for both fundamental science and technological applications. When attempting to use single-wall carbon nanotubes for electronic devices on a pre-patterned microelectrode array, one encounters difficulties in controlling the alignment and the contact of the tubes to the pre-patterned electrodes. Recently, AC-dielectrophoresis was demonstrated as a useful tool to deposit and align single bundles of SWNTs [1-3].

In the present work we demonstrate a fast and site-selective deposition of *individual* SWNTs from a surfactant stabilized tube suspension by using dielectrophoresis.

Here we focus on scanning electron microscopy (SEM) and atomic force microscopy (AFM) techniques to detect the presence of a single tube connecting between pairs of electrodes. Transport measurements on these individual tubes with correlation to local Raman Spectroscopy are currently under way and will be published elsewhere.

CP723, *Electronic Properties of Synthetic Nanostructures*, edited by H. Kuzmany et al.
© 2004 American Institute of Physics 0-7354-0204-3/04/$22.00

EXPERIMENTAL

Single wall carbon nanotubes were synthesized using the pulsed laser vaporization (PLV) technique. As-grown SWNTs (10 mg) were suspended in 10 ml D_2O with 1 weight-% sodium dodecylbenzene sulfonate (NaDDBS). The suspension was further sonicated, centrifuged and diluted to a concentration of a few ng/ml of individual SWNTs.

An array of sub-micron electrodes was prepared on a thermally oxidized silicon substrates using standard electron-beam lithography and lift-off technique. Prior to tube deposition, the structure was bonded to a chip carrier and wired with two bonds wires to a frequency generator as schematically shown in figure 1. This setup was shown to be efficient in depositing simultaneously tubes on many pairs of electrodes, due to capacitive coupling between the floating electrodes and the ground [1].

A drop of the nanotube suspension (100 µl) was applied to the chip via a pipette immediately after switching on the generator. A typical time for deposition was a few minutes, followed by methanol rinsing, drying the surface with a stream of nitrogen gas and then turning off the generator. Finally, the sample was annealed for 2 hours at 200°C in air and subjected to SEM and AFM characterization.

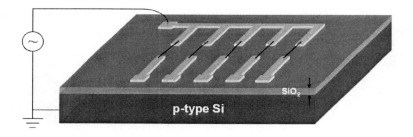

FIGURE 1. Electrode wiring scheme (not to scale). The electrodes are ~30 nm thick, 200 nm wide and have a gap of 1 µm on a p-type silicon substrate with 600 nm thermally oxidized SiO_2. Gold was used as the top electrode material (~30 nm) with titanium as an adhesion layer (~2-3 nm). Black lines represent tubes after deposition.

RESULTS AND DISCUSSION

Figure 2 shows two examples of a single tube connected between a pair of electrodes, using the experimental setup described above. The deposition was performed with an applied voltage of $V_{rms} = 2$ V at a frequency of $f = 300$ KHz for a duration of 6 minutes. The individual tubes were then trapped between the electrodes with precise and reproducible alignment. This effect has been shown before on bundles of SWNTs [2] and in analogy to the former case, we ascribe the movement of individual tubes to the interaction of the external electric field with the field-induced dipole moment of the tube. The tubes align along the electric field lines and are carried along the field gradient towards the area between the electrodes. As a consequence, the deposited tube bridges the electrode pair as a straight line.

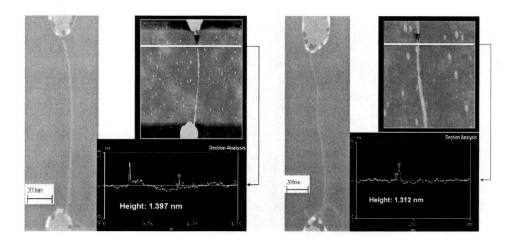

FIGURE 2. SEM (left image in each panel) and AFM images of tubes trapped between electrodes, using dielectrophoresis. The height of the tube in the left and the right panel is shown to be ~1.4 nm and ~ 1.3 nm, respectively, which implies the presence of an individual tube. The PLV tubes used in these experiments are known to be 1.2 - 1.4 nm in diameter. Dots in the AFM topography images reveal residual surfactant molecules, on and in the vicinity of the tube.

Careful inspection of the AFM images reveals the presence of residual surfactant molecules, on and in the vicinity of the tubes. The thickness varies abruptly along the tube, an observable fact that can not be attributed to small tube segments that bundled along the tube. Moreover, this phenomenon was not observed with tube deposited from N, N, dimethylformamide (DMF) solution. Hereafter, we attribute the dots in the vicinity of the tube to be residual surfactant molecules as well. These residual molecules were still present after rinsing with methanol and annealing the sample. We anticipate a challenge to remove the surfactant molecules entirely.

We performed the AFM characterization in the tapping mode, which turned out to be unusually sensitive to the oscillation amplitudes of the tip and the tip to surface distance. We attribute this to residual surfactant molecules on the tube surface, and their influence on the water layer formation in air. As a consequence the tip may sample two different interfaces; one between tube and water and another between water and air. This difficulty was circumvented by moving the tip close to the surface and tapping at small tip oscillation amplitudes, which apparently favors sensitivity to the tube rather than to the water layer (see Figure 3). This observation was not seen in the case of tubes in DMF whereas in the case of surfactant suspensions, the surfactant apparently enhances water adsorption from air, which results in a significantly thicker water layer on surface relative to a non surfactant solution.

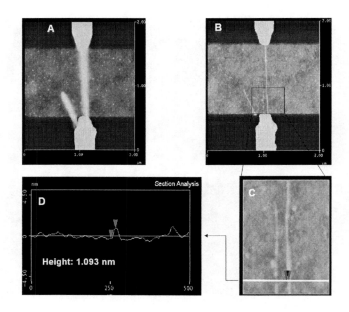

FIGURE 3. AFM images of a single tube taken at different tip amplitudes and distances from the surface. The scan displayed in panel A, which was taken with a tip being more distant from the surface than in panel B, resulted in a misleading topography. Panel C shows a higher magnification of the squared area shown in panel B. The line scan in panel D indicates the height of the tube from surface, ~1 nm, implying the presence of a single tube. Dots at the AFM topography images reveal residual surfactant molecules on and in the vicinity of the tube.

SUMMARY

We have demonstrated the ability to site-selectively deposit individual tubes via the AC-dielectrophoresis technique. AFM and SEM images demonstrate the presence of single tubes trapped between the electrodes. Investigations on the influence of residual surfactant molecules on the contact resistance as well as correlations between transport measurements, local Raman spectroscopy and AFM characterization are currently in progress.

REFERENCES

1. Krupke R., Hennrich F., Weber H. B., Beckmann D., Hampe O., Malik S., Kappes M. M., Löhneysen H. V., *Nano Lett.*, **3**, 1019-1023 (2003).
2. Krupke R., Hennrich F., Weber H. B., Kappes M. M., Löhneysen H. V., *Appl. Phys. A* **76**, 397-400 (2003).
3. Chen X. Q., Saito T., Yamada H., Matsushige K., *Appl. Phys. Lett.*, **78**, 3714-3716 (2001).

Carbon Nanotubes: Can they become a microelectronics technology?

Wolfgang Hoenlein, Franz Kreupl, Georg S. Duesberg, Andrew P. Graham, Maik Liebau, Robert Seidel, Eugen Unger,

Infineon Technologies AG, Corporate Research, 81730 Munich, Germany

Abstract. Carbon nanotubes (CNTs) have a large variety of properties that make them attractive for applications in microelectronics. A comparison of carbon nanotube field-effect transistors with silicon MOSFETs shows that CNT devices outperform state-of-the-art silicon transistors. A silicon technology review gives the benchmark for an assessment of the current CNT technology and identifies the growth and placement procedures that are not yet sufficient for industrial applications. Finally, the vertical CNT transistor concept is introduced which deals with the technological problems and opens a new route for 3D integration.

INTRODUCTION

Carbon nanotubes (CNTs) can be fabricated in both metallic and semiconducting configurations [1] in which the current transport is ballistic giving rise to mobilities much higher than in silicon. Further, the absence of scattering in the perfect crystal-like one dimensional structure of the CNTs allows current densities well in excess of metals [2]. Due to their wire-like structure, carbon nanotubes can be used to fabricate transistors as well as interconnects, both of the basic building blocks of a microelectronics technology. In this context we discuss the possible applications of single-walled carbon nanotubes (SWCNTs) as transistors. Most of the conclusions, however, may also be adopted for the case of interconnects if semiconducting SWCNTs are replaced by metallic multi-walled nanotubes (MWCNTs).

COMPARISON OF SWCNT FIELD-EFFECT TRANSISTORS WITH SILICON MOSFETS

Since their first appearance CNT field-effect devices have made tremendous improvements with respect to the most important parameters used to characterize silicon MOSFETs. In Table 1 state-of-the-art silicon devices are compared with CNT-FETs that have been published recently [3]. It is evident that carbon nanotube transistors are comparable or even superior in all individual parameters except for the gate length. The fabrication of very short transistors, however, is not expected to be a roadblock. It can also be seen from Table 1 that tremendous achievements have been made in the last few years indicating that CNTFET development is still in its early

CP723, *Electronic Properties of Synthetic Nanostructures*, edited by H. Kuzmany et al.
© 2004 American Institute of Physics 0-7354-0204-3/04/$22.00

stages. It should be also noted that CNTFET devices are mostly p-type after growth but can be transformed to n-type behavior by annealing in vacuum or doping [4]. Unfortunately, not much data has been published on n-type devices up to now. It is also important to note that all of the parameters for CNTFETs have been virtually upscaled, which means that the objects of investigations were CNTFETs consisting of single SWNTs. In order to compare them to the much larger silicon MOSFETs, a number of identical, equally spaced CNTs that are separated by the nanotube diameter and form a grid completely filling up the space between source and drain contacts is assumed. However, no such arrangement has been realized up to now. Moreover, even for single nanotubes, no placement method is available that achieves reproducible deposition of single CNTs. In order to benchmark the CNT technology more in detail we have to identify the most prominent features of the state-of-the-art silicon technology.

Table 1. Comparison of state-of-the-art CNTFETs and silicon MOSFETs [3].

	p-CNT FET 260 nm (1 V) Wind (2002)	p-CNT FET 1.4 μm (1 V) Rosenblatt (2002)	p-CNT FET 3 μm (1.2 V) Javey (2002)	p-CNT FET 300 nm (0.6 V) Javey (2003)	p-CNT FET 200 nm (1V) Seidel (2003)	MOSFET 100 nm (1.5V) Ghani (1999)	FinFET 10 nm (1.2V) Yu (2002)
Drive Current Ids (mA/μm)	2.142	2.99	3.5	14	4-14	1.04 nFET 0.46 pFET	0.450 nFET 0.360 pFET
Transconductance (μS/μm)	2284	6666	6000	1470	260	1000 nFET 460 pFET	500 nFET 450 pFET
Subthreshold slope (mV/dec)	130	80	70	150-170	700-1100	90	125 101
On-Resistance (Ohm /μm)	660	360	342	17- 54	71-250	1442 nFET 3260 pFET	2653 nFET 3333 pFET
Gate-length (nm)	260	1400	2000	300	150 - 200	130	10
Normalized gate-oxide (1/nm)	4/15= 0.26	80/1 = 80	25/8 = 3.12	4/67= 0.059	4/200 = 0.020	4/2 = 2	4/1.7 = 2.35
Mobility (cm²/(Vs))	--	1500	3000	4000	--	--	--
I_off (nA/μm)	7	--	1	0.5	1	3	10

SUCCESS FACTORS OF SILICON TECHNOLOGY

Since the advent of large scale integration in the early seventies, silicon technology has pushed the limits of functionality per chip area to dizzying heights. The key to this successful development was the steady miniaturization of transistors and interconnects without changing the basic processing sequence of depositing a layer, covering it with a resist, patterning with lithography and etching the lithographically defined structure into the underlying layer. Due to highly parallel processing steps many identical devices can be fabricated at the same time. The basic element of this technology is the semiconducting field-effect transistor made on the surface of a single crystal silicon wafer taking advantage of the perfect Si-SiO$_2$ interface. Thus, it was not the material with the best mobility (e.g. GaAs) that made it to a mature technology, but the best overall package with respect to integration.

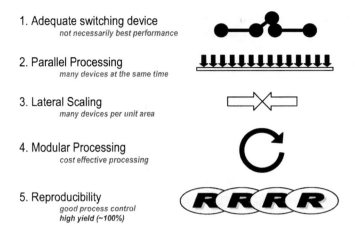

1. Adequate switching device
   not necessarily best performance

2. Parallel Processing
   many devices at the same time

3. Lateral Scaling
   many devices per unit area

4. Modular Processing
   cost effective processing

5. Reproducibility
   good process control
   high yield (~100%)

Figure 1. The key success factors of silicon technology

Finally, tremendous processing control and reproducibility have to be achieved in order to obtain reasonable yields at the end of a highly complex processing sequence. Figure1 summarizes the success factors of silicon technology. We will now address the growth and deposition methods for carbon nanotubes and assess them with respect to the silicon technology requirements.

CARBON NANOTUBE GROWTH

Basically two methods for fabrication of carbon nanotubes are known. First, nanotubes can be produced externally in a reactor by evaporation of carbon electrodes in an arc discharge [5] or by laser ablation [6] of a carbon containing target in a furnace, both operating at elevated temperatures. The soot in the reaction tube is then collected and dissolved in a suitable solvent. After cleaning procedures that reduce the amount of unwanted amorphous carbon, the nanotubes have to be sorted with respect to semiconducting and metallic properties. Some procedures for this separation have been described in the literature [7] but no comparable yield data is available. For applications in microelectronics it is important that with *ex situ* production methods it is possible to sort the nanotubes with respect to length, diameter and conductivity type. It is also very important to increase the purity of the solutions: For microelectronic applications the yield from liquid deposition has to be close to 100% with respect to conductivity type. For the diameter dependent energy gap the distribution should be approx. 1 eV +/- 10%. So far no solution with these values has been produced.

The second method is the *in situ* growth of CNTs using catalysts [8]. The catalyst is deposited as a thin layer, e.g. by evaporation and treated in a way that it breaks up into small clusters that correspond to the size of the nanotubes required. The catalyst material is typically derived from the iron group elements (Fe, Ni, Co) or alloys of these elements. The CNTs are then grown from the catalyst particles in CVD-type process using a carbon containing feedstock gas (e.g. methane or acetylene). Preferential growth of multi-walled and single-walled nanotubes can be achieved by choosing the proper particle size, composition and the growth temperature. However, the high yield production of a certain species of defined carbon nanotubes (semiconducting with defined energy gap, metallic) has not been achieved up to now. Obviously, the control of the catalyst size, morphology and processing conditions is the key to a reproducible high yield production of CNTs. For microelectronic applications the relevant investigations of the catalyst mediated deposition yield that have been published recently are summarized in Table 2 [9]. Considerable progress has been made but no thorough understanding of the growth process has been reported and the yields are still not sufficient for microelectronics.

Table 2. Yield of semiconducting CNTs from different deposition techniques (after ref. [9])

Laser Ablation	$30 \pm 6\%$
HiPCO	$61 \pm 7,6\%$
CVD	62,5%
PECVD	$89,3 \pm 2,3\%$

CARBON NANOTUBE PLACEMENT

The exact placement of the nanotubes is an essential step in building a nanotube technology. As mentioned previously, only parallel placement methods will be acceptable for microelectronics. Thus, we have to exclude serial nanotube processing methods, such as manipulating nanotubes with AFM or similar tools in a sequential manner. Rather we expect that methods similar to optical lithography will be used for patterning of nanostructures in the future.

For the deposition out of the liquid phase we are faced with the problem that a number of nanotubes are deposited in a random way within the window defined by the minimum lithographic length. Given that the cleaning and sorting procedures have supplied us with the right species we are now confronted with the fact that the diameter of the nanotubes is much smaller than lithographic dimensions, even in ten years from now. Thus, the number of nanotubes that form a given device will not be defined. One way to achieve this is to encapsulate nanotubes by a spacer that keeps them at a certain distance to each other, preferably in the range of their diameter. In this way the gate field can protrude through the asperities giving sufficient gate control. Spacers can be made of large molecules attached to the CNT scaffold or molecules wrapped around the CNTs like DNA [10]. Further, self-organized alignment of the tubes would help to form a grid in a way schematically shown in Figure 2. If the number of tubes that fit into a minimum lithographic length is not too large, this technique could also control the number of tubes per device. The lack of control over the individual CNT placing procedure, however, renders deposition from the liquid phase less suited to the needs of microelectronics.

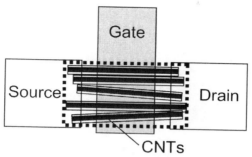

Figure 2. Schematic representation of a planar CNTFET made by liquid phase deposition of coated carbon nanotubes. Coating acts as a spacer to keep the tubes away from each other. The dashed line represents the window into which the nanotubes have to be deposited. If the window dimensions are small enough a reproducible number of tubes can be placed.

Catalyst-mediated deposition relies on the availability of properly deposited catalyst particles. If the catalyst is deposited in the form of a thin layer, lithography can be applied to pattern the catalyst layer in order to define the places where the nanotubes grow.

Figure 3. Selective CVD-growth of many entangled multi-walled CNTs on top of a patterned catalyst layer (not visible). The contour of the CNT-blocks can be maintained to even larger aspect ratios.

In Figure 3 we show an example where the selective growth of a large number of multi-walled nanotubes from such patterned catalyst regions has been obtained [11].

Due to entanglement of the individual tubes, the contour of the catalyst layer is preserved even if the height of the blocks exceeds the base length. In Figure 4 a much smaller number of tubes was grown in a template that supports the tube structure. Beyond the supporting structure no entanglement is visible. This structure is the basic element of a nanotube via, where nanotubes are used to connect two different metallic layers [12]. For this application the targeted growth of only metallic nanotubes is not necessary, because the multi-walled tubes always contain metallic shells, if the number of shells is large enough.

Figure 4. Catalyst mediated CVD-grown MWCNTs in a SiO$_2$ template on top of a metal layer. The catalyst layer has been deposited only at the bottom of the hole. This structure can be used as a via between two metal layers [12].

Figure 5 shows the template growth of a single multi-walled tube, which fits to the diameter of a nanohole [13]. It can be deduced that control of the template and of the catalyst diameter enables a certain degree of control over the growth of the nanotube. By choosing a smaller diameter of the nanohole it should also be possible to grow SWCNTs which may be exploited for transistor devices. Figure 6 shows the growth of a single SWCNT out of a nanohole that is still too wide [14]. Most probably the hole is V-shaped and the diameter shrinks at the bottom to form a smaller catalyst particle.

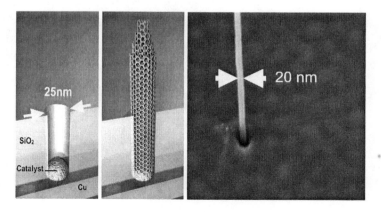

Figure 5. Catalyst mediated CVD-growth of a single MWCNT schematically (left) and from a 25nm nanohole in SiO$_2$ (right). The tube diameter fits to the hole diameter.

Template growth thus opens a route for exploiting microelectronics patterning procedures even for the control of single SWCNTs. However, 100% yield for the desired species at the desired places has also not yet been achieved with template-type CVD-growth.

Figure 6. Catalyst mediated CVD-growth of a single SWCNT from a template nanohole in SiO_2. The tube is vibrating in the electron beam of the SEM.

THE VERTICAL CNT FIELD EFFECT TRANSISTOR

Under the assumption of complete control over the catalyst mediated CVD process we proposed a concept for a vertical CNT Field-Effect Transistor (VCNTFET) [15]. In Figure 7 (left) a structure is shown where the nanotube is grown in a template that has an additional third gate electrode separated from the tube by a thin gate dielectric. The gate now surrounds the nanotube completely giving optimum gate control. Moreover, this device has the smallest footprint conceivable, as source and drain electrodes are situated on top of each other. Gate width and distance to source and drain electrodes are no longer defined by lithography but by the thickness of deposited layers which can be controlled very easily down to nanometer dimensions.

Figure 7. Schematic representation of a vertical CNTFET (VCNTFET) (left). Source and drain electrodes are on top of each other. The gate electrode surrounds the tube and is separated by a thin dielectric. Single VCNTFET modules can be built in parallel to deliver higher output currents (right).

The VCNTFET concept combines well known interconnect techniques like the damascene process with the free placement in space for transistors which are normally bound to the surface of the silicon wafer in the state-of-the-art microelectronics approach. Thus, modular 3D structures can be developed opening a new route for large scale integration. Furthermore, by placing many identical VCNTFETs close to each other as shown in Figure 7 (right), transistors for different output currents can be designed. Prerequisite for such devices is the feasibility of sub-lithographic structures that can only be created by self-organization. An example would be the self-adjustment of nanoholes in a PMMA resist that can be used as etching mask for a template [16].

CONCLUSIONS

CNTFET development has made tremendous advances. CNFET devices consisting of single SWCNTs have properties that outperform silicon MOSFETs. However, growth and placement techniques, when compared with the silicon technology, are not yet able to fulfill the requirements of a large scale integration technology. Methods have to be developed that deliver nanotubes with defined properties at a yield very close to 100%. Furthermore, placement procedures have to be refined to cope with the requirements of the microelectronics technology, e.g. parallel processing and reproducibility on the nanometer scale. For the catalyst mediated deposition the concept of vertical CNTFET (VCNTFET) is proposed that enables yet unattained packing densities and modularity for 3D architectures. The formation of the sub-lithographic template structures necessary for the VCNTFET concept can be achieved by self-organization.

ACKNOWLEDGMENTS

This work has been supported by the German Ministry of Science and Technology (BMBF) under Contract No. 13N8402.

REFERENCES

1. R. Saito, G. Dresselhaus, and M. S. Dresselhaus, Physical Properties of Carbon Nanotubes, *Imperial College Press* (1998).
2. B. Q. Wei, R. Vajtai, and P. M. Ajayan, *Appl. Phys. Lett.*, **79** (8), 1172 (2001).
3. S. J. Wind, J. Appenzeller, R. Martel, V. Derycke, and Ph. Avouris, *Appl. Phys. Lett.*, **80**, 3817 (2002).
   S. Rosenblatt, Y. Yaish, J. Park, J. Gore, V. Sazonova, and P.L. McEuen, *Nanoletters*, **2**, 869 (2002).
   A. Javay, H. Kim, M. Brink, Q. Wang, A. Ural, J. Guo, P. McIntyre, P. McEuen, M. Lundstrom, and H. Dai, *Nature Materials*, Vol. **1**, No. 4, 241-246 (2002).
   A. Javay, J. Guo, Q. Wang, M. Lundstrom, and H. Dai, *Nature* **424**, 654-657 (2003).
   R. Seidel, M. Liebau, G. S. Duesberg, F. Kreupl, E. Unger, A. P. Graham, W. Hoenlein, and W. Pompe, *Nanoletters* **3**, 965-968 (2003).

T. Ghani, S. Ahmed, P. Aminzadeh, J. Bielefeld, P. Charvat, C. Chu, M. Harper, P. Jacob, C. Jan, J. Kavalieros, C. Kenyon, R. Nagisetty, P. Packan, J. Sebastian, M. Taylor, J. Tsai, S Tyagi, S. Yang, and M. Bohr, *IEDM Tech. Dig.*, 415 (2002).

B. Yu, L. Chang, S. Ahmed, H. Wang, S. Bell, C. Yang, C. Tabery, C. Ho, Q. Xiang, T. King, J. Bokor, C. Hu, M. Lin D. Kyser, *IEDM Tech. Dig.*, 251 (2002).

4. V. Derycke, R. Martel, J. Appenzeller, and P. Avouris, *Appl. Phys. Lett.*, **80** (15), 2773-2775 (2002).
5. C. Journet, W. K. Maser, P. Bernier, A. Loiseau, M. L. delaChapelle, S. Lefrant, P. Deniard, R. Lee, and J. E. Fischer, *Nature* **388**, 756-758 (1997).
6. A. Thess, R. Lee, P. Nikolaev, H. J. Dai, P. Petit, J. Robert, C. H. Xu, Y. H. Lee, S. G. Kim, A. G. Rinzler, D. T. Colbert, G. E. Scuseria, D. Tomanek, J. E. Fischer, and R. E. Smalley, *Science* **273**, 483-487 (1996).
7. R. Krupke, F. Hennrich, H. von Lohneysen, and M. M. Kappes, *Science* **301**, 344-347 (2003).
8. J. Kong, H. T. Soh, A. M. Casell, C. F. Quate, and H. Dai, *Nature* **395**, 878 (1998).
9. Y. Li, D. Mann, M. Rolandi, W. Kim, A. Ural, S. Hung, A. Javay, J. Cao, D. Wang, E Yenilmez, Q. Wang, J. F. Gibbons, Y. Nishi, and H. Dai, *Nanoletters* **4**, 317-321 (2004).
10. M. Zheng, A. Jagoda, E. D. Semke, B. A. Diner, R. S. McLean, S. R. Lustig, R. E. Richardson, and N. G. Tassi, *Nature Materials* **2**, 338-342 (2003).
11. G. S. Duesberg, A. P. Graham, M. Liebau, R. Seidel, E. Unger, F. Kreupl, and W. Hoenlein, *Diamonds and Related Materials* **13**, 354-361 (2004).
12. F. Kreupl, A. P. Graham, G. S. Duesberg, W. Steinhogl, M. Liebau, E. Unger, and W. Hoenlein, *Microelectronic Engineering* **64**, 399-408 (2002).
13. G. S. Duesberg, A. P. Graham, M. Liebau, R. Seidel, E. Unger, F. Kreupl, and W. Hoenlein, *Nano Letters* **3**, 257-259 (2003).
14. R. Seidel, G. S. Duesberg, E. Unger, A. P. Graham, M. Liebau, and F. Kreupl, *Journal of Physical Chemistry B* **108**, 1888-1893 (2004).
15. W. Hoenlein, F.Kreupl, G. S. Duesberg, A.P. Graham, M. Liebau, R. Seidel, E. Unger, *Materials Science & Engineering C* **23**, 663-669 (2003).
16. T. Xu, J. Stevens, J. Villa, J.T. Goldbach, K. W. Guarini, C. T. Black, C. J. Hawker, and T. P Russel, *Adv. Funct. Mater.* **13** No. 9, 698-702 (2003).

Suitability of carbon nanotubes grown by chemical vapor deposition for electrical devices

B. Babić, J. Furer, M. Iqbal and C. Schönenberger

Institut für Physik, Universität Basel, Klingelbergstr. 82, CH-4056 Basel, Switzerland

Abstract. Using carbon nanotubes (CNTs) produced by chemical vapor deposition, we have explored different strategies for the preparation of carbon nanotube devices suited for electrical and mechanical measurements. Though the target device is a single small diameter CNT, there is compelling evidence for bundling, both for CNTs grown over structured slits and on rigid supports. Whereas the bundling is substantial in the former case, individual single-wall CNTs (SWNTs) can be found in the latter. Our evidence stems from mechanical and electrical measurements on contacted tubes. Furthermore, we report on the fabrication of low-ohmic contacts to SWNTs. We compare Au, Ti and Pd contacts and find that Pd yields the best results.

The present work is structured in two main sections. The first is devoted to our results on carbon nanotubes (CNTs) grown by chemical vapor deposition (CVD) emphasizing on the problem of CNT bundling, which occurs during growth. The second section discusses our results on the contacting of CVD-grown tubes using the metals Au, Ti and Pd.

SUPPORTED AND SUSPENDED CARBON NANOTUBES PREPARED BY CVD

The full control and understanding of structural and electronic properties of carbon nanotubes remain a major challenge towards their applications in nanoelectronics. Today, there exists several different production methods of carbon nanotubes (CNTs). Among them, chemical vapor deposition (CVD) emerged [1, 2, 3] as the most prominent one for the investigation of the electronic and electromechanical properties of CNTs. The most important advantages of the CVD method are that CNTs can be grown at specific locations on the substrate and at lower temperatures with simpler equipments as compared to the arc discharge and laser ablation methods. However, CNTs grown with this method vary in a quality and display a rather large dispersion in diameter which might be a severe problem for potential applications. Following the published recipes, we found that CVD grown CNTs differ dramatically if they are grown supported on a substrate or suspended over structured slits. This suggests that the nanotube-substrate interaction plays an important role in the final product in addition to growth parameters and catalysts.

CP723, *Electronic Properties of Synthetic Nanostructures*, edited by H. Kuzmany et al.
© 2004 American Institute of Physics 0-7354-0204-3/04/$22.00

Growth method

Two types of catalysts are used for the growth of CNTs. The first catalyst, which we will name catalyst 1, is similar to that described in Ref. [2]. The catalyst suspension consists of 1 mg iron nitrate seeds ($Fe(NO_3)_3\cdot9H_2O$) dissolved in 10 ml of isopropanol. The other catalyst, which we will call in the rest of the paper catalyst 2, has been prepared similar to that described in Ref. [3]. To 15 ml of methanol, 15 mg alumina oxide, 20 mg $Fe(NO_3)_3\cdot9H_2O$ and 5 mg $MoO_2(acac)_2$ are added. Both suspensions are sonicated for 1 hour, stirred overnight and sonicated every time for at least 20 min before deposition on the substrate [5]. A drop of the suspension is placed on a bare substrate surface or on a substrate with predefined structured areas by electron-beam lithography (EBL) or optical lithography in the corresponding resist. After spinning at 2000 r.p.m. for 40 sec, the substrate is baked at 150 °C for 5 min, followed by lift-off. The CVD growth of CNTs is performed in a quartz-tube furnace between $750-1000\,°C$ at atmospheric pressure using different gases. For catalyst 1 we used a mixture of either ethylene or methane with hydrogen and argon with respective flow rates of 2, 400, and 600 cm^3/min [4]. For the catalyst 2, we have used a mixture of methane and argon with respective flow rates of 5000 and 1000 cm^3/min [4]. During heating and cooling of the furnace, the quartz tube is continuously flushed with argon to reduce the contamination of the CNTs and to avoid burning them once they are produced.

Results and Discussion

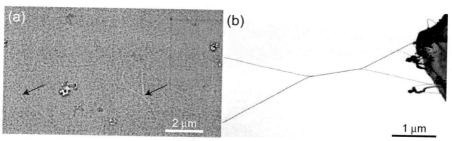

FIGURE 1. SEM images of CNTs grown from catalyst 1. In (a) the CNTs were grown on a Si/SiO_2 substrate at $T=800$ °C. The arrows point to visible branches. (b) Typical CNT network, grown over structured slits at $T=750$ °C. Note, that CNTs can bridge very large distances.

Carbon nanotubes which are grown at the same temperature but with the two mentioned catalysts on thermally oxidized silicon substrates show similar characteristics. In both cases there is a profound temperature dependence. At relatively low temperatures (750-850 °C) predominantly individual MWNTs or ropes of SWNTs are obtained with high yield. At intermediate temperatures (850-975 °C) individual SWNTs are grown with a typical diameter of 2 nm or thin bundles of SWNTs, but with less yield than at lower temperatures. At high temperatures (>1000 °C), the substrate and the CNTs are often found to be covered with an additional material, which is most likely amorphous carbon. Carbon nanotubes used in transport measurements have been solely produced at the intermediate temperature range. Fig. 1a shows a scanning electron microscope (SEM) image of CNTs grown from catalyst 1 on a Si/SiO$_2$ substrate.

For the purpose of mechanical and electromechanical studies, CNTs have been grown over structured slits patterned in Si$_3$N$_4$, an example of the outcome is shown in Fig. 1b. It is expected that for sufficiently long CNTs thermal vibrations should be readily observed with transmission and scanning-electron microscopy (TEM and SEM) [6, 7, 8]. This holds only, however, for 'small' diameter tubes, because the vibration amplitude is strongly reduced with increasing diameter d according to $(\sim 1/d^2)$. Only individual SWNTs are expected to show a substantial vibration amplitude which could be observed in SEM. We suggest this as a simple check to distinguish individual from bundled SWNTs. Fig. 1b shows a representing SEM image of suspended CNTs spanning over long distances ($L > 1\,\mu$m). None of the visible 'strings' display observable vibrations. This is not surprising considering the observed CNT branches. Clearly, in this case the CNTs must be bundled. This bundling increases the wider the slit is resulting into complex (but marvellous looking) spider webs. Further details on the search for vibrating suspended tubes can be found in Ref. [8]. We argue that in the absence of a support and at the relatively high temperature CNTs may meet each other during growth. The likelihood is increased if growth proceeds in 'free' space over a large distance. Once they touch each other they stick together due to the van der Waals interaction leading to a bundle.

In contrast, the growth on a substrate is different, as the tubes interact with the substrate rather than with each other. Hence, bundling is expected to be reduced. This is confirmed in AFM images, provided the catalyst density is low. However, there are bundles as well, which is evident from the observed branches visible in the AFM image of Fig. 1a (arrows). Even at locations where bundling is not apparent, one can still not be sure that such a nanotube section corresponds to a single-wall tube. Usually this is checked by measuring the height in AFM, but this can be misleading too, because the diameters of CVD-tubes can vary a lot, over $1 - 5$ nm as reported by Ref. [9]. We confirm this with our own measurements. Further insight into the question of bundling of CVD-grown CNTs can be obtained from electrical characterizations, which we report next.

Carbon nanotube devices

We have produced CNT devices on chip following two strategies. In the first method the substrate is covered with a layer of polymethylmethacrylate (PMMA) in which windows are patterned by electron-beam lithography (EBL). Next, the catalyst is spread from solution over these patterned structures, after which the PMMA is removed with acetone, leaving isolated catalyst islands ($5 \times 10\,\mu m^2$) on the surface. The substrate with the catalyst is then transferred to the oven where CVD growth of CNTs is performed. From the catalyst islands, CNTs grow randomly in all directions, but because of the relatively large distance between the islands ($5\,\mu$m) just one or a few CNTs bridge them usually. An atomic force microscope (AFM) image in phase mode with several CNTs growing from the catalyst islands is shown in Fig. 2a. An individual SWNT bridging the

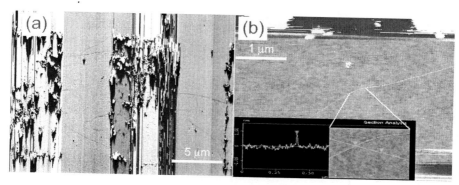

FIGURE 2. (a) Phase image recorded by tapping mode AFM, showing CNTs grown from the patterned catalyst islands and bridging between islands. (b) Topography image of an individual SWNT grown between the catalyst islands recorded by tapping mode AFM. Inset: Height measurement on the line cut (white line) for the SWNT shown in (b). The height measurements yield for the diameter $d = (1.2 \pm 0.2)$ nm for this particular tube.

catalyst islands is shown in Fig. 2b. Metal electrodes (Au, Ti, Pd) are patterned over the catalyst islands with EBL, followed by evaporation and lift-off. The alignments during the EBL structuring have been done corresponding to chromium markers [10]. SEM and AFM images of contacted individual CNTs are shown in Fig. 3a and 3b.

In the second method we spread the (diluted) catalyst over the entire substrate at low concentration. The density is chosen such that at least one CNT grows inside a window of size $10 \times 10 \, \mu m^2$. After the CVD process a set of recognizable metallic markers (Ti/Au bilayer) are patterned, again by EBL, see Fig. 3c. Using AFM in tapping mode, a suitable CNT with an apparent height of less than 3 nm is located with respect to the markers. In the final lithography step, electrodes to the selected CNT are patterned by lift-off.

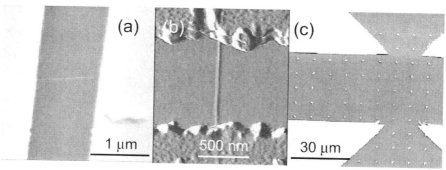

FIGURE 3. (a) SEM image of a SWNT contacted with a Ti/Au bilayer. (b) AFM image recorded in tapping mode of a contacted individual SWNT. (c) SEM image of a set of Ti/Au markers which are used to register the contact structure to the SWNTs selected before by AFM.

Room temperature characterization

Once the samples are made, it is common practice to distinguish semiconducting and metallic CNTs by the dependence of their electrical conductance (G) on the gate voltage (V_g), measured at room temperature $(T \approx 300\,\text{K})$. This, however, cannot be considered as a proof that an individual SWNT has been contacted, because it is not well understood how the linear response conductance is altered if more than one tube is contributing to electrical transport. Even if measurements were performed on ropes of SWNTs, the measured signatures agreed quite well with the behavior expected for a SWNT [11, 12, 13]. This has been attributed to a dominant electrode-CNT coupling to one nanotube only. This scenario may be true in exceptional cases, but one would expect that the majority of measurements should display signatures that arise from the presence of more than one tube. We have recently observed Fano resonances which we attribute to the interference of a SWNT which is strongly coupled to the electrodes with other more weakly coupled ones [14].

Assuming that all chiralities have equal probability to be formed in growth, 2/3 of the SWNTs are expected to be semiconducting and 1/3 metallic. From the measured response of the electrical conductance to the gate voltage (back-gate), $\approx 60\,\%$ of the devices display metallic (the conductance does not depend on the gate voltage) and $\approx 40\,\%$ semiconducting behavior. Based on our assumption the larger fraction of metallic gate responses points to the presence of bundles or multishell tubes. If there are on average 2 or 3 tubes per bundle, which are coupled to the electrodes approximately equally, the probability to observe a semiconducting characteristic would amount to $(2/3)^2 = 44\,\%$ or $(2/3)^3 = 30\,\%$. Hence, we can conclude that the bundle size is very likely small and close to 2 on average.

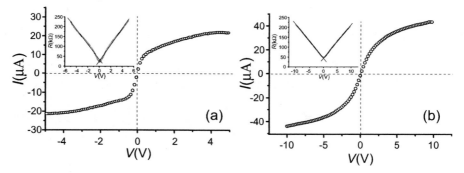

FIGURE 4. Typical $I-V$ characteristics at high bias voltage for CNT samples with a contact spacing of $1\,\mu$m. The insets show $R \equiv V/I$ versus V and fits to Eq. 1 for positive and negative V (lines). (a) The extracted mean value for the saturation current for this device is $I_0 = 24.3 \pm 1.2\,\mu$A which suggests transport through an individual SWNT. (b) A higher saturation current of $I_0 = 59.3 \pm 2.1\,\mu$A is found in this device suggesting transport through $2-3$ CNT shells.

A powerful method to characterize contacted CNTs is to perform transport measurements in the nonlinear transport regime (high bias). As previously reported by Yao *et al.* [15] the emission of zone-boundary or optical phonons is very effective in CNTs at high fields. This effect leads to a saturation of the current for an individual SWNT at $\approx 25\,\mu A$. High bias I/V characteristics are shown in Fig. 4. Fig. 4a corresponds to an individual SWNT. The saturation current can be extracted from the relation for the electrical resistance $R \equiv V/I$ [15]

$$R = R_0 + V/I_0, \qquad (1)$$

where R_0 is a constant and I_0 is the saturation current. The dependence of $R(V)$ versus the bias voltage V is shown in the insets of Fig. 4 with corresponding fits to Eq. 1. Because the saturation current is relatively well defined, its measurement allows to deduce the number of participating CNTs. Whereas Fig. 4a corresponds to a single SWNT, two nanotubes seem to participate in transport in the example shown in Fig. 4b. This result is consistent with the one above and points to the presence of more than one tube. This saturation-current method works for SWNTs but also for multi-wall CNTs [16]. One can therefore not distinguish whether one deals with two tubes in a rope or with one double-wall CNT.

LOW-OHMIC CONTACTS

It is well known that physical phenomena explored by electrical transport measurements (especially at low temperatures) dramatically depend on the transparency between the contacts and the CNT. At low energies, the electronic transport through an ideal metallic single-wall carbon nanotube (SWNT) is governed by four modes (spin included). In the Landauer-Büttiker formalism [17] the conductance can be written as

$$G = T \cdot 4e^2/h, \qquad (2)$$

where T is the total transmission probability between source and drain contacts. For low transparent contacts ($T \ll 1$) the CNT forms a quantum dot (QD) which is weakly coupled to the leads. Charge transport is then determined by the sequential tunnelling of single electrons (Coulomb blockade regime). If the transmission probability is increased (for which better contacts are required), higher-order tunnelling processes (so called co-tunnelling) become important which can lead to the appearance of the Kondo effect. This phenomenon was first reported by Nygård et al. [18]. At transparencies approaching $T \approx 1$ we enter the regime of ballistic transport where residual backscattering at the contacts leads to Fabry-Perot like resonances [19]. Good contacts with transparencies close to one are indispensable for the exploration of superconductivity [20], multiple Andreev reflection [21] or spin injection [22] in CNTs. Nevertheless, modest progress has been made so far on the control of the contact resistances between CNTs and metal leads. Annealing is one possible route, as proposed by the IBM group [23] and we confirm their results here. We compare in the following Ti, Au and Pd contacts.

Comparison between Ti, Au and Pd contacts

In the ideal case of fully transmissive contacts, a metallic SWNT is expected to have a conductance of $G = 4e^2/h$ (two modes), which corresponds to a two-terminal resistance

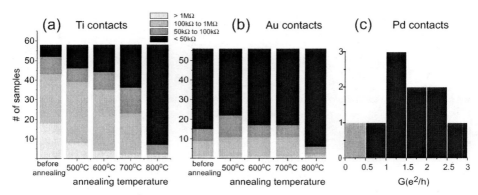

FIGURE 5. Comparison of the two-terminal resistance R at room temperature of CNT devices which were contacted with different metals: (a) Ti, (b) Au and (c) Pd. Post-annealing has been done in vacuum ($< 10^{-5} mbar$) in case of Ti and Au. In (a) and (b) the evolution of R for a large number (≈ 55) of samples as a function of annealing temperature is displayed in the form of a histogram. The representation for Pd (c) is different: the conductance $G = 1/R$ of 10 samples are compared, out of which only one has a resistance $R > 50 k\Omega$, corresponding to $G < 0.5 e^2/h$.

of 6.5 kΩ. In case of contacts made by Ti, Ti/Au or Cr on as grown SWNTs, most of the devices show resistances in the range between 100 kΩ to 1 MΩ. In contrast, Au contacts are better, because the measured resistances range typically between 40 kΩ and 100 kΩ. Even for the highest conductive sample the transmission probability is rather small and amounts to only $T \approx 0.16$ (per channel).

To lower the contact resistances we added an annealing step to the process, which was motivated by the work of R. Martel *et al.* [23]. We have performed annealing on more than 50 samples in a vacuum chamber fitted with a heating stage at a back-ground pressure of $< 10^{-5} mbar$. The resistance is first recorded on as prepared devices. Then, they are annealed with temperature steps of $100\,°C$ for 5 min starting at $500\,°C$. The results for titanium and gold contacts are shown in Fig. 5a and 5b, respectively.

In agreement with previous work [23] we find a pronounced resistance decrease for Ti contacts, if annealed at $800\,°C$. It was suggested by R. Martel *et al.* [23] that the origin of the resistance decrease is the formation of titanium carbide (Ti$_x$C) at temperatures over $700\,°C$. In contrast to Ti contacts, we do not observe a dramatic change in the sample resistance versus annealing temperature in case of Au contacts. This suggests that unlike Ti on carbon no chemical reactions take place between Au and carbon even at temperatures as large as $800\,°C$. We have also compared annealing in vacuum with annealing in hydrogen within the same temperature window (not shown). The outcome in terms of resistance change is comparable to the vacuum results provided that $T < 700\,°C$. At temperatures above $\approx 700\,°C$ the majority of the devices display a short to the back-gate. We think that the reducing atmosphere is very effective in partially etching the SiO$_2$ at these high temperatures.

Finally, we have also studied as-grown Pd contacts, which were recently reported to lead to contacts that are lower ohmic than Au [9]. In our own work (Fig. 5c) we have indeed found independently of Javey *et al.* [9] that palladium makes excellent contacts to CNTs. There is no need for an additional post-growth treatment [14]. Metallization of CNT devices with Pd is the preferred method, because it yields low-ohmic contacts without an additional annealing step. Careful transport studies of Pd contacted SWNTs show Coulomb blockade, Kondo physics and Fano resonances [14]. The observed resonances suggest that even in nanotubes, which look at first sight ideal, interference with additional transport channels may appear. The only plausible explanation for this observation is the existence of other tubes, hence a bundle or multishell nanotube.

CONCLUSION

Many applications of carbon nanotubes (CNTs) require to reproducibly place and contact single *small* diameter tubes. This is important, for example, for the realization of mechanical resonators [8], for field-effect transistors with reproducible characteristics and for fundamental studies of electron transport. One approach is to start from a powder of CNTs which is obtained, for example, in arc-discharge or laser-evaporation. Because these methods yield bundles of dozens of tubes, individual CNTs can only be obtained by rigorous ultrasonics and separation in an ultracentrifuge in the presence of a surfactant. If the ultrasonic step is too rigorous, the CNTs are cut into short pieces. Spreading and contacting of single tubes is possible. However, one has to bear in mind that these CNTs are covered by a surfactant which is likely to affect the fabrication of low-ohmic contacts. Moreover, the surfactant may carry charge which dopes the CNTs. In contrast to this approach, chemical vapor deposition (CVD) yields tubes in a very direct way immediately on the chip and without a surfactant, which makes this approach very attractive. Whereas a profound comparison of the quality in terms of the number of defects between these two major classes of CNTs is not yet established, the degree of bundling can be compared today. If grown by CVD on a surface at relatively high temperature and with a low catalyst density, apparently single-wall CNTs can be grown, though with a much larger spread in diameter as compared to e.g. the laser method. Although, the tubes appear to be single, as judged from SEM and simple tapping-mode AFM in air, we find in a number of different experiments clear signs for the presence of more than one tube. Measured saturation currents are often larger than the value expected for a single tube. Suspended tubes, even if no bundling is apparent in SEM in the form of branches, do not thermally vibrate as expected for a typical SWNT [8]. And finally, the presence of interference effects in transport (Fano resonances) point to additional transport channels that are likely due to additional shells or tube [14]. The results presented in this work show however, that the number of tubes can be small, e.g. 2-3. This gives hope that with refined catalysts, the controlled production of single tubes should be possible. In addition, we have demonstrated that relatively low-ohmic contacts can be achieved either with Ti, if an additional annealing step is used, or by Au and Pd without any additional treatment. Out of these three materials, Pd yields the best contacts (lowest contact resistance).

ACKNOWLEDGMENTS

We acknowledge contributions and discussions to this work by T. Y. Choi (ETHZ), J. Gobrecht (PSI), and D. Poulikakos (ETHZ). Support by the Swiss National Science Foundation, the NCCR on Nanoscience and the BBW is gratefully acknowledged.

REFERENCES

1. Kong, J., Cassel, A. M., and Dai, H., *Chem. Phys. Lett.*, **292**, 567–574 (1998).
2. Hafner, J. H., Bronikowski, M. J., Azamian, B. R., Nikolaev, P., Rinzler, A. G., Colbert, D. T., Smith, K. A., and Smalley, R. E., *Chem. Phys. Lett.*, **296**, 195–202 (1998).
3. Kong, J., Soh, H. T., Cassel, A. M., Quate, C. F., and Dai, H., *Nature*, **395**, 878–881 (1998).
4. Calibrated massflow controllers were used, except for the large flow rate of methane of $5000\,cm^3$/min, which was adjusted and meseared with a floating ball meter calibrated to air. The correction factor for methan is ≈ 1.4, i.e. the actual flow was 1.4 times larger.
5. Two different substrates are used: Si/SiO_2 or $Si/SiO_2/Si_3N_4$ heterostructure.
6. Treacy, M. M. J., Ebbesen, T. W., and Gibson, J. M., *Nature*, **381**, 678 (1996).
7. Poncharal, P., Wang, Z. L., Ugarte, D., and Heer, W. A., *Science*, **283**, 1513 (1999).
8. Babić, B., Furer, J., Sahoo, S., Farhangfar, S., and Schönenberger, C., *Nano. Lett.*, **3(11)**, 1577 (2003).
9. Javey, A., Guo, J., Wang, Q., Lundstrom, M., and Dai, H., *Nature*, **424**, 654 (2003).
10. Chromium is used because it could sustain high temperature of CVD process with no apparent diffusion.
11. Tans, S. J., Devoret, M. H., Groeneveld, R. J. A., and Dekker, C., *Nature*, **394**, 761–764 (1998).
12. Bockrath, M., Cobden, D. H., McEuen, P. L., Chopra, N. G., Zettl, A., Thess, A., and Smalley, R. E., *Science*, **275**, 1922–1925 (1997).
13. Nygård, J., Cobden, D., and Lindelof, P. E., *Nature*, **408**, 342 (2000).
14. B. Babić *et al.* (to be published).
15. Yao, Z., Kane, C. L., and Dekker, C., *Phys. Rev. Lett.*, **84**, 2941 (2000).
16. Collins, P. G., Arnold, M. S., and Avouris, P., *Science*, **292**, 706–709 (2001).
17. Büttiker, M., *Quantum Mesoscopic Phenomena and Mesoscopic Devices in Microlectronics*, Kluwer Academic Publishers, New York, 2000, pp. 211–242.
18. Nygård, J., Cobden, D., and Lindelof, P. E., *Nature*, **408**, 342 (2000).
19. Liang, W., Bockrath, M., Bozovic, D., Hafner, J. H., Tinkham, M., and Park, H., *Nature*, **411**, 665–669 (2001).
20. Kasumov, A. Y., Deblock, R., Kociak, M., Reulet, B., Bouchiat, H., Khodos, I. I., Gorbatov, Y. B., Volkov, V. T., Journet, C., and Burghard, M., *Science*, **284**, 1508 (1999).
21. Buitelaar, M. R., Belzig, W., Nussbaumer, T., Babić, B., Bruder, C., and Schönenberger, C., *Phys. Rev. Lett.*, **91**, 057005 (2003).
22. Jensen, A., Nygård, J., and J.Borggreen, *World Scientific*, pp. 33–37 (2003).
23. Martel, R., Derycke, V., Lavoie, C., Appenzeller, J., Chan, K., Tersoff, J., and Avouris, P., *Phys. Rev. Lett.*, **87**, 256805-1 (2001).

A Few Electron-Hole Semiconducting Carbon Nanotube Quantum Dot

Pablo Jarillo-Herrero[1,2], Sami Sapmaz[1], Cees Dekker[1], Leo P. Kouwenhoven[1,2] and Herre S.J. van der Zant[1]

[1]Department of Nanoscience and the [2]ERATO project on Mesoscopic Correlations, Delft University of Technology, PO Box 5046, 2600 GA, The Netherlands

Abstract. Carbon nanotubes are ideal systems to explore the physics of 1-dimensional materials. At low temperatures carbon nanotubes exhibit a variety of phenomena, such as Coulomb blockade[1-4], quantum interference[5] and the Kondo effect[6]. So far these observations have been made mainly in metallic nanotubes. Of particular interest is the study of electronic transport in nanotube quantum dots, where information about the quantized energy levels can be obtained. Despite some studies at intermediate temperatures[7], semiconducting nanotube single quantum dots have been proven difficult to realize at low temperatures preventing, for example, the observation of the electronic spectrum. Here we show that semiconducting individual single-walled carbon nanotubes can behave as fully coherent single quantum dots operating both in the few-electron and few-hole regime. We find that the discrete excitation spectrum for a nanotube with N holes is strikingly similar to the corresponding spectrum for N electrons. The data indicate a near-perfect electron-hole symmetry as well as the absence of scattering in semiconducting carbon nanotubes.

The fabrication of the devices follows those of refs [8,9]. Basically, HiPco nanotubes[10] are deposited from a suspension onto an oxidized Si substrate. Individual nanotubes are located with an atomic force microscope and electrically contacted by source and drain Cr/Au electrodes (Fig. 1a). The highly doped Si serves as a backgate. We then suspend the nanotubes by etching away part of the SiO_2 surface[9]. We generally find that removing the nearby oxide reduces the amount of potential fluctuations (i.e. disorder) in the nanotubes, as deduced from transport characteristics.

A low-temperature measurement around zero gate voltage (Fig. 1b) shows a large zero-current gap of about 300meV in bias voltage, reflecting the semiconducting character of this nanotube. The zigzag pattern outside the semiconducting gap is due to Coulomb blockade. These Coulomb blockade features are more evident in Fig. 1c, where a high-resolution measurement of the differential conductance shows the semiconducting gap with the first two adjacent Coulomb blockade diamonds. These correspond to a single electron (right) and a single hole (left), demonstrating that semiconductor nanotubes can operate as single electron transistors in the single electron and single hole regime[1].

CP723, *Electronic Properties of Synthetic Nanostructures*, edited by H. Kuzmany et al.
© 2004 American Institute of Physics 0-7354-0204-3/04/$22.00

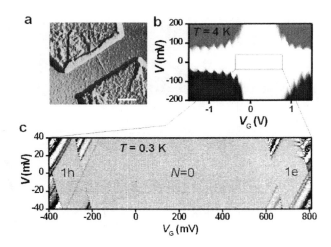

FIGURE 1. a, AFM picture of the device. b, Current vs bias and gate voltage. c, Conductance vs bias and gate voltage, showing the single electron and single hole region.

Fig. 2a shows the filling of the QD, one by one, up to hole number 20. The region for the first 2 holes is enlarged in Fig. 2b. The regularity in the Coulomb diamonds indicates a nanotube that is free of disorder. A closer inspection shows that the size of the Coulomb diamonds varies periodically on a smooth background as the hole number increases (Fig. 2c). The alternating, even-odd pattern in this addition energy, E_{add}, reflects the subsequent filling of discrete orbital states with two holes of opposite spin[1,4].

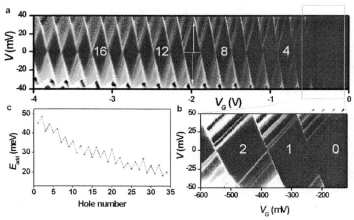

FIGURE 2. a, Stability diagram in the few-hole region. b, Region corresponding to 0-2 holes. Excitation lines can be seen. c, Addition energy as a function of hole number.

The additional discrete lines outside the Coulomb diamonds running parallel to its edges correspond to transport through excited states[1], as for instance indicated by arrows in Fig. 2b. This excitation spectrum can be measured both for electrons and holes. Fig. 3 shows this comparison for the Coulomb diamonds corresponding to 1-2 holes and 1-2 electrons. Remarkably, the excitation spectrum is nearly identical. The observation of electron-hole symmetry poses severe restrictions on the QD system: the effective masses for holes and electrons should be equal and the QD should be free of disorder. Scattering by negatively charged impurities, for example, is repulsive for electrons but attractive for holes, so it would break electron-hole symmetry. A symmetric band structure has been theoretically predicted for graphite materials and carbon nanotubes[11]. In contrast, the absence of scattering has come as a positive surprise.

The observation of the quantum properties of semiconducting carbon nanotubes opens the way to both fundamental and applied research on one of the potentially most relevant materials in the technology at the nanoscale.

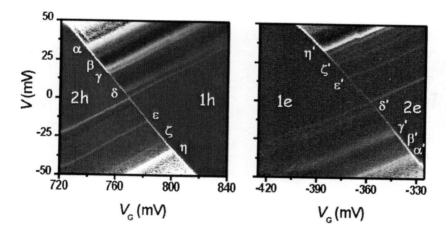

FIGURE 3. Stability diagram showing the symmetric spectrum for electron and holes.

ACKNOWLEDGMENTS

We thank R. E. Smalley and coworkers for providing the high-quality HiPco nanotubes, and S. De Franchesci, J. Kong, K. Williams, Y. Nazarov, H. Postma, S. Lemay and J. Fernández-Rossier for discussions. We acknowledge the technical assistance of R. Schouten, B. van der Enden and M. van Oossanen. Financial support is obtained from FOM. The details of this paper can be read in the May 27 issue of Nature (http://www.nature.com/nature).

REFERENCES

1. Kouwenhoven, L. P., Austing, D. G. & Tarucha, S. *Reports on Progress in Physics* **64**, 701-736 (2001).
2. Tans, S. J. *et al. Nature* **386**, 474-477 (1997).
3. Bockrath, M. *et al. Science* **275**, 1922-1925 (1997).
4. Cobden, D. H. & Nygård, J. *Phys. Rev. Lett.* **89**, 046803-1–046803-4 (2002).
5. Liang, W. *et al. Nature* **411**, 665 (2001).
6. Nygård, J., Cobden, D.H. & Lindelof, P.E. *Nature* **408**, 342 (2000).
7. Park, J. & McEuen, P. L. *Appl. Phys. Lett.* **79**, 1363-1365 (2001).
8. Fuhrer, M. S. *et al.* Science **288**, 494-497 (2000).
9. Nygård, J. & Cobden, D. H. *Appl. Phys. Lett.* **79**, 4216-4218 (2001).
10. Bronikowski, M. J., Willis, P. A., Colbert, D. T., Smith, K. A. & Smalley, R. E. *J. Vacuum Sci. Technol. A* **19**, 1800-1805 (2001).
11. Dresselhaus, M. S., Dresselhaus, G. & Eklund, P. C. *Science of Fullerenes and Carbon Nanotubes* (Academic Press, San Diego, 1996).

Electrically Driven Vaporization Of Multiwall Carbon Nanotubes For Rotary Bearing Creation

A. M. Fennimore, T. D. Yuzvinsky, B. C. Regan and A. Zettl

Physics Department, University of California, Berkeley, CA 94720 USA
Materials Sciences Division, Lawrence Berkeley National Laboratory, Berkeley CA 94720 USA

Abstract. We have previously reported on the creation of nanoscale rotational actuators based on multiwall carbon nanotubes. During the fabrication of these devices, we torsionally sheared the outer walls of the MWCNT to form a rotational bearing. We have designed an alternate technique for forming a rotational bearing geometry using electrically driven vaporization (EDV) of multiwall nanotube shells. While applying this technique, we have discovered an interesting failure mode.

INTRODUCTION

Investigating the exact behavior of nanoscale systems is often quite difficult. Easily accessible imaging techniques such as optical or scanning electron microscopy may not offer high enough resolution, while techniques that do (scanning probe or transmission electron microscopy) are limited in what geometries and materials they can examine (such as planar, conductive or electron-transparent substrates). Components integrated in multi-planar devices on silicon wafers can be particularly hard to image. We recently reported on one such device, a rotational actuator mounted on a multiwall carbon nanotube bearing (see Figure 1) [1]. A gold plate attached to the outer walls of a suspended multiwall nanotube was torqued about the nanotube by electric fields until rotational freedom was achieved. From the lack of restoring force, we determined that one or more outer shells of the multiwall nanotube had failed and were rotating about an inner core. We here explore an alternate (and hopefully highly controlled) method for bearing creation: electrically driven vaporization to selectively remove the outer walls of the multiwall carbon nanotube. In principle, the controlled removal of these walls creates a desirable geometry in which the behavior of the bearing can be easily characterized. Furthermore, with no outer walls extending past the edges of the rotor, the rotor should be able to slide along the inner core, creating a combination of linear and rotational bearing. This opens the door for investigation into chiral mismatch between the inner and outer walls of a multiwall nanotube and its effect on linear translation between the two [2].

CP723, *Electronic Properties of Synthetic Nanostructures*, edited by H. Kuzmany et al.
© 2004 American Institute of Physics 0-7354-0204-3/04/$22.00

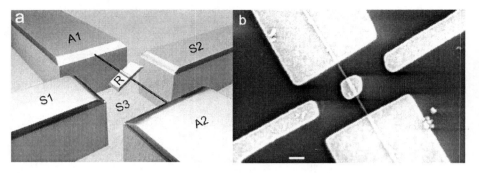

FIGURE 1. Rotational actuator mounted on a MWCNT. The gold rotor paddle can be rotated by applying voltages to the three stator electrodes (two surface stators and the conducting back gate). Artist's conception **(a)** and SEM image **(b)**. The scale bar is 300nm.

METHODS

Electrical driven vaporization (EDV) of the outer walls of a multiwall carbon nanotube (MWCNT) was first discovered by Cumings *et al* in 2000 [3]. By passing current through a MWCNT, they were able to instantaneously vaporize several walls, corresponding to a step up in the resistance of the MWCNT. Related work refining this technique to presumably step-by-step single wall vaporization has been reported by Collins *et al* [4], in which regularly spaced current steps corresponding to discrete steps in the thinning of the nanotube were presented. This technique has also been applied for use in sharpening STM/AFM tips [5] and in the fabrication of a torsional nanotube device with an architecture similar to ours [6].

We find that good electrical contact to the nanotubes is necessary for controlled vaporization to take place. Attempts made on devices with high resistance (>50 kΩ) result in breakage of the nanotube with no intermediate thinning observed. Devices with resistances lower than 10 kΩ reliably achieved stepwise current decays at constant bias voltages. These stepwise current decays sometimes exhibited current steps of equal magnitude (on the order of 10-20 µA, varying from device to device), but were often found to vary greatly (in the range of 5 to 25 µA) on a single device (see figure 2). The exact mechanism underlying these steps is still unknown.

Our first attempt to use EDV in our devices consisted of passing current from one anchor to the other, in the hope that sections of the outer walls would be removed on both sides of the rotor. We found, however, that once a shell failed on one side of the rotor (determined by scanning electron microscope (SEM) imaging of thinning of the MWCNT), all subsequent vaporization would happen on the same side, with no apparent failures occurring on the other side. This could not be remedied by reversing the bias, and would continue all the way to complete breakage of the nanotube.

We were able to vaporize sections of the MWCNT on both sides of the rotor, however, by making electrical contact to the center of the MWCNT and passing current through each side separately. The contact was made by adding an extra lithography step in the device fabrication, during which a thin strip of metal is evaporated to form a bridge between the stator electrodes and the rotor (see inset of figure 2). This contact, however, must later be removed for the device to be able to function. We therefore used Al or Ti, both of which, due to their very high etch rates in hydrofluoric acid, quickly disappear in the subsequent buffered hydrofluoric acid etch used to undercut and suspend the device. Due to the propensity of Al for oxidation we found Ti to be the ideal metal for this temporary contact.

Once a device was contacted with a Ti short we were able to pass current from the stators to either anchor in turn. It proved difficult, however, to induce equal amounts of damage on both sides; the resistance was rarely the same on both sides, often requiring different voltages and currents to begin the current cascades, and sometimes the nanotubes would completely fail without showing any steps at all. When they did occur, the cascades were sometimes difficult to controllably stop. Upon testing the devices *in situ* in an SEM, we found that many would have significantly reduced torsional spring constants. They would not, however, exhibit free bearing behavior – they would eventually break without showing the freedom of motion seen in the torsionally freed devices. We surmise that one side of the tube had been rotationally freed while the other remained as a torsional spring.

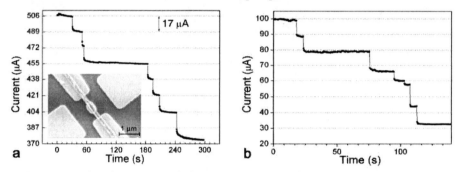

FIGURE 2. Cascades of current steps during EDV. **(a)** Some devices had remarkably equally-sized steps. **(b)** Many, however, showed a large variation in step size. The inset shows a scanning electron micrograph of a device with a Ti bridge connecting the stators to the rotor.

Despite these difficulties we were able to shed more light on the bearing nature of our devices. We repeatedly saw one particularly interesting failure mode. Instead of snapping at some point along its length (as was seen, for example, in devices freed by reactive ion etching [7]), the MWCNT would telescope out, dropping the rotor to the underlying surface (telescoping behavior in MWCNTs was first reported by Cumings & Zettl in 2000 [8]). The result of one such failure is shown in Figure 3. We were able to extend the MWCNT even further by attracting the rotor to the two side stators. Other devices failed similarly, some combining telescopic extension with rotation of the paddle. We submit two possible explanations for this failure mode. EDV may be able to remove internal, unexposed shells, in which case we are seeing the result of a break in the inner core near to the EDV-induced gap in the outer walls. This is contrary to the results of Collins *et al*, as they could correlate each current step with a thinning of the MWCNT (implying that each subsequent exposed wall is being removed). We find it more likely that the inner core is indeed decoupled from the outer shells and free to move, both linearly and rotationally, and we simply removed too many shells, making the exposed core too flexible to support the rotor.

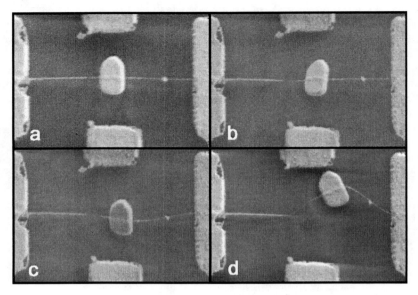

FIGURE 3. Telescopic failure of a MWCNT that has undergone EDV on both sides of the rotor paddle. The images are in sequential order, showing increasing extension: **(a)** No voltages applied. **(b)** Rotor pulled down towards substrate (voltage applied to back gate). **(c)** Rotor pulled towards lower stator. **(d)** Rotor pulled towards upper stator (though hard to see, the nanotube is still intact).

The difficulty of inducing equal damage and failure of the same shells on both sides of the rotor suggests this method requires additional refinement for creating reliable rotational bearings. Additional work on applying EDV to nanotube bearings and other geometries is in progress.

ACKNOWLEDGMENTS

This research was supported by the Office of Science of the U.S. Department of Energy. We thank Noah Bodzin for assistance with graphics.

REFERENCES

1. Fennimore, A.M. *et al*, *Nature* **424**, 408-410 (2003)
2. Saito, R. et al, Chem. Phys. Lett. 348, 187-193 (2001)
3. Cumings, J., Collins, P. G., and Zettl, A., *Nature* **406**, 586 (2000).
4. Collins, P.G., Arnold, M.S., and Avouris, P., *Science*, **292**, 706-709 (2001).
5. Nakayama, Y. and Akita, S., *New J. Phys.*, **5**, 128.1-128.23 (2003)
6. Bourlon, B., *et al*, *Nano Lett*, **4**, 709-712 (2004)
7. For more information see the report on reactive ion etching, also in these proceedings
8. Cumings, J., and Zettl, A., *Science* **289**, 602-604 (2000)

Conducting transparent thin films based on Carbon Nanotubes – Conducting Polymers

N. Ferrer-Anglada[a,c], V. Gomis[a], Z. El-Hachemi[b], M. Kaempgen[c] S. Roth[c]

[a] Departament de Física Aplicada, Universitat Politècnica de Catalunya, J. Girona 3-5, 08034 Barcelona (Spain)
[b] Department de Química Orgànica, Universitat de Barcelona, Diagonal 647, 08028 Barcelona, Catalonia (Spain)
[c] Max-Planck-Institut für Festkörperforschung, Heisenbergstraße 1, D-70569 Stuttgart, Germany

Abstract. The present work reports on the characterization and optimization of thin transparent and electrically conducting films (from 120 to 180 nm thick) based on single walled carbon nanotubes (CNT) and conducting polymers, polypyrrole (PPy) or polyaniline (PA). We obtained a number of different CNT-PPy doped with PTS or PF_6 and CNT-PA under different parameters (electrodeposition time, density current or voltage) and analyzed the required properties, electrical conductivity and transparency, and other significant properties: Raman Spectroscopy, and AFM, from which we can estimate the film thickness. The electrochemical conditions for the polymer thin film deposition were studied in order to improve their conductivity and transparency.

Compared to the well known transparent conducting oxides like ITO, the best of our composite thin films are from 10 to 100 times less conductive and highly transparent. As a great possibility, these conducting films could be prepared on a flexible substrate with a continuous deposition procedure.

INTRODUCTION

Conducting polymers have been intensively studied during the last decades, due mostly to their useful and smart applications. Polyaniline and polypyrrole are between the more interesting conjugated organic conducting polymers [1,2]. Polypyrrole (PPy) or polyaniline (PA) samples obtained electrochemically can be made using a variety of procedures each of which generates a material with a different electrical behavior that can be related to structure [3].

Besides, from their discovery in 1991 [4], carbon nanotubes (CNT) have attracted much attention due to their unique mechanical and electrical properties. CNT show a very high electrical conductivity, as high as 2000 S/cm in bucky paper [5]. For this reason, different composites using CNT have been performed, with the aim of improving some particular properties. Recently O. Hjortstam et al. [6] estimated that a room temperature resistivity 50% lower than Cu is achievable in carbon nanotube-metal composites.

CP723, Electronic Properties of Synthetic Nanostructures, edited by H. Kuzmany et al.
© 2004 American Institute of Physics 0-7354-0204-3/04/$22.00

EXPERIMENTAL RESULTS

We prepared electrically conducting thin films, quite transparent (70% to 95%) and enough conducting that can be used as electrodes. A thin network of Single Walled Carbon Nanotubes (CNT) was deposited on a transparent substrate. As deposited, the network is conducting enough to be used as an electrode in order to grow electrochemically a conducting polymer on it. We essayed the deposition of doped polypyrrole (PPy) and polyaniline (PA) conducting thin films.

Thin films of PPy and PA respectively (CNT-PPy and CNT-PA) have been obtained on a thin network of Single Walled Carbon Nanotubes close to percolation threshold, CNT-PA was described elsewhere [7]. CNT-PPy was grown galvanostatically from an acetonitrile solution 0.1 M of tetraethylammonium p-toluensulphonate (PTS) and 0.05 M Pyrrol [8] to obtain PTS-doped polypyrrole, or using an acetonitrile solution 0.06 M tetramethylammonium hexaflourophosphate and 0.06 M pyrrol in order to obtain PF_6 doped polypyrrole. Both at different density currents j, from 0.08 mA/cm^2 to 1 mA/cm^2. The proportion of the conducting polymer may vary depending on the electrodeposition time from t = 7 min to 60 min. From AFM (see Figure 1) we obtained information about the granular structure and the connectivity of the sample, and estimated the thickness of the thin film (from 120 – 180 nm). It can be seen that at the beginning the polymer grows mostly on the carbon nanotubes, showing its granular structure. When obtained at low density current the polymer films are quite uniform, covering the all surface.

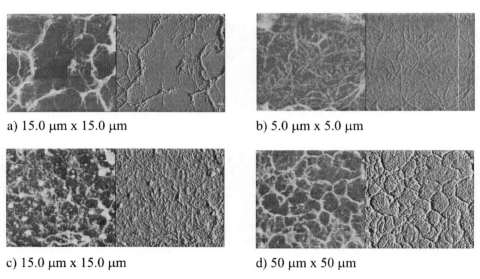

a) 15.0 μm x 15.0 μm   b) 5.0 μm x 5.0 μm

c) 15.0 μm x 15.0 μm   d) 50 μm x 50 μm

FIGURE 1. AFM image of: a) transparent CNT network on glass; b) Network of CNT-PA on quartz; c) CNT-PPy doped with PTS on glass, eletrodeposition time 33 min, density current j = 0.08 mA/cm^2; d) CNT-PPy doped with PF_6 on glass, eletrodeposition time 12 min, density current j = 0,05 mA/cm^2.

Different electrochemical conditions for the polymer deposition were studied in order to improve their conductivity and transparency. The composite films were characterized by AFM, optical and Raman spectroscopy, and electrical conductivity. The Raman spectra for CNT-PPy doped with PF$_6$ are shown in Figure 2, whereas CNT-PPy doped with PTS spectra are shown on Figure 3, for different films. In both figures spectra a) corresponds to pristine CNT network. We can see the characteristic bands of doped polypyrrole [9] superimposed to the CNT bands. Looking at the characteristic breathing mode for CNT, the lines appearing near 200 nm, they disappear completely for some films (Figure 2: c), e); Figure 3: c), d), f)) indicating that the CNT are completely covered by the polymer, as it is also observed by AFM.

The experimental results of the best conducting samples are summarized on Table 1, the conductivity values are estimated taking the thickness as 150 nm.

TABLE 1. Resistance per square R$_{sq}$, Electrical conductivity σ and transparency (transmission %T), of single walled carbon nanotubes (CNT), carbon nanotubes-polyaniline (CNT-PA, from Ref. [7]) and carbon nanotubes-polypyrrole (CNT-PPy) thin films, with PTS or PF$_6$ doped polypyrrole, obtained at different parameters: density current j, deposition time t.

	CNT	CNT-PA	CNT-PPy PTS	CNT-PPy PTS	CNT-PPy PTS	CNT-PPy PTS	CNT-PPy PF$_6$	CNT-PPy PF$_6$
j (mA/cm²)	-	at 700 mV	0.6	1.0	0.24	0.09	0.15	0.05
t (min)	-	20–45	6	7	10	30	10	12
R$_{sq}$ (kΩ)	6.6	2.5	1.1–7	1.5–10	1.5–10	3.5–15	19–30	110
σ (Scm⁻¹)	20	55	60–10	44–7	43–7	19–10	3.5	0.60
%T	90	62–90	75	45–75	75–83	85	75–80	77–90

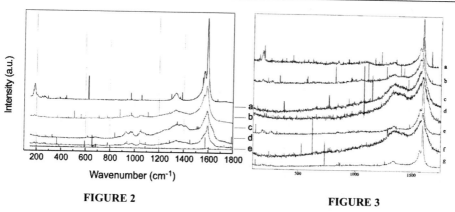

FIGURE 2   **FIGURE 3**

FIGURE 2. Raman spectra with a 514.5 nm excitation laser line, of transparent thin networks of: a) CNT composites where we can see the radial breathing mode (RBM) below 300 nm and the high-energy modes (HEM) near 1590 nm; compared to spectra of thin networks of CNT-PPy: PF$_6$, obtained at: b) density current j = 0.15 mA/cm²; t = 10 min. c), d), e) lower density current (j = 0.05 mA/cm²); t = 12 min.

FIGURE 3. Raman spectra with a 514.5 nm excitation laser line, of transparent thin networks of CNT-Ppy: PTS composites: a) pristine CNT network as reference; different density current j (mA/cm²) and time t were used: b) j = 0.60, t = 6 min; c), d) j = 0.08, t = 33 min; e) j = 1.0, t = 7 min; f) j = 0.09, t = 30 min; g) j = 0.24, t = 10 min.

CONCLUSIONS

We prepared single walled carbon nanotubes – polyaniline and single walled carbon nanotubes – polypyrrole composites as electrically conducting thin films (thickness 120–180 nm), quite transparent (70% to 90%) that can be used as transparent electrodes. The conductivity of the composite films is better than the carbon nanotubes network, due probably to a best contact between nanotubes. Compared to the measured conductivity on thick, self-standing polymer, the composite films show similar conductivity values. Raman spectra are very sensitive to the polymer deposition, either the Raman breathing mode (RBM) and high energy mode (HEM) change in intensity and shape. The RBM decreases or practically disappear when the characteristic bands of doped polypyrrole increase their intensity, indicating that CNT are covered by PPy. The highest conductivity is obtained on carbon nanotubes – polyaniline, or on the carbon nanotubes – polypyrrole films that are obtained for polypyrrole deposited at high density current (between 0.25 and 1mA/cm2) and short time (<10 min), for which the measured room temperature conductivity is 60 S/cm and the transparency near 75%. The conductivity value are form 100 to 10 times lower than the commonly used transparent oxides such as ITO, but can be enough for some applications, particularly for electromagnetic shielding. As a major advantage, the film can be deposited on any material and any shape of surface.

ACKNOWLEDGMENTS

This work was partially supported by the MCYT (Spain), No. MAT2002-04264-C04-01. N. Ferrer-Anglada wish to thank the MECD of the Spanish Government for a mobility grand, Ref. PR2002-0050, for a Research stage at Max Planck Institute for Solid State Research, Stuttgart, Germany.

REFERENCES

1   Skotheim, T. A., Elsenbaumer, R. L., and Reynolds, J. R. (Eds), *Handbook of Conducting Polymers*, 2nd Edition, Marcel Dekker, New York, 1998.
2   Kaiser, A.B., *Rep. Progr. Phys.* **64**,1-49 (2001) 1-49; Rogers, S. A., and Kaiser, A. B., *Curr. Appl. Phys.* **4**, 407-410 (2004).
3.   Ribó, C. M., Anglada, M. C., Hernandez, J. M., Chaibí, A., Ferrer-Anglada, N., and Movaghar, B., *Synth. Met.* **97**, 229-236 (1998).
4.   Iijima, S., *Nature* **354**, 56-58 (1991).
5.   Ajayan, P. M., and Zhou, Q. Z.,*Top. Appl. Phys.* **80**, 391 (2001).
6.   Hjortstam, O., Isberg, P., Söderholm, S., and Dai, H., *Appl. Phys. A* **78**, 1175 (2004).
7.   Ferrer-Anglada, N., Kaempgen, M., Skákalova, V., Dettlaff-Weglikowska, U., and Roth, S., *Diamond and Related Materials*, **13**, 256-260 (2004).
8.   Gomis, V., Ferrer-Anglada, N., Movaghar, B., Ribó, J. M., El-Hachemi, Z., and Jhang, S. H., *Phys. Rev. B* **68** 115208 (2003).
9.   Furukawa, Y., Tazawa, S., Fujii, Y., and Harada, I., *Synth. Met.* **24**, 329-341 (1998).

Fluorine Effect on the Binding Energy of Nitrogen Atoms Incorporated into Multiwall CNx Nanotubes

L.G. Bulusheva[1], A.V. Okotrub[1], E.M. Pazhetnov[2], A.I. Boronin[2]

[1]*Nikolaev Institute of Inorganic Chemistry SB RAS, pr. Ak. Lavrentieva 3, 630090 Novosibirsk, Russia*
[2]*Boreskov Institute of Catalysis SB RAS, pr. Ak. Lavrentieva 5, 630090 Novosibirsk, Russia*

Abstract. Fluorination of multiwall CN_x nanotubes has been shown to change the N 1s line of X-ray photoelectron spectrum (XPS). *Ab initio* Hartree-Fock calculations on various models of fluorinated nitrogen-doped graphite fragments were performed to reveal fluorine effect on the binding energy of N 1s core level. Comparison between experimental and theoretical data indicated that the most probable sites for fluorine attachment are (1) carbon atoms bonded to three-coordinated nitrogen, (2) pyridinic nitrogen atoms, and (3) carbon atoms located para and ortho to the pyridinic nitrogen.

INTRODUCTION

Nitrogen doping of carbon nanotubes can be a way for modification and controlling of their electronic structure. Actually, multiwall CN_x nanotubes are characterized by metallic conductivity and high values of field emission current [1]. X-ray photoelectron spectroscopy (XPS) and electron energy loss spectroscopy (EELS) showed the nitrogen, incorporated into graphitic layers, is in two different electronic states at least [2]. The high-energy spectral maximum is obviously attributed to three-coordinated nitrogen atoms replacing carbon ones within the graphite sheets, while the low-energy maximum is assigned to pyridinic nitrogen or nitrogen atoms bonded to sp^3-hybridized carbon [2–4]. STM images of the surface of CN_x nanotubes had exhibited large holes, which were suggested to be caused by the presence of pyridine-like islands [2]. Nitrogen located at the zigzag edge of vacancy acts as an acceptor lowering the Fermi level energy for carbon nanotube [5].

Carbon nanotubes surface is rather inert towards various reagents and nitrogen doping could affect its chemical activity. Thus, three-coordinated nitrogen was shown by the results of quantum chemical calculation to increase the reactivity of carbon atoms occupied the ortho- and para-positions [6]. Fluorination is most extensive method for sidewall functionalization of carbon nanotubes covalently attaching up to 50% of fluorine atoms [7]. Fluorination of multiwall CN_x nanotubes has been found to have significant effect on the N 1s line due to oxidation of nitrogen atoms with lower binding energy [8].

The purpose of the present work is quantum-chemical study of fluorine influence on the 1s binding energy of nitrogen incorporated into graphitic layers.

EXPERIMENTAL

CN_x nanotubes were obtained using a chemical vapor decomposition (CVD) process described in details elsewhere [9]. Catalytic Co/Ni nanoparticles were prepared in the result of thermal decomposition of solid solution of Co and Ni bimaleates. CN_x nanotubes grew via pyrolysis of acetonitrile in an argon flow (3 l/min) at 850°C and atmospheric pressure. Fluorination of CN_x nanotubes was carried out following a procedure previously applied to the arc-produced multiwall carbon nanotubes [10]. The sample placed in a teflon flask was held in a vapor over a solution of BrF_3 in Br_2 for 7 days. Thereafter the flask content was dried by a flow of N_2 up to the termination of Br_2 evolution. X-ray photoelectron spectra (XPS) of N 1s core level of the samples were measured using VG ESCALAB spectrometer with the AlKα line.

COMPUTATIONS

Geometry of models was relaxed in Hartree-Fock self-consistent field using 3-21G basis set within the quantum chemical package Jaguar [11]. This approach calculates ground state of a system while the XPS lines arise in the result of sample ionization. Thus, for the XPS spectra interpretation, the calculated N 1s level energy (E^{N1s}) should be corrected to take into account the relaxation for the ionized state. The correction factor was determined from the correlation dependence between the experimental and theoretical values for ten nitrogen-containing molecules [12]. The theoretical binding energies E^{BE} were computed using the derived linear dependence $E^{BE} = 75.69 + 0.78271 E^{N1s}$.

RESULTS AND DISCUSSION

N 1s XPS spectra of pristine sample and fluorinated one are compared in fig. 1. The spectrum of CN_x nanotubes exhibits two maxima A and B centered at 399.1 eV and 401.3 eV respectively. Fluorination of the sample causes maximum A shift to the increase of binding energy that provides an asymmetric shape of N 1s line. To interpret the experimental data, the results of quantum-chemical calculation on models presented in table 1 were used. The model 1 is a graphite fragment with central carbon atom substituted for nitrogen one. The models 2, 3, and 4 were produced by removing the central atom from the model 1 and substituting carbon atoms at the zigzag sites of vacancy boundary for one, two, and three nitrogen atoms. The hydrogen atoms were bonded to the edges carbon atoms of models 1–4. The table 1 presents the theoretical N 1s binding energies E^{BE} computed for the considered models. Fitting of these values to the experiment (fig. 1(a)) revealed that two models with different location of nitrogen in the graphite network are sufficient for N 1s spectrum interpretation. The maximum A can be assigned to the model 4 incorporated three pyridinic-like nitrogen atoms at the vacancy boundary. The maximum B is related to three-coordinated nitrogen (model 1). The values of E^{BE} computed for models 2 and 3 fall between main maxima and, hence, these models are not required for interpretation of N 1s spectrum

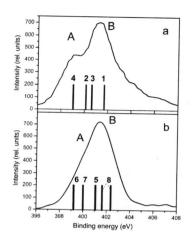

FIGURE 1. N 1s spectra measured for CN_x nanotubes (a) and fluorinated sample (b). The vertical lines correspond to the N 1s core level binding energies calculated for the models presented in table 1.

of CN_x nanotubes.

Effect of fluorination on the electronic state of nitrogen incorporated into the layers of carbon nanotubes was examined using the selected models 1 and 4. We found the fluorine does not approach to the three-coordinated nitrogen owing to large negative charges on atoms. Carbon atom, linked with three-coordinated nitrogen, has a positive charge. This site is attractive for fluorine attachment (model 5) resulting in lowering of the binding energy of N 1s core level by 0.6 eV. Electronic state of such kind of nitrogen corresponds to the maximum B of the fluorinated CN_x spectrum (fig. 1(b)).

TABLE 1. Theoretical energy of N 1s core level E^{BE} for nitrogen-contained graphite fragments (models 1-4) and those fluorinated (models 5-8).

Model	E^{BE} for N 1s (eV)	Model	E^{BE} for N 1s (eV)
1	401.7	5	401.1
2	400.6	6	399.2
3	400.1	7	400.0
4	399.0	8	401.6 402.4

The E^{BE} value for three-coordinated nitrogen is slightly affected by fluorination of the rest of the neighbors and amounts to 399.9 eV.

Fluorination of the model 4 is the most favorable for the carbon atom located para to the pyridinic nitrogen (model 6). As the result the N 1s binding energy slightly increases compared to that for the non-fluorinated fragment (model 4) and could be attributed to the low-energy intensity of the spectrum (fig. 1(b)). Attachment of the fluorine atom to the carbon one located ortho to the pyridinic nitrogen (model 7) results in the value of E^{BE} being equal to 400.0 eV. One can see the models 6 and 7 are necessary for interpretation of line A in the N 1s spectrum of the fluorinated sample. We found the pyridinic nitrogen can be also fluorinated that initiates formation of a five-membered ring with a bond between nitrogen atoms (model 8). Fluorination of pyridinic nitrogen has the greatest effect on the E^{BE}. The values for nitrogen atom linked to fluorine and atoms from the N-N bond constitute 401.6 eV and 402.4 eV (table 1) and correspond to the spectral intensity in the high-energy region.

In summary, the XPS of N 1s core level of CN_x nanotubes and quantum-chemical calculations on model structures showed the fluorination is the most preferable for following sites: (1) carbon atoms ortho located to three-coordinated or pyridinic nitrogen, (2) carbon para located to pyridinic nitrogen, and (3) pyridinic nitrogen.

ACKNOWLEDGMENTS

We thank the INTAS (project 01-254) and the RFBR (project 03-03-32286) for financial support.

REFERENCES

1. Golberg, D., Dorozhkin, P. S., Bando, Y., Dong, Z.-C., Tang, C. C., Uemura, Y., Grobert, N., Reyes-Reyes, M., Terrones, H., and Terrones, M., *Appl. Phys. A* **76**, 499-507 (2003).
2. Terrones, M., Ajayan, P. M., Banhart, F., Blasé, X., Carroll, D. L., Charlier, J. C., Czerw, R., Foley, B., Grobert, N., Kamalakaran, R., Kohler-Redlich, P., Ruhle, M., Seeger, T., and Terrones, H., *Appl. Phys. A* **74**, 355-361 (2002).
3. Casanovas, J., Ricart, J.M., Rubio, J., Illas, F., and Jimenez-Mateos, J. M., *J. Am. Chem. Soc.* **118**, 8071-8076 (1996).
4. Souto, S., Pickholz, M., dos Santos, M. C., and Alvarez, F., *Phys. Rev. B* **57**, 2536-2540 (1998).
5. Zhao, M., Xia, Y., Lewis, J.P., and Zhang, R., *J. Appl. Phys.* **94**, 2398-2402 (2003).
6. Stafstrom, S., *Appl. Phys. Lett.* **77**, 3941-3943 (2000).
7. Mickelson, E. T., Huffman, C. B., Rinzler, A. G., Smalley, R. E., Hauge, R. H., and Margrave, J. L., *Chem. Phys. Lett.* **296**, 188-194 (1998).
8. Okotrub, A. V., Maksimova, N., Duda, T. A., Kudashov, A. G., Shubin, Yu. V., Su, D. S., Pazhetnov, E. M., Boronin, A. I., and Bulusheva, L. G., Full.erenes, *Nanotubes, and Carbon Nanostructures* **12**, 99-104 (2004).
9. Kudashov, A. G., Okotrub, A. V., Yudanov, N. F., Romanenko, A. I., Bulusheva, L. G., Abrosimov, O. G., Chuvilin, A. L., Pazhetnov, E. M., and Boronin, A. I., *Phys. Solid State* **44**, 652-655 (2002).
10. Yudanov, N. F., Okotrub, A. V., Shubin, Yu. V., Yudanova, L. I., Bulusheva, L. G., Chuvilin, A. L., Bonard, J.-M., *Chem. Mater.* **14**, 1472-1476 (2002).
11. *Jaguar 3.5*, Schrodinger, Inc., Portland, OR, 1998.
12. Kudashov, A. G., Okotrub, A. V., Bulusheva, L. G., Asanov, I. P., Shubin, Yu. V., Yudanov, N. F., Yudanova, L. I., Danilovich, V. S., Abrosimov, O. G., *J. Phys. Chem. B* (to be published).

A

Aboutanos, V., 469
Achiba, Y., 222
Ajami, D., 19
Albertini, D., 465
Albrecht, P. M., 173
Alig, I., 478
Allioux, M., 485
Aoyagi, Y., 103
Apih, T., 298
Arčon, D., 298, 302, 427
Ardavan, A., 255
Arenal, R., 293
Arnold, K., 116

B

Babić, B., 574
Baibarac, M., 201
Baltog, I., 201
Balzano, L., 27
Bando, Y., 229
Barišić, N., 107
Bartsch, K., 36
Beckmann, D., 561
Béguin, F., 133, 460, 532
Behr, G., 141
Bernier, P., 45, 181, 238, 469
Bhattacharyya, A. R., 478
Biedermann, K., 285
Biró, L. P., 40, 149, 389
Bizot, H., 465
Björk, M. T., 449
Blacher, S., 133
Blau, W. J., 544
Blinc, R., 298, 302
Bocharov, G. S., 528
Böhme, T., 334
Bohnen, K.-P., 289
Bolton, K., 87, 364
Bonnet, P., 465
Borondics, F., 137
Boronin, A. I., 595
Borowiak-Palen, E., 141, 285
Bouchiat, H., 103

B (continued)

Briggs, G. A. D., 255
Brintlinger, T., 520
Britz, D. A., 255
Brown, C. M., 8
Brown, E., 91
Buleon, A., 465
Bulusheva, L. G., 595
Burghard, M., 99, 168, 415
Bystrzejewski, M., 95

C

Cançado, L. G., 372, 407
Cantoro, M., 81, 445
Carella, A., 503
Carvalho, A. C. M., 347
Castiglioni, C., 334, 359
Cech, J., 478
Cevc, P., 298, 302, 427
Chapman, B., 99
Chauvet, O., 201, 465
Chen, Z., 69, 137
Chernozatonskii, L. A., 351
Chiu, P.-W., 556
Chou, L. J., 53
Chueh, Y. L., 53
Cobas, E., 520
Cohen, S. R., 306
Colli, A., 445
Colliex, C., 293
Costa, P., 278
Couteau, E., 107
Cravino, A., 548
Csanyi, G., 81
Czumińska, K., 95

D

Danno, T., 431
Davy, J., 465
Deblock, R., 103
de Jonge, N., 485
Dekker, C., 583
Delpeux, S., 133
Dennis, T. J. S., 255

Deppert, K., 449
de Souza, M., 157
Dettlaff-Weglikowska, U., 189
DiCarlo, A., 355
Di Donato, E., 334, 359
Ding, F., 364
Dinse, K.-P., 3, 259
Dipasquale, M., 524
Di Zitti, E., 524
Dolinšek, J., 298
dos Santos, M. C., 347
Doytcheva, M., 485
Dresselhaus, G., 157, 372, 407
Dresselhaus, M. S., 157, 372, 407
Dubosc, M., 473
Dudkin, S. M., 478
Duesberg, G. S., 503, 516, 536, 540, 565
Dunsch, L., 242, 247
Duppel, V., 415

E

Eklund, P., 87
Eletskii, A. V., 528
El-Hachemi, Z., 591

F

Fantini, C., 157, 372, 407
Fennimore, A. M., 512, 587
Feringa, B. L., 498
Ferrari, A. C., 81, 445
Ferrer-Anglada, N., 591
Fink, J., 57, 205, 217
Finnie, P., 111
Fonseca, A., 40
Fórró, L., 61, 65, 107, 427
Frackowiak, E., 460, 532
Friedrichs, S., 278
Fuhrer, M. S., 520
Furer, J., 574

G

Gaál, R., 61, 107
Gabric, Z., 536
Gallop, J. C., 91

Gao, C., 193
Gao, Y. H., 229
Gartsman, K., 306
Gatti, F., 524
Gembus, A., 259
Gemming, S., 368
Gemming, T., 141, 285
Glapiński, J., 95
Glasgow, D. G., 455
Glerup, M., 45, 469
Godon, C., 201, 465
Golberg, D., 229
Gomis, V., 591
Gommes, C., 133
Goze-Bac, C., 181, 238
Graham, A. P., 503, 536, 565
Green, M. L. H., 278
Grubek-Jaworska, H., 95
Grüneis, A., 372, 407
Grupp, A., 12
Guéron, S., 103
Gusenbauer, A., 548
Gyulai, J., 149

H

Haddon, R. C., 197
Hansen, A. E., 449
Hao, L., 91
Hartschuh, A., 163
Hasi, F., 234, 251, 273
Hassanien, A., 145
Haufe, O., 12
Hauptmann, J. R., 552
Hecht, M., 12
Hennrich, F., 49, 116, 330, 561
Herges, R., 19
Herrera, J. E., 27
Heyning, O. T., 45
Hoenlein, W., 503, 565
Hofmann, S., 81, 445
Holzinger, M., 469
Holzweber, M., 234, 251
Homma, Y., 111
Hornbostel, B., 473, 478
Hsu, W. K., 53
Hu, H., 197
Hu, J. Q., 229
Huczko, A., 95

Hulman, M., 278
Hummelen, J. C., 315

I

Iqbal, M., 574
Ishii, H., 222
Itkis, M. E., 197
Ito, T., 326
Ivanovskaya, V., 306
Iwasa, Y., 326

J

Jagličić, Z., 298, 427
Jančar, B., 302, 423
Jánossy, A., 259
Jansen, M., 12
Jarillo-Herrero, P., 583
Jaud, J., 503
Jensen, A., 552
Jesih, A., 302
Jiang, J., 372, 407
Jin, Y. Z., 53
Johnels, D., 238
Johnson, Jr., A. T., 129
Jonkman, H. T., 315
Jorio, A., 157, 372, 407
Jost, O., 57, 185

K

Kaempgen, M., 591
Kaiser, A. B., 99
Kaiser, M., 485
Kalbáč, M., 242, 247
Kalenczuk, R. J., 141, 285
Kamarás, K., 8, 137, 197
Kammermeier, S., 19
Kanai, M., 255
Kane, C. L., 402
Kaplan-Ashiri, I., 306
Kappes, M. M., 49, 116, 561
Kasumov, A. Y., 103
Kasumov, Y. A., 103
Kataura, H., 217, 222, 234, 238, 242, 247

Kato, T., 3
Kavan, L., 242, 247
Kawaguchi, J., 431
Kawamura, M., 103
Kern, K., 168, 415
Khlobystov, A. N., 255
Khodos, I. I., 103
Kienle, L., 415
Kim, B. M., 520
Kleinsorge, B., 81
Klinov, D. V., 103
Klupp, G., 8
Knupfer, M., 141, 205, 217, 285
Kobayashi, T., 103
Kociak, M., 103, 293
Kodama, T., 103
Kong, H., 193
Kouwenhoven, L. P., 45, 583
Kramberger, C., 234, 251, 263, 268
Krause, M., 247
Kreupl, F., 503, 536, 565
Kroto, H. W., 53
Krstić, V., 121
Krupke, R., 49, 561
Kubozono, Y., 326
Kürti, J., 343, 377
Kuzmany, H., 213, 234, 251, 259, 263, 268, 273, 278

L

Lacerda, R. G., 485
Lahiff, E., 544
Lake, M. L., 455
Lambin, P., 389
Lange, H., 95
Larsson, M. W., 449
Launay, J.-P., 503
Leahy, R., 544
Leao, J., 8
Lebedkin, S., 116, 289
Lee, S. T., 445
Lefebvre, J., 111
Lefrant, S., 201
Leonhardt, A., 36, 478
Li, Y. B., 229
Liebau, M., 503, 516, 536, 565
Lifshitz, Y., 445
Lillo-Rodenas, M., 460

Lindelof, P. E., 552
Liu, X., 57, 205
Liu, Z. Y., 32
Lobach, A. S., 209
Löhneysen, H. v., 49, 561
Loiseau, A., 293
Lota, G., 532
Lota, K., 532
Lu, Y., 32
Lyashenko, D. A., 490
Lyding, J. W., 173

Nejman, P., 95
Nemanic, V., 435
Nemes, N. M., 8
Neugebauer, H., 548
Nikolaev, A. V., 339, 393
Nikolou, M., 137
Niyogi, S., 197
Novosel, B., 423
Novotny, L., 163
Nygård, J., 552

M

Macfarlane, J. C., 91
Machón, M., 381
Maggini, M., 548
Magrez, A., 61
Mannsberger, M., 234, 251
Marcoux, P. R., 65
Márk, G. I., 389
Mårtensson, T., 449
Maser, W., 149
Matarredona, O., 27
Maultzsch, J., 153, 330, 381, 397
Mayer, A., 389
McRae, E., 133
Meden, A., 423
Medjahdi, G., 133
Mehring, M., 12, 181, 315
Meixner, A. J., 163
Mele, E. J., 402
Mertig, M., 185
Mevellec, J. Y., 201
Meyer, J. C., 540, 556
Michel, K. H., 339, 393
Mihailovic, D., 419, 423, 435
Mikó, C., 61, 107
Milas, M., 107
Milne, W. I., 485
Minett, A. I., 544
Monteferrante, M., 355
Morton, J. J. L., 255
Mrzel, A., 419, 435

O

Obergfell, D., 540, 556
Obraztsov, A. N., 490
Obraztsova, E. D., 209
Okada, Y., 431
Okotrub, A. V., 595
Ordejón, P., 381, 397
Oron, M., 561
Osváth, Z., 149

P

Paillet, M., 540
Pal, A. F., 528
Panjan, P., 419
Panthöfer, M., 12
Paris, M., 465
Payne, M. C., 81
Pazhetnov, E. M., 595
Pejovnik, S., 423
Pernbaum, A. G., 528
Pető, G., 149
Petrushenko, Y. V., 490
Pfeiffer, R., 234, 251, 263, 268
Pichler, T., 141, 205, 213, 217, 285
Pichugin, V. V., 528
Pimenta, M. A., 157, 372, 407
Pinto, N. J., 129
Piscanec, S., 445
Ploscaru, M., 435
Podio-Guidugli, P., 355
Podobnik, B., 419, 435
Pompe, W., 57, 185, 503
Ponikvar, M., 423
Ponomareva, I. V., 351
Popov, V. N., 263

N

Nagy, J. B., 40
Nakatake, M., 222
Namatame, H., 222

Porfyrakis, K., 255
Possamai, G., 548
Poteau, R., 503
Pötschke, P., 473, 478

R

Rabe, J. P., 334
Rachid Babaa, M., 133
Rafailov, P. M., 153
Rapenne, G., 503
Rauf, H., 213, 217
Regan, B. C., 587
Reich, S., 330, 381, 397
Renker, B., 289
Repetto, P., 524
Requardt, H., 397
Resasco, D. E., 27
Ricci, D., 524
Rikken, G. L. J. A., 121
Rinzler, A. G., 69, 137
Ritschel, M., 478
Robertson, J., 81, 445
Röding, R., 238
Romero, H., 87
Rosén, A., 87, 364
Rosentsveig, R., 306
Roth, S., 181, 189, 473, 478, 516, 540, 556, 591
Rubio, A., 293
Rümmeli, M. H., 141, 285

S

Sadowski, J., 552
Saito, R., 157, 372, 407
Samaille, D., 469
Samsonidze, G. G., 157, 372, 407
Samuelson, L., 449
Sansalone, V., 355
Sapmaz, S., 583
Sariciftci, N. S., 548
Satanovskaya, O. P., 490
Scherer, W., 315
Schlecht, U., 99, 415
Schlögl, R., 32
Schmid, M., 181
Schneider, M. A., 168

Schönenberger, C., 574
Schreiber, M., 368
Schweiss, P., 289
Seidel, R., 503, 536, 565
Seifert, G., 306, 368
Seifert, W., 449
Seo, J. W., 61, 107, 427
Severin, N., 334
Shchegolikhin, A. N., 209
Shimotani, H., 326
Shinohara, H., 326
Shiozawa, H., 222
Simon, F., 213, 234, 251, 259, 263, 268, 273
Sioda, M., 95
Sippel, J., 69
Skákalová, V., 189
Škarabot, M., 427
Skipa, T., 289
Sköld, N., 449
Sokolov, V. I., 209
Solomentsev, V. V., 209
Souza Filho, A. G., 372, 407
Staii, C., 129
Steinmetz, J., 469
Stephan, O., 293
Stoll, M., 153
Su, D. S., 32
Sundqvist, B., 238
Suzuki, S., 222
Szabó, A., 40
Szabó, I., 149
Szostak, K., 460

T

Taeger, S., 185
Takata, M., 326
Tang, C. C., 229
Taniguchi, M., 222
Taninaka, A., 326
Tanner, D. B., 137
Tapasztó, L., 389
Täschner, C., 478
Taverna, D., 293
Telg, H., 330
Tenne, R., 306
Teo, K. B. K., 485
Teresi, L., 355

Thelander, C., 449
Thomsen, C., 153, 330, 381, 397
Tibbetts, G. G., 455
Tokumoto, M., 145
Tommasini, M., 334, 359
Tsukagoshi, K., 103

U

Umek, P., 298, 302, 427, 435
Unger, E., 503, 536, 565
Uplaznik, M., 419, 435

V

van Delden, R. A., 498
van der Veen, M. H., 315
van der Zant, H. S. J., 583
Van Haesendonck, C., 40
Vengust, D., 435
Venturini, P., 423
Verberck, B., 339
Vértesy, G., 149
Vitali, L., 168
Volkov, A. P., 490
Volkov, V. T., 103
Volodin, A., 40
Vrbanic, D., 419, 423, 435

W

Wågberg, T., 238
Wagner, H. D., 306
Wagnière, G., 121

Wallenberg, L. R., 449
Walton, D. R. M., 53
Watts, P. C. P., 53
Weber, H. B., 561
Weidinger, A., 315
Weinberg, G., 32
Woo, Y., 516
Wu, Z., 69

Y

Yan, D., 193
Yang, Sha., 556
Yang, Shi., 556
Yoshioka, H., 222
Yuzvinsky, T. D., 512, 587

Z

Zajec, B., 435
Zakhidov, A. A., 490
Zalar, B., 298
Zapien, J. A., 445
Zerbetto, F., 439
Zerbi, G., 334, 359
Zettl, A., 512, 587
Zhao, B., 197
Zheng, L., 27
Zheng, M., 75
Zhu, Y. C., 229
Zhu, Y. Q., 53
Zhu, Z. P., 32
Zólyomi, V., 343, 377
Zorko, A., 427
Zukalová, M., 242, 247
Zumer, M., 435

RETURN TO: PHYSICS LIBRARY

351 LeConte Hall   510-642-3122

LOAN PERIOD 1 **1-MONTH**	2	3
4	5	6

ALL BOOKS MAY BE RECALLED AFTER 7 DAYS.
Renewable by telephone.

DUE AS STAMPED BELOW.

FORM NO. DD 22
500   4-03

UNIVERSITY OF CALIFORNIA, BERKELEY
Berkeley, California 94720–6000